Handbook of VLSI Chip Design and Expert Systems

A. F. Schwarz
*Faculty of Electrical Engineering
Delft University of Technology
Delft, The Netherlands*

ACADEMIC PRESS

This book is printed on acid-free paper.

Academic Press Limited
24–28 Oval Road
London NW1 7DX

United States edition published by
Academic Press Inc.
San Diego, CA 92101

Copyright © 1993 by
ACADEMIC PRESS LIMITED

All Rights Reserved

No part of this book may be reproduced in any form
by photostat, microfilm or any other means
without written permission from the publishers.

A catalogue record for this book is available from the British Library

ISBN 0-12-632425-5

Printed and bound in Great Britain by Hartnolls Ltd, Bodmin, Cornwall

£70-

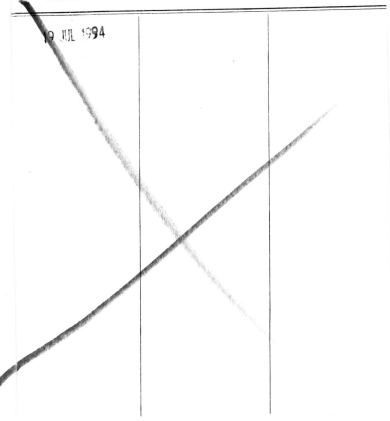

PREFACE

This book consists of three interrelated parts: (1) Computer-Aided Design of VLSI Digital Circuits and Systems (CADCAS), (2) Artificial Intelligence (AI) and Expert Systems, and (3) Applications of AI / Expert Systems to CADCAS. The field of CADCAS has experienced a remarkable growth and development during its short existence. It primarily depends on a set of well-formalized algorithms, which means that the many procedures for performing subtasks in the design cycle follow fixed preprogrammed sequences of instructions. The ultimate aim in CADCAS is to automate the design process so that human-introduced errors are avoided or at least minimized.

As system complexity increases, chip design becomes more and more a multidisciplinary enterprise. The production of a complex chip is the result of individual contributions from many experts, including system design experts, logic designers, circuit analysts and designers, device engineers, technologists and application-specific designers. Designing and producing complex chips requires a hierarchical top-down approach with several levels of abstraction, starting from the highest system level going down to the lowest fabrication level. Each expert working at any level uses a specific data description and representation associated with this level. Chip production requires the interdisciplinary teamwork of specialists, implying adequate communication among specialists and unambiguous translation of descriptive languages between the subsequent levels of the design cycle. This integration problem appears to be intractable when exclusively algorithmic methods are used. Knowledge-based methods may offer a solution to this problem.

In the past decade, progress in artificial intelligence, with an emphasis on expert systems, has been overwhelming. In addition to the algorithmic approach as applied in CADCAS, expert systems provide us with a knowledge-based approach to problem solving. This approach, which is based on heuristics and rules contained in a knowledge base, has been adopted in a great variety of applications, e.g., in medical diagnosis. Though CADCAS as well as expert systems are growing to maturity, each in its particular domain, their integration (that is, the application of expert systems in CADCAS) is still in its infancy. Books on pure CADCAS are available in abundance; so are books on AI and expert systems and its applications. However, publications on applications of expert systems in CADCAS are scattered among a multitude of monographs, specialized journals and conference proceedings. Books published so far are restricted to specific subtasks of the chip design cycle, for example, either for high-level design, routing or testing. The present book is an attempt to encompass, in a single monograph, all important issues emerging when expert systems are applied to microelectronic chip design.

The main objective of writing this book is to present the reader with the use of expert systems in every possible subtask of VLSI chip design as well as in the interrelations between the subtasks. Usually, the entire design cycle consists of a set of algorithmic procedures whose sequential steps are controlled by the human designer who also performs the intelligent tasks which cannot be done by the computer. A typical session of designing microelectronic chips consists of running a set of automatic programs, knit together by a human designer sitting in front of a workstation. Knowledge-based tools can be used to replace the human designer in performing the heuristic and intelligent decisions to be made. This is implemented by storing non-algorithmic design expertise acquired from recognized human design experts into the knowledge base of expert systems.

Only recently have heuristic knowledge-based methods been considered applicable to CADCAS. In my view, future CADCAS procedures will be mainly algorithmic, while heuristic AI methods will be used as a complement to the algorithmic methods in those parts of the design cycle which lend themselves very well to a heuristic approach. Future design systems will involve both algorithmic and heuristic methods. Generally speaking, expert systems in CADCAS can be applied both to specialized subtasks and to the integration aspect of the design cycle. The different character of each subtask requires a different kind of expert system. For example, an expert system used for high-level design is different from an expert system used for routing or testing. The kind of expert system to be used depends on the subtask for which it is used. Therefore, this subtask has first to be defined, the algorithmic approach should be outlined, and then the question can be answered as to whether an expert system can be used. A desirable goal of using expert systems is to shift the burden of the remaining human contributions to the computer so that the entire design cycle is fully automatic. To this end, an integration of the algorithmic and the essentially different knowledge-based procedures into one coherent system is necessary. This dynamic area on the frontier of microelectronics research is not yet well established with many problems requiring further study.

CADCAS and expert systems have grown so rapidly that it is difficult to keep abreast of the latest developments in the field. There is a need for emphasizing fundamentals in order to give the book a lasting value. At the same time, a presentation of the state of the art (though it changes in the course of time) is necessary to give the reader a sufficient amount of knowledge as to where we are now so that we can assess future developments on a reasonable basis. To that end, this book gives extensive surveys of representative expert systems for CADCAS purposes as published in the literature.

This book is intended to be an introductory guide through the maze of articles published on the use of expert systems in VLSI chip design. Readers who want to probe deeper into a subject will benefit from the extensive list of references at

the end of the book. In summary, the main ingredients of this book are as follow:
- CADCAS is discussed focusing on the most important methods which will influence future developments. I will not restrict myself to only one specific subtask of CADCAS, but place particular emphasis on the integration of the separate subtasks.
- A review of the growing field of expert systems will be given as an introduction in order to help the reader to conceive the possible uses of expert systems to CADCAS. Due attention will be paid to the programming languages and to the personal computers which can be used in expert systems.
- Applications of expert systems in specific subtasks as well as in the integrated CADCAS system will be described with illustrative examples.

This scheme of organization is quite natural considering the fact that the AI approach complements rather than replaces the algorithmic approach to CADCAS. The objective of the book is to give the reader
- a basic conceptual and methodological understanding of all that is involved in chip design and a clear insight into fundamental principles in CADCAS so that possible future trends can be evaluated.
- sufficient knowledge of AI, expert systems and personal computers to be able to understand their use in CADCAS.
- an insightful grip and knowledge as to where, why and how expert systems can be employed in subtasks of CADCAS and in the integrated chip design system.

The targeted readership consists of three categories:
- Electronics engineers and computer scientists involved in any subtask of CADCAS (chip design) in need of more insight into the interrelations of the various subtasks in the design cycle.
- AI scientists and expert system designers who are interested in the applicability of expert systems to CADCAS and, last but not least,
- university professors and undergraduate or graduate students who want to acquaint themselves with the latest developments in this highly active research field.

Acknowledgments

This book should be considered as a tribute to all researchers in the field of CADCAS and AI whose work has brought the field to a remarkably high level of sophistication. In describing their methods and tools, I have carefully referenced the original publications, wherever appropriate.

I am deeply indebted to Professor Patrick Dewilde and his staff of the Delft University's Network Theory Section, who provided the stimulating environment in which I could work on the book. I have greatly benefited from the critical reviews and comparative studies of recent developments in CADCAS and expert systems undertaken by students. Erik Platzbecker was particularly helpful whenever text-processing problems occurred.

I am obliged to the publisher, Academic Press, especially to editor James Gaussen, who handled the final editing and production process. I wish to express my appreciation to the staff of the Production Department of Academic Press who transformed my camera-ready copy into this fine book.

Last but not least, my special thanks go to my wife, Nicoline, for her understanding and patience during years of composing and editing of the text.

Dolf Schwarz

CONTENTS

PREFACE	i

Chapter 1 VLSI CHIP DESIGN	1
1.1 COMPUTER-AIDED CIRCUIT AND SYSTEM DESIGN	1
1.1a Impacts of the Rapidly Changing Technology	1
1.1b The Hierarchical Approach to Chip Design	6
1.1c Design Methodologies	11
1.2 CADCAS TOOLS IN CHIP DESIGN	18
1.2a Design, Simulation and Testing	18
1.2b Layout Design	22
1.3 COMPUTERS AND ARTIFICIAL INTELLIGENCE	25
1.3a Digital Computers	25
1.3b The AI Approach to Computer-Aided Chip Design	31
Chapter 2 PROGRAMMING TOOLS IN CHIP DESIGN	37
2.1 PERSONAL COMPUTERS	37
2.1a PC Hardware	37
2.1b PC Software	43
2.2 OPERATING SYSTEMS	49
2.2a Operating System Principles	49
2.2b Operating Systems for Personal Computers	54
2.3 DATABASE SYSTEMS AND COMPUTER NETWORKS	62
2.3a Design Databases and Data Models	62
2.3b Database-Management Systems	68
2.3c Computer Networks	75
Chapter 3 NEW DEVELOPMENTS IN PROGRAMMING	81
3.1 FOURTH-GENERATION LANGUAGES	81
3.1a Five Generations of Programming Languages	81
3.1b Fourth-Generation Programming	85
3.2 NON VON NEUMANN ARCHITECTURES	91
3.2a Computer Architectures	91
3.2b Non Von Neumann Computers	100
3.2c Processor-Array Architectures	104
3.3 ARTIFICIAL INTELLIGENCE	108

3.3a Heuristic Methods of Solving 108
3.3b Expert Systems, Powerful AI Applications 110
3.3c Fifth-Generation Languages 114
3.4 LISP 117
3.4a Basics of LISP 117
3.4b LISP Implementations and Machines 122
3.5 LOGIC PROGRAMMING 124
3.5a Predicate Calculus 124
3.5b Prolog 127
3.5c Prolog Implementations 138
3.6 THE OBJECT-ORIENTED ENVIRONMENT 147
3.6a Object-Oriented Programming 147
3.6b Smalltalk 150
3.6c Object-Oriented Databases 154

Chapter 4 EXPERT SYSTEMS 159
4.1 EXPERT SYSTEM FUNDAMENTALS 159
4.1a Expert System Architecture 159
4.1b Problem Solving Strategies 163
4.1c Dealing with Uncertainty 169
4.2 KNOWLEDGE REPRESENTATION 172
4.2a Knowledge Acquisition 172
4.2b Production Rules 175
4.2c Semantic Networks and Object-Oriented Representations 178
4.3 KNOWLEDGE PROCESSING 183
4.3a Production Systems 183
4.3b Logic Programming and Object-Oriented Systems 187
4.3c Shells and Tools 190
4.4 PROGRAMMING ENVIRONMENTS 194
4.4a Knowledge-Based Management 194
4.4b AI Architectures 198
4.4c Linking Declarative and Procedural Languages 201
4.5 APPLICATIONS OF EXPERT SYSTEMS 202
4.5a Types of Expert System 202
4.5b An Anthology of Expert Systems 205
4.5c Expert Systems in Electronic Engineering 209

Chapter 5 VLSI SYSTEM DESIGN AND EXPERT SYSTEMS 221
5.1 LOGIC-CIRCUIT DESIGN 221
5.1a Logic Modules 221

5.1b Synthesis of Sequential Circuits	225
5.1c Data Processor and Controller Design	232
5.2 HIGH-LEVEL SYSTEM DESIGN	241
5.2a VLSI Design Aspects	241
5.2b Intelligent High-Level Synthesis	247
5.2c Automated Systems for High-Level Logic Synthesis	258
5.2d Expert Systems for Logic Circuit and System Design	269
5.3 ASIC DESIGN	285
5.3a ASIC Design Methodologies	285
5.3b Silicon Compilers	291
5.3c Expert Systems for ASIC Design	297
Chapter 6 DESIGN VERIFICATION AND EXPERT SYSTEMS	**303**
6.1 SYSTEM SIMULATION	303
6.1a High-Level Simulation and Design	303
6.1b Hardware Verification and Expert Systems	306
6.2 LOGIC SIMULATION	310
6.2a Gate-Level and Functional Simulation	310
6.2b Timing Verification	319
6.2c Switch-Level and Timing Simulation	323
6.2d Expert Systems for Digital-Circuit Verification	329
6.3 CIRCUIT SIMULATION	334
6.3a Circuit-Element-Level and Mixed-Mode Simulation	334
6.3b Expert Systems for Analog Circuits	340
Chapter 7 VLSI TESTING AND EXPERT SYSTEMS	**349**
7.1 FAULT DIAGNOSIS	349
7.1a Fault Models	349
7.1b Testability Measures	353
7.1c Expert Systems for Fault Diagnosis	356
7.2 TEST GENERATION	364
7.2a Path Sensitizing and D Algorithm	364
7.2b Alternative Methods for Test Generation	372
7.2c Fault Simulation	377
7.2d Expert Systems for Test Generation	382
7.3 DESIGN FOR TESTABILITY	389
7.3a Scan-In/Scan-Out Methods	389
7.3b Built-In Testing	392
7.3c Expert Systems for VLSI Design Testability	395

Chapter 8 LAYOUT DESIGN AND EXPERT SYSTEMS — 407
8.1 PLACEMENT AND ROUTING — 407
8.1a Partitioning and Placement — 407
8.1b Global Routing — 410
8.2 LAYOUT DESIGN OF BLOCK STRUCTURES — 416
8.2a Block Structures and Symbolic Layout — 416
8.2b Heuristic Approaches to Floorplanning — 420
8.2c Expert Systems for Layout Design — 428
8.3 LOCAL ROUTING — 444
8.3a Algorithms for Local Routing — 444
8.3b Expert Systems for Local Routing — 455
8.4 LAYOUT VERIFICATION — 462
8.4a Design-Rule Checking and Circuit Extraction — 462
8.4b Expert Systems for Layout Verification — 464

Chapter 9 MODERN DESIGN METHODOLOGIES — 467
9.1 HUMAN-COMPUTER INTERACTION — 467
9.1a Interactive Graphics Tools — 467
9.1b Intelligent User Interfaces — 471
9.2 WORKSTATIONS FOR CHIP DESIGN — 475
9.2a Engineering Workstations — 475
9.2b Advanced CADCAS Workstations — 477
9.3 PROBLEM-SOLVING TOOLS — 480
9.3a Machine Learning — 480
9.3b Neural Networks — 485
9.3c VLSI Applications of Neural Networks — 496
9.3d Toward the Sixth Generation — 503

REFERENCES — 515

INDEX — 567

Chapter 1
VLSI CHIP DESIGN

1.1 COMPUTER-AIDED CIRCUIT AND SYSTEM DESIGN

1.1a Impacts of the Rapidly Changing Technology

Introduction
The impressive advances in digital computers and communication engineering have been made possible by the various achievements in electronic design and semiconductor technology. The first three generations of computers are usually associated with the changing electronic hardware components, which were successively vacuum tubes, discrete transistors and integrated circuits. The advent of the integrated circuit (IC), which ushered in the *microelectronics* era, has initiated the emergence of *chips* with ever-increasing integration densities. In about two decades, the manufactured chips have reached the density of *Very-Large-Scale Integration* (VLSI), which implies more than 40000 transistors per chip. At present, a chip may contain millions of transistors.

The growing density in a chip goes hand in hand with an increase in the complexity of the system functions which a chip can perform. Systems have gradually become more comprehensive in scope and more complex in conception and design. As systems become larger and more complex, they strain the capabilities of the designers in that the immense masses of detail that the design emcompasses can no longer be comprehended. As a consequence, we witness changes in some of the older concepts and methods of digital-circuit design, while new approaches to solving the complexity problem emerge. The fourth generation of computers, which was developed in the 1980s, is characterized by an abundant use of VLSI and distributed computing systems. The combined use of microelectronics, artificial intelligence and automatic development tools will beget the fifth-generation computer, which is expected to flourish in the 1990s.

The emergence of complex logic systems requires a *systems approach* to design in addition to the structural *circuit-design approach*. This systems approach, which stresses the functional behavior of a system rather than its structure, has led us to consider higher levels of abstraction, such as the register-transfer level and the system level.

Design Automation
The increasing complexity of logic systems forces us to search for methods and tools which automate the design process. This has led to the concept of design automation. Strictly speaking, *design automation* (DA) implies the automatic generation of an electronic circuit or system, given a specification of design requirements. However, except for restricted applications and layout structures, a fully automatic design procedure from initial design concept to the final chip production is unattainable, if ever desirable. For this reason, design automation will be defined here as the art of utilizing the digital computer to automate as many steps as possible in the design of integrated circuits and systems. Design automation can be identified as *Computer-Aided Design of Circuits and Systems* (CADCAS) with automation in mind [Sch87].

Design automation exploits the basic capabilities of computers to perform complex calculations and to handle huge amounts of data with a high accuracy and speed. These capabilities are enhanced by new data-storage techniques and the use of interactive graphics systems, which facilitate human-machine communication. The primary incentive of design automation is the reduction of the required investments entailed in the design efforts:
a. Substantial reduction in the turnaround time, i.e., the time between design specification and product delivery.
b. Reduction in errors made during the entire process from initial design to product manufacture.
c. Fewer demands on the designer's skills. Human experience can be exploited for improving complex designs in an interactive way.
d. Reduction in production costs and increased manufacturing yield and hence a higher probability of initial success.

Disadvantages of design automation are the relatively inefficient usage of silicon area and the lower chip performance compared to full-custom design.

Papers on virtually all aspects of design automation have been presented at several conferences. Starting in 1964, the Design Automation Conference, sponsored by the ACM and the IEEE Computer Society, is held annually in the USA at the end of June. The Proceedings of this conference give a good picture of how design automation has evolved through the years. Similar conferences are being held in the USA as well as in Europe. Several books and survey articles on computer-aided design of VLSI chips have been published [Ein85, Fic87, Sch87, Dil88, DiG89, Sha89].

Expert Systems
It is generally agreed that human intervention in appropriate stages of the design cycle will always be desirable. Chip design involves two main approaches to problem solving: (a) the algorithmic approach, and (b) the knowledge-based approach. The algorithmic approach, substantiated by design automation or

1.1 COMPUTER-AIDED CIRCUIT AND SYSTEM DESIGN

CADCAS, primarily depends on well-formulated algorithms. This means that the procedure of solving a problem follows a fixed preprogrammed sequence of steps, ultimately leading to the solution. The aim is to automate the design process as much as possible so that human-introduced errors are avoided or at least minimized.

The second approach involves those activities which are not amenable to algorithmic methodology and which therefore have been relegated to human intervention. These activities are based on knowledge (in the form of heuristics and rules) derived from many years experience acquired from a selected group of human experts. Only recently have attempts been made to implement hardware systems which can take over the activities thus far performed by human designers. The knowledge in such systems is stored in a knowledge base, while a reasoning procedure is incorporated into the system. Such a system is now known under the name *expert system*. A main objective of this book is to survey the possibilities of using expert systems in the design of VLSI chips. It must be emphasized that expert systems are to be used as complements rather than as replacements of human designers.

Rapid Changes in VLSI Technology

Complex system design requires adequate computer-aided design tools. When the IC technology changes (which it does at a dramatic pace), the design tools will have to be changed accordingly. The advances of VLSI technology and the exploitation of design automation tools have made possible the manufacture of digital systems with smaller sizes and with much lower power consumptions and production costs and higher speeds than ever before. The main problem that designers of complex systems have to face can be phrased as follows: Given a rapidly changing technological environment, how can we develop and maintain design tools which remain useful and at least easily updatable as technology progresses?

Without special measures being taken, such as the use of CADCAS and DA tools, the chip-design time will increase exponentially as circuit complexity increases. In the literature, many solutions to the complexity problem in the area of circuit or system design, simulation, testing and layout are attempts to keep up with the rapid changes in technology. The most important *impacts of VLSI technology* on the design of complex digital systems will be summarized below:

a. *New Design Methodologies.* To cope with the increasing complexities in VLSI sytems, new system-design methodologies must be developed. Major issues are the hierarchical approach to system design (involving the partitioning of system design into smaller, more manageable parts), a structured method of software development and the use of an integrated design system with updating facilities. There is an urgent need for technology-independent CADCAS or DA tools for complex system designs.

b. *Regular Structures and Design Automation.* Increasing system complexities can be handled by using regular system structures. This facilitates the exploitation of design automation, which in turn is evolving as a major contributor to further technological advances. Design automation and VSLI systems usually involve predefined structures, notably regular structures, such as RAMs, ROMs, PLAs and gate arrays. An advantage of regular structures is the efficiency of design and the ease of design testability and verification.

c. *Hardware-Software Tradeoffs.* The decreasing costs of hardware components, made possible by VLSI technology, force the designers to consider tradeoffs between the high performance and speed of hardware implementations and the programming flexibility of software to perform various tasks. As a consequence of the increasing costs of software development, the designer is compelled to use more sophisticated software-development tools. On the other hand, the decreasing costs of hardware encourage the development of new multiprocessor system architectures, different from the Von Neumann architectures. This causes a movement of software functions into hardware.

d. *Emergence of New Products.* A milestone in the history of IC technology is the fabrication of a complete computer on a chip. The dramatic reduction in physical size, power consumption and production cost of LSI and VLSI chips (including microprocessors) has made new products proliferate, thereby opening the door to new markets. There is a growing demand for applications in the home, factory, office and just about everywhere else.

e. *Revival of Customized Designs.* Many semiconductor manufacturers used to be reluctant to produce low-volume customized ICs because of low profits. The explosion of new application areas for LSI and VLSI chips and the advances in CADCAS and DA systems for IC design have led to reduced development costs and time scales for design. As a consequence, semicustom designs or Application-Specific Integrated Circuits (ASICs) are now showing significant cost benefits, even at low-volume production. Many approaches to custom design with a short turnaround time have been proposed with great success.

f. *New Computer Architectures.* The increasing complexity of problems encountered in advanced applications has an impact on the computer architecture. The trend is to diverge from the general-purpose computer toward a cooperative set of processors each dedicated to specific tasks and from the traditional Von Neumann machine toward parallelism, both in function and control.

g. *Expert Systems.* Non-algorithmic tasks in problem solving, hitherto a domain of human experts, are being performed by expert systems. Proposed applications of expert systems in VLSI chip design is the main topic of the

1.1 COMPUTER-AIDED CIRCUIT AND SYSTEM DESIGN

present book.

h. *Neural Networks.* While problem solving using traditional computers is based on exploiting the inherent properties of computers (e.g., high-speed processing of computational tasks), there is a trend to let hardware systems perform many intelligent tasks in various applications by using neural networks, which mimic the structure and performance of the human brain. Possible applications of neural networks in chip design are being investigated. At least, the problem of how to implement them as VLSI chips falls within the scope of this book.

i. *Need for a New Breed of Chip Designer.* A complete logic system can be integrated in the form of a very small physical chip, which is actually a microelectronic circuit capable of performing logic functions. Due to this fact, the borderline between system designers and circuit designers is becoming blurred. A modern circuit or system designer is necessarily concerned with (and hence should develop a basic foundation of knowledge in) all aspects of the design cycle from the highest system level down to the fabrication level. This statement does not suggest that everyone involved in chip design and manufacture must do everything. The need for system designers, circuit designers, device engineers and chip manufacturers who are expert in their own fields remains. The point is that a basic knowledge of the entire design cycle facilitates intelligent and effective interactions between various specialists and a better appreciation of each of the other's problems.

j. *Integrated Workstations for Chip Design.* The fast rate of change in semiconductor technology has created a scarcity in chip designers who are proficient in a multidisciplinary role. As a consequence, there is a trend to develop powerful workstations which integrate the whole spectrum of the design cycle and expert systems which incorporate design knowledge and experience from expert human designers. Though such systems are required to be technology-independent, regular updates will remain necessary and chip designers as indicated in i. are well qualified to exploit such systems for increasing the design quality and productivity. Computer communication networks allow designers to access remote databases and to exchange design information with colleagues working on the same project.

Consequences of the VLSI Explosion

The tremendous impacts of VLSI technology outlined above pose substantial problems to chip designers. The more rapid the technological changes are, the more imperative it is that we should master basic principles which have a lasting value. Fortunately, we may still rely on a number of principles and concepts, which have withstood the ravages of time. The advent of new technologies does not automatically make all the existing methods and tools of design obsolete. A new VLSI tool is often either an extension or an adjunct to an existing design tool.

For all that, considerable ingenuity must be employed in developing a methodology which is able to cope with the rapid changes in technology. Design systems must be updated on a continuing basis to accomodate changes in chip technology, customer requirements and critical time schedules. A major goal in IC design and manufacture is to produce the largest possible percentage of error-free chips that will work upon first operation. With this in mind, the ability to develop adequate tools for design verification and testing may be a rate-limiting factor constraining the potential growth in the complexity of logic systems [Mea80].

The scarcity of skilled personnel able to cope with the rapidly changing technology has forced different companies to join efforts with the objective of advancing the state of the art in chip technology by improving the planning and coordination of system research and development. In the USA, the Semiconductor Research Cooperative proposed by the Semiconductor Industry Association coordinates university research, while the Microelectronics and Computer Technology Corporation supports joint efforts among industrial companies. In Japan, the Ministry of International Trade and Industry provides financial support to research projects in the microelectronics field. In Europe, the Commission of the EEC (European Economic Community) provides finances for launching innovative projects, in particular through the Microelectronics Program and ESPRIT (European Strategic Progamme of Research and Development in Information Technology).

1.1b The Hierarchical Approach to Chip Design

From SSI to VLSI

There has been tremendously fast progress in microelectronics starting with MOS technology and later followed by bipolar technology. This has led to the development of more complex digital systems with increased device density in a chip. Since digital systems are composed of logic gates, the *integration density* of integrated circuits is usually expressed in terms of *equivalent gates per chip*, where each equivalent gate (EG) is assumed to contain four transistors. Based on the integration density, the following classification of IC technologies is commonly used:

SSI (Small Scale Integration), up to 10 EG/chip;
MSI (Medium Scale Integration), between 10 and 10^2 EG/chip;
LSI (Large Scale Integration), between 10^2 and 10^4 EG/chip;
VLSI (Very Large Scale Integration), more than 10^4 EG/chip.

As the integration density increases, more and more functions can be

1.1 COMPUTER-AIDED CIRCUIT AND SYSTEM DESIGN

incorporated into one chip. This has led to the introduction of standard modules in more advanced forms of integration: SSI → MSI → LSI → VLSI. Consequently, circuit primitives can be defined according to the level of integration density. When an interconnected set of transistors, which are primitives in a transistor circuit, is used to form a logic gate, and several types of logic gates are defined, it is expedient to base a logic-circuit design on logic gates than on transistors. Logic gates form a higher level of primitives than transistors. Similarly, interconnected sets of gates form functional units which can be taken as primitives for designing a logic system. The development of higher-level primitives leads to the problem of modeling and simulation at different levels of detail.

Complex high-density LSI and VLSI circuits require a hierarchical approach to design which implies simulation at various levels. The levels of detail (or abstraction) in simulation are directly related to the model primitives which are selected.

Hierarchical Levels of a Complex Logic System
A complex logic system requires a top-down hierarchical design approach, starting at a high system level and working down to the mask-artwork and fabrication level. At the higher levels, most emphasis is placed on the *behavioral* aspects of the target system, i.e., the ultimate hardware realization of the system. As one proceeds to lower levels, more and more *structural* information is added to the design in progress, until at the end the design can be put into fabrication.

With a view to Fig. 1.1, let us describe the various abstraction levels which are encountered during the entire design cycle.

System Level or Algorithmic Level
This level serves to specify the behavioral and performance requirements of the system according to customer needs. System analysis is related to the applications for which the design is intended. A feasability study is conducted taking into account present-day physical constraints. The global behavior, particularly the information flow in the target system, is evaluated and a suitable design methodology is selected aimed at an optimal solution in terms of performance and throughput. The system level is partitioned in terms of *system primitives*, such as central processing units, memory modules, buffers and peripheral devices. Parameters of interest to be studied include information rates, memory capacities and system costs.

Register-Transfer Level
At this level, *data flow* and *control flow* are of interest. The register-transfer primitives are registers which are operated upon to perform all kinds of data processing ranging from simple data transfer to complex logic and arithmetic operations. This level deals with detailed data path and control algorithms

without emphasizing the constraints imposed by a particular implementation technology. The functional design of the system can be simulated, thereby enabling microprograms etc. to be evaluated.

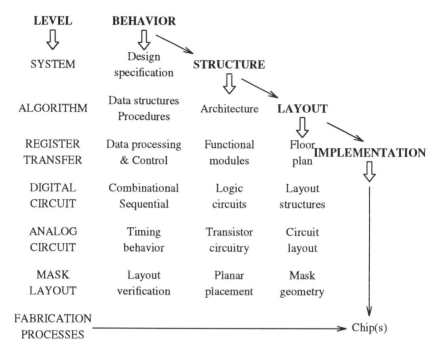

Fig. 1.1 Hierarchy of a complex logic system

Digital-Circuit Level
The construction of the target system from gates and flip-flops is emphasized. At this stage, the structure of the circuit implementation becomes visible. The choice of the implementation technology (bipolar, NMOS, CMOS, etc.) may be made at this level, provided that one is not restricted to a fixed technology at the outset. For the purpose of simulation at this level, models of gates and flip-flops can be defined with different levels of accuracy. The *Switch Level*, relevant to the MOS technologies, can be considered as a logic-gate level as far as the logic values (e.g., 0, 1 and *U*) are concerned. At the same time, the switch level can be considered as a form of circuit-element level, where all circuit elements (usually transistors) are represented by very simple models: voltage-controlled switches.

Analog-Circuit Level
The target system has taken the form of a transistor circuit in the selected technology. The circuit primitives are transistors, capacitors, resistors, etc. In an

1.1 COMPUTER-AIDED CIRCUIT AND SYSTEM DESIGN

integrated circuit, the main primitives are transistors. Hence, it is important to be thoroughly acquainted with the behavior of transistors in a circuit. A characteristic feature of this level is that the transistors can be represented by very accurate models. The major limitation of circuit simulation is the size of the network that can be handled (at most a few hundred transistors). When the circuit to be simulated is large, some form of partitioning into suitable subcircuits is necessary. The main advantage of circuit simulation is that extremely accurate transistor models can be used, taking into account the effect of any desirable parameter. As a result, realistic simulations are possible. When we want to simulate large circuits, we may employ macromodels which are models in a simplified form.

Physical-Device Level
At this level, the physical properties of the transistor as a semiconductor device are studied. In particular, the effect of physical contraints on the electrical behavior of the transistor is of great importance. Simulation at the device-physics level allows one to study the effect of physical and geometric parameters of the transistors on the terminal behavior of the transistors or transistor circuits. This level is intimately related to the layout design and the fabrication technology.

Mask-Artwork and Fabrication Level
It is assumed that the target system is a transistor circuit, that is, we do not consider mechanical peripheral devices which are also parts of the complete digital system. To allow the transistor circuit to be implemented in the desired integrated-circuit form, a set of masks is required. The mask artwork which defines the set of masks is obtained as the result of the layout design. The symbolic layout level may precede the geometric mask-artwork level. To verify the correctness of the layout, this layout is thoroughly examined to check for violation of the specified mask-design.

Top-Down Design
When a very complex system is to be designed, it is natural to adopt a *top-down approach*, which proceeds from the top level with system simulations down to the implementation level, resulting in the ultimate fabrication of the desired integrated circuit.

The top-down approach implies partitioning a large problem into manageable subproblems which can be solved in an appropriate manner. Let us illustrate this approach by considering the design of a complex digital logic system. In general, the design of a complex system proceeds in a top-down manner, after which the implementation takes place in a bottom-up manner. This is clarified by the flowchart of Fig. 1.1. This chart demonstrates the hierarchical character of the design cycle. The different levels of hierarchy allow a complex system to be designed in a systematic way that is amenable to human understanding and compatible with computer processing capabilities.

The hierarchical levels in the lefthand column of Fig. 1.1 are in conformity with the top-down approach that a human designer uses. The other columns show the corresponding software and hardware features which are relevant at each particular level. The *design specification* includes the definition of the data or signal processing function to be realized. The upper two levels (system level and register-transfer level), which deal with system functions without considering physical implementation details, will be discussed in Chapter 5. A large part of this book will be concerned with the lower levels, which are distinguished from the upper two levels by the presence of a physical structure, as an interconnection of functional units (functional level), as a logic-gate circuit (gate level) or transistor circuit (circuit-element level), or as a pattern of mask primitives (implementation level). The hierarchical top-down approach focuses the design successively on the system behavior, the data and control flow at the register-transfer level, the logic-circuit behavior, the analog-circuit performance and finally the mask artwork, which defines the planar mapping of the complete transistor circuit.

A software capability (e.g., simulation) is provided for each stage of design. The software tool associated with a particular level is tailored to the distinctive requirements and desirable design verifications which are characteristic of this level. Note that simulation at intermediate levels or at multiple levels can be successfully accomplished. For example, a timing simulation is a simulation mode which attempts to combine the speed of gate-level simulation and the accuracy of simulation at the circuit-element level (see Subsection 6.2c). Mixed-mode simulators have been developed which permit simultaneous simulation of different parts of the circuit at different levels (logic, timing and circuit-element levels).

The top-down approach may include *backward loops* which represent repeating design cycles. For example, when the design of the complete mask artwork is finished, useful information regarding the corresponding logic circuit, including calculated logic delays, can be extracted from it. The extracted circuit can be submitted to a logic or timing simulator, or even for better accuracy to a circuit simulator. This circuit extraction allows the simulation of the actual circuitry, including parasitic and other device parameters which are revealed during the mask-artwork analysis.

The hierarchical top-down approach which passes through all design stages, as indicated in Fig. 1.1, is not pursued in all cases of complex design. When we restrict ourselves to special structures, it is possible to move from the logic-function specification directly to the mask artwork, sometimes referred to as *silicon compilation* (Subsection 5.3b).

In addition to the required simulations for verifying the correctness of a circuit design, we have to consider the *testing problem* in order to facilitate system maintenance in the field. Just like the simulation problem, the testing problem should be solved in a top-down manner. Functional tests at successively lower levels may detect the location of a faulty component. VLSI chip design, design verification, testing and layout design will be dealt with in Chapters 5 through 8.

1.1c Design Methodologies

Introduction

By a *design methodology* we may include the whole body of selected procedures or techniques and the various hardware and software tools employed in solving a design problem. The advances in device technology, software engineering and system architectures as well as the available design tools and automatic or interactive design facilities have generated a plethora of system-design methodologies which present an overwhelming problem to the novice in the field. This is serious, since there is no single design approach that can be applied to all possible design cases with optimal results. This situation cries out for developing a classification of design methods and tools so that each individual design approach can be evaluated in the proper perspective.

To avoid ambiguity, let us first introduce frequently used terminology associated with the design of complex chips.

Customer, Designer and Manufacturer

When chips are to be manufactured, we may identify three distinctive persons or groups of persons, which we simply call the customer, the designer and the manufacturer. The *customer* wants a chip to be designed and manufactured. The design activity required to meet the customer's specification is called *customization*. The *designer* performs the circuit and system design according to the customer's requirements. Finally, we need a chip *manufacturer* to implement the circuit and system in the form of the desired chip. The three categories (customer, designer and manufacturer) need not reside at different locations. In big companies, such as IBM, AT&T Bell Labs and Philips, all three categories are employed in-house. The customer and designer may be identical so that we have the customer/designer on the one hand and the chip manufacturer on the other hand.

For reasons of profit, semiconductor companies were formerly mainly concerned with the manufacture of standard chips, ranging from simple off-the-shelf chips to microprocessors and large-scale memories. Gradually, these companies are producing more and more final products, such as digital watches, video games, educational toys and even microcomputers. At the same time, system companies (such as IBM, Honeywell, NCR, Burroughs, Sperry Univac and Control Data Corporation), which were founded as mainframe-computer designers, are now equipped with extensive in-house chip-manufacturing facilities. The above trend of bringing together the system-design and chip-manufacturing activities within one company is sometimes called *vertical integration*. The increasing demand for custom designs has created the opposite trend, which will be discussed below.

Silicon Foundries
The past decade has shown a trend toward transferring the fabrication process to separate semiconductor companies which commit themselves exclusively to the manufacture of the chips, which are the physical implementations of customized designs. These companies are called *silicon foundries*. The idea is that one company (or customer/designer) is engaged in the IC design process resulting in the complete set of masks, which is used by another company (the silicon foundry) to produce the desired chips. The reason for the emergence of silicon foundries is obvious. A system designer cannot afford to set up a fabrication line which requires a large starting investment in the equipment needed along with the required skills and experience.

Leaving the chip manufacture to a silicon foundry is not devoid of difficulties. The mere fact that the system company (customer/designer) and the silicon foundry are separate organizations may bring up several questions. In what form should the designer submit the chip-design specifications (e.g., layout-description files for pattern generation) to the chip manufacturer? To what extent is the manufacturer willing to provide detailed information on the processing technology, including the latest developments? In any case, the system company and the silicon foundry should reach agreement on several aspects of design and implementation. Particularly, the performance of an IC design must be accurately verified before committing it to fabrication. For this purpose, many silicon foundries make circuit-simulator models and parameters for their technologies available to the customer. Design verifications must be performed with the proper device parameters and process specifications pertaining to the implementation technology which is supported by the silicon foundry. The layout descriptions must meet the design rules imposed by the technology. The large capital involved in chip production justifies testing the prototype circuit thoroughly. To this end, a special integrated transistor circuit is usually fabricated on the same wafer as the customer's prototype circuit.

There is a wide variety among silicon foundries as to the implementation technology, the integration density, the system performance and the minimum amount of chips that can be fabricated. Some silicon foundries accept small-volume orders, while others require a minimal level of commitment. Important issues in dealing with silicon foundries are the the implementation technology

1.1 COMPUTER-AIDED CIRCUIT AND SYSTEM DESIGN 13

used, the production costs, the simulation facilities and post-fabrication services.

Automated Design

From the viewpoint of design automation, we can distinguish manual design and automated design. *Manual design* proceeds by human methods and the layout design is carried out as a handcraft. The experience of the chip designer plays an essential role. In *automated design*, automatic synthesis methods are employed and the masks are produced automatically by appropriate layout design progams. Manual design is aimed at achieving compact layouts, which may have an irregular structure. It produces the best results in terms of high performance and silicon area utilization. The handcrafted design is profitable for either high-volume production of standard circuits or full-custom designs which require high circuit performance. When the circuit becomes too complex or when a small turnaround time is required, we have to resort to using automatic design tools. Automated design typically differs from manual design in that the layout structure is restricted to standard cells, PLAs, gate arrays or other structures which are amenable to easy automation of the design process. Regular layout structures tend to alleviate the interconnection problem.

Application-Specific Integrated Circuits

An IC design is referred to as *(semi)custom design* or *Application-Specific Integrated Circuit* (ASIC) *design* when the IC package is designed according to the particular specification of an individual customer. As mentioned earlier, the customer and chip designer may be the same person or group of persons. The custom design approach may take several forms. A simple approach is to utilize a manufacturer's standard chips, which are interconnected on a thin-film or thick-film substrate and encapsulated in a single package.

In a full-custom design, all the steps of the design cycle are dictated by the customer's requirements, which are unique in terms of the hardware implementation and the circuit performance. Full-custom design is the right choice when a high circuit performance is mandatory, either in terms of a very high operating speed, an extremely low power consumption or any other desirable property required by the particular application. Special care is exercised with respect to the layout structure and the diagnosability. The use of manual layout techniques imposes strong demands on the ingenuity and the experience of the chip designer. In order to achieve the optimum compromise between the high circuit performance of a full-custom design on the one hand and the constraints imposed by the implementation technology on the other hand, there should be a close liaison between the customer/designer and the chip manufacturer.

When the number of chips required is low, the long design time inherent in the full-custom approach may make the total design cost prohibitive. There is a considerable need for customized designs which must be completed on a rigid time scale. A small turnaround time can be achieved by using *automatic design tools*. A custom design which makes use of automatic design tools is often referred to as *semicustom design*. ASIC design is usually connected with semicustom design rather than with full-custom design. Let us look closer at the trend toward using automatic design tools, which is characteristic to ASIC design.

Selecting a Design Methodology
Which actual design strategy to choose in a particular design case depends on various factors, including the kind of application, the production volume, the implementation technology and the CADCAS tools which are available. Manual layout, standard-cell, gate-array and block layouts can be employed for virtually all kinds of applications. The PLA structure is useful for implementing combinational logic and finite-state machines and hence is suitable for implementing the control path of the system. The standard floorplan technique in the form of a silicon compiler is a special form of the hierarchical block design approach, which is useful for designing microprocessors.

The advent of the microprocessor has opened up two *design style* options:
1. The *hardware-oriented* design style, which involves the design of a logic system and the implementation of dedicated hardware components required to construct the logic system.
2. The *microprocessor-based* design style, which uses a general-purpose data processor (notably a microprocessor) as the central component of the system so that the design can be completed only by developing the hardware interfaces and the software needed for the system to perform the prespecified system function.

Option 1. can be divided into two different design approaches:
a. The *distributed* or *hard-wired* custom design style, which emphasizes manual procedures to achieve a high circuit performance.
b. The *structured* semicustom design styles, which imply the use of structured layouts and automatic methods for achieving short turnaround times, even at the expense of the circuit performance and chip-area utilization.

Option 2. with the microprocessor as the basic component is flexible in that different tasks can be performed merely by changing the controlling software.

An alternative classification of design styles may be based on the *system architecture*. Among other styles, we may discern:
a. The *distributed* composition of logic modules, which are SSI, MSI or LSI building blocks.
b. The *bus-oriented* system architecture, which uses registers and other system components for data processing and storage, all of which are attached to buses which function as communication media between the different parts of the system.

c. *Parallel* and *pipeline* architectures, which use a number of processors, as opposed to the normal Von Neumann architecture.

The ultimate design style to be selected must provide an efficient solution tailored to the particular customer's requirements. The total design costs may include the costs for conventional programming and microprogramming, the design costs for hard-wired methods or the costs for using automatic design tools when using standard cells, gate arrays or PLAs. A special methodology which has gained a widespread popularity in the past decade will be described below.

CADCAS and Expert Systems

The increasing complexity of microelectronic chips has stimulated designers to develop automatic tools which perform the various tasks involved in chip design [Sha89]. Undoubtedly, these tools facilitate and speed up the design process, at least in a large number of individual design steps in the hierarchical design cycle as shown in Fig. 1.1.

A characteristic feature of these tools is that they execute a fixed algorithm, that is, given the specific method for solving the problem at hand, a prescribed procedure of executing a sequence of instructions is followed, leading to a fixed solution, if there is one. In fact, the design specification (including performance criteria and design constraints) define a potential solution space, within which a particular design is selected after performing a number of trade-offs between counteracting design parameters (e.g., operating speed and power dissipation). In the hierarchical design cycle, a number of intelligent decisions have to be made by the human designer. It is the aspiration of the chip designer community to conceive a design methodology that, given a design specification, produces an optimal design without any human intervention. This implies that all tasks involved in the entire design cycle, including the intelligent decision making, must be executed solely by the computer.

In recent years, much effort has been spent to the development of expert systems (see Chapter 4), which solve problems by utilizing expert knowledge (in the form of facts and rules) captured in the knowledge base of the system. Expert systems employ reasoning methods in the way humans do. While various tasks are performed by executing algorithms, intelligent tasks not amenable to algorithmic problem solving can be relegated to expert systems. Although some human intervention can never be dispensed with, the introduction of expert systems into the set of CAD tools will bring the design process nearer to the point of full automation. The use of expert systems not only reduces the turnaround time, but also explains and justifies why a particular solution is proposed.

An introduction to expert systems will be given in Subsection 3.3b and Chapter 4. The applications of expert systems in VLSI chip design, verification, testing and layout design are dealt with in Chapters 5 through 8, after a brief introduction of conventional algorithmic design. It should be emphasized that expert systems are not to replace conventional algorithmic approaches to design, but rather as a complement to them.

Implementation Technologies

A rich variety of implementation technologies and circuit-design techniques are available to the integrated-circuit designer. The decision as to which implementation technology or logic family will be chosen for the circuit implementation is governed by the intended applications of the integrated circuit. In some cases, the designer is committed to using the implementation technology available within the company. Let us consider some technologies.

Bipolar technology. The propagation delay of a logic gate is proportional to the node capacitances and is inversely proportional to the transconductance of the gate. The exponential current-voltage relationships in a bipolar transistor lead to a high transconductance and the attendant high current-drive capability. A bipolar transistor can thus provide a larger current per unit active area than a corresponding MOSFET device can. This explains the good high-speed quality of bipolar transistors. Bipolar technology is primarily suitable for analog-circuit applications due to the high single-stage gain-bandwidth product. A major problem with Si bipolar devices is the power dissipation [Nin86].

The TTL (Transistor-Transistor Logic) technology has long been a standard technology for LSI circuits. ECL (Emitter Coupled Logic) is still being used due to its high-speed properties.

MOSFET technology. The simplicity of MOSFET circuitry and fabrication along with its amenability to further size reduction makes MOSFET technologies very well suited to VLSI fabrication of logic circuits and memories. The high input impedance of a MOSFET gate alleviates the loading problem, allowing a large fanout to be used. Especially, silicon-gate MOSFETs [Mye86] hold promise for the future due to the reduced threshold voltages and speeds as high as 100 MHz. The ease of designing MOSFET circuits has given impetus to the rapid development of LSI and VLSI systems. MOSFET technology is responsible for several landmarks in the history of microelectronics: the pocket calculator, the MOS memory and the microprocessor. The first commercially available microprocessor (manufactured in 1971) was the four-bit Intel 4004. MOSFETs are divided into N-channel and P-channel transistors.

CMOS technology. Complementary-MOS technology is acquiring a growing stake in the world markets for microelectronic chips. These are some of the most striking features of CMOS technology:

a. CMOS chips are praised for their extremely low power consumption. At moderate speeds, the power dissipation is less than a microwatt per gate. CMOS is a good choice when low standby power is imperative, notably in battery-operated applications.

1.1 COMPUTER-AIDED CIRCUIT AND SYSTEM DESIGN

b. CMOS inverters have a nearly ideal DC transfer characteristic. CMOS noise immunity is very much better than those of other technologies.
c. The above properties are valid for a wide operating range of power supply voltages, i.e., CMOS circuits do not need expensive close-tolerance power supplies. The body effect, which degrades the characteristics of non-complementary static inverters, is of no account in CMOS inverters.
d. The fact that CMOS inverters are ratioless facilitates the design of CMOS circuitry. The term "ratioless" is related to the length-width ratios of the transistor channels.

As drawbacks, CMOS circuits require more complex processing and more layout area than NMOS circuits and, in addition, CMOS circuits has the possibility of latch-up.

BiCMOS technology. Merging a CMOS structure with a bipolar transistor leads to the BiCMOS technology which is successfully employed in the design of high-speed static RAMs and gate arrays. This technology offers the system designer the low-noise characteristics of bipolar devices at high speeds, while maintaining active CMOS power-dissipation levels [Cor87].

GaAs technology. Most existing devices are based on the use of silicon (Si) as the semiconductor material. For high speeds, gallium arsenide (GaAs) may offer an attractive alternative [MIL86]. Electrons move faster through GaAs crystal lattice than through Si crystal. Depending on conditions, the electron mobility is 6 to 10 times higher in GaAs than in Si. Since chip density of GaAs technology is rapidly increasing, GaAs becomes a severe contender of silicon. The saturated drift velocity is about twice as high as in Si. On the other hand, the relative effect of interconnection capacitances in GaAs technology on the propagation delays seems to be more severe than in Si technology. The very high intrinsic bulk resistivity of the GaAs substrate minimizes parasitic capacitances and permits easy electrical isolation of multiple devices on a GaAs integrated circuit chip. GaAs technology will prevail at frequency levels greater than 100 MHz. GaAs chips have been employed in the Cray-3 supercomputer.

Though in principle any technology can be used for different applications, a specific technology is particularly suitable for some class of applications: ECL for mainframes, BiCMOS for Superminis, CMOS for minis and personal computers, and GaAs for high-speed applications.

Since integrated circuits started to emerge, feature sizes have shrunk continuously. Submicron technology has already emerged. With minimum feature size, chips are vulnerable to defects and improving the reliability becomes an important consideration.

1.2 CAD/CAS TOOLS IN CHIP DESIGN

1.2a Design, Simulation and Testing

Analog Circuits

Analog circuit design has a long history, starting with the use of vacuum tubes in the early twenties. At the present time, microelectronic circuits employ bipolar, NMOS, CMOS or BiCMOS technology. It is apparent that circuit design exploits the vast knowledge gained in the past. Therefore, new circuit designs are modifications of existing designs.

Although the bulk of this book is focused on digital circuits and systems, the significance of analog circuits should not be underestimated for two reasons:
1. Analog integrated circuits are being widely used as low-cost and high-quality components in various applications, notably in signal processing, including channel filters, analog-to-digital and digital-to-analog converters, modulators, speech processors, operational amplifiers and equalizers.
2. Even digital circuits exhibit an inherently analog behavior if exact timing analysis of digital signals is performed. Though binary logic is based on the use of two distinct signal levels, the transition from one signal level to the other occurs continuously rather than abruptly.

The field of analog circuits and signal processing is changing dramatically for two reasons:
a. VLSI is now maturing, with emphasis on submicron structures and on sophisticated applications that combine digital and analog circuits on a single chip. Examples are advanced systems for telecommunications, robotics, automotive electronics, image processing, and intelligent sensors.
b. The rapid technological developments that are leading toward single-chip, mixed analog/digital VLSI systems require design strategies that bridge the gap between classical analog design and VLSI. MOS switched-capacitor and continuous-time circuit designs, on-chip automatic tuning, and self-correcting and self-calibrating design schemes are but a few examples of these emerging technologies.

Smart and dependable CAD tools are needed for designing analog and mixed analog/digital VLSI, mixed-mode simulators, analog layout tools, automatic synthesis tools, and the development of analog-hardware languages.

In early days of electronics history, the design of an electronic circuit is intimately related to its hardware implementation. The relatively simple circuitry made it possible to verify design correctness by electrical measurements directly on the circuit nodes, followed by modifications, if necessary. Integrated analog circuits have the characteristic that only input nodes and output nodes are available for applying input signals and observing the output responses. Since the internal nodes are usually not accessible for measuring purposes, design verification must be carried out on a realistic circuit model, which is appropriately analyzed. The combination of circuit modeling and the subsequent analysis is called *simulation*. Several circuit simulators have been developed, including SPICE2, ASTAP and NAP2 [Sch87]. Most analog circuit simulators are based on the same principle. The circuit is assumed to be composed of an interconnection of circuit components (transistors, resistors, capacitors, etc.).

Unlike design automation of digital systems, which has grown to maturity, analog design automation has made little progress. The reason is that analog circuits are less amenable to design automation than logic circuits. To minimize design time delays, there is a strong need for improved CAD tools that support analog circuit design. One practical approach to analog circuit design is the use of basic subcircuits and amplifiers as building blocks to construct more cpmplex circuits [All87]. Several articles on automated analog-circuit design have been published. See also Subsection 6.3b.

Though analog design automation has by far not reached the advanced stage of digital design automation, analog system design environments are beginning to emerge. An example is IDAC3 (Interactive Design of Analog Circuits, third version) of the Swiss Center of Microelectronics, Neuchâtel [Deg88]. Its key features are multifunctionality, advanced modeling, new simulation approaches, general sizing algorithms, hierarchy and interactive layout-generation capabilities. A workstation-based environment for the fabrication process and the device level is the Process Engineer's Workbench of Carnegie-Mellon University, Pittsburgh [Str88].

Digital Logic Circuits and Systems

Increasing the accuracy of analog-circuit performance requires better control of physical processes. Digital systems can be made more accurate simply by increasing the number of bits for representing parameter and data values. A higher transistor density in digital chips results in falling prices and better performance of the system. Digital chips are able to encompass more and more complex functions on one chip.

A key aspect of VLSI designs is the computational complexity which increases as the system size increases. Some deterministic computational algorithms are of *polynomial-time complexity*, that is, such algorithms can be performed in a computational time $T = c\,n^k$, where c and k are constants. Unfortunately, many VLSI design problems are non-deterministic and substantially more complex. They are said to be *NP complete*. Appropriate heuristics are used to solve NP-complete problems. The classical solution to controlling computational complexity of a design problem has been to identify subsets of the problem which are computationally manageable.

A VLSI design cannot be committed to chip fabrication, unless the design has has been verified to be correct. In special cases, a formal design procedure can be chosen such that design correctness is ensured (see Subsection 6.1b). In most cases, particularly when complex systems are considered, a wide range of simulation tools at different levels of abstraction must be used to verify the correctness of the design (see Chapter 6).

Design of complex logic systems comprises a top-down sequence of steps going down to lower levels. The number of steps to be considered depends on the complexity of the target system (i.e., the system to be designed). To ensure a satisfactorily working final product, the correctness of the design must be verified at each design level. Complex modern chips have the characteristic that signals can be controlled or observed at only a small percentage of the circuit nodes. Computational aspects of VLSI design have been discussed by Ullman [Ull84].

In all phases of the design cycle prior to the prototype realization, design verification relies on some form of simulation. Referring to the discussion of the hierarchical top-down design cycle given in Subsection 1.1b, different types of simulation must be applied, each associated with a specific level of abstraction. Some remarks follow below (see also Fig. 1.1).

Any system design starts with a design specification which specifies the complete set of requirements that the target system must satisfy. At the system level, the design specification is examined to see if the requirements are consistent and practically realizable. Proposals are made as to which feasible solutions may be considered. It is appropriate to introduce three distinguishable, though related, system levels which emerge when either the system behavior, the architecture or the instruction level is emphasized.

The numerous human manipulations at each simulation level and the necessary switching from one level to the next one in the top-down approach is susceptible to errors. The fragmentation of a design into disjunct simulation stages ignores the fact that we are dealing with one and only one system, whichever simulation level is being considered. It would be convenient to have a single simulation package which could be used throughout all phases of the hierarchical design cycle. Successful solutions have been achieved in combining two adjacent levels. Design systems which unify both the behavioral and structural modes of higher-level simulation include SCALD, SARA and SABLE (see Subsection 6.1a).

At lower levels (logic, switch, circuit), multi-level simulators are useful (see Subsections 6.2c and 6.3a). Though ideal combinational logic is free from timing problems, realistic logic circuits propagate logic signals in a nonzero finite time. This phenomenon may produce malfunctioning of the logic circuit. To avoid such undesirable effects, an appropriate timing analysis is necessary. A large variety of CADCAS tools have become available to the VLSI chip designer [Fic87, Rub87, Sch87].

A very important design parameter is the *design turnaround time*, i.e., the

1.2 CADCAS TOOLS IN CHIP DESIGN

time that elapses from the start of the design project to the shipping of the final product. This parameter is of great importance since an early product release on the market gives the producer an edge on the competitors.

ASICs are final products which usually are produced in small volumes. The smaller the volume, the larger the design effort per chip will be. To overcome this problem, automatic design tools and standardized layout structures are used in ASIC design (see Section 5.3).

Simulation and Testing

The objective of simulation is to detect and eliminate *design faults* due to incorrect design. System or circuit simulation may reveal inadequacies or inadvertences of the designer, static or dynamic hazards in binary sequences and other undesired effects, which cause a circuit behavior to deviate from the design specification.

Unfortunately, in spite of all possible care being bestowed on design and simulation, hardware faults due to *physical defects* (e.g., mask defects and manufacturing process flaws) will occur in the hardware implementation of the logic circuit. Since these faults cause malfunctioning of the circuit, knowledge of their origin is essential to permit design modification or, at least, a checking capability in the maintenance field. Hence, when a fault occurs anyway, we must be able to detect the presence of the fault and, if desired, to pinpoint its location. This task is accomplished by testing the circuit (usually the prototype chip). System maintenance draws heavily upon the testing capability of the logic system.

Types of Testing

The faults which are relevant in logic-circuit testing can be divided into *parametric faults*, which bring about changes in analog-circuit parameters (such as propagation delay, rise and fall times of logic transitions) and *logic faults* which cause changes in the logic function (such as circuit lines stuck at logic 0 or logic 1). There are three methods of testing a logic circuit:

a. Parametric testing for checking the presence of parametric faults.
b. Functional testing for analyzing logic faults in a circuit.
c. Clock-rate testing for checking clocked logic circuits at clock frequencies simulating the normal-rate operation of the circuit.

The need for designing reliable systems is beyond doubt. For the sake of regular checks of working circuits, a system must be testable, that is, one must be able to detect and localize faults. At larger gate densities per chip (accompanied by an increase of the ratio of the numbers of internal nodes and accessible output nodes), the testability becomes a problem of increasing complexity. To ensure the testability, the design process must be performed with the requirement of testability in mind.

When a system in the field is tested, an important point of consideration is

whether or not the system is repairable. Two types of testing can be distinguished:
a. Go/no-go testing, when non-repairable components are rejected in cases of circuit malfunctioning.
b. Diagnostic testing, with the purpose of detecting or localizing the fault. Correcting the circuit design is needed to eliminate the fault.

Testing is so important that it must not be handled as an afterthought. Instead, the testing should be taken into consideration during the design cycle. An important property of a VLSI circuit is its testability. *Design for testability* means that a design is aimed at an easily testable circuit. Testing issues include test generation and fault simulation (Section 7.2). A circuit may be said to be testable if a suitable test pattern applied to the circuit input is able to detect a fault by observing the circuit output.

1.2b Layout Design

The Layout of an Integrated Circuit

When the circuit designer has completed the circuit design, including all necessary design and test verifications at various levels in the design hierarchy, the *integrated circuit* (IC) has to be laid out as a planar structure suitable for chip production. Layout design takes up a large portion of the total design cycle from concept to the ultimate IC chip. Therefore, it is imperative to pay due attention to it.

The layout-design technique to be used depends on the functional and structural properties of the desired analog or logic circuit and on the device technology employed. For example, linear bipolar circuits and digital MOS circuits require different approaches to layout design. In general, the following aims are taken into consideration:
a. An integrated circuit is produced that electrically satisfies the design specifications within specified tolerances.
b. A minimum area is used.
c. Circuits are manufactured with a high yield of production.

Large-volume high-yield production is essential to the economic success of integrated circuits. Usually, a large number of integrated circuits is produced simultaneously on one slice of silicon, called a wafer. This wafer must undergo over 100 individual processing steps, after which the IC chips are cut from it. The production of many identical circuits by a sequence of simultaneous operations on a single wafer makes the technology attractive from the point of view of economy and reliability.

1.2 CADCAS TOOLS IN CHIP DESIGN 23

Layout Design is Mask Design
A silicon (or other semiconductor) integrated circuit is a two-dimensional assembly of electronic devices, such as transistors, diodes, resistors and capacitors, and their interconnection wires (with isolation barriers in bipolar technology). All devices are arranged on the surface of a silicon wafer, each one having a structure that extends into the body of the wafer. The *planar process* which is based on this planar arrangement has become the basic process in IC manufacture.

The essence of layout design is to assign to each electronic device or interconnection wire of the circuit a restricted area of the wafer, usually extending to two or three layers. A manufacturing process may involve the disposition or removal of material at selective areas of the wafer, i.e., a particular process should be effective over specified areas of the wafer and not effective in the remaining areas. To accomplish the separation between effective and non-effective areas, a mask is needed. For the purpose of IC production, a set of masks is required, each of which has a specific function in the sequence of manufacturing processes (diffusion, metallization, etc.).

A *mask* is a square or rectangular plate with a pattern of *opaque* and *transparent* regions. The transparent regions allow the electrical properties to be altered in selective parts of the wafer where the desired components of the integrated circuit must be formed. The smallest geometric figures which outline the opaque or transparent regions of a mask are called *mask primitives*. Most mask primitives are rectangles or polygons. In some cases, circular primitives are used. The complete set of mask primitives of the two *colors*, opaque and transparent, is called *mask artwork*. A point of a mask is either in a region of one of the two colors or on the boundary of two regions of different colors. Depending on the technology, a distinct number of equal-sized masks is required for the fabrication of IC chips.

The desired structure of each device or interconnection segment in the chip is reflected in the transparent/opaque patterns at the assigned locations of one or two masks. The actual integrated circuit is implemented by using the masks one by one in the sequential steps of the production process. This shows the indispensability of masks in IC production and, in effect, layout design boils down to mask design.

Intermediate Mask-Description Languages
In order to define the complete mask artwork unambiguously, a suitable graphics language has to be used. The complete artwork description can be established as a *layout-design file*. This file must specify the geometrical dimensions and the locations of all layout primitives along with the associated mask types (diffusion, metallization, etc.). From the layout-design file, a *pattern-generation file* (PG file) can be derived by using a suitable data-conversion program. The PG file is

used to control a pattern generator which is located in a mask-making firm. The working copies of the production masks which are produced may be sent to a manufacturing firm for the ultimate chip production. Usually, however, the mask-making facility and the fabriaction line are located within the same building.

The layout-design file, applied above to pattern generation, can also be used for other purposes, such as the production of hard copies (check plots), design-rule checking and alternative pattern generation. For this reason, the layout-design file is often referred to as as *intermediate file* (IF). This intermediate file serves as an intermediate between the circuit designer on the one side and the layout-checking and chip-fabricating facilities on the other (Fig. 1.2).

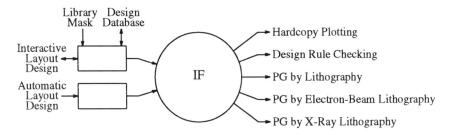

Fig. 1.2 Uses of the intermediate file

The importance of the intermediate file suggests the use of a unified graphics language for describing a standard intermediate file in any IC design cycle. Such a unified language would provide a standard interface between a designer using any design technique and a fabrication facility using any pattern-generation equipment. Hence, the language is used independently of the particular design method, and of the fabrication process and device technology employed. The designer may use an automatic program based on standard cells, an interactive graphics layout procedure, or a batch-type input for the layout. At the output, the IF can be used to derive a machine-readable code of the chip geometry that serves as an unambiguous input to a chip-fabrication facility.

The IF is not a design database but rather an interchange format to couple the IC designer to the chip manufacturer. Many graphics languages have been proposed to function as the intermediate language described above. A number of commands offer the designer a wide variety of functions, including the description of geometric shapes, such as rectangles and straight-line or circular paths with a specified path width.

In *CIF* [Mea80], the mask artwork is expressed as a collection of primitives which includes boxes, wires, round flashes and polygons of arbitrary shapes. Groups of such primitives may be combined to form a macro ("symbol" in CIF terminology) which can be placed anywhere on the chip plane upon a call

1.2 CAD/CAS TOOLS IN CHIP DESIGN

command. The symbol may be mirrored, rotated and shifted at the time of the call. A CIF file contains commands, delimited by semicolons, and terminated by an end marker. The unit of distance is 0.01 μm. CIF uses signed and positive numbers in the range $[-2^{24} + 1, 2^{24} - 1]$. A direction is specified by a pair of integers representing the X and Y coordinates.

There are four or five different *approaches to automated design* based on the *layout structure*: the standard-cell approach, the gate-array approach, the use of programmable logic arrays or other programmable technologies, and the hierarchical block design approach. Subsection 5.3a on ASIC design methodologies will further elaborate on these approaches to VLSI chip design.

1.3 COMPUTERS AND ARTIFICIAL INTELLIGENCE

1.3a Digital Computers

Basic Components of a Computer

A distinguishing feature of a digital computer is that *data* (as the computer-manageable expression of essential information) and *instructions* (i.e., the sequence of operations which the computer is requested to carry out) are generated, operated upon, transmitted and represented as a sequence of bits. The processing unit of a computer processes information in blocks of bits known as *words*. The *word length* is the number of bits contained in one word. Microprocessors have evolved from 8-bit to 16- and then to 32-bit word lengths. Higher word lengths have several advantages, including a higher precision, increased functionality and wider accessibility of existing applications.

A program prescribes the execution of a sequence of instructions. An instruction consists of two parts of a word: an operator and an operand. The operator specifies the specific operation to be performed by the central processor (e.g., ADD or JUMP), while the operand specifies the data or address in the memory to be operated upon.

Conventional computers are of the *Von Neumann type*, which means that all instructions are executed one at a time. Such a computer is composed of three major components: the Central Processing Unit (CPU), the memory and the Input/Output (I/O) Processor (Fig. 1.3). These components are interconnected by means of one or more buses. A *bus* is a set of electrical connection wires through which information can be conveyed in parallel from one unit to another unit in specific time slots.

A program for solving a problem is presented at the input of the computer via some peripheral device, such as an alphanumerical keyboard. At the output of the computer, the result is made visible on an appropriate peripheral device, such as a graphics display screen, a printer or a plotter. The *I/O processor*, which functions as the interface with the outer world, has two specific tasks: buffering and data conversion. Buffering provides the communication between the relatively slow I/O peripheral devices and the very fast processing unit. The different code representations of information data in different peripheral devices and the word organization in the memory require data conversions when data are transferred into and from the memory.

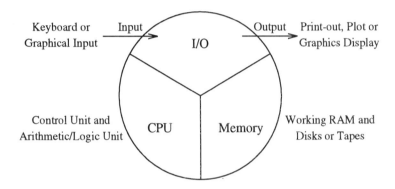

Fig. 1.3 Basic organization of a Von Neumann computer

A *command language* allows the user to present external inputs or jobs to the computer system. Such a language tells the system what activities must be executed as a part of the dialog with the user and messages related to the user requests are returned to the user. Modern user interfaces involve the use of a graphical input device (e.g., a mouse), which controls a menu or other graphical information displayed on the screen. For our purposes, the programs presented by the user to the computer are meant to aid chip design and are called *application programs*, usually written in a high-level programming language, such as Pascal or C.

The *memory* may be viewed as a collection of information data stored in locations. Each location is associated with a unique address by means of which the data can be written into or read from the location. The main memory which is directly connected to the central processing unit is usually a high-speed random-access memory (RAM) with a capacity sufficient to store the working program (instructions and data) and all system software necessary to execute a specific problem-solving task. In addition, disks and tapes are used to store complete programs, files of data and other additional information which may be called upon, whenever needed. During computer runs, bulk transfers of data to and from the main memory may be necessary.

1.3 COMPUTERS AND ARTIFICIAL INTELLIGENCE

At the heart of a computer system is the *Central Processing Unit* (CPU) which controls the information flow and performs the actual data processing within the computer. The CPU has two components: the control unit and the arithmetic/logic unit. The *control unit* performs the basic supervision and synchronization of all other units of the computer, including the time multiplexing of the buses. The control unit decodes the instructions that it receives from the memory and ensures that the appropriate units of the computer operate in the proper sequence as specified by the instructions in the working program. Many computer programs require a large number of operations to be repeated many times. The *Arithmetic/Logic Unit* (ALU) performs the necessary arithmetic and logic operations on the operands specified by the instruction sequence at a high speed. Registers are used for storage of the instructions currently being executed, its operands, the address of the next instruction and other information needed by the CPU. Registers with capacities equal to the computer word length actually function as small temporary memories for the CPU.

Input/output devices are quite slow compared to the electronic speed at which the CPU operates. The resulting idle time of the CPU is reduced by loading more than one program into the memory at a time (*multiprogramming*). Multiprogramming is made possible by introducing an interrupt by a higher-priority request. When an interrupt occurs (say, because of a request of a peripheral device), the execution of the current program is suspended and another program is started or continued. When the requested operations have been completed during the interrupt, the interrupted program can be resumed. The main function of multiprogramming is to keep the CPU and I/O devices as busy as possible. By using interrupts, the processor handles one particular job, while many other jobs are performing I/O operations. When more than one processor is used to fulfill the processing task, we have a multiprocessor system.

The *operating system* of a computer is a set of system programs that functions as an interface between the human user and the computer hardware. It has a supervising and managing task which includes the allocation of CPU time and memory, and the control of peripheral devices, files and system software. It allows several users to share the same computer in an efficient and non-interfering manner. It schedules user programs for execution in an order that optimizes the utilization of the system resources. These resources comprise not only hardware units, such as the CPU, the memory and I/O units, but also system software. In addition to the various kinds of translator, such as compilers and assemblers, system software includes all programs which control hardware resources as well as such programs as program testing aids and text editors.

The Evolution of Computer Design
The extremely high transistor density in VLSI chips allows one to implement

complete systems involving arithmetic and memory functions on a single chip. The many decisions to be made in designing a very complex VLSI system depend on many factors:
a. Cost of system design, development and maintenance by a judicious trade-off between realizing a function in hardware and in software.
b. An effective exploitation of VLSI technological advances in parallel hardware architecture.
c. Flexibility and modifiability of the system design.
d. A high quality as to system performance, throughput, power dissipation, operating speed, testability and reliability.
e. A user-friendly interface accomodating the particular application.

One of the criteria for selecting a computer for a specific application is the *instruction set*, i.e., the unique set of basic operations that can be executed by a particular computer. This set includes instructions for performing data transfer (add, subtract, etc.) and logic (AND, OR, etc.) operations, branching operations (conditional or unconditional jump, subroutine call and return), operations on stacks, and status and I/O operations.

The trend of computers has always been to handle more complex problems [Sta87]. Early computers perform very simple operations, resulting in an enormous amount of traffic in the data bus by the frequent fetching instructions and data from the memory. By allowing instruction sets involving quite complex operations (including building and accessing stacks, arithmetic expression evaluation, etc.), the amount of bus traffic is alleviated. This has led to CISC (Complex Instruction Set Computer) architectures, which need massive hardware to decode these instructions.

The development of computer architectures are closely connected with the complexity of the instruction sets. A computer needs a *compiler* to convert the code written in a particular high-level language into machine-executable code. In effect, complex instructions are broken down into basic instructions from the instruction set. Unfortunately, there are several alternative ways to partition complex instructions. Moreover, some complex instructions may be used so rarely that they cost more to implement than they are worth. The increasing complexity of instructions makes the partitioning process more difficult. This has led researchers to propose RISC (Reduced Instruction Set Computer) architectures, which eliminate complex instructions in the interest of speed and simplicity. RISC's instruction set is restricted to a small set of judiciously selected instructions (see Subsection 3.2a). A precise distinction between CISC and RISC architectures is difficult to make. At any rate, a wide variety of RISC architectures have been developed [Gim87].

The availability of ROMs (Read Only Memories) allows a more efficient execution of arithmetic and logic operations within the ALU. In *microprogramming*, sequences of microinstructions are programmed by single microprogram instructions. Such sequences are stored in a ROM and are executed when they are called upon. For long programs utilizing complex functions recurrently, the programming task is considerably simplified. A *microcontroller*, which is a dedicated processor with its own microprogram stored in a ROM, reads the CISC instructions from main memory and manipulates them as data.

The leading computer manufacturers in the USA are IBM, DEC and Hewlett-Packard. IBM has announced the IBM System/390 mainframe line. Major features are new hardware technologies, enhanced operating systems and the powerful ES/9000 processor series. The channel architecture ESCON (Enterprise Systems CONnection) extends the reach and range of computer systems with high-speed, fiber-optic technology. IBM and Apple Computer have agreed, in cooperation with Motorola, to jointly develop and market new technologies and software.

The greatest increase in transistor density can be realized in memories, due to their regular layout pattern. The rapid decline of hardware costs has led to the increasing use of large numbers of identical processors, contrary to the Von Neumann computer, which contains a single central processing unit, implying that machine operations can only be executed one at a time. A multiprocessor system can increase its throughput by exploiting several forms of parallelism for efficient data handling. This has led to the development of *supercomputers* for rapidly solving large-scale scientific problems. See Subsection 3.2a.

The Impact of VLSI on Computer Design

Over the years, the performance of digital computers have been upgraded by

a. *technological advances* in the speed and reliability of logic devices and logic circuits, and
b. new *system architectures*.

Technological advances include the increasing number of logic functions which can be performed in a chip, the reduction in heat dissipation and the drop in production costs. A computer or system *architecture* will be defined here as the whole of hardware and software organized in such a way that the system properly performs the tasks for which it is designed.

There is an interaction between technological advances and innovative system architectures. Firstly, there is a trend toward integrating more and more identical functions, such as memories and parallel processors, on the same chip. This provides the technological basis for developing parallel computer architectures. The capability of manufacturing cost-effective VLSI hardware has brought forth a myriad of new architectures for either general-purpose or special-purpose system applications. On the other hand, it is up to the system architect to investigate all possible forms of parallel processing and to suggest the manufacture of the required processing units for implementing a particular architecture [McC87].

The desire to integrate more and more different functions on the same chip may impose the requirement of combining different technologies in a compatible manner. To avoid the use of a number of chips for implementing a very large system with the concomitant delays between the chip packages, *wafer-scale integration* may be employed. This implies that the complete circuit covers the entire surface of a silicon wafer, unlike the usual procedure in which a wafer contains a number of identical chips. When very-high-speed operation is at a premium, gallium arsenide can be used as the basic semiconducting material instead of silicon.

When the extreme demands for fast computations (e.g., in real-time applications) are to be met, the required throughputs cannot be effectively obtained merely by technological advances. We have to design new architectures which exploit the presence of the various parallelisms inherent in the computational algorithms for many practical problems. Cost-effective design is achieved by using high-performance LSI or VLSI system modules which are easily adaptable to the algorithmic implementation of various computational tasks.

Software Engineering

Software engineering is concerned with software systems and encompasses all activities relating to the more practical aspects of computer engineering. The *life cycle* of a software system consists of several phases: analysis of software-system requirements and specifications, preliminary and detailed design, coding and debugging, testing and optimization, field operation and maintenance. The main objectives of software engineering are to improve the quality of the software, to maximize the productivity of software engineers and to minimize the cost of all phases of the software life cycle. Characteristic measures which must be minimized include programming time (i.e., the time it takes to get the programs running), debugging time, speed of execution and memory use for the programs and data, cost of maintenance and testing. When different objectives are conflicting, it is inevitable to assign priorities, which will affect the produced software system.

Primary considerations in designing and implementing software systems are: *reliability* (which is prerequisite for high-quality programs), *flexibility* (the capability of applying the programs to a wide variety of problems), *adaptability* and *expandability* (the ease with which a program can be modified and expanded), *portability* (the ability to use the programs in different computer environments) and *user convenience* (which must not be underestimated, although it is not a technical attribute). An important tool to meet these goals is the use of a suitable programming methodology, which enhances the comprehensability, testability and reliability of the software system. Besides the

1.3 COMPUTERS AND ARTIFICIAL INTELLIGENCE

wide range of problem-solving programs, we need special-purpose progams, such as system programs (compilers, operating systems, etc.).

The concept of applying software tools during the software life cycle has led to CASE (Computer-Aided Software Engineering). CASE, which is rapidly evolving during the 1990s, includes everything from design methodologies to application generators and maintenance tools [Fis88, McC89].

1.3b The AI Approach to Computer-Aided Chip Design

Traditional Chip Design

Design is a creative process. A design cycle of a complex logic system involves a sequence of subtasks, each of which requires a complete set of activities at a specific level of abstraction. A short account of such a hierarchical design is given in Subsection 1.1b (see also Fig. 1.1). The human designer has developed automated tools to alleviate the specific subtasks of the chip design. The subsequent levels of design have different essential features: system behavior, realization as a logic system and an electrical circuit, and a geometric layout of the masks needed for the ultimate chip fabrication. Developments in the past have been focused on the automation of most of the activities at the individual design levels. However, linking the subsequent levels has traditionally been a task of the human designer and hence we have an interactive design system.

In an interactive design cycle, the human designer takes an important part in making decisions from alternative subsolutions, at least when a transition takes place from one level to the next lower level. The traditional design cycle is illustrated in Fig. 1.1. The human designer uses automatic tools at each level. Results from a specific design level are transferred to the next level top-down, until the final chip is obtained. The human designer functions as a central controlling and monitoring agent taking care that, given the design specification, the design process proceeds according to the requirements. This is done by intervening in the design process when a change of level takes place. Bottom-up steps are needed when at any level some requirements are not satisfied. In an endeavor to automate the whole design process, a first step is to develop tools which enable a smooth automatic transition from one level to the next. This leads to multilevel simulation and design procedures.

Artificial Intelligence and Expert Systems

There is no consensus among scientists and engineers about what is meant by artificial intelligence. *Intelligence* has to do with several human properties, such as the ability to apprehend existing knowledge and solve new problems and hence create new knowledge by reasoning and other means associated with human skill.

Artificial Intelligence (AI) refers to the trend to let computers perform activities which require intelligence as ascribed to human beings. Heuristics play an essential part in AI. Whereas algorithmic methods are deterministic of nature, following a prefixed sequence of steps leading to the solution, heuristic methods imply exploratory problem-solving techniques, in which rules of thumb and empirical knowledge from past experience form basic ingredients. Major applications of AI include vision systems, robotics, natural language processing and, last but not least, expert systems. For our purpose, *expert systems* will be studied in relation to their use in computer-aided chip design.

A general feature of AI applications is the manipulation of an appropriate amount of knowledge stored in a knowledge base. Particularly in expert systems, knowledge is the mainstay of the business. Expert systems are therefore said to be knowledge-based systems with a humanlike behavior. Knowledge can be represented as facts and rules.

Conventional algorithmic programs usually deal with computation-intensive problems. Efficient programs have relieved humans from dull, repetitive work. Features of expert systems, which are absent in algorithmic programs, include the ability to deal with uncertainty (solutions are given with some measure of probability), the productions of multiple solutions (different alternative solutions) and the interactive way of the design process (including an explanation of how the expert system has reached a certain conclusion). Though particular subtasks in chip design can be solved effectively through the use of an algorithmic procedure, which yields a unique solution, many other subtasks have a nondeterministic character and needs a more heuristic approach. The latter class of subtasks imply uncertainty, choices between alternative solutions and other points, which can best be handled by expert systems.

An essential feature of an expert system is that the knowledge is restricted to a well-defined domain of application. A *domain* is the subject area of the specific application, for which an expert system is designed. The diversity of subtasks in chip design necessitates a division in a large number of domains, for each of which a separate expert system is needed. Artificial intelligence and expert systems will be discussed in greater detail in Section 3.3 and Chapter 4.

Algorithmic versus AI Approaches to Design
A wide variety of computer-aided design (CADCAS) tools at virtually every abstraction level in the design hierarchy have been developed for the convenience of the chip designer. These tools perform specific design tasks automatically by using algorithmic techniques. This design automation has released designers from computation-intensive and time-consuming tasks. As a result, the designer can manage large amounts of design information and concentrate on knowledge-intensive and innovative tasks. The human designer uses heuristics which tell when and how to use them in order to avoid too much

1.3 COMPUTERS AND ARTIFICIAL INTELLIGENCE

expensive processing and the combinatorial explosion. The next step toward a higher level of design automation is to relegate these tasks to expert systems [Beg84, Hol87]. In many applications, where low-level human decisions at present still have to be made, completely automated design systems will evolve. Trends to incorporate expert systems into computer programs are intended to aid the designer.

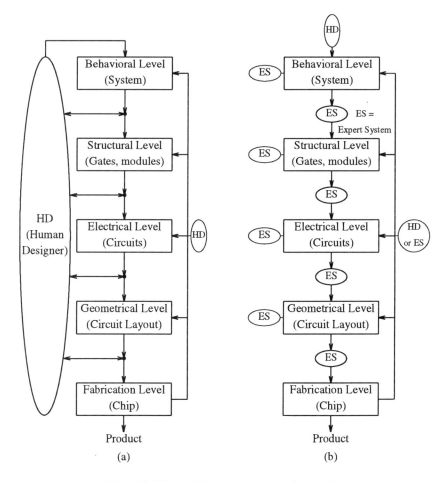

Fig. 1.4 Hierarchical top-down design cycle

The following points are meant to give an introductory exposition of the relative differences between AI methods and traditional methods.
a. Traditional computational methods are based on implementing an algorithm. AI methods abundantly use heuristics. An algorithm is defined as a fixed

sequence of instructions whose execution leads to the desired solution in a straightforward manner. Heuristics comprise indications or clues as to how to solve a problem often in the form of rules of thumb, based on previous experience on a specific domain of experience.

b. The knowledge needed to solve a problem in a given domain is concentrated in a knowledge base in contrast with the data scattered throughout a traditional program.

c. Computational methods use imperative languages. AI problems require declarative languages. Imperative languages are needed for the implementations of the algorithm associated with the computational problem. On the other hand, the reasoning associated with AI problems requires a declarative language, enabling reasoning processes. Imperative and declarative languages are described in later sections.

d. Computational problems deal with quantitative parameters leading to a unique solution, whereas AI problems involve symbolic and qualitative parameters, often entailing approximate and multiple solutions. Solving AI problems generally implies a search in a very large space containing possible solutions.

e. In principle, computational algorithms manipulate fixed data. Expert systems are essentially based on knowledge rather than data. For a chip design, a solution space can be defined. A solution space, if it exists, consists of a virtually infinite number of solutions.

f. Expert systems work in an interactive user-friendly environment.

The AI Approach to Design

AI expert systems are based on expert knowledge, that is, all forms of information as acquired from human design experts in a specific domain of knowledge. The total amount of information is stored in a knowledge base and forms all the knowledge needed to solve a problem.

One could wonder if a complex algorithm for solving a linear programming problem (or any other optimization problem) could be considered as a knowledge-based expert system. As a matter of fact, an optimization program can be viewed as a result of human effort, stored as a piece of knowledge which can be used to solve specific domain of applications. We would say that such a program can be seen as an expert system. However, the optimization program represents a fixed algorithm with a predefined sequence of instructions. This kind of knowledge can be stored in a database. The knowledge associated with knowledge-based expert systems is usually of a more general nature than the above predefined fixed knowledge, consisting of fixed data. The knowledge stored in the knowledge base of an expert system goes beyond the fixed knowledge which can be defined as data, as indicated above. In addition to data, knowledge comprises a set of heuristic rules. Moreover, the data stored in a

1.3 COMPUTERS AND ARTIFICIAL INTELLIGENCE

knowledge base may be given with a probability factor. Essential in expert systems is that during the solution process several decisions must be made, based on specific conditions.

Knowledge-Based System for Chip Design

Let us now examine where and when AI or expert systems can be applied in chip design. Let us consider the problem of designing a complex digital system with a conventional architecture. The traditional approach, illustrated in Fig. 1.1, comprises a sequence of subtasks at several levels. The activities at each level comprises a set of automated processes which may form a coherent whole.

Traditionally, there is a general aim to automate as many design subtasks as possible. To this end, algorithmic methods have been developed. However, most subtasks belong to the class of hard combinatorial problems. This means that the problems are computationally too difficult, that is, they go beyond the polynomial-time complexity, which implies a calculable computational cost. A problem has a polynomial-time complexity when the number of operations is proportional to n^c, where c is a finite constant and n is the parameter which represents the complexity of the problem.

Example: The computational cost of the Gauss-elimination and LU-factorization algorithm for solving systems of linear equations is proportional to n^3, where n is the number of equations (or unknowns). Hence, such an algorithm is said to be polynomial-time complex.

The complexity becomes more serious as chip sizes increase. A way out of this problem is to use expert systems, in which heuristic methods are used abundantly. Since CADCAS tasks generally consist of subtasks of different characters, a collection of different expert systems is needed. This is indicated by ES on the lefthand side of Fig. 1.4b. Though attempts have been made to embrace several levels into an automatic procedure, the total design cycle may be considered to be composed of individual automatic procedures at different levels. As Fig. 1.4a indicates, the transition from one level to the next lower level involves an exploration of possibilities, which is a task reserved to a human designer. This transition often requires a decision as to which of several alternative procedures must be selected. For example, at some time in the design process one of the available architectures, data processor and control styles, ASIC layout structures (standard cell, gate arrays, etc.) must be selected. Such decisions require intelligence and is reserved to the human designer to execute. When the specification at a particular level is not met, a backtrack to a higher level must be performed.

The question arises whether these intelligent tasks could be executed by expert systems. The transformation from a specific design level to the next one implies making decisions from different alternatives. For example, in the behavior-structure transformation, one could consider the question whether a parallel or a series structure should be selected, when it comes to the optimal choice of the architecture. In general, in a level-level transformation, selections between alternatives have to be made. These decisions, traditionally assigned to human designers, could be carried out by expert systems. As Fig. 1.4b indicated, an expert system (ES) performs the transformation from a level to the next lower one. A transition at higher levels requires a different expert system as one at lower levels. Hence, each level transformation requires a different expert system. The design process involves successive transformations of specifications from one domain of knowledge to another. Expert systems work with knowledge in a specific domain and hence should be suitable for performing the intelligent tasks of different character.

The knowledge-based approach to chip design consists of a sequence of problem-solving expert systems which invoke automatic procedures for solving subproblems, where necessary. The design cycle requires a number of different expert systems, each of which embodies an appropriate knowledge base relevant to the subproblem to be handled. The big problem in knowledge-based chip design is to decide which subtasks should be executed by an efficient algorithmic program and which subtasks by a knowledge-based solving procedure. Generally speaking, those subproblems which involve selections from different alternatives or which need heuristic procedures are amenable for use in expert systems, whereas those subproblems for which highly efficient algorithms have been successfully developed can be used as automatic subroutines which can be invoked by an expert system, where necessary. An intelligent knowledge-based CADCAS system consists of a coherent set of cooperating expert systems, each of which can be invoked whenever appropriate [Che83].

Heuristic methods and expert systems which can be used at any stage of the design cycle will be discused in Chapters 5 through 9.

Chapter 2

PROGRAMMING TOOLS IN CHIP DESIGN

2.1 PERSONAL COMPUTERS

2.1a PC Hardware

Characterization of Microprocessors

The microprocessor is the most sophisticated of the building blocks developed for logic systems. It constitutes the key component of a microprocessor-based signal-processing system or the Central Processing Unit (CPU) of a personal computer (PC). Let us now examine which characteristic features distinguish different types of microprocessor from one another.

A *word* is a ordered set of n bits which can be manipulated in a computer in a single step. The number n is called the *word length*. The longer the word length, the more data can be handled in a single instruction and the more powerful and sophisticated the computer will be. For that reason, personal computers are often classified according to their word lengths. An *n-bit microprocessor* refers to the data path from memory to the CPU of n bits wide. In principle, the word length determines the potential applications of the microprocessor-based system. For example, 4-bit microprocessors are used in games or simple control applications (e.g. traffic light, cash registers, etc.), while 16-bit or 32-bit microprocessors permit such applications as tools for computer-aided design.

The *memory size* of the RAM directly determines the number of lines of the address path. For example, a 16-bit microprocessor with 24 address lines from the control unit to memory allows 2^{24} = 16 Mwords of memory to be addressed directly.

Another distinguishing feature of microprocessors is the *instruction set*, i.e., the set of possible basic instructions that can be executed by the microprocessor. Finally, a microprocessor is judged by the operating speed, which is determined by the *clock frequency*.

Summarizing, a microcomputer is specified by the word length (of the data path), the RAM size in kbytes or Mbytes (corresponding to the number of lines in the address path), the instruction set and the clock frequency.

Microprocessors and Personal Computers

The development of ever more advanced microprocessors is reflected in the availability of ever more powerful personal computers. The evolution of the microprocessor and its use in personal computers will be outlined below.

The first microprocessor, released in 1971, was the 4-bit microprocessor Intel 4004, implemented in PMOS technology. The PMOS 8-bit microprocessor Intel 8008 with an address range of 16 kbytes was superseded in 1974 by the NMOS Intel 8080, which has an address range of 64 kbytes and an instruction set of 78. From that moment on, 8-bit microprocessors were primarily based on NMOS technology, the most important being Intel 8085, MOS Technology 6502 and Zilog Z80. The 6502 was implemented in the personal computers: Pet, Vic, Acorn, Commodore 64/128, Apple and Atari. The Z80 with an instruction set of 158 was implemented in Sharp MZ, Tandy 4, Sinclair ZX, Philips MSX and Schneider.

The first available 16-bit microprocessor was the Texas 9900. Many personal computers are based on 16-bit microprocessors, such as Intel 8086, Intel 8088, Zilog Z8000 and Motorola MC68000. The 8086 is implemented in Tulip, Sharp MZ-5600 and Olivetti. The 8088 is implemented in IBM PC, Philips PC, Sharp PC, Tandy 1000, Toshiba and Zenith. The powerful Intel 80286 is implemented in the IBM AT (July 1984), Texas and IBM System/2, Model 50 (1987). Another powerful device is the Motorola 68000, which is implemented in the Apple Macintosh, Atari ST, Commodore AMIGA and Hewlett Packard 200/PC.

Representative examples of 32-bit microprocessors are Intel 80386 and Motorola 68300. Intel's 80386 has been introduced in IBM System/2, Model 80, Compaq Deskpro 386 and Mitsubishi MP386. Intel's 80486 is a more powerful version of its predecessor, the 80386.

In the more powerful personal computers, a coprocessor is used to assist the CPU chip, particularly for performing driving-point calculations very fast.

Configuration of a Personal Computer
The central working horse of a personal computer is the *microprocessor*, which is responsible for all the data processing and control operations. The microprocessor has been discussed earlier (Subsection 1.3a). In addition to the microprocessor unit, a personal computer is provided with at least one input device (notably a keyboard) and at least two output devices (a display monitor and a printer or a plotter). See Fig. 2.1.

The complete set of hardware units as indicated in Fig. 2.1 is not sufficient for solving problems. For this purpose, specific software is needed. All the software needed for solving a specific problem must first be loaded in a RAM. This software includes the required operating system and relevant application programs. The RAM, which is the working memory of the computer, usually resides in the computer housing which also contains the microprocessor. Setting a computer at work requires a number of files to be transferred to the RAM.

Input Devices
Every personal computer is invariably provided with a keyboard. The letter keys

2.1 PERSONAL COMPUTERS

resemble those of a conventional typewriter. In addition, the keyboard includes functional keys and a set of special-purpose keys.

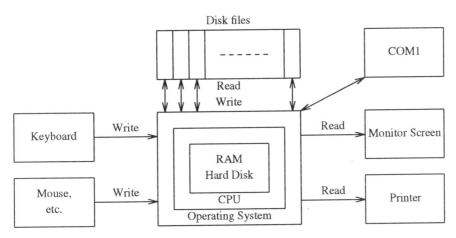

Fig. 2.1 Configuration of the personal computer

To move around the display screen we use a *cursor* in the form of a small symbol (e.g. an underlining dash or an arrow) to indicate the point on the screen where text and graphics can be manipulated from an input device. For example, the cursor must be moved to specific places on the screen where characters or words are to be entered or deleted during a word processing session. The cursor can be moved to a desired point on the screen by pressing the appropriate arrow keys on the keyboard. Cursor movements are possible in horizontal or vertical directions, that is, up, down, left and right.

Easy and fast movements of the cursor can be achieved by using a *mouse*. On the bottom side of this palm-sized input device is a round ball, which can be rolled around on a flat table top. As the mouse is moved in the direction you want the cursor to go, the movement of the ball is translated via horizontal and vertical gears into a shift of the cursor in the desired direction. A mouse has one, two or three buttons on its top surface. Pressing a button once or twice (referred to as clicking or double-clicking) is used to select and run or cancel a particular option within a program.

Personal Computers
During many years, IBM has set the trend for personal computers with the PC-DOS operating system. Many IBM-compatible computers have been put on the market with MS-DOS as the operating system equivalent to PC-DOS. Three types of IBM computers were trendsetting: IBM PC (the first IBM personal computer), IBM XT (requiring DOS, version 2.00 or later) and IBM AT (requiring DOS, version 3.00 or later). IBM PC and XT are based on the use of

the Intel 8088 microprocessor, while the AT uses Intel 80286. The latter microprocessor is also used in the Model 30 and 50 of the new IBM OS/2 series of personal computers. The more powerful Model 80 of this series uses Intel 80386.

Whereas many personal computers are clones of the IBM computer using Intel 8088, some other computer firms pursue a different course with Motorola 68000 microprocessors. First introduced in the Apple Macintosh and soon followed by Atari and Commodore Amiga, a new user-friendly kind of personal computing becomes the vogue, characterized by WIMP features. WIMP stands for Window, Icons, Mouse and Pull-down (or Pull-up) menus. Fast movements of the cursor for editing of text and graphics can be easily performed by using the mouse. The mouse also allows the user to make drawings on the screen. A *help* option provides answers to standard queries. *Scrolling* is the up, down or sideways movement on the screen of a document being edited and displayed.

Icons are pictorial symbols on the screen identifying a file or a specific program that can be selected by clicking the mouse, while the cursor is pointed at the relevant icon. A file can be deleted by pointing the cursor at the icon representing this file and dragging it over to the waste bin icon, while holding down one of the mouse buttons. WIMP allows the user to access files and directories through manipulation with the mouse rather than through a series of learnt key strokes. Usually, a menu bar along the top of the screen displays a number of headings representing menu titles. A menu can be pulled down by moving the cursor to a specific heading and clicking the execute button on the mouse. The pull-down menu contains a list of alternative commands from which one can be selected.

Windows are discrete areas of the screen which independently can accept and display information. Using two or more windows means that several displays are available simultaneously on one display screen. Several windows may be displayed side by side or may be partly or entirely overlapping. Each window represents a document being processed or a programming task in progress. Switching from one window to another is easily accomplished by clicking the mouse.

The IBM OS/2 series computers also have WIMP features by using special software, such as GEM and Windows.

Disk Storage Devices

Files (containing programs and data) can be permanently stored in disks. Personal computers use two types of disk:
a. *Diskettes*, i.e., separate disks which can be put into or removed from a disk drive mounted on the computer.
b. A *hard disk*, i.e., a fixed disk which is permanently mounted within the computer housing.

2.1 PERSONAL COMPUTERS

Most diskettes are of two sizes: 5¼ inch and 3½ inch diameter. The 5¼ inch floppy disk is a flat circular piece of plastic coated with a magnetic oxide material. The newer 3½ inch disks are housed in protective, hard plastic shells. A diskette which can be kept in a box fits into a disk drive built into the computer housing.

A diskette cannot store any file, unless it has been formatted by a program contained in the operating system. This FORMAT program writes appropriate data onto the diskette, defining concentric rings on the surface, called tracks, while each track is broken into sections, called sectors. A sector is the smallest block of storage on a diskette. Most diskettes are now double-sided (meaning that both sides can be used for data storage) and double-density (i.e., 48 tracks per inch).

A hard-disk drive, originally developed by IBM under the name Winchester drive, is a closed unit containing the hard disk which is not easily removable. Winchester drives have storage capacities ranging from 5 to 60 Mbytes and a data transfer rate of 5 Mb/s. Hard disks can be used for storing the operating system and other frequently used programs, such as a word processor.

A personal computer is started by loading the operating system (e.g., MS-DOS) into the system's memory, either from the hard disk or from a DOS system diskette.

Display Systems

The two types of display, monochrome and color, have unique features. The monochrome type can only be used in the text mode, but the resolution is better than in color. Several display attributes allow the characters to be displayed with underline, reverse video, highlighting and blinking. The screen has a uniform coating of phosphor. Dots which are hit by an electronic beam are illuminated to produce an image corresponding to a particular text or graphics display. Since the illuminated dots disappear after a short time, the screen is scanned (50 to 60 times a second) by sweeping across and down in a zigzag pattern to maintain the desired image.

Monochrome (amber or green) monitors use a single gun to illuminate the uniform coating of phosphor on the screen. A high-resolution color monitor uses three guns, each firing and illuminating dots corresponding to a primary color (red, green or blue). All three dots illuminated simultaneously will produce a white dot. By varying the intensity of each beam these dots manage to produce various shades of colors in addition to the three primary ones.

The cathode-ray tube (CRT) display described above is driven by an interface board inside the PC known as the "video display adapter". Through a Motorola 6845 CRT controller chip, this adapter generates the signals for the CRT to display characters and graphics on the screen. Popular adapters are the Monochrome Adapter and Color Graphics Display Adapter (CGA) from IBM and the Enhanced Graphics Adapter (EGA). The IBM Monochrome Adapter is intended for use specifically with the IBM Monochrome Display Monitor and other compatible monitors. It features a display resolution of 720 horizontal and 350 vertical dots. This resolution is far superior than that of the CGA for displaying text.

In contrast, the IBM CGA offers several display modes and can be used with many different types of monitor. Two display modes of CGA are text and graphics. In the graphics mode, the CGA offers medium resolution graphics at 320 horizontal and 200 vertical dots and high resolution graphics at 640 horizontal and 200 vertical dots. The more dots the finer the picture appears. The CGA offers several choices of colors. In the text mode, you can have up to sixteen foreground colors and eight background colors per character, and one of sixteen border colors. In the graphics mode, you can have up to sixteen background and foreground colors in low resolution, and up to sixteen background and four foreground colors in medium resolution graphics modes. The high resolution graphics mode, however, only offers black and white colors for foreground and background. The CGA also offers considerable flexibility in the type of monitor that can be used with it.

Printers

A printer has the task to produce text and graphics on a hard copy. A short survey of printers is given below. A 9-pin-head *dot-matrix printer* is a low-resolution printer which allows graphics to be printed. All characters and figures in dot-matrix printers are made up of dots. The 24-pin dot-matrix printer has a near-letter-quality. The printing quality is the *resolution*, that is, the number of dots per inch (dpi) in horizontal and vertical directions. *Daisy-wheel printers* offer a high resolution for text, but are not suitable for printing graphics. *Ink-jet printers* are based on spraying a fine stream of wet ink on those places of the paper as to produce the image to be copied. A characteristic of this type of printer is the high printing speed.

Thanks to the very high resolution, the *laser printer* has become an essential part of a Desktop Publishing System (see Subsection 2.1b). This printer includes a Raster Image Processor (RIP) which allows both text and graphics to be printed together. Typical laser printers have a resolution of about 500 dpi. This resolution is far short of the performance of professional typesetters (e.g., Linotronic). However, laser printers with ever higher resolution will become available. How a laser printer works may be described as follows.

According to the dot pattern of the image to be processed and copied, an on-and-off pulsed laser beam is directed on a rapidly revolving hexagonal mirror and reflected on a large photosensitive drum. As the drum rotates, the electronically charged tiny spots attract fine toner powder from a toner cartridge. When the drum is pressed on the paper on which the copy is wanted, exposure to heat will ultimately produce a clear image of text and graphics on the paper. The print engine of the laser printer determines the dot resolution of the image, printing speed, the paper handling and the life span of the printer, while the printer controller determines the command sequences and gives functionality to the print engine.

Scanners
When a photograph or other pictorial material is to be included in a document, it has to be digitized, that is, the image must be converted into a bit map, a collection of bits that form the image. Such a bit map is obtained by a scanner which scans the image line by line. The bit map can be stored on a disk and called up for placement on a page. The first scanners were suitable for line art only. At present, they can be used for photographs, even in color.

2.1b PC Software

Text Processing
One of the most frequently used software programs undoubtedly deal with *text processing* or *word processing*. A *text processor* enables one to create a written text on paper with facilities which go far beyond the typing process using a traditional typewriter. A word processor can store blocks of text for future recall and modification and even check the spelling. Word processing has three major elements: file handling, editing and printing.

A central function of each word processor is *editing*, which includes entering text for the first time, inserting text within an existing document, deleting or changing text, moving or copying blocks of text and searching for or replacing (groups of) characters. A special command can be given, when a deviation is desired from the default line spacing, line length, margin and tab settings.

Hundreds of word-processing software packages have been released in the past. Text-processing software for PCs is available from a large number of vendors. The capabilities of this software range from simple text processors to complex programs. Text processors have become popular in large offices and have gained widespread approval. The quality of the printed documents mainly depends on the output printer used. Dedicated text-processing systems include Wang, IBM's Display Writer and Digital's DECMate.

Wordstar was long considered a sort of standard text-processing software. Blocks of text, indicated by its begin KB and its end KK, can be copied or moved to a new position in the document. *Wordperfect* 5, a contender of Wordstar, is now one of the most popular text processors. In addition to text processing at a high level, both Wordstar and Wordperfect provide such facilities as combining functions, spelling control, a list of synonyms and working with

text columns. Advanced text processors, particularly those used in Desktop Publishing systems, have special sophisticated features:
a. A font-selecting feature allowing appropriate titles and headlines to be printed with different formats. A font identifies a specific typeface (boldface, italics, underlining, etc.) and size (often specified in pointsize).
b. Printed letters must be proportional, that is, wide letters (such as W) and narrow letters (such as i) take up proportional spaces on the line. Kerning means that spaces between particular letters (e.g., W and A) are allowed to overlap.
c. Controlling the widths of lines allows the creating of two or more columns on a page. Column formatting requires that the text on the bottom line of a column is continued on the top line of the next column.

Desktop Publishing

The traditional procedure of producing a document, e.g. a book, consists of several phases. The manuscript, written by one or more authors, should be presented in as clean a form as possible, preferably typewrittten using double space on one side of sheet only. This manuscript, i.e., the text along with high-quality graphics originals, including line drawings, graphs and photographs, is submitted to a publishing company whose editor sees to it that all this material is well suited to further handling by the production department. The manuscript which is provided with correction marks and other indications necessary for the printing process is sent to the make-up expert or typesetter who does the make-up of the pages, which includes placing the text and graphics in a way satisfying the wishes of the author. Originally, typesetting has the literal meaning of setting or organizing individual letters upon a page. At present, typesetting uses a professional system of making up the page layout prior to the printing process. A set of proofs is sent to the author for proof reading after which the corrected proofs are printed in a type-setting machine. This procedure, which involves a close cooperation between the author, the editor and the printing office, takes about eight months for a scientific book of conventional size (about 300 pages).

Desktop Publishing (DTP) is an in-house facility centered around a personal computer and appropriate DTP software to do all preparatory text and graphics manipulations, printing and even typesetting activities within the walls of the own firm [Bat87, Lan87]. The main motivation is the saving in time and money by the reduction in turnaround time to obtain preprinting proofs. The manual procedure of cutting out pieces of text and graphics from various sources and then pasting the results into a suitable format is replaced by a screen-based cut-and-paste software program. The camera-ready copy produced on a computer can be sent to a typesetting bureau either physically on disk or electronically via a modem.

A characteristic feature of a DTP system is *WYSIWYG* (What You See Is

2.1 PERSONAL COMPUTERS

What You Get), that is, everything will be printed exactly as it appears on the screen, except that the print quality ultimately depends on the resolution of the printer. WYSIWYG allows text and graphics to be arranged to the full satisfaction of the user. A page, or a portion of it, can be displayed on the screen of the monitor. Portions of a page can be displayed on a larger scale. Another typical feature of a DTP system is the use of user-friendly software with WIMP facilities, as discussed in Subsection 2.1a. The use of windows, a mouse, icons and pull-down menus facilitates the programming task of the user considerably.

In order to allow the printer to adapt to the WYSIWYG features, we must use a *page-description language*, which describes the composition of a page on the screen which must be handled by the Raster Image Processor. By using such a language, a personal computer can control type styles and sizes, draw lines and arcs and fill blank spaces with special text or pictures. Typical page-description languages are Adobe's *Postscript* [Ado85] and Xerox's *Interpress*. Whereas these languages describe only a single page at a time, there are *document-composition languages*, like ImageGen's DDL, which conveys the format of a full document to the printer.

Modern text processors, including Wordperfect 5 and Wordstar, have WYSIWYG features.

The components of a DTP system are shown in Fig. 2.2. A central component is a suitable desktop personal computer at which the larger part of the preparatory work is carried out, including word processing, preparing the graphics part and page make-up.

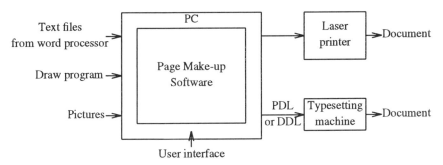

Fig. 2.2 Desktop publishing system

The input ingredients which fill up a page of a document originate from several sources: text files produced by word processors, structured graphics, such as flow charts, technical diagrams, form, etc., data information from database and spreadsheets and, occasionally, photographs or other digitizable copies which can be stored as a file by using a scanner. The printing device associated with desktop publishing systems is at least a laser printer with a minimum resolution of 300 dpi (dots per inch).

There are many possibilities of constructing a DTP system, ranging from simple low-cost solutions to a complete system producing high-quality printed documents. When a low budget cannot afford the purchase of an expensive typesetting machine, the digitized output of a manuscript can be stored on a diskette, which is delivered to a professional type house for further handling. This output may be transmitted to the type house over telephone lines. A factor determining the print quality is the *resolution* of the print. For newsletters, reports and other small documents, a laser printer with a resolution of 300 dpi will produce a satisfactory print quality. For a high quality as required for scientific books, a typesetter, such as Allied Linotype's Linotronic 100 or 300 typesetters, must be used, providing a resolution of up to 2540 dpi.

An important feature of a DTP system is the use of a Raster Image Processor (RIP), interfacing the personal computer and the laser printer. This processor treats both text and graphics to be printed as images which are made up of dots. These dots are scanned line by line by the RIP, stored in the printer memory and reproduced on the printing page. The text is composed with a word processor. Graphics is produced by software, specifically developed for creating drawings and schematic diagrams. Photographs and other pictorial information can be digitized by a scanner.

When the text and all pictorial material are available, *page make-up* software can be used to position columns, headlines, text, illustrations and their captions, on the page as wanted by the author. The most important page make-up programs are the Aldus PageMaker and Xerox's Ventura Publisher.

Apple's Desktop Publishing System

Several dedicated desktop publishing systems are available from some manufacturers, including Apple Computer, Hewlett-Packard and Xerox. To give an idea of a real DTP system, let us outline the Apple Macintosh system (Fig. 2.3).

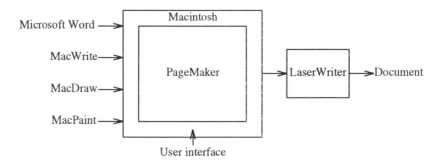

Fig. 2.3 Macintosh Desktop publishing system

2.1 PERSONAL COMPUTERS

The main computer is the Apple Macintosh II with a 20 Mbyte hard disk. The word processor used is Microsoft *Word*, which offers more advanced facilities than MacWrite. In addition to the standard features known with the normal high-standard word processors, Word includes font and style menus to create a wide variety of typefaces in different sizes (from 9 to 24 points) and styles (roman, bold, italic, underline, outline, shadow, superscript and subscript). The text in final form with the desired format can be stored in memory from which it can be retrieved when needed.

MacDraw is a graphics package for line drawings, composed of varying thickness, boxes, circles, arcs, polygons and a choice of hatches or fill patterns, including various shades of grey. *MacPaint* allows freehand black-and-white drawings, facilitated by the use of a mouse, available icons and pull-down menus. Several type fonts and styles, lines of various widths are available and geometric manipulation, such as magnifications, mirror images and movements across the screen are possible.

The page-makeup program PageMaker is used for final WYSIWYG page composition by manipulating the text composed by Word and the graphics created with MacPaint, and occasionally, a picture digitized by a scanner. Reduction in size to see the whole page and enlargements of portions of a page for closer inspection are possible. PageMaker permits up to 128 pages to be stored in a file. This means that inserting additional lines of text on a page will cause shifting the text in the subsequent pages. *MacAuthor* is a package that combines word processing with page layout facilities.

When the make-up has been to the user's satisfaction, the page can be printed on the LaserWriter, which incorporates the page-description language PostScript. Camera-ready copy produced by the LaserWriter can be sent to a typesetter. The AppleTalk network can be used for the data transfer, with a Raster Image Processor inserted between the Macintosh computer and a Linotronic typesetter.

The success of the Apple-Macintosh desktop publishing system has inspired other computer manufacturers and software producers to design easy to use systems with WIMP and WYSIWYG features. MS-DOS imitations of the Macintosh environment include Digital Research's *GEM* software and Microsoft *Windows*. GEM is an easy to use method of dealing with the screen and keyboard. It includes several programs, such as GEM Desktop, Draw, Paint, Graph and WordChart. Microsoft Windows is supplied with its own suite of application software: Windows Write, Draw and Paint.

Multi-User Desktop Publishing Systems

Besides the above desktop publishing systems based on dedicated personal computers, a more powerful computer can be used as the host computer for multi-user purposes. Several terminals are connected to the host computer via a local-area network. A frequently employed operating system for multi-user systems is UNIX. An example of a multi-user publishing system is AT&T's documenter's workbench with its standard UNIX screen editor *vi* and its major formatter *troff*, which was designed to produce high-quality output on a typesetter [Chr87]. Several powerful preprocessors extend the capabilities of *troff* by providing layout facilities for equations (*eqn*), tables (*tbl*), graphics (*pic*) and graphs (*grap*). By using a utility program, the *troff* document can be converted to *PostScript* format and then printed on a PostScript laser printer. This camera-ready copy eliminates the need of extensive re-editing, typesetting and proofreading on the part of the publishing house.

Another UNIX-based publishing system is Interleaf Workstation Publishing Software from Interleaf Corporation. This system is implemented on the IBM6150 UNIX computer, and the DEC, Sun and Apollo workstations. It offers a WIMP and WYSIWYG environment.

Spreadsheets and Decision Support Systems
In many cases, there is a need for making proper decisions based on results obtained by manipulating and evaluating available data. This need has led to the development of Decision Support Systems (DSSs). The simplest form of a DSS is the *spreadsheet*, which enables one to carry out matrix-oriented calculations. The basic form is a matrix of r rows and c columns of cells. Usually, a spreadsheet tool supports arithmetic functions, trigonometric expressions, conditional expressions and miscellaneous functions.

The decision support or strategic policy in a spreadsheet program can be determined by applying the *what if* or the *goalseeking* statement. At the *what if* question, numbers and formulas are introduced and then it is observed what is the effect of choosing specific numbers on the other numbers. *Goalseeking* is the opposite concept. Specifying the desired final result, it is investigated which parameters must be changed in order to achieve this result. Though spreadsheets were particularly developed for financial applications, such as bookkeeping, they can also be used for solving simple electrical circuit problems [Bre87].

Representative examples of spreadsheets are Visicalc and Microsoft's Excel. Spreadsheets have a number of drawbacks:
a. The maximum number of rows and columns is fixed.
b. Many spreadsheets are not portable: they are not suitable for mainframes, minis as well as personal computers.
c. The documentation capabilities of most spreadsheet software are poor.
d. It is difficult to change the computation rules.
e. It is not possible to protect a part of the working space from unauthorized use.

A more sophisticated form of spreadsheet is the Decision Support System. A DSS is suitable for users who

2.1 PERSONAL COMPUTERS

- want to communicate with other files,
- can at least utilize the capabilities of a spreadsheet,
- make high demands on the portability between users and computer systems,
- want flexibility of using the software on mainframes as well as personal computers, and
- want a high degree of flexibility in realizing applications.

A Decision Support System generally consists of three parts: a database, a model library and a software system.

Integrated Software Packages
End users will increasingly be given facilities to format their documents into meaningful graphics displays, reports and intelligent spreadsheets. Software packages for personal computers have been developed in a wide variety: text processing, performing calculations, carrying out administrations, etc. Transferring data from one package to another package (e.g., when tables or spreadsheets must be inserted in a text) is usually not an easy task. A way out of this problem is provided by bringing together various software packages into one integrated set. A typical example of such a set is Symphony [Cob84], in which text processing, spreadsheet, database, communication and graphical processing have been put together. Other integrated software packages include Lotus 1-2-3, version 3, and SuperCalc 5.

The Wordperfect Corporation offers a family of the following software packages: Wordperfect for text processing, DrawPerfect for graphics presentation, PlanPerfect for spreadsheet operations, DataPerfect for database applications and WordPerfect Office for network use.

2.2 OPERATING SYSTEMS

2.2a Operating System Principles

What is an Operating System?
An *operating system* is a specialized software environment in which other software operates. It is a set of utility programs which provides the interface between the computer system and its users. The operating system sees to it that shareable and nonshareable resources are allocated to the users and other parts of the computer system such that their utilization is optimized. This task can be done by using *multiprogramming*, i.e., several jobs (user programs) may reside concurrently and compete for using the various resources.

The efficiency of an operating system can be measured by the *throughput*, i.e., the total amount of work which can be processed per unit time. The inevitable differences in the speeds of input/output devices, memory devices and the CPU require that the data processing be scheduled so as to maximize the throughput. A computer user is interested in the *turnaround time*, i.e., the time it takes for a programming job to be processed. Some operating systems handle the various jobs with different priorities.

From the viewpoint of a user, an operating system provides an interface to the computer hardware which makes it easier to code, test and execute computer programs for solving particular problems. *Systems programming* is concerned with the design, implementation and maintenance of the programs that constitute the operating system. Systems programs include assemblers, compilers, linkers, loaders, debuggers and command-language interpreters. An operating system may serve special purposes, such as text editing, program development, real-time process control, data communication and file management, or it may be a general-purpose system providing a wide range of services for a varied user community.

In terms of the way in which access to a mainframe computer may be available, we may have batch access (for processing one program after the other) or time sharing (multi-access system for shared use among many operators, permitting concurrent interaction with the computer system). A stand-alone facility is one that is dedicated to a single user and is physically accessible through a control panel. It may provide either interactive facilities similar to those in a time-sharing system or a batch facility.

File Systems

Files are collections of related records which are stored in sequential or random-access storage devices (tapes, disks, etc.). An appropriate number of files can be organized as a *file system*. A *file-management system* provides the service of managing the creation, insertion, modification, retrieval and deletion of the individual files in the file system. The files are cataloged by their file names in a *file directory*. Usually, a designer is given a separate storage device, such as a floppy disk on which a private file can be stored. In time-sharing systems, private files are stored in a large storage device on behalf of the individual designers. Such files can be retrieved by supplying the file name which was assigned when the file was created.

File access is concerned with the manner in which the files are read or written. A text editor uses line numbers as keys. In general, files in RAMs (random-access memories) can be accessed by their file names or record keys to identify the memory address. A file system typically contains modules which are organized hierarchically. Then, accessing a file requires a tree search combined with linear searches, scanning the main directories and proper subdirectories until the desired file is found. In order to reduce the number of disk accesses and

2.2 OPERATING SYSTEMS

thus decrease the average transaction processing time, records are arranged into blocks.

Fundamental Concepts
Primarily, an operating system is a collection of systems software, such as assemblers, compilers, linkers, monitors, input/output controllers, command-language interpreters, memory-management routines and file-management utilities. Besides the systems software, an operating system is often considered to include those hardware components, such as I/O and memory devices, which are intimately related to the systems software. Important issues in operating systems are the protection of an individual user from either malicious or inadvertent actions of other users and the managing of the overhead involved in the concurrent execution of diverse processes.

Crucial to the understanding of an operating system is the concept of *process*, which can be defined as a set of states of a program in execution or the activity resulting from the execution of the program. Concurrent processing needs appropriate techniques of synchronization. In concurrent programming, a *monitor* is a shared data structure and a set of functions that access the data structure to control the synchronization of concurrent processes. When a process desires access to the shared variables, it must call the monitor, which either grants or refuses access. A process may have to wait on one or more events to be true.

Many operating systems are organized as a collection of separate system modules, providing different services under the supervision of a monitor which keeps control of the status of each system resource and user program. An alternative organization of an operating system is based on a *kernel* or *nucleus*, which is a small monitor whose function is to support multiprogramming, i.e., the concurrent execution of several system modules or user programs. An essential part of a *kernel system* is formed by interrupt handlers, one for each kind of interrupt that can occur. Routines are required for process management and synchronization, I/O device handling, memory management and protection mechanisms.

Examples of Operating Systems
The IBM 360-370 mainframes run a number of operating systems, notably OS-MVS, OS-VSE and VM/370. These systems support virtual memory so that a user's program can be partitioned, allowing segments of the program to be swapped between the primary and secondary memories. Digital Equipment Corporation provides at least three operating systems for the PDP-11 minicomputers: RT-11, RSTS and RSX-11 (registered trademarks of DEC). RT-11 can serve two jobs with a priority granted to one of the jobs. RSTS is a time-sharing system serving a number of users with different computing needs. RSX-11M-PLUS (a superset of RSM-11M) is a disk-based, multi-user operating

system that provides a multiprogramming environment using a priority-structured, event-driven scheduler. VAX computers are run under the VMS operating system [Mil92].

The THE (Technische Hogeschool Eindhoven) System is an operating system developed by a small group led by Dijkstra [Dij68]. It is based on a kernel which implements processes and semaphores. All other modules implement processes which are synchronized by means of semaphores. A semaphore is a data structure which provides a means to suspend execution of a process until certain conditions are satisfied. MULTICS (Multiplexed Information and Computing Service), written primarily in PL/I, was developed by MIT, Cambridge (Mass.), in cooperation with General Electric and Honeywell. Its principal goal was to implement a time-sharing system providing the utility of various computer resources to a number of remote users with special attention to security and privacy users. The UNIX system, which in some respects is influenced by the design of MULTICS, will be described in greater detail.

The UNIX System
The name UNIX is a trademark of AT&T Bell Laboratories for a family of powerful, interactive, multi-user operating systems along with a rich variety of utility software. UNIX, developed and maintained by AT&T Bell Labs, and licensed by Western Electric Company, has gained an enormous popularity, notably at universities all over the world. One of the major properties of UNIX is its portability. Originally developed for DEC PDP-11 minicomputers, UNIX is now in use in a wide range of computers, from mainframes to microcomputers. The UNIX kernel and the utility programs are written in C [Ker88]. Detailed information concerning the UNIX system can be found in several textbooks.

An important feature of UNIX is its hierarchical file system supporting demountable modules. The files are of three types: ordinary files, special files and directories. An *ordinary file*, identified by a name of maximally 14 characters, is a string of characters, e.g., a program to be executed. a *special file* is read and written just like ordinary files except that it is related to some physical storage medium, such as the primary memory or any secondary memory device. *Directories* (containing lists of files) are organized in a hierarchical structure, with names assigned to ordinary files, special files or other directories. Input and output file names can be passed as parameters to functions so that the user need not be aware of the physical locations of the sources and destinations of the files. At the root of the tree structure is the file-manager module, which defines directories and implements operations. Normally, system users do not have access to the root.

The UNIX kernel schedules processes, allocates memory and supervises data transfer between the main memory and the peripheral devices. Each user process executes commands or programs which may use the file system. When a user

2.2 OPERATING SYSTEMS

(that is, someone who is allowed access to the system) logs in by entering an authorized name, the current directory associated with this name is accessible to the user. Each user can create additional directories, which become subtrees under the user's name. Files within the tree structure are named by indicating the path to be followed in the tree. For example,

/ user / dir1 / ... / dir$n$ / filename

describes the top-down path which must be taken from the current-directory user name to the desired file name. Each transition to a lower level in the tree is indicated by a slash /. The *shell* is a command interpreter which acts as an intermediary between the user and the UNIX system. A UNIX function is invoked by using a simple command, which consists of the program name followed by arguments. The shell transfers control to the appropriate file and, as soon as the command execution has been completed, it writes a dollar symbol to indicate its readiness to accept the next command. Program inputs and outputs are normally in standard input-source and output-destination files respectively, which may be the user's terminal. UNIX's transfer capabilities extend to compatible file, memory device and interprocess I/O processes. Commands may be strung together with a vertical bar, indicating a pipeline of different processes to be succesively executed in different files. UNIX commands also provide multitasking, with a priority order assigned to different processes. For example, disk events have a higher priority than character I/O events.

The widespread acceptance of the UNIX operating system has led to the adoption of UNIX as a standard operating system for a wide range of computers, though UNIX is most powerful from the supermicro to the small mainframe level. A large number of languages, including C, FORTRAN77, RATFOR, BASIC, APL and Pascal, are supported. UNIX has spawn a host of utility tools for software development, including about 200 commands (or programs). All these features provide a welcome software-development environment which enhances the programmer's productivity. UNIX has powerful facilities for the text processing prose (including spelling and formatting checks), mathematical equations and statistical tables with the capability of linking to the phototypesetter, which provides high-quality master documents for reproduction. The text editor is very useful in on-line software development.

Two other utility programs worth mentioning are LEX and YACC, which in combination with the available C compiler can perform useful compiling tasks. LEX is used to generate a lexical-analysis routine and produces a stream of tokens as output. YACC (Yet Another Compiler Compiler) is a parser generator which takes a syntax specification as input and produces either a C or a RATFOR code of a parser subroutine. The compiled parser subroutine transforms the stream of tokens delivered by a lexical analyzer into a sequence of appropriate actions.

Several versions of UNIX have been introduced: AT&T Bell Labs version V, the Berkeley variant 4.2/4.3, IBM AIX (Advanced Interactive Executive) and DEC's ULTRIX. The 4.2 BSD version of the University of California, Berkeley, is the first operating system to focus on superior communications and networking capabilities. A Microsoft PC variant of UNIX is XENIX. Two groups (AT&T Co and Sun Microsystems Co on the one hand, and the Open Software Foundation on the other) are trying, each in its own inimitable manner, to improve the user interfaces of the UNIX operating system.

2.2b Operating Systems for Personal Computers

Introduction
Every personal computer requires an operating system that takes care of the information transport between the keyboard, the monitor (display unit), the disk drives and all other peripheral devices. An operating system enables the computer user to create, print, copy or otherwise manage files. It also serves as a program-development environment.

Each personal computer operates in conjunction with a specific operating system. For that reason, personal computers are often classified on the basis of the associated operating system. This is important since application software is written with a specific operating system in mind and hence can be used by personal computers which run on this operating system. Personal computers which run on similar operating systems can handle the same application software and hence these computers are said to be *compatible* with each other. For example, MS-DOS computers are said to be IBM-compatible, since the operating system MS-DOS is similar to PC-DOS, which is the operating system of IBM personal computers XT and AT.

CP/M
CP/M (Control Program for Microcomputers), implemented by Digital Research, was the first widely used operating system for microcomputers and is still being used for 8-bit personal computers.
CP/M-86 is a single-user 18-bit version of CP/M and hence is usable for 18-bit personal computers.
Concurrent CP/M-86 is a multitasking version which can run up to four programs simultaneously.

MS-DOS
PC-DOS for the IBM personal computers (XT and AT) and its companion operating system *MS-DOS* (Microsoft DOS) for IBM compatibles have been the most popular operating systems of 16-bit personal computers for many years. Both systems were developed by Microsoft Corporation. An interesting fact to note is that Microsoft also wrote ROM *BIOS* (Basic Input and Output) which can

2.2 OPERATING SYSTEMS

be considered as a small operating system built into the computer hardware. Microsoft's BASIC is included as part of the DOS system. BIOS is hard coded as a ROM module of the PC (like the ROM BASIC module). When the system is switched on, BIOS carries out a number of tests to see if there are hardware faults in some components of the computer, such as the time registration of the built-in time clock. BIOS routines provide the working interface between the operating system, or an application program, and the I/O devices, such as the display monitor, keyboard, disk drives, printer and modem.

PC-DOS and MS-DOS are virtually identical. Below, general common features of (PC and MS) DOS will be summarized. The first version 1.0 of DOS was released in August 1981. Improved versions have been released, the last one being DOS 3.2. This version is suitable for use in a token ring network and supports the use of 3.5-inch disks.

DOS (Disk Operating System) has all the standard tools for disk and file management (the manipulations of data stored in files). A *file* contains related information on a storage device. Typical file operations include creating files, writing to files, reading from files, modifying and appending data in existing files. Data files are usually maintained on disk storage devices. A file is identified by two parts: a *filename* (up to eight characters long) and an *extension* (up to three characters long), separated by a period. Common DOS file extensions are:

- ·COM   Command file for executing a DOS command
- ·EXE   Executable program file
- ·SYS   File containing DOS system information
- ·BAT   Batch files with DOS commands

The on-screen appearance of the DOS prompt (A >, B > or C >) indicates that the operating system has been loaded and the computer is ready to receive and execute the commands. It is assumed that the system is equipped with two floppy disk drives (referred to as A and B) and a hard disk drive (referred to as C). The default drive is the one that shows up in the DOS prompt (A >, B > or C >). This drive is also the one for which the commands are meant. When a change of the default drive is wanted, the new default drive is to be specified. For example,
C > A:
indicates that the current default drive C is to be changed to A. The next DOS prompt will then be: A >.

DOS commands, file names, etc., can be entered either in uppercase or in lowercase characters. In what follows, they will be shown in uppercase. Some commands have shorter and longer forms. For example, MD or MKDIR for the Make Directory command, and CD or CHDIR for the Change Directory command.

DOS commands always follow the DOS prompt A >, B > or C >, depending on the current default drive. A text following the command REM is considered as remarks or comments. Frequently used DOS commands are listed below. The comments on these commands are enclosed by the symbols /* and */. Note that these symbols are no DOS standards.

CLS /* Clear the screen */

DIR B: /* Display the directory of the disk in B. If the parameter B: is omitted, the directory of the default disk is displayed */

DIR B: /P /* Display the directory of disk B one screenful (i.e., 24 lines) at a time. The next page is displayed on typing any key of the keyboard */

TYPE \ FILE1 /* Display FILE1 on the screen */

TYPE \ FILE1 | MORE or MORE < FILE1 /* Display FILE1 one screenful at a time */

PRINT \ FILE1 /* Print this file on the printer */

DISKCOPY A: B: /* With an original disk in drive A and a blank formatted disk in drive B, copy all the contents of the original disk to the blank disk. When only one floppy disk drive is available, the user is asked to place both disks alternatively in this drive */

A > COPY PROJ.TXT C: PR.T /* Copy the file PROJ.TXT on the disk in drive A to the hard disk C under the file name PR.T */

COPY C: * ·EXE A: /V /* Copy all files with the extension EXE from the hard disk C to the disk in drive A. The symbol * is a wild card character denoting any string which in this case is followed by ·EXE. The symbol /V means that the copy must be checked after copying has finished */

COPY FILE1+FILE2+FILE3 NEWFILE /* Create a new file NEWFILE by combining the files FILE1, FILE2 and FILE3 */

ERASE B: FILE1 or DEL B: FILE1 /* Erase file FILE1 from the disk in drive B */

ERASE *.* or DEL *.* /* Erase all files on the disk in the default drive */

RENAME OLDF NEWF /* Change the file name OLDF into NEWF */

CHKDSK B: /F /* Check the directory and file allocation table of disk B. The option /F corrects and displays detected errors */

RECOVER FILE1 /* Recover file FILE1 with bad sectors or a damaged directory */

BACKUP and RESTORE are available for protecting data in the hard disk.

2.2 OPERATING SYSTEMS

The control key CTRL of the keyboard in combination with one of the letter keys corresponds to a specific command. For example,

CTRL-C   /* Instruction to end the program */
CTRL-S   /* Stop scrolling */
CTRL-P   /* Stop output to the printer */

Tree Structure of DOS Directories

Peripheral devices of a microcomputer system can be divided into input devices and output devices. Figure 2.1 shows a general configuration to illustrate a number of possible flows of data from input devices to output devices. Central to this system is the microcomputer and its associated operating system, which makes the system ready to perform its tasks. The RAM is the memory device which stores all data files during a computing session, needed to perform a specific task.

The default input device is the keyboard. In user-friendly systems, a mouse or other devices may be used as input devices. The default output device is the monitor with its display screen. The default situation in running a program occurs when the computer user enters all commands, programs, input data and goals through the keyboard, while the monitor displays these data and the required results on the display screen.

Usually, additional I/O devices constitute essential parts of a computer system. A printer, used as an output device for providing printed output, is connected as a parallel device. COM1 is the first of the serial-type outputs of the computer. It can be used as an input or an output device. Disks (in drives A, B or C), which contain data files, can also be used as input or output devices. Data can either be read from or written into a disk. An output can successively be sent to different output devices, e.g., first to the screen, then to a disk and finally to a printer. The assignment of different output devices as the write device is known is *redirection*.

DOS provides several features, including redirection, filters and pipes. *I/O redirection* enables us to select a variety of operations to perform on files as we make up a command. A *filter* is a program that reads data from a standard input device, modifies it in some way and then writes it out to the standard output device. A *pipe* is a cascade of filters which processes the data in different ways during the data flow from input to output.

Many personal computers include a hard-disk facility. Three steps must be performed to prepare a hard disk so that DOS can be started from it:
1. Use the FDISK command to identify the hard disk to DOS.
2. Use the FORMAT command to format the hard disk.
3. Use the COPY command to copy the DOS files from the DOS system diskette to the hard disk.

Management of files in a hard disk is obtained by a hierarchical tree-structured organization. Each directory may point to subdirectories as well as to files. The first directory, called *root directory*, is typically used to point to several subject-specific areas of the disk. Each of these subject-specific areas has its own directory. Sometimes, these areas are themselves subdivided into smaller groups of files. Figure 2.4 shows a simple example of the tree structure of directory.

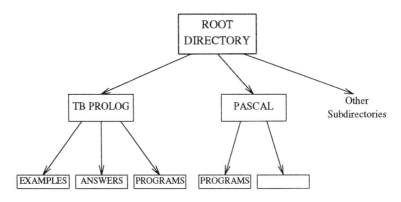

Fig. 2.4 Directory tree structure

The process of finding a file boils down to following a path from the root directory (or the current directory) down to the required file. This is symbolically indicated by a sequence of directory names separated by the \ symbol.

Example: The file PROLOG.HLP is found by writing
C: \TBPROLOG/PROLOG.HLP

Creating subdirectories in a directory structure of a fixed disk is done with the MD command (Make Directory).

Example: MD \TBPROLOG
             MD \TBPROLOG\PROBLEMS

By default, a subdirectory is created in the current directory. The current directory is the one we are working in; if a command is typed, it will address that directory. When DOS is started, the default directory is the root directory of the default drive. By specifying a complete directory path, a new subdirectory will be created anywhere you want.

Accessing files and programs from the various subdirectories can be done by changing the current directory, specifying the path to a file or using the PATH command. The CD command (Change Directory) is used to change the current directory of the specified or default drive.

2.2 OPERATING SYSTEMS

Example: CD [path]

If the backslash \ is used as the first character, DOS starts its search in the root directory. Otherwise, it refers to a subdirectory of the current directory.

Example: CD \TBPROLOG\PROGRAMS
            CD \PROGRAMS

To check if the current directory has been changed, the CD command can be used without any path specification. DOS will respond with the name of the current directory.

It is sensible to make working copies of disks containing new software. Before copying, the destination disk must first be formatted. Assuming that the DOS operating system is operative and the disk has been placed in drive A, the format command is

FORMAT A: /S

where the /S option causes the operating system files to be copied to the disk so that they can be used to start the computer. Any information previously contained on the disk will be destroyed.

When two disk drives A and B are available, the contents of an original disk placed in drive A can be copied on the blank formatted disk placed in drive B. The command is

DISKCOPY A: B:

When only one disk drive is available, the system will prompt the user to place the original disk and the blank disk alternatively in this drive until all data have been copied.

When the contents of a new disk is to be copied on the hard disk of a computer, we have to take into account the hierarchical tree structure of the directory. Suppose that the file to be stored in the hard disk has the filename NEWPROG and the original disk has been placed in disk drive A. Then, the successive commands to be given following the system prompt C > are

MD \NEWPROG /* A subdirectory under the name NEWPROG is created */

CD \NEWPROG   /* This command changes the current directory to the subdirectory NEWPROG */

COPY A: *.* C: /V /* All files from the disk in drive A are copied to the directory C: \NEWPROG. The symbol *.* means that all files have to be copied. The option /V means that the copy has to be checked for correctness */

When the file NEWPROG is to be called again, the following commands are needed:

CD /NEWPROG /* This command brings us to the NEWPROG directory */

NEWPROG   /* The file NEWPROG is called and loaded to the RAM */

The TREE command helps us in keeping track of tree-structured directories.

TREE B: /F | MORE /* Display all the directories on disk B. The vertical bar | is the symbol used to indicate piping. In this case, the program TREE is followed by the filter program MORE, which displays the text one screenful at a time. The option /F displays all the files in the root directory and in all subdirectories */

Other standard filters in addition to MORE are FIND and SORT.
DIR | FIND "01-06-92" | SORT > DIRS /* All lines containing the string 01-06-92 are sorted alphabetically by the SORT program and placed in a file named DIRS. The > symbol is used to denote redirection */

There are three external DOS commands: LINK, EXE2BIN and DEBUG. The LINK command is used to link program sections produced by compiling source programs written in different languages (such as Pascal, C and Prolog) or produced by assemblers. The result is a file with an EXE extension. EXE2BIN converts EXE files in a more efficient format. DEBUG helps the user detect and correct errors in programs.

Though not comparable to a dedicated word processor, DOS includes a program EDLIN, which allows one to write and print short documents, such as memos or lists.

DOS-5

DOS-3 is limited to using memories with a maximum of 640 kbytes. Expanded memory in blocks of 64 kbytes can be used to surpass the 640-kbyte limit up to 1 Mbytes. *DOS extenders* on 80286, 80386 and 80486 computers provide programs access to memories up to 4 Gbytes. No special software is needed to run standard DOS-4 oriented programs.

The latest improved version of DOS is MS-DOS-5, which will be incorporated in new IBM-compatible personal computers. MS-DOS-5 has a user's shell, which can be operated via a menu interface, similar to Windows. It allows the user to switch to another application without leaving the current application. The new software has new programming features, including an improved editor, a help facility, an UNDELETE function (which allows files wiped out unintentionally to be restored) and an UNFORMAT facility (which undo the effects of the dangerous FORMAT command). MS-DOS-5 is supplied with the Q-BASIC interpreter.

MS-Windows

A new concept in user-friendly interfacing is the *window*, which allows us to view parts of a large imagined two-dimensional space, e.g., containing a large printed data file. *Windows* is the answer of software manufacturers to the demand of clients to make operating systems more user-friendly. Windows

2.2 OPERATING SYSTEMS

divide the display screen into several subwindows, enabling the system to carry out simultaneously different tasks, such as file editing, electronic mail, program compiling, etc. [PEt88].

Windows was first used in the Apple MacIntosh personal computer. Because of the great success, this concept is copied on other systems. Mostly, Windows is generated from the application. A disadvantage was that each application had its own way to produce Windows. An attempt for standardization is X-Windows (see Subsection 9.1b). The mouse, icons and pull-down menus now constitute daily tools of the PC user.

MS-Windows 3.1 employs a DOS extender and is a recent attempt of Microsoft to give the MS-DOS system a more user-friendly appearance. Users of MS-Windows are no more confronted with the meaningless prompt A > or C >. Like the usual practice with Apple MacIntosh computers, Windows 3.1 enables MS-DOS users to control the computer by finger tips. The system requirements for Windows 3.1 are: 286 or higher with 1 Mbytes (for standard mode), 386 or higher with 2Mbytes (for enhanced mode), a mouse and a supported monitor. Multimedia extensions require a sound board and/or a CD-ROM drive. With Windows/386, specifically developed for 32-bit machines, the MS-DOS world has been provided with a powerful tool, which can compete with computers, such as Macintosh II.

New Developments

OS/2 is an operating system which was recently developed by IBM for the IBM computers, Models 40, 50, 60 and 80 [Let88]. OS/2 was announced by Microsoft and IBM as a successor of MS-DOS. One aim is the effective integration of mainframes, minis and PCs. Many powerful features have been added to MS-DOS: the 640-kbit limit for the internal memory is shifted to a maximum of 16 Mbits, while several tasks can be executed simultaneously. A user-friendly graphics interface using a mouse is provided.

OS/2 and UNIX operating systems (see Subsection 2.2a) provide the computer the power of a minicomputer or a workstation. They have complete multitasking and/or multi-user possibilities and can be grouped to large computer networks. However, it is not easy to develop applications that exploit these improved features.

Sales of MS Windows 3.1 show a remarkable growth. By far most of PC operating systems will be Windows. Also, applications of Windows expand enormously. Applications range from scalable fonts to Word or Turbo Pascal for Windows and Excel to multimedia support. Microsoft and Digital are cooperating in marketing Windows, which was developed by Microsoft. Apple and IBM have joined hands to design a new operating system.

2.3 DATABASE SYSTEMS AND COMPUTER NETWORKS

2.3a Design Databases and Data Models

Databases for Computer-Aided Design

A *database* is an organized, updatable repository of interrelated data which can be manipulated so that it can be used for the purpose of data retrieval and decision making. The software needed to perform the data-managing task is called the *database-managing system* (DBMS). The principles of DBMSs have been treated in a multitude of books [Dat86, Kor86, Elm89].

The problem of data handling in VLSI chip design is becoming increasingly important. The use of databases in a CADCAS environment is a natural evolution. However, CADCAS databases are more complicated than commercial databases, such as inventory, personnel and payroll databases. A CADCAS system employs a database-managing system as a design tool in conjunction with application programs, by which we mean the programs for simulation, testing, layout and other tasks to be performed during the design cycle. The ensemble of database, database-managing system and application programs is called a *database system*.

Since in all hierarchical stages of the design cycle we are dealing with one and the same circuit or system, we have to do with common data, e.g., the module interconnections of the circuit at the gate and circuit-element level and even in the circuit layout. Therefore, it would be effective to store these common data once in a storage system and utilize it whenever needed by any application program in the design cycle. The solution to this problem is furnished by the database system. The database in this system is an integrated collection of interrelated data stored together on disks, drums or other secondary storage media without unnecessary redundancy, accessible by several application programs and users. The problem of *data redundancy* is the problem of ensuring that each item of information appears only once in a collection of master files. When an item of data appears in more than one file or in more than one place in a single file, there is not only a waste of storage space, but also a high risk of inconsistency in the stored data, in particular when data files must be updated frequently.

As in file systems, database systems contain files for collecting data. However, a database system is distinguished from the file system in that it incorporates relations between data in different files. Hence, while file systems may be efficient in simple design systems, a database system is needed when there are numerous and complex relationships between data which are subject to change. A database used as part of a CADCAS system will be called a *design database*. All relevant data in the hierarchical approach is organized and structured in the design database in such a way that it can be accessed by all the

2.3 DATABASE SYSTEMS AND COMPUTER NETWORKS

application programs involved in the design cycle. The computer can be used as an aid in database design.

The ultimate goal of using a design database is the achievement of a coherent design-automation system by unifying existing CADCAS tools for the design and manufacture of IC chips with an updating capability to cope with technology changes. Database systems have evolved from simple data-handling systems to sophisticated database systems. Several implementations of design databases have been developed.

A Database System as a Tool for Automated Design
The data contained in a design database may be divided into two classes: design data and library data. Design data are data that pertain to software or hardware which is to be designed and implemented. This data may be input data entered into the database through some high-level user-oriented language or may be output data which are design results after the input data have been processed. Design data may include descriptions of logic and circuit interconnections, parts lists, tests from test generators, placement and routing data, mask artwork and pattern-generator data.

Library data are fixed stored data which are retrieved in a read-only mode. They may be shared by the users of the system to supply additional information for their application programs. Examples are technological parameters, mask-design rules and symbol descriptions for graphics applications. Library data may also include complete descriptions of standard cells or other subcircuits which have already been designed thoroughly and may be used as building blocks for constructing more complex IC structures. Such data may include pin descriptions, electrical parameters (such as power rating and tolerances), logic information (such as truth tables) and other relevant data. As technology progresses, more of such building blocks can be stored as library data. Updating library data is a specific task assigned to a database administrator, not to the chip designer. We may also distinguish between basic data (i.e., input-specification data supplied by the designer and library data needed by an application program) and derived data (i.e., outputs of application programs, such as generated test sets and mask-artwork data).

A query language is usually provided to allow the designer to define accesses interactively for interfacing with the database in a user-friendly manner. The main requirement of databases is that of *data independence*, which implies that the application programs are by no means affected by any changes in the actual file-storage organization. In effect, the application programs are isolated from the storage structure.

Data Models
Depending on the particular design task to be performed, the data provided by a user or an application program must be organized in an appropriate data

structure, which is the user's view of what the database is. In database terminology, this data structure is called the logic data model or *data model*, for short. Each data model implies a unique data representation and a means of manipulating this representation. In this subsection, we review the most important data models that have been proposed so far. In the next subsection, we discuss the ways in which the database system manipulates these data and stores them in the storage medium.

The smallest unit of data in the database is a field which contains a data item. A coherent collection of data items forms a *record*, which is usually represented as a node in the graph representation of the data model. A node is the basic unit that can be accessed. Any particular record, called a *record occurrence*, is an instance of one of a fixed number of *record types*. These record types, their names, formats and other characteristics are defined by a Data-Definition Language (DDL).

The numerous kinds of relationships which may exist among the data in different CAD problems do not permit the definition of one specific data model that would be most efficient in all practical cases. Three basic data models (the hierarchical, network and relational models) have become classic in that they are well known in the database community.

Hierarchical Data Model

The *hierarchical data model* has a tree structure. The root, which is the single highest-level node in the graph, is the only access point in the tree. All nodes, except for the root, have only one higher-level node (superior) and zero, one or more lower-level nodes (subordinates). The tree can be traversed downwards via the root (master key) to reach any subordinate node in the tree. In this way, a user can quickly trace a path through the tree to find the required information.

The tree structure of the hierarchical model reflects the hierarchy that we encounter in the top-down IC-design approach. The different levels of complexity in the design cycle, such as system, functional, gate, circuit and layout level. Another example of the tree structure is found in graphics systems, where we may wish to display either detailed portions or less detailed higher-level portions of a design. A drawback of the hierarchical model is that it is not suitable for representing many-to-many relations that we encounter in IC design. If these relations were to be implemented as hierarchical structures, redundancy would be created with a danger of inconsistency. The hierarchical data model can be considered as a special case of the network model to be discussed below.

CODASYL Network Data Model

The *network data model* is based on the 1971 CODASYL (Conference on Data Systems Languages) report in which standardized data definition and manipulation languages are proposed. Instead of hierarchical nodes, a node in the network model may represent an owner (superior) or a member

2.3 DATABASE SYSTEMS AND COMPUTER NETWORKS

(subordinate). A particular owner may have any number of members, while a member may have any number of owners.

The fundamental relationships in the network model is the set of records which are related to one another. A set type consists of an owner and one or more members (a one-to-many relation). A member in one set may be an owner in another set. Owners (each corresponding to a specific set type) are directly addressable via a principal key. Members can only be accessed via an owner. Network data models are usually represented by linked lists or rings, each one representing a set as defined above.

Relational Data Model

In the hierarchical and network models introduced above, relations between records are given by a pointer structure. In the relational data model, introduced by Codd [Cod70], all records in the database are viewed as being represented by two-dimensional tables. The actual relationships between data are implied by the characteristics of such a table. The main features and usual terminology in the relational approach are clarified in Fig. 2.5.

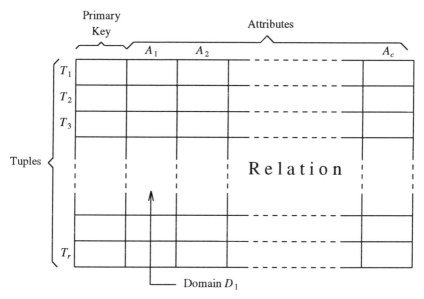

Fig. 2.5 Relation represented by a table

A complete table with r rows and c columns is referred to as a *relation*. Each row represents a *tuple*, which combines all related data of each record in a specific order. The columns represent the *attributes* A_j, $j = 1 \ldots c$, which specify the different types of data items contained in a record. The set D_j, $j = 1 \ldots c$, of allowable values of A_j in a column j is referred to as the *domain* of the attribute

A_j. To represent the data model unambiguously, the ordering of the columns is prefixed, unless each column is given a distinct attribute name. All rows should be different by definition. Each tuple (row) is addressed by a unique *primary key*, which is provided by the first column of the table.

A relational database is formed by the complete set of relations or tables. An advantage of the relational approach is that it is founded upon the mathematical theory of relations, which provides a formal basis for handling problems containing many-to-many relations between entities. Let c sets D_j, $j = 1 \ldots c$, of the attributes A_j, $j = 1 \ldots c$, be given. Then R is a relation on these sets if it is a set of all c-tuples $\{a_{i1}, a_{i2}, \ldots, a_{ic}\}$, $i = 1 \ldots r$, where the cardinality r is the number of the set members (rows or tuples), which is usually variable. The number of columns c is called the degree of the relation. One domain (column) contains the primary keys, which are used to identify the disjunct c-tuples (rows).

A relation is said to be *normalized* if each entry in the table is allowed to take only one value (not a set of values). Normalization is the process which classifies relations, records, relationships, etc., into groups by the characteristics that they possess. Depending on whether certain rules are met, five normal forms can be defined. The successive normal forms correspond to improved performance in certain operations, particularly in updating.

The database-managing system performs well-defined mathematical operations on the tables, which are expressed as normalized relations. Operators, such as PROJECT, JOIN and UNION, allow the user to extract the desired data or merge data together to form new relations. These operations correspond to the traversals through the hierarchical or network structures. The table-oriented structure used in the relational approach is suitable for display on visual-display units. The advantage of the relational model is obvious. Besides the solid mathematical basis, there is a large flexibility in manipulating the relations and data. However, apart from the unfamiliarity with relational algebra, efficient implementation in computer-aided design applications is not so easy and the operations in relational algebra are often time consuming.

Alternative Data Models

For solving real-world problems, it is useful to take the concepts of *entity* and *relationship* (between two or more entities) as basic primitives for data models. Figure 2.6 contains the usual terminology of corresponding characteristics of the CODASYL network model and the relational model. In what follows, we report briefly on alternative data models which have been proposed in recent years. Most of these models have the objective to add semantic information to a data model, notably to the relational model.

The concepts of entity and relationship have led Chen [Che76] to introduce the *entity-relationship model* (ER model). For each relationship, there is a

2.3 DATABASE SYSTEMS AND COMPUTER NETWORKS

relationship relation which specifies the primary key values for each participating entity in a relationship. The general ER model can be applied to several research areas in engineering and science. International conferences have been devoted to systems analysis and design, information modeling and analysis, software engineering and knowledge representation. The ER model permits a top-down approach to data modeling. First, the entities and the relationships between them are selected. Then, attributes (properties) are assigned to these entities and relationships in such a way that a set of fully normalized tables is obtained. The motivation for using the ER model as the basic data model in a database system is the need for a modeling tool that can specify user needs and requirements in a natural way.

Real-World Concepts	CODASYL Network Model	Relational Model
Entity type	Record type	Relation
Entity (instance)	Record (occurrence)	Tuple
Relationship	Set (type)	Common Domain
Attribute	Data item	Attribute, Domain

Fig. 2.6 Comparison of the network and relational models

Borkin [Bor80] introduced *semantic data models* in which the structures and operations performed on them explicitly represent certain types of real-world information. Two models, the graph-data model and the semantic relation-data model have been introduced to study the equivalence properties of the two data models. The *structural data model*, described by Wiederhold [Wie83], is based on the relational model, which is augmented with a formal definition of relationships between files, called *connections*. Semantic concepts can be described in terms of five relation types and three conenction types.

Fagin *et al.* [Fag82] consider the entire database as a single relation, while all existing relationships are descibed through a family of dependencies. This has led to the *universal-relation data model* in which, unlike the relational model, the user no longer has to allocate attributes to appropriate distinct relations. This model facilitates the formal analysis of data semantics. Major efforts are being made to develop a complete set of modeling procedures for database design.

In the *functional data model*, the relationships are formally expressed by functions, which can be implemented directly. The interest in functional models is in line with the increasing efforts in upgrading functional techniques for programming and verification.

Data Models for VLSI Chip Design

A database system provides a useful means to integrate a variety of CADCAS tools around a common pool of data. Given a large number of proposed data models, we face the problem of which model should be chosen for use in VLSI chip design. In the past, relational database systems have been widely used for several applications. Unfortunately, relational base systems have a number of serious limitations [Hor91]. These limitations are mainly due to the stringent record-based representation, as illustrated in Fig. 2.5. For example, each record of a certain type is assumed to be composed of the same fields, while each field must be from the same domain in all the records.

Designing a VLSI chip involves large volumes of design data of different structures, several views and perspectives to the data and many design tools at different levels are available to process the data. Recently, a new class of database systems, the *object-oriented (OO) database system*, has emerged as a powerful contender in database technology. Among the many advantages of OO systems are the modularity and easy extendability of the OO database. The versatile concept of *object* makes OO database systems particularly useful in VLSI chip design [GUp91]. Object-oriented database systems will be discussed in Subsection 3.6c.

2.3b Database-Management Systems

Database Management

Database-Management Systems (DBMSs) are cost-effective tools for organizing and maintaining large volumes of data. A DBMS is the software package that acts as an intermediary between human users or application programs requiring access to the database and the physical data which are stored in the database of a database system. A number of important features related to a DBMS will be summarized below:

a. The DBMS provides mechanisms for the data-manipulating operations: data insertion, deletion and retrieval.
b. Data from a related set of application programs are analyzed and organized in the database such that redundant items are eliminated.
c. The DBMS should be compatible with available programming languages, operating systems and user-written programs.
d. The DBMS takes care that different application programs can be modified or updated without requiring changes in the storage medium.
e. Means is provided for monitoring the usage and performance of the database system.
f. In addition to the batch mode, a time-sharing facility allows users to interact online from remote terminals.
g. *Data privacy* is ensured by protection mechanisms which prevent unauthorized access to data or unauthorized types of usage.

2.3 DATABASE SYSTEMS AND COMPUTER NETWORKS

h. The DBMS provides *data and system integrity*, that is, all data-managing operations mentioned in a. are executed correctly.
i. In the case of hardware or software failures, recovery of stored data is effectuated by providing a backup facility.
j. The *Database Administrator* (that is, a person in charge of the updating and maintenance task of the DBMS) monitors the system performance and may advise application programmers and users.

Query Languages

To make access to files easier, a DBMS provides a *query language* to express query operations on files. Query languages differ in the level of detail they require of the user, with systems based on the relational data model generally requiring less detail than languages based on other data models.

Query-by-Example (QBE), developed by IBM, Yorktown Heights, contains a number of features not present in relational algebra or calculus. QBE is designed to be used through a special screen editor that helps compose queries. A key on the terminal allows the user to call for one or more *table skeletons* to be displayed on the screen. The user then names the relations and attributes represented by the skeleton, using the screen editor.

SQL (SeQueL), developed by IBM, San Jose, combines features from known abstract query languages and is the most commonly implemented relational query language. SQL is a declarative language with assertions and triggers to specify when (*if* condition) an integrity constraint should be run and what (*then* condition) is to be done. Triggers in SQL are demons, which initiate an action as soon as a condition is TRUE. SQL is used in a large number of commercial database systems.

Data Structures and Storage Structures

The organization of the database must be tailored to the needs of current and prospective users of the database system. The problem of which data must be included in the database, which interrelationships among data exist and which operations must be performed on the data depends strongly on the intended applications of the system. Below, we discuss the intermediary role that the DBMS plays between the user or application programs and the storage medium for the database. DBMS software provides tools for storing and accessing data in such a way that the user needs not be concerned with the physical location of data or with the intricacies of the particular storage device on which it is held.

A distinction should be drawn between the *data structure* (as seen by the application progams) and the physical *storage structure* (the way in which data are stored in a storage medium, e.g., a tape or a disk). Data structures and storage structures are two different concepts, the former being determined by the applications and the latter being governed by the optimal storage capabilities of a database system. Therefore, a conversion from the data structure (also called the *user's view*) to the physical storage structure is needed. Note that the user's view can be converted to different physical storage structures.

A DBMS distinguishes several levels of abstraction (also calles *views*) used in describing databases. Different views can be conceived, starting from the outside world (users, applications) to the actual implementation of data storage within the physical storage medium. A definition of a data structure associated with a specific view is called a *schema*. A description of a schema is given by a *data-definition language* (DDL). In fact, DDL is a notation for describing the types of entities, and relationships among types of entities, in terms of a particular data model. A *data-manipulation language* (DML) is a facility for expressing queries and operations on the schemas.

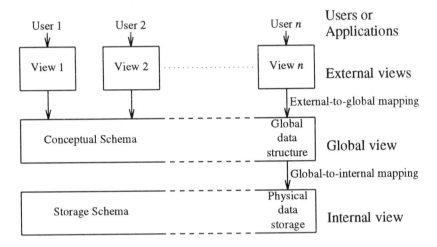

Fig. 2.7 Architecture of a DBMS

As depicted in Fig. 2.7, there are three views of the data in a DBMS:
1. The *external view* is different for different users or applications. A suitable DML or DDL (e.g., SQL) can be used to define the external view associated with the requirements of a specific user or application.
2. By means of an external-to-global mapping, each particular view is converted into a conceptual schema, which provides a *global view* of the database in terms of common record layouts, data-item definitions, and set membership and ownership.
3. A global-to-internal mapping is needed to define the ultimate physical storage structure (storage schema) of the data. This *internal view* specifies where records should be placed, clustered, scattered etc. across physical files and how they should be physically accessed, such as by using an index or a serial search.

2.3 DATABASE SYSTEMS AND COMPUTER NETWORKS

DDL descriptions of the three views are usually held in a central repository, such as a directory or dictionary in association with the database.

Data Independence

Database systems must cope with the rapid changes in technology. It may be desirable to replace the storage medium by newer and better hardware implementations. Frequently, the scope of the database must be extended since new records or data types and new relationships among data must be stored in the storage medium. All these changes cause changes in the storage structure of the database. One of the major requirements of a database system is that the storage structure and its associated access methods can be changed to meet any desirable purpose without affecting existing application programs. In other words, changes in the storage structure do not require reprogramming the user's software.

Since the storage structure and the access methods must be independent of the application programs which make use of the data in the database, this independence is referred to as *data independence*. This implies that either the data or the application programs which use them may be changed without changing the other. Data independence is achieved by a mechanism, called the *database schema* (in CODASYL terminology). Such a schema contains a complete description of all logic data elements and its relationships in the database. It is defined by the use of a high-level data-description language which specifies each data element type, its characteristics, relationships to other data, the record structure, etc. The description is compiled into an intermediate form for use by the DBMS. Since the schema defines everything in the database explicitly, its construction is a very critical task. It requires an adequate knowledge of how application programs view the data as well as the characteristics of the DBMS.

Client/Server Architecture

In this architecture, the user software (client) is logically (and frequently physically) separated from the data storage and access facilities (server). Client (front-end) and server (back-end) are connected via a network (see Fig. 2.8). A client/server architecture allows several users at terminals or workstations to access the server, which has a high processing power and contains a suitable database.

A *transaction* is a single execution of a program required by a user of an information system. This program may be a simple query expressed in one of the query languages or an elaborate host language program with embedded calls to a query language. A transaction reads and writes data to and from the database into a private workspace, where all computations are performed. In particular, computations performed by the transaction have no effect on the database until new values are written into the database.

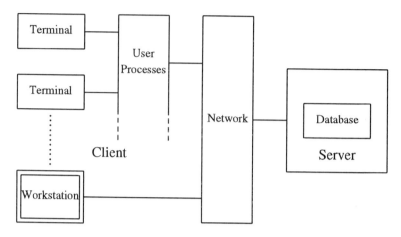

Fig. 2.8 Client/server architecture

A single-phase transaction consists of a single output message and a reply. A multi-phase transaction consists of a series of logically connected message-reply pairs.

Several independent executions of the same program may be in progress simultaneously; each is a different transaction. Transaction processing systems enable multiple users to concurrently access common resources in an efficient manner and control a number of processes as a single service. A Transaction Processing Monitor controls facilities, including a database management system, allocated store and application code.

Transactions are required to have recovery atomicity and concurrency atomicity [Lef91]. Recovery atomicity means that either all of the operations that constitute the transaction are performed or none are performed. This implies that possible software or hardware faults do not corrupt the database with incorrect results. Concurrency atomicity assures users that concurrent execution of transactions of other users does not affect their own applications. So, transaction processing systems let each user think himself or herself as the sole user of an idealized, failure-free database system.

Commercial Database Systems

A brief discussion of a few popular database systems follows below.

INGRES

Fourth-generation languages and Relational Database-Management Systems have been combined in INGRES (INteractive Graphics and REtrieval System) to support the entire application-development trajectory [Sto76]. INGRES can be applied in the most frequently occurring development methods from design to product release of a specific application. It supports database development,

2.3 DATABASE SYSTEMS AND COMPUTER NETWORKS

including database creation, in a dynamic way. It simplifies the access to data for application developers as well as for end users.

INGRES/MENU is the coordinating tool which allows one to select between the other development tools. The creation of applications involves the use of ABF (Application-By-Forms), the fourth-generation language, and the Frame and Procedure concept. Query and updating of tables are performed with the QBF (Query-By-Form) tool. Selecting, adding and updating the database are carried out by the database language SQL, which *de facto* is a standard query language for relational databases. Techniques for designing visual screens are provided by VIFRED (VIsual FoRms EDitor). RBF (Report-By-Forms) is used for designing and producing reports.

dBase III and dBase IV

DBMSs have also been developed for personal computers. In general, such DBMSs are intended for single-user systems. On the one hand, access and control features need not be so elaborate as in multi-user systems. On the other hand, DBMSs for PCs may have special facilities, such as the use of integrated software, including spreadsheets and desktop publishing.

dBase III, developed by Ashton Tate, Culver City, is a widely used relational database system for personal computers [Sim87, Jon88]. It is a versatile storing and look-up program with which data can be stored, retrieved, corrected and expanded. dBase is a programming language which allows one to write menus, own commands and new programs. dBase III has the following properties:

a. It can store 128 fields and 4000 bytes per record over a billion records per file, with 10 files open simultaneously.
b. It has a data catalog and a fast indexing and sorting mechanism.
c. It can exchange files in a number of formats, including ASCII, DIF and Lotus 1-2-3.

The dBase II Assistant is a sophisticated front-end to dBase which, by means of a series of drop-down menus, enables the user to create, update and link databases and then process queries to extract data and to generate reports.

dBase IV is a database-management system [LeB89]. It differs from dBase III in that it employs a fourth-generation language and provides extensive options for building end-user applications. By using advanced techniques, dBase IV is capable of handling large amounts of data. For example, dBase IV can be applied for bookkeeping, personnel, client and stock administration. It is a powerful tool for anyone who wants to create and maintain data banks. dBase IV has the following features:

- multi-user features, such as record locking and rollback,
- connectivity with minicomputers and mainframes,
- dBASE/SQL for data query and manipulation, and

– up to 99 files can be open simultaneously.

FoxPro 2 (which is now marketed and supported by Microsoft) is a contender of dBase IV (now marketed and supported by Borland International). Other DOS database packages include Borland's Paradox, Microrim's R:Base, DataEase from DataEase International, Nantucket Software's Clipper, WindowBase from Software Products and Alpha Four from Alpha Software [Gre92].

Database Machines

Database management is a software activity and therefore suffers from the drawbacks inherent in developing large software packages at increasing complexities of the database. Particularly the storage, retrieval and management of large databases require the execution of a large number of instructions. When a normal Von Neumann computer is used, much time is spent in interpreting data-management instructions. The CPU-memory bus tends to become a bottleneck which may degrade the quality and speed of database accesses to an intolerable extent. Most database-management operations can be executed in parallel.

By relegating the tedious and time-consuming functions of database management to dedicated hardware and leaving program preparation to a Von Neumann computer, a substantial reduction in execution times can be achieved. This has led to the development of the *database machine*. A database machine may be viewed as a backend machine, which implies that it is added to the host computer for the purpose of performing specific functions by a hardware implementation.

A database machine may contain multiple hosts and multiple backends. It can be implemented for very large databases. Berra [Ber82] discussed six different architectures for implementing a database machine. Hsiao *et al.* [Hsi83] described several architectures of database machines which are being developed in different countries.

As in hardware-implemented CADCAS tools, the decline of hardware costs in VLSI technology is an impetus for developing hardware implementations for DBMSs. According to Hsiao [Hsi83], database machines may be classified on the basis of their applications, the technology and the architecture. Applications of database machines may be globally divided into those for text search and retrieval and those for formatted database management. The class of applications affects the architecture. As the state of the technology evolves, it will affect the hardware/software allocation of the specific database-management tasks to the backend machine and the host computer.

Database Computers (which include intelligent secondary devices, database filters, associative memory systems, multiprocessor database computers and text processors) have been discussed by Su [Su88].

2.3c Computer Networks

Distributed Computing

Ever since the 1970s, computer users have felt a growing need for overall accessibility of a number of remotely located host computers, which can be connected by a switching network. The main goal of such a *computer network* is the sharing of programs, data and hardware resources [Bla89, Tan89]. The price-performance characteristics of hardware and software indicate that it is more economical to achieve a high performance and throughput by utilizing dedicated minicomputers in a network rather than increasing the size and complexity of one mainframe computer.

Distributed computing implies the collective use of a number of autonomous computers which collaborate in varying degrees for solving a variety of problems. The distributing feature applies to the processing and control as well as to the database. Separate computers may perform specialized processing functions or the processing task may be distributed over multiple computers located at different sites.

Since the telephone network, which was designed for voice communication, was already available, it was soon employed as the transmitting medium for data communication. The necessary conversion of the digital data signals into analog signals, and vice versa, has been accomplished by using a *modem* (modulator-demodulator). Besides the telephone network, other dedicated networks have been developed for data communication.

Distributed Database Management

Since data communication over long distances is expensive and limited in speed, there is a trend toward introducing distributed database management such that specific data is distributed to locations where usage is highest. Very large databases may be dispersed over a number of linked computer installations. Purposefully introducing redundancy to avoid excess amounts of data transfer and to improve the performance and reliability may cause additional problems on data updating and other operations. Special-purpose hardware tailored toward distributed database management would be very useful.

A *distributed database* consists of a collection of *nodes* (or *sites*), each of which represents one computer and its associated secondary storage devices. Possibly, some nodes have little or no secondary storage, and other nodes represent only secondary storage, with the minimal computing power needed to store and retrieve data. Some pairs of nodes are connected by links, allowing data or messages to be transmitted from one of the pairs of nodes to the other, in either direction. For example, a collection of workstations on a local-area network could hold a distributed database, with part of the data at each workstation. The workstations are the nodes and there is a "link" between every pair of nodes, because the network permits a message to be sent directly from any node to any other.

We may distinguish between three basic configurations of the database: the central database, the satellite database and the distributed database. In the *central database*, all design data resides at a single site, usually on a single machine. In the *satellite database*, a central computer is augmented with satellite computers, each of which may copy data from the central computer. To reduce the redundancy introduced in the satellite database, the *distributed database* has gained increased importance. It stores data physically in the computer where it is needed most, while part of the data may be replicated at multiple sites. There is a trend toward establishing national and international computer networks which will enable organizations to transmit data between remote sites at high speeds.

Network Topologies
Communication networks have increased in size and complexity. In a typical network configuration, a *hierarchy* of computer resources can be identified in the network. For example, a mainframe host computer with large processing and storage capacities can be used as the root of a tree network. A number of minicomputers is connected to the mainframe. The leaves of the tree represent intelligent or "dumb" terminals and other I/O devices. An *intelligent terminal* is usually a standard I/O terminal equipped with data-processing and storage devices, such as microprocessors, RAMs and floppy-disk drives. Some specialized functions which can be performed by such a terminal are text editing, preprocessing and local data storage.

Depending on the available computers and the interconnecting medium, including switching devices, several network topologies have been proposed for implementing the computer network. A survey of interconnection-network topologies was given by Feng [Fen81]. The main goal of a computer network is to make every computer resource, program and data available to other computers and users without degradation.

A *topology* indicates how the computers and other devices are interconnected, not the way in which the data are transmitted. Each computer may be provided with a number of user terminals. The *distributed network* provides direct connections between pairs of computers via switches, subject to priority constraints. The switches see to it that the correct data are transmitted at the correct times.

In the *star network*, all computers are connected to a central device M, that is, all traffic between computers is performed via M. Control may be concentrated in M or may be distributed among M and the computers C. In the *ring network*, the switching nodes and the associated computers are connected in a loop structure. Data provided with an address can be passed from node to node along unidirectional links until it reaches the node for which it is intended. In the *bus structure*, several computers are connected via cable taps or connectors to a

2.3 DATABASE SYSTEMS AND COMPUTER NETWORKS

time-shared bus.

Standards and Network Protocols
Internetworking is concerned with the methods of interconnecting individual communication networks with different characteristics [Spr91]. In this way, a larger user community can be served from a larger potential of computer resources. A major driving force in the evolution of telecommunication networks has been the development of standards for ISDN [STa88]. ISDN (Integrated Services Digital Network) has been viewed as a goal for future telecommunication networks, providing flexibility, user control and responsiveness, thus benefiting both the network providers and users. The main advantage of ISDN is that many different types of telecommunication services can be offered by using a single digital network and one type of equipment with considerable savings in investment and maintenance costs. Audio, video, computer and other data can be transported fast and efficiently from one user to one or more other users. The power of ISDN is that two different types of channels are available to the network. A basic connection offers a user two B channels and one D channel. The B channels are used for user information (e.g., speech or computer data). The D channel is primarily for signalling information (necessary for constructing the connection), but can also be used for data communication between users.

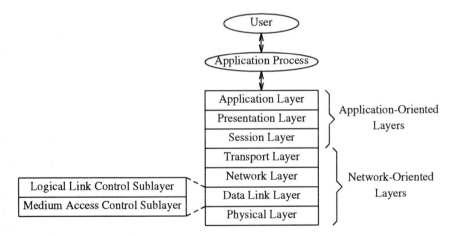

Fig. 2.9 OSI model of communication protocols

This structure gives ISDN considerably more flexibility than the existing telephone network. For example, User-to-User Signalling enables one during a conversation to transfer a document via the D channel. Another example is the Closed User Group, where a specific group of users is protected, that is, the users in the group are able to communicate with each other, while the communication possibilities with the outside world is limited.

What receives much attention is the OSI (Open Systems Interconnection) model of communication protocols from the International Organization for Standardization. OSI is the international standard for creating networks from heterogeneous hardware and facilitates openness between computer systems. The OSI functions involved in the interchange of information among very different end systems, such as computers, workstations and Local-Area Networks, are divided into seven layers (bottom-up: physical, data link, network, transport, session, presentation and application). See Fig. 2.9.

Wide-Area Networks

The advantages provided by multicomputer networks have stimulated the development of networks which encompass areas thousands of kilometers wide, crossing several different countries. A well-known example of such a wide-area network is ARPANET, developed by the Advanced Research Projects Agency of the United States. The primary aim of this network is to unite the computing resources of American universities and corporations working on common scientific problems. In addition to the continental United States, satellite channels allow ARPANET to extend its working area to the Hawaiian Islands and Western Europe.

EURONET was set up by the European Economic Community to provide user access to scientific and technical documentation in various countries in Western Europe. Other wide-area networks include TELENET (of the Communications Corporation in the USA), DATAPAC (Trans-Canada Telephone Company) and TRANSPAC (of the Ministry of Communications in France).

Local-Area Networks

Improving communications among PC users is a key challenge to extending the benefits of personal computing, not the least for users working on the same VLSI design project. A *local-area network* (LAN) is distinguished from other types of computer network in that communications are confined to a moderately-sized geographic area, such as a single office building, a warehouse or a scientific center [Sta91]. The localized area allows one to use coaxial cables with a high data rate and a low bit-error rate. Fiber optics offer a very-high-performance alternative to coaxial cables. High-performance local-area networks using complex autonomous protocols can be viewed as a generalized alternative to the hierarchical tree structure mentioned earlier. Each computer in a local-area network is potentially capable of communicating with any other device in that network.

A local-network standard is needed to enhance the communication compatibility between computing resources of different manufacturers. The IEEE Local-Network Committee established one such standard for the purpose

2.3 DATABASE SYSTEMS AND COMPUTER NETWORKS

of data communication at data rates of at least 1 Mbit/s within a local area with data paths up to 4 kilometers. In fact, the IEEE 802 standard addresses three different kinds of problems: CSMA/CD (Carrier Sense Multiple Access with Collision Detection), token passing in a token ring and a bus.

A well-known example of a local-area network is ETHERNET, initiated by Xerox Corporation and later developed in cooperation with DEC and Intel. ETHERNET was intended for combining minicomputer- and microcomputer-user systems, disk drives, printers and copiers. It has a bus structure, much in the same way as cable television is implemented. The communication medium is a coaxial cable consisting of one or more segments up to 500 m long. Each segment can handle up to 100 user systems via adapters stations. Two successive segments are connected by repeaters (amplifiers). The System 8000, developed by Xerox Corporation, includes a database containing up to 10000 pages of text, a laser printer and adapters for connecting various computers and other hardware equipment.

Local computer networks of personal computers combine the advantages of personal computers (e.g., low cost per user) with those of local-area networks (access to shared resources).

Interconnecting LANs

Designers working on the same project (e.g. a large VLSI design project) must be able to interchange in real time their views and ideas about problems and intermediate design results. For this purpose, a suitable LAN interconnection network is needed.

Though LANs may be classified according to their topology, access protocol and transmission medium, the data rate appears to be a useful criterion. According to the data rates, the LANs can be classified as low and medium speed (up to 20 Mb/s), high speed (50 - 150 Mb/s), supercomputer (~ 800 Mb/s) and ultragigabit speed (> 10^9 Mb/s) [Chl90]. The transmission medium may be copper wire, copper cable (32 or 64 parallel connections), optical fiber [Pal88] or wireless. The supercomputer and ultragigabit speed LANs are to be developed in the nineties. Conventional LANs are low- or medium-speed copper-wire-based ETHERNET or token ring or wireless LANs. A simple interconnection medium is the multiplexer.

High-speed local-area networks in the 50- 150 megabit-per-second range are becoming available, complying with the Fiber-optic Distributed Data Interface (FDDI) and Distributed Queue Dual Bus (DQDB) standards [Kap91]. A current development is the interconnection of LANs as metropolitan-area networks and the connection to wide-area networks.

There are three major methods for interconnecting two high-speed LANs, depending on the protocols of the LANs to be interconnected:

a. *Bridges* interconnect LANs with identical protocols for the two lower layers of the OSI model. One bridge is needed for interconnecting two nearby LANs. When LANs are at some distance from each other, the interconnection is made through two remote bridges connected by a leased line.
b. *Routers* can be used when similar or dissimilar LANs with three identical OSI layers must be interconnected.
c. A *gateway* is used for linking dissimilar LANs that employ different high-level protocols, such as DECnet and SNA (Systems Network Architecture). Gateways perform the necessary adaptations so that information can be forwarded from one network to another.

Different LANs can be interconnected across wide areas with a packet-switched data network, Internet services or dedicated digital lines [Kap91]. *Packet switching* is a store-and-forward method of transmission. The messages are partitioned into data packets (e.g. of 400 bits). Each switching network contains switching centers (denoted by nodes in a graph) whose functions are to transmit the packets correctly from source to destination. A switching node receiving a packet places it in a queue. When the packet reaches the head of the queue, the switching node examines its destination address, selects the next switching node on the route and places the packet in an output queue after having notified the sending node of the receipt of the packet. This process continues, until all packets of the same message have reached the destination node, where they are reassembled and sent to the destination computer. A large variety of (distributed control) switching mechanisms is possible: time multiplex, space/time switch, bus or serial link. The network graph may be a star, a ring or a mesh network of multistage switches and computers. Although centralized control is easier to implement, decentralized control has the advantage that the failure of one node does not necessarily disable the entire network.

Since partitions of a message are transmitted simultaneously, savings in transmitting time are achieved. Short messages may be grouped into a single packet. Proper packet sequencing and correct information transfer are ensured by letting the nodes of the network perform the required checks. Packet switching makes efficient use of the available bandwidth of the interconnection media, while the bandwidth is distributed among multiple users in a fair manner. A negative aspect of packet switching is that much time may be spent in reassembling packets when some packets arrive out of sequence, while the necessary headers contaning the required information, such as destination address, packet-sequence number, packet size and priority, produce an overhead which tends to be dominant as the packets are taken smaller.

Chapter 3
NEW DEVELOPMENTS IN PROGRAMMING

3.1 FOURTH-GENERATION LANGUAGES

3.1a Five Generations of Programming Languages

Evolution of Programming Languages
The impressive advances in digital computers and communications technology have been made possible by the various achievements in the *electronics* field. The first three generations of computers are usually associated with the changing electronic hardware components, which were successively vacuum tubes, discrete transistors, and integrated circuits. The fourth generation of computers, which have been developed in the 1980s, is characterized by an abundant use of VLSI and new computer architectures. The combined use of advanced microelectronics, artificial intelligence and automatic programming tools will beget the fifth-generation computer, which is expected to flourish in the 1990s [Fei83].

Examples: Typical examples of computers of the five generations are:
First generation: IBM 650.
Second generation: IBM 7094, Control Data's CDC-6600.
Third generation: IBM 360 and 370, Cray 1, Control Data's Cyber 205.
Fourth generation: ILLIAC IV, Cray X-MP, Cray 2, Control Data's Cyberplus.
Fifth generation: Expert systems.

In order for a computer to perform a specific computing or other data-processing task, we need a programming language. The growing integration density in a chip goes hand in hand with an increase in the complexity of the system functions which a chip can perform. Concurrently, the programming task of the chip designer becomes more and more complex. To face the growing complexity, the programming task has to be simplified. The evolution of programming languages is characterized by an increasing programming ease and user-friendliness [Set89]. Whereas the first computers could only be programmed by professional programmers proficient with the in-and-outs of the computer, the fifth-generation computer is supposed to be a system that can be put to use by persons without any knowledge of computer hardware.

The following five generations of programming languages can be identified.

First Generation

The first generation of languages comprises the *machine languages* associated with the earliest computers. A machine language belonging to a specific computer can be directly recognized by the computer as a sequence of executable instructions and data. Each instruction coded in a machine language specifies a predefined hardware operation and the location of an operand. For example, the addition of two numbers A and B and the storage of the result C using a single-accumulator, single-address machine consists of the three instructions:

a. Load the accumulator from location A.
b. Add to the accumulator from location B.
c. Store the result in location C.

Instructions and data coded in a machine language are composed of sequences of zeros and ones, which are far from the way humans communicate with each other.

A slight improvement is achieved by assembly languages which represent the second generation of programming languages.

Second Generation

An *assembly language* is closely related to the machine language of a specific computer, except that easy-to-remember mnemonic codes are used instead of the zero-one sequences to represent the basic operations which must be performed to indicate the data and to specify the memory locations. The available mnemonic codes are provided by the instruction set associated with the particular processor of the computer. The above example of addition might be coded in assembly language as

$$\text{LOAD } A$$
$$\text{ADD } B$$
$$\text{STORE } C$$

The use of symbolic addresses rather than the physical machine addresses eliminates the need of changing the assembly program, when changes in the storage medium are made.

The program in assembly language must be translated by an *assembler* into the appropriate bit sequences of the machine language. A dedicated ROM could be used to convert the mnemonic code into the machine-interpretable code. Both machine and assembly languages depend on the particular machine which is used.

Third Generation

Programming in machine or assembly language is very cumbersome, tedious and prone to human errors. To facilitate the programming task of a human programmer, *high-level languages* have been conceived. The principal feature of these third-generation languages is that an instruction, which requires several

3.1 FOURTH-GENERATION LANGUAGES

machine operations to be performed, can be coded as a single statement without bothering about machine details, such as the actual memory addresses. The above example of addition might simply be written as the statement:

$$C = A + B$$

Since high-level languages allow programmers to formulate the problem in a readable form, the progammer is able to concentrate on how to solve the problem in an optimal way without losing time in error-prone coding. The introduction and use of high-level languages creates the need for compilers which translate the high-level code into the required machine-executable code. The availability of compilers makes the programs *portable*, which means that the program can be run on different computers.

In the past, many third-generation languages have been developed, the most important being FORTRAN, ALGOL, PL/1, Pascal and, above all, C.

Fourth Generation
The first computers were designed for computational purposes, that is, for solving numerical problems. Gradually, the computers are more and more used for manipulating all kinds of data. Although third-generation languages give an improvement compared to languages of the first and second generation, they are not suitable for solving numerical as well as non-numerical problems with the complexity encountered today. The vast amount of code needed for solving current problems requires much effort on the part of the professional programmer. Modification or debugging complex programs takes a large amount of time.

It is desirable to be able to instruct computers much more easily and more quickly than we have been able in the past. Processing speed of computers is increasing steadily and hence will increase the throughput.

Fifth Generation
There is a continuing need for storing and manipulating non-numerical data. Many problems require some reasoning in the solution procedure. Languages which declare all the information necessary to solve the problem, but leaves to the programming system with its inherent rules and relationships of the declared facts the task of finding the way of how to solve the problem, are called *declarative languages*. Advances in Artificial Intelligence have led to the development of expert systems which are useful in particular applications. Fifth-generation languages, such as LISP, Prolog and Smalltalk, will be discussed in Sections 3.4 through 3.6. The main representatives of fifth-generation computers are expert systems, which are dealt with in Chapter 4.

Procedural versus Nonprocedural Languages
It is expedient to distinguish between procedural and nonprocedural languages. In principle, we may state that a program written in a procedural language

specifies *how* a given problem is solved, whereas a nonprocedural language can be used to specify *what* problem has to be solved but not in detail how.

Third-generation languages are in essence procedural. For solving a given problem using a *procedural language*, we need a unique algorithm, i.e., a well-defined sequence of instruction steps leading to the solution of the problem being considered. Running the program, which is the coded form of an algorithm, means that a fixed *procedure* must be followed, that is, in solving a specific problem, the same sequence of instructions is executed. For this reason, procedural languages are often referred to as *imperative* languages. Procedural languages are particularly suitable for solving numerical problems.

For many classes of problems, a *nonprocedural language* is more suitable. For example, suppose a database has to be searched for acquisition of certain information. When different questions are posed, different paths may be followed to arrive at the solution. Nonprocedural languages are primarily applied to non-numerical problems.

Professional Programmer, Systems Analyst and End User
It is useful to divide all persons who in some way have to do with application software into three classes: professional programmers, systems analysts and end users. In the beginning, computers were exclusively programmed by *professional programmers*, who simultaneously were *end users*. The programmers of the first and second generation were of necessity familiar with the technical and operational aspects of the computer. From the time third-generation languages (3GLs) came into use, professional programmers and end users gradually become two separate classes of persons. In special cases, someone in need of a specific 3GL program writes and also uses this program so that the programmer and the end user are the same person. In many other cases, the complete code of an application program has been written by a professional programmer, and this program, provided with special commands, is invoked by an end user for solving some problem. For example, the circuit simulator SPICE, written by a professional programmer, can be used by a circuit designer for verifying if a circuit performs as desired. By means of a user's manual, the end user specifies the circuit to be analyzed and the desired responses. In such cases, the end user is the person who applies the program written by the professional programmer. Computer programs have become attractive for use by designers without having a detailed knowledge of the computer hardware.

When a large software project is to be developed, a *systems analyst* may function as an intermediary between the end user and the professional programmer. In such cases, the systems analyst performs an analysis of the project in consultation with the end user or with a potential group of end users in mind. Then the problem is defined and instructions are passed to one or more professional programmers who write the code.

3.1 FOURTH-GENERATION LANGUAGES

As personal computers become available to an increasing number of people, it would be undesirable for the end user to turn for help to a professional programmer every time a programming problem arises. Therefore, it is imperative that the end user is able to program as many applications as possible. A first prerequisite would be that the software created by the professional programmer be available to the end user. A primary requirement for such software is the ease of use, which can be guaranteed by using fourth-generation languages.

3.1b Fourth-Generation Programming

3GLs versus 4GLs

The use of computers, notably personal computers, is increasing at a rapid rate. Advances in microelectronics have led to improved processing speed. The growing demand for application software would create the need of ever more professional programmers. When the third-generation languages (3GLs) would be used, an intolerable amount of programming effort, and consequently of professional programmers, would be needed. The solution to this problem lies in the use of fourth-generation languages (4GLs).

The objective of the fourth generation of languages is to alleviate the coding task of the professional programmer on the one hand, and to enable the end users to design their own applications on the other. In short, the fourth generation has two objectives:
a. To provide an increased productivity for the professional programmer.
b. Powerful capabilities for end users to program their own applications.

Properties of Third-Generation Languages

Third-generation languages are high-level languages which are founded on a theoretical basis. The formal definition of a high-level language, and hence the program statements, must conform to a well-defined set of rules. Important attributes of a programming language are its syntactic structure and the semantics of the statements. A bulk of knowledge must be transferred to compilers or interpreters which are needed to translate the source programs into machine-executable code. Much effort has been made in the construction of formal grammars. Automatic software tools for constructing compilers range from scanner generators (which apply regular-expression-based techniques) to compiler-compilers (which produce a translation from a source code into a target code [Tra80]). Advances in software engineering and the development of software tools were directly related to third-generation languages.

Third-generation languages can be pretty well defined. They have the following features:

a. A third-generation language is meant for the traditional Von Neumann computer, that is, instructions are executed one by one.
b. As a consequence of a., 3GLs are procedural, implying a sequential execution of the instructions contained in the program.
c. A basic operation in 3GLs is the assignment of a value to a global memory. Destructive statements, such as $X = X + 1$, are valid.
d. The language has a syntax, which comprises the set of grammatical rules that the statements of a program must satisfy. The usual notation is known as the Backus Normal Form or the Backus-Naur Form (BNF).
e. A program written in a high-level language actually consists of a finite string of characters over some alphabet. The semantics comprise a set of interpretation rules which define the meaning of each statement so that the compiler can generate the appropriate machine instructions and data. The programmer is confined to using a strictly defined set of instructions.
f. Usually, a compiler is needed for translating the source code into the machine-executable code. Some languages, such as BASIC, needs an interpreter for this translation.
g. The input is usually entered through commands on a keyboard. The usual output is a print-out.

Properties of Fourth-Generation Languages

The way fourth-generation languages have been introduced is rather elusive. The problem is that there is no unambiguous definition of fourth-generation languages (4GLs). These languages differ in many respects from 3GLs:

a. A 4GL gives a substantial improvement in *programmer productivity* compared to 3GLs. For solving many problems, a 4GL program needs significantly fewer lines of program code than a 3GL would need. A 4GL is easy to learn and to work with. Development time scales are reduced.
b. A 4GL permits easy data entry, called screen painting, into the screen by means of a pointing device, in particular a mouse. Screen painting means that commands and data can be entered onto the screen by a pointing manipulation, e.g., by using a mouse, rather than defining cartesian coordinates, as usual in 3GLs. The use of help screens, menus and icons on the screen facilitates the *human-computer interface*. A 4GL allows one to write an application direct onto the screen.
c. Fourth-generation languages are primarily designed for *interactive operation*. In addition to sequential parts of the code, 4GLs employ a diversity of mechanisms, such as screen interaction, computer-aided graphics and filling in forms or panels. End users are capable to carry out their own software development.
d. Fourth-generation languages provide on-line identification of errors, which can easily be corrected. Applications can be easily modified. Software

3.1 FOURTH-GENERATION LANGUAGES

maintenance workloads are substantially reduced.
e. Most 4GLs are dominantly *nonprocedural*. Some 4GLs are primarily procedural.
f. An important step forward in data processing and the development of 4GLs is the *database management facility*. Fourth-generation languages are usually based on database technology, notably with relational databases. A data dictionary ensures that all commonly used information will be used in a consistent, irredundant manner.
g. There is a diversity of 4GL types, ranging from simple query languages to complex integrated software packages.

Professional Programming of Application Software
Fourth-generation languages are usually designed either for professional programmers or for end users. Languages for programmers require the memorization of mnemonics and formats. The programmer should also be familiar with the hardware performance of the computer. A primary requirement to languages for end users is the user-friendliness. Many fourth-generation languages and software for end users have appeared in the past. They have been developed by commercial enterprises or software houses, but often also by individuals. Unfortunately, most of these languages have been created without a theoretical foundation. The explosion of 4GL development activity has also held off the definition of a standard for 4GLs.

In advanced 4GL software systems, we may distinguish four classes of problem area:
a. A database back-end, which contains all the information needed for solving the problem under consideration. Usually, the database is of a distributed nature.
b. The data communications and networking, which link together the distributed databases.
c. Data processing, which deals with all the data manipulations needed for solving the problem.
d. End-user facilities, which is the problem area directly touching the user.

It would be useful to have a universal language that could handle each one of the four problem areas. Fourth-generation languages are designed to be useful in any, if not all, of these problem areas. At present, the distinct requirements in these four areas are so diverse that a range of programming tools must be developed to fulfill the overall programming task. Much of the work, at least the simple tasks, can be done nonprocedurally. However, 4GLs should combine the best of procedural and nonprocedural capabilities fitting to the specific application. Of the 4GLs introduced in the past, only the best ones and those supported by large organizations will survive. Since no universal 4GL is available to handle all problem areas, a range of 4GLs will be used to cover all these areas. To ease the programming task, computers handle a large part of the actual coding phase, which previously was the domain of the professional programmer.

In the usual system prevailing for many years, the operating system fills the role of the interface between the user and the application programs on the one hand and the computer on the other. Future systems will allow the user to interface directly with the computer via the graphics display screen. It is desirable to have common data structures and user interfaces. A window-management system addresses databases in mainframes or in local files. Many 4GLs are built upon a database and its data dictionary. The *data dictionary* does not contain merely simple data. It stores also information which is useful in several application areas. This includes validity rules, security control, authorizations to read or modify data, report formats, dialog structures, and many other things.

Substantial processing is needed to translate the fourth-generation application language to machine-executable language or to carry out a user-computer dialog. A dialog must be designed in terms of a subsecond response time. The language may need substantial access to a dictionary or encyclopedia. This has become possible by the increase of processing speed and the availability of low-cost memories. For simple applications (e.g., queries, report generation or spreadsheets), simple nonprocedural languages give much faster results than procedural languages. When the level of complexity is high, development effort will rise and a procedural 4GL may be necessary to reduce the programming effort.

The professional programmer in charge of writing the 4GL code for end users need to take the users into account. Useful basic operations include help operations, cursor movement and editing. Therefore, the programmer should be proficient in dealing with the complex possibility of screen interaction. A programmer must have a thorough understanding of the programming languages used, in particular how to exploit their capabilities and overcome their weaknesses.

A strategy for designing application software for a given class of end users comprises several steps. First, the system functions and objectives are defined. The hardware configuration and the data structures are selected. To meet user needs, an appropriate human-computer interface must be designed. Given the system specification, the program code is written along with the necessary testing routines. Finally, a well-written user's manual must be provided. Bug-free code should be generated from high-level expressions, while debugging problems should be minimized.

Application Software Should Be User-Friendly

Professional programmers should bear in mind that the software packages to be

3.1 FOURTH-GENERATION LANGUAGES

developed are to be used by end users who are supposed to have no knowledge of computer hardware or software programming. End users should be able to have computers at their command with a minimum effort and to design their own particular applications without the help of a professional programmer. The end-user languages should be easy to learn without requiring the memorization of mnemonics, formats and complex constructs.

User-friendly software implies that it is easy to use and that it has attractive features, for example, because the human-computer interface is guided with menus and detailed on-screen explanations. A primary requirement for end-user languages and software packages is that the users be acquainted with the software capabilities quickly so that they may obtain valuable results without much expenditure of time. Many end users are not willing to take a special training course to acquaint themselves with the capabilities of the application software. Rather than struggling through a big user's manual to find out which command codes should be used for executing specific tasks, the end users should be able to communicate with the computer in an easy way, preferably with a simple visual programming interface.

An important feature is the use of *on-screen graphics facilities*. According to the saying that "a picture is worth a thousand words", the graphics display is a powerful human-computer communicating medium. An invaluable controlling input device is the *mouse,* with which a cursor can be positioned at any point of the screen by moving the mouse appropriately over a flat desk surface. By means of a mouse, the user may select commands and operands to execute an operation.

One of the features which make 4GLs software attractive is the use of *windows* and *menus.* The computer displays lists of options in the form of text codes. When an option is selected by pointing the cursor at it, a pull-down menu becomes visible. Tracing the list of menu options, one of the options is selected. Frequently, the screen displays *icons,* which represent specific files. Clicking a mouse on an icon may open a file, for example, by displaying a window containing a number of icons, each of which represents a subfile. In this way, the user can quickly access files and select certain commands and operands to execute an operation. A powerful improvement is the capability of *dialog interaction* with which users respond to questions displayed on the screen. A good software product encourages the user to respond to clear messages to take action at the keyboard or the screen. Word processing and other software packages incorporate a set of *help* pages on the screen that are available during editing.

A useful feature of many 4GLs is the availability of a *default option,* which will automatically be selected by the computer in case the user does not make a choice.

Example: A default format for a report is preselected by the computer. When the user wants something different from the default option, a definite choice can be made from the menu on the screen or by typing the appropriate commands on the keyboard.

The above facilities make it easy for the end user to exploit the capabilities of the 4GL software directly, even when after a long period of doing other things, the software must be used again. In the case of complex software packages, it is sensible to learn an appropriate subset of the package which can be taught quickly. When this subset has been mastered, the user is more receptive to learning more advanced subsets.

Some Applications of Fourth-Generation Languages
The numerous aspects of 4GL programming have led to the development of a wide variety of languages and applications. Unlike the 3GLs, the 4GLs have been developed by a large variety of developers. Some 4GLs are designed only for a restricted class of applications. Others can handle a wide range of applications. In order to be able to select the proper language that fits the application we have in mind, we should be acqainted with the 4GL packages that are available. Fourth-generation languages designed for a specific application are often easy to use and provide quick responses. In such cases, it is not sensible to use complex tools for simple tasks. Let us consider a number of applications of 4GLs.

The simplest *query languages* allow stored records to be printed or displayed in a suitable format. Examples of such languages are QBE (Query-By-Example) and Info Center/1. Some languages permit users to create their own files, which means that new data can be entered, existing data can be updated and queries in the database are possible. End-user database languages are available, which differ by their syntax and structure.

Frequently, we need a facility for extracting information from a file or a database. Such a facility which displays or prints the data in a suitable format is called a *report generator.* In many cases, substantial arithmetic or logic has to be performed on the data. Report generators are either an extension of database query languages or completely independent of database or query facilities.

In order to get an attractive graphics presentation of the results of data processing, *graphics facilities,* including drawing aids, are available. Graphics output revealing relationships of data and effects of parameter changes are very useful. Together with color plotting machines, the output represents user-friendly means.

Decision making needs the analysis of consequences of certain decisions. *Decision-support languages* can be used for this purpose. Applications include financial or investment analysis and business planning. Examples are Lotus 1-2-3 and Multiplan.

3.1 FOURTH-GENERATION LANGUAGES

In many applications, routine data processing is frequently used. For this purpose, we need software which is often called *application generators*. These facilities can speed up application development greatly. Some application generators can generate only a part of an application, and routines in a procedural third-generation language are called. Application generators usually generate preprogrammed packages which run a certain type of application. A wider application range can be achieved by using parameterized application packages. Selecting suitable parameters enable one to modify the application.

To avoid misinterpretation, formal methods of representing the specifications for programs have been developed. The tools for this purpose are called *specification languages*. A specification language is more general than an application generator. It has the ability to specify all types of applications.

Application generators are designed for specific applications. There is a need for end users to develop their own application programs. For this purpose, *application languages* can be used.

3.2 NON VON NEUMANN ARCHITECTURES

3.2a Computer Architectures

Pipelining
Suppose that a recurrent process consists of a sequence of n distinct subprocesses. When only one suitable processor P is available, all the distinct subprocesses must be handled one after the other by P. *Pipelining* is a technique that can speed up such a process. It refers to the parallel, synchronized execution of these subprocesses by using several dedicated processors. When n processors are available, the n subprocesses can be handled by these processors so that an overlapped execution of the consecutive subprocesses is possible. For example, a floating-point addition may consist of the four operations: align exponents, add mantissas, normalize and round off the result.

Let a process be partitioned into three sequential phases, denoted by q, r and s respectively, and assume that each phase is completed in one clock cycle. Fig. 3.1a shows the sequential case in which a complete three-phase process must be completed before the next process can be handled. When three processors (Q, R and S) are available, each of them can be dedicated to perform each of the three phases q, r and s respectively, of a process. The successive processes P_1, P_2 and P_3 can be executed in the way indicated in Fig. 3.1b. The pipelining process is schematically depicted in Fig. 3.1b.

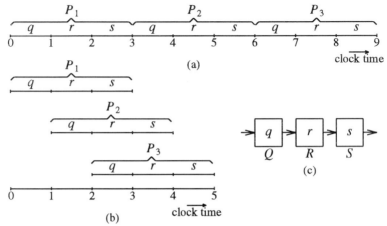

Fig. 3.1 Principle of pipelining

The following remarks are in order. In the sequential case (Fig. 3.1a), a single processor has to perform all three different phases of a process. When pipelining is used, the three processors, which operate concurrently, are implemented as specialized hardware units dedicated to perform a specific task, resembling the practice of assembly lines in an industrial plant. What is very important in pipelining is that the processing time can be considerably reduced. Some practical considerations have been ignored above. For example, interface latches must be used between clock cycles to hold the intermediate results. By predefining a sequence of repetitive operations, the pipeline principle can be applied in arithmetic processes, such as in the multiplier/accumulator structure. Pipelining can be applied to the extent that concurrent execution of different stages of a computation can be identified. The throughput may be badly affected when the subprocesses take different times. By partitioning a repetitive process of clock period T into n subprocesses, the clock period of each subprocess is determined by the longest subprocess, which is larger than T/n. In order to obtain a greater speedup of system operation, various forms of parallel processing, including pipelining, can be combined in the same system.

The optimal situation arises when all the subsequent phases in a process require the same time, that is, a particular processor does not have to wait for the next subprocess to be executed. Such an ideal situation is not likely to occur in practical cases. For example, in instruction sequencing, three distinct tasks must be performed sequentially in handling each instruction:
1. The memory address that contains the instruction is determined.
2. The instruction is fetched from memory.
3. The instruction is executed.

Let the times required to perform each of these tasks be t_1, t_2 and t_3

3.2 NON VON NEUMANN ARCHITECTURES

respectively. Depending on the changing technology used through the years, different imbalanced values of t_1, t_2 and t_3 required different solutions for optimizing the pipelining process [Kri91]. The use of cache memories has greatly simplified the instruction sequencing requirements of the first task. In RISC architectures, the executions are simple so that the execution time t_3 can be made small enough to approach that of t_1 and t_2.

RISC Architecture

Systems on a single chip fall into special-purpose and general-purpose processors. *Special-purpose processors* usually perform a specific task, often on behalf of a host computer. Original Von Neumann computers were Complex Instruction Set Computers (CISCs).

A typical example of an early single-chip *general-purpose processor* is the RISC-I 32-bit microcomputer [Pat85]. The design of RISC (Reduced Instruction Set Computer) was led by the argument that a very short design time is possible, when the instruction set is restricted to a small set of judiciously selected instructions, so that a simple architecture tailored to the efficient execution of the instruction set can be chosen [Sta90, Tab90]. RISC-I, once developed as part of the graduate curriculum of the EECS Department of the University of California, Berkeley, was designed at two levels: a low-level mask description and a high-level functional description. The most difficult parts of the logic design appeared to be the timing and the miscellaneous gates for driving the control lines.

High-performance commercial RISC architectures include the SPARC (Scalable Processor ARChitecture) for SUN workstations, the R3000 line of Digital Equipment Corporation and Silicon Graphics, the Precision-Architecture line of Hewlett-Packard, RISC system/6000 AIX of IBM, machines based on Motorola 68000 series and workstations with the Intel 860 series as the kernel. Unfortunately, the RISC architectures are not compatible. It is claimed that the SPARC architecture is amenable to implementation in different technologies (ECL, CMOS, GaAs) and hence supports an open system approach to computing. SPARC is based on the Motorola 68040 architecture. It combines RISC and CISC techniques with the CPU, the Floating-Point Unit and the cache memory integrated in one chip. At a clock rate of 33 MHz, SPARC can operate 26 Mi/s and 4.5 Mflop/s.

Other general-purpose systems on a single chip include the MIT SCHEME-79 [Sus81], the Stanford MIPS [Hen82] and the IBM 801 [Rad82]. A characteristic feature of the SCHEME-79 chip is that its machine instructions closely match the LISP programming language.

Multiprocessor Systems

Under this heading falls a wide variety of computer systems, with the common feature that more than one processor (usually a large number of processors organized in a regular structure) are used in the system. Novel forms of multiprocessor computer architectures have been made possible by the advent of VLSI technology. Cost-effective implementation of multiprocessor systems is particularly feasible through the use of many repetitions of identical processor elements with local communications, that is, each element is only connected to adjacent elements, which simplifies the routing problem.

In multiprocessor design, the communication aspects play an important role. Communication and the related control problems can be decomposed into three related issues: (a) the interconnection topology between processors and memory, (b) the communication protocol, and (c) the synchronization of data and control signals.

Multiprocessor systems may be special-purpose or general-purpose systems. Typical examples of special-purpose systems are the Geometry Engine [Cla80] and systolic arrays. The Geometry Engine is a 12-chip system which performs three of the very common functions of computer graphics. Systolic arrays and other array structures will be discussed in Subsection 3.2c. The use of many processors in a general-purpose system can be exploited to maximize the throughput of scientific computations which contain a lot of parallelism. The various parallel architectures which have been proposed differ by the way in which the different forms of parallelism (that is, all kinds of simultaneity in computer operations) are implemented by several processors [Kri89, Sto90].

Supercomputers
Stimulated primarily by the requirement for high-speed complex computations and large-scale simulations of distributed parameter systems, major computer manufacturers undertook to develop computer systems, now known as *supercomputers*. A feature is the use of pipelining and a large number of processors. Today's supercomputers typically have a performance of hundreds of Mflops (millions of floating-point operations per second) with a word length of 64 bits and a main memory capacity of millions of words.

There is such a wide diversity of computers designed with different goals that comparison of their performance is a delicate matter. A useful measure of *computer performance* is the *number of instructions* which can be processed per second. The measuring units are Ki/s (thousands of instructions per second) or Mi/s (millions of instructions per second). Present-day supercomputers for scientific applications are measured at least by Mflops (millions of floating-point operations per second). The Intel Touchstone Delta System at Caltech has a performance of 8.6 Gflops.

A technological measure of computer performance is the *clock period*. For example, Cray-1 has a clock period of 12.5 ns. The number of useful arithmetic operations performed per second depends on the gate delays, which have shown gradual improvements in the past. Probably a more important technological measure, which influences the computer performance, is the *access time* to the

3.2 NON VON NEUMANN ARCHITECTURES

main memory for instructions and operands. The *memory bandwidth* is measured by the number of memory words that can be accessed (either FETCH or STORE) per unit time.

There is no consensus about the term *supercomputer*. Initially, supercomputers were scientific computers which could be specified in terms of Mflops or even Gflops (G = 10^9). Typical examples of supercomputers are the various Cray systems and the CDC CYBER 205 family. Nowadays, the term supercomputer is used for the most powerful existing high-performance computers which are applied in sophisticated problems, including signal and image processing. Supercomputers need not be general-purpose computers. They may be dedicated to solving special classes of problems, including large-scale vector and matrix computations in various branches of engineering. Thanks to their impressive computing capability, supercomputers are sometimes called *number crunchers*.

A large machine containing a number of processors may be driven by a common control unit. Typical examples of such machines include the ILLIAC IV (which contains an array of 64 floating-point processors), the ICL Distributed Array Processor DAP and the Goodyear STARAN IV (both of which contain arrays of bit-serial processors). To allow different functional units to operate simultaneously on different data, a small number of different functions, such as logic, addition and multiplication, may be replicated and the control unit expanded accordingly. A typical machine based on this multifunctional principle of parallelism is the CDC 6600. Typical examples of high-speed supercomputers are Cray X-MP [Aug89], NEC, ETA and Fujitsu [Hwa89]. IBM, Cray and Digital have committed themselves to offering a massively parallel computer in the near future.

Besides parallel processing, which allows an array of identical processing elements to operate independently on their own individual data streams, pipelining is extensively used at the instruction and arithmetic levels.

Classification of Computer Architectures
In the past, many viable computer architectures were proposed to replace the traditional Von Neumann architecture. A general classification scheme based on flow of control and flow of data is due to Flynn [Fly72]. By considering *Single* or *Multiple Streams* of *Instructions* or *Data*, Flynn introduced four categories of computer systems:
SISD: Single Instruction Stream / Single Data Stream,
SIMD: Single Instruction Stream / Multiple Data Stream,
MISD: Multiple Instruction Stream / Single Data Stream, and
MIMD: Multiple Instruction Stream / Multiple Data Stream.

The well-known paradigm for the SISD machine is the traditional Von Neumann computer. A SIMD machine usually contains a synchronous array of processors under the supervision of one control unit. Typical examples are vector processors and array processors. In an MIMD machine, several processors and memories obey their own instruction streams on their own data. The use of a large number of processors makes pipelining viable in SIMD and MIMD machines.

The above classification scheme, though still in use, is not adequate to classify the multitude of different computer architectures. Baer [Bae83] discussed several possible classification schemes. He argued that giving a satisfactory taxonomy of computer systems is a formidable task because of the diversity and constant evolution of computer architectures and the continuing blending of the best characteristics of different classes. A simple classification scheme stems from Händler [Han77] who based the classification on the triplet (k, d, w), where k denotes the number of control units, d is the number of ALUs controlled by one of the k control units and w is the word length or the number of bits handled in one of the d ALUs. Three additional quantities k', d' and w' can be used to indicate the degree of pipelining at the processor control unit, arithmetic/logic unit and the bit level respectively.

A useful classification of processor architectures has been adopted in the ACM Computing Reviews Classification System. This scheme distinguishes three main classes: Single-Data-Stream Architectures (including the Von Neumann architecture, MISD, pipeline, SISD and MISD processors), Multiple-Data-Stream Architectures (including parallel, pipeline, SIMD, MIMD, associative, array and vector processors and interconnection architectures) and other architecture styles (such as high-level-language, data-flow, adaptable and capability architectures and stack-oriented processors).

Treleaven [Tre82] proposed a classification of parallel models of computation by the way in which the execution of computations is initiated (control mechanisms either by availabilty or by need) and by the way in which computations manipulate their arguments (data mechanism either by value or by reference). This leads to four classes of parallel modes:

a. Control by availability, data by value (e.g., data-flow computers).
b. Control by availability, data by reference (e.g., multiple form of control flow).
c. Control by need or demand, data by value (e.g., string reduction).
d. Control by need or demand, data by reference (e.g., graph reduction).

The classes a and b imply that data are available before a computation can be executed, reason why they are said to be *data driven*. In contrast, the classes c and d are *demand driven* since instructions are executed whenever needed.

Treleaven [Tre83] proposed still another classification scheme which is based on control flow and the different types of *data handling*. He considered five classes of architectures succinctly denoted by: control flow, data flow, reduction, actor and logic. In an *actor computer*, each instruction is object-oriented (see

3.2 NON VON NEUMANN ARCHITECTURES

Subsection 3.6a) and executed on the arrival of a *message*, which causes the instruction to change its state. Actor instructions may be control driven, data driven or demand driven. An example of an actor computer is ACT2 (a combination of ACT1 and OMEGA) of MIT. The *logic architecture* is based on predicate logic.

An architecture not mentioned before is that of the *recursive computer*, which is a sophisticated Non Von Neumann computer, proposed by Barton of Burroughs Corporation and Wilner of Xerox Corporation. Instructions are handled by a special recursive machine language. The architecture is based on either one or a mixture of control flow, data flow, reduction, actor or logic. For example, the Xerox recursive computer is an actor machine.

Shankar and Fernandez [Sha89] proposed a classification of computer architectures based on the instruction set. The following groups can be identified: CISC, RISC, independent instruction unit and data part of the processor, language oriented, object-oriented, and microprogrammable instruction set.

Functional Languages versus Imperative Languages
The traditional high-level languages, such as ALGOL and Pascal, which are directly associated with the Von Neumann model of a digital computer will be referred to as *imperative languages*. These languages suffer from several drawbacks which can be traced to the Von Neumann architecture. Successive assignment statements in the instruction sequence of a program in an imperative language cause changes in a state space which contains all possible states of variables in the program. A variable, in the terminology of imperative languages called an *identifier*, may have different meanings during the program execution. For example, the sum of a series $A(i)$, $i = 1 \ldots 4$, can be programmed as the repetitive summation:
$$S := S + A(i),$$
where the "variable" S on the lefthand side takes on intermediate values and has the meaning of the actual sum
$$S = A(1) + A(2) + A(3) + A(4)$$
only after the last summation with $i = 4$ has been completed.

The above example is an indication of the ubiquity of *assignment statements*, which require repetitive (fetch and store) memory accesses. An imperative language is not suitable for parallel execution of several tasks and is disposed toward side effects. A *side effect* is an undesirable effect that accompanies a specific computer operation. For example, a change in a parameter of a function may change a global parameter, in addition to returning a value. The change in meaning of a local or global variable, as exemplified above, may be considered as a side effect. Another example of an undesirable effect is the access of improper data in memory resulting from aliasing or improper timing. Due to the connection between the CPU and the primary memory inherent in Von Neumann computers, imperative languages are not suitable for constructing large programs from existing smaller ones. Since an imperative program does not lend itself easily to mathematical characterization of the program, the verification of this program is very difficult.

To overcome the shortcomings of imperative languages, we need new languages which support Non Von Neumann architectures. To this end, Backus [Bac78] and others proposed a new class of languages which have now become known as *functional languages* or *applicative languages*. Functional languages are based on the mathematics of functional algebra and eliminate the need for such notions as state and assignment. For that reason, they are sometimes called *zero-assignment languages*. Functional languages may be viewed as mathematical objects. They describe computations in terms of expressions and function definitions rather than commands and procedures. Every occurrence of a variable in a given context of a functional program has the same meaning. As a result, proving program correctness is considerably facilitated since advantage can be taken of transformation of expressions rather than assertions on a usually complex state space. The capability of formally manipulating descriptions and deriving proofs by transformation of expressions is referred to as *referential transparency*. Pure LISP can be considered as a forerunner of functional languages.

Functional languages are suitable for handling parallel evaluation of subexpressions. Programming a parallel machine may be pursued along two different lines:

a. Extend an imperative language with constructs to express parallelism.
b. Use a functional language, which takes full advantage of the inherent parallelism contained in the expression structures of the functions involved in an algorithm.

Though functional programming has its advantages compared to imperative languages (higher-level programming, freedom from side effects, enhanced system modularity, the natural exploitation of concurrency and the potential of easier program verification and debugging), it has its shortcomings. For example, functional programming is not effective for describing inherently sequential instructions (e.g, I/O operations). Handling array structures may also be troublesome in a functional language.

Procedural and Concurrent Languages

The main features of a procedural language are assignment as the basic action and control structures for the sequential execution of statements stored in a global memory. The procedural control structure, known in third-generation languages such as BASIC and C, can be extended to parallel control structures, defined in concurrent languages [And83]. Such languages are based on

3.2 NON VON NEUMANN ARCHITECTURES

processes, plus communication and synchronization mechanisms.

A *process* is an independent program consisting of a private data structure and sequential code that can operate on the data. Concurrently executing processes operate on their own private data and only interact with one another using the communication and synchronization mechanisms. The *communication* mechanism is the way in which processes communicate data among themselves. The *synchronization* mechanism is the way in which processes enforce sequencing restrictions among themselves.

Concurrent Languages

Besides the control structures inherent in Von Neumann machines, and which are consequently based on sequential execution of the instructions, various control structures have been proposed which permit parallel or concurrent programming. Conventional third-generation languages are not suitable for distributed information systems environments. Coupled to the development of multiprocessor and multicomputer systems has been the introduction of the concept of concurrency in the programming language to recognize the fact that most computing applications are inherently parallel in nature [Alm89, And91]. Concurrency implies the use of processes, shared variables, traffic signals (pipes, semaphores) and monitors.

Brinch Hansen [Bri75] devised the language Concurrent Pascal, which is suitable for programming monitors for use in operating systems. Another proposal to add concurrency to a Pascal-like language has led to the language MODULA [Wir77, Wir88]. The basic structuring unit in MODULA is the module, which is used both as a form of abstract data type and for the concurrent execution of processes.

Parallel Computer Architectures

There is no single method of implementing parallel processing in the new generations of parallel computers. A wide range of parallel computer systems have become commercially available [Tre91]. However, the rate at which parallel computer technology changes is phenomenal.

Two approaches to exploiting parallelism in computers can be distinguished:
1. Different parallel RISC architectures for general-purpose applications.
2. Systolic and waveform array architectures for special-purpose applications.

The trade-off between flexibility in implementing the applications with general-purpose machines and the performance gain in special-purpose hardware determines the choice of the desired architecture.

Multiprocessor systems may be *shared-memory systems* (in which processors exchange data and synchronize through a global address space, accessible by all processors) or *distributed-memory systems* (in which each processor can access its own memory and communicates by sending messages to other processors). Array architectures will be described in Subsection 3.2c. A set of major computer models of computation for general-purpose applications is given in Fig. 3.2, along with the corresponding programming style.

Model of Computation	Programming Style	See Subsection:
Control Flow	Procedural	3.2b
Data Flow	Single-Assignment	3.2b
Reduction	Applicative	3.2b
Logic	Predicate Logic	3.5a,3.5b,3.5c
Object-Oriented	Object-Oriented	3.6a,3.6b
Cellular Array	Semantic Network	4.2c
Rule-Based	Production System	4.3a
Neurocomputer	Neural Network	9.3b

Fig. 3.2   Models of parallel computation

The various models of parallel computation will be discussed in the subsections indicated in Fig. 3.2.

3.2b Non Von Neumann Computers

Parallel Computers

The Von Neumann computer contains a single processing unit. As a consequence, programming languages are designed so as to carry out only a single arithmetic or logic operation at any one time and the programmer is to specify the appropriate sequence of instructions. The availability of more than one processing element allows two or more operations to be executed simultaneously. This is referred to as *parallel processing*. A parallel program is a decomposition of a given program into a set of parallel subprograms, each of which is executed in a separate processor. A strong incentive for developing parallel machines is the fact that a large subset of computational problems, including those encountered in VLSI chip design, are inherently parallel in nature.

Every action of the processor in a standard Von Neumann computer must be preceded by a READ from memory and followed by a WRITE to memory. This I/O inefficiency, often referred to as the *Von Neumann bottleneck*, is a significant barrier in attempts to speed up algorithms in such a computer. Parallel algorithms designed for multi-processor architectures are meant to avoid the Von Neumann bottleneck. By using parallel processing and extensive pipelining, the ratio of computation time to time for I/O traffic can be considerably increased.

3.2 NON VON NEUMANN ARCHITECTURES

A short survey of some important classes of parallel computers is given below.

Control-Flow Computers

The main feature of control-flow computers is the use of a procedural language. Explicit flows of control cause the subsequent instructions to be executed. Processing modules, called at the same time, are executed in parallel and the control module waits until all processing procedures are complete before continuing. Procedural languages for parallel processing may be classified by the nature of their communication mechanism into shared memory, such as in Concurrent Pascal or ADA [Bar82], or on message passing, such as in OCCAM [INM84], which is used in INMOS Transputer Systems [Bar83].

By using OCCAM, a programming language developed by the INMOS Group of Companies [INM84], an application can be described as a collection of processes which operate concurrently and communicate through channels. An OCCAM process describes the behavior of one component of the hardware implementation, while each channel represents a connection between components.

A suitable component to implement an OCCAM process is the *transputer*, which is a programmable VLSI device with communication links to other transputers or to peripheral devices [INM88, Gra90]. A transputer includes both a processor and a local memory element. The transputer has a modular RISC architecture. A transputer system consists of a number of interconnected transputers, each executing an OCCAM process and communicating with other transputers. The IMS T800 transputer [Hom87] is a scientific computer which forms the basis for a powerful supercomputer.

A process executed by a transputer may itself consist of a number of concurrent processes implemented by sharing the processor time between the concurrent processes. To effectively implement simple procedural programs, a transputer is fairly conventional except that additional hardware and microcode support the OCCAM model of concurrent processes.

Data-Flow Computers

Considerable attention is being given to the class of *data-flow computers*. In such computers, there is a data flow from one statement to another in such a way that the execution of instructions is initiated by the presence of input data, indicated by *data tokens*, after Petri-net terminology. A processing element will perform the required arithmetic operations on input data as soon as data tokens appear at all of its inputs. The resulting data and the associated data tokens are passed to the next processing element, while the data tokens at the inputs of the preceding processing element are no longer required and hence can be removed from the inputs. The use of data tokens implies that control flow goes hand in hand with data flow.

Data flow can easily be represented by a *data-dependence graph* or *data-flow graph*, which consists of nodes (indicated by bubbles) and directed edges. A *node* represents a processor element which can execute an instruction or an operation, notably a function which maps inputs to an output. An edge directed from a producing node A to a consumer node B indicates that a data token is conveyed from A to B as soon as A has completed its task. The data-flow principle will be illustrated by an example.

Example: The simple arithmetic computation:
$$Z = A * B * C * D * E * F,$$
can be broken down into the simpler multiplications: (1) V = A * B; (2) W = C * D; (3) X = E * F; (4) Y = V * W; (5) Z = X * Y. When, in addition to registers for storing intermediate results, only one multiplier can be used, Z can be found in five sequential multiplication steps. When two processors are available for parallel processing, three multiplication steps will suffice: (1) V = A * B, W = C * D; (2) X = E * F, Y = V * W; (3) Z = X * Y. The flow of data from input to output makes the designation *data-flow computer* obvious.

Data flow implies that instructions or operations (multiplications in the above example) are *data driven*, i.e., an instruction can only be executed when all the input data are available. This principle is useful for implementing algorithms with a high degree of parallelism [Sha85]. Data-flow computers (e.g., Manchester, MIT) use a single-assignment language, which means that data actually flow from one function entity to another or directly support such flowing semantics. Examples of single-assigment languages include ID and LUCID [Wad85].

Reduction Computers

Wheras a processing element in a data-flow computer performs work as soon as data at its inputs are *available*, a *reduction computer* lets a processor perform a task whenever data are *needed*. Which data are needed depend on the current result of a computation and the next computational step to be taken. Since the sequencing of instruction executions is governed by demands for operands, a reduction machine is said to be demand-driven. Demands may return both simple and complex arguments (such as functions), as inputs to a higher-order function.

In traditional Von Neumann computers, variables are used to store intermediate results in a memory during a computational process. In the reduction computer, just as in the data-flow computer, an intermediate result is transmitted directly from its point of origin to the point of destination, thus eliminating the need for accessing addressable cells in a memory. The program structure, instructions and arguments in a reduction computer take the form of expressions with prespecified definitions. During the computational process, the

3.2 NON VON NEUMANN ARCHITECTURES

original expression is subjected to a sequence of reduction steps in line with a predefined set of computational rules.

According to the way in which instructions share data, two types of reduction can be distinguished: string reduction and graph reduction. *String reduction*, which implies the scanning of a string for reduction purposes, is based on a *by-value* data mechanism, that is, an instruction accessing a particular definition takes and manipulates an instance as a value for the computational process, which uses a graph representation, shares arguments through pointers in a *by-reference* data mechanism.

In a reduction machine, processors break up an expression of the programming language into subexpressions, and the expression is reduced by meaning-preserving transformations to the final result. Successively, a part of the expression is reduced by the substitution of a more compact term, until the expression is no longer reducible. A simple example is the expression (6 * (3 + 5)), which is replaced by (6 * 8), which in turn is replaced by 48. Berkling [Ber77] developed a string-reduction machine consisting of a reduction unit (comprising four different PLAs), a set of 4-kbyte pushdown stacks (for storing and transmitting intermediate results) and a one-byte-wide bus system for communication between the various units. While an expression stored on a stack is transmitted to another stack, reducible subexpressions are substituted for appropriate simple terms.

Mago [Mag79] of the University of North Carolina developed the *cellular reduction machine*, which has a binary tree organization. The tree consists of the root (which interfaces with the user), leaf cells (the most primitive bottom-level cells) and tree cells (which lie between the root and the leaf cells). Cells are activated as they receive demands for results needed to evaluate an expression. The expression submitted to the root is broken down in the tree cells and distributed downwards, until the leaf cells can evaluate subexpressions. Intermediate results are sent upwards to be manipulated by the tree cells and then sent back down again for further processing by the leaf cells. While sweeping up and down the tree, a larger and larger part of the expression is evaluated. Ultimately, the final solution of the problem is available at the root of the tree. The leaf cells are processors with associated storage, while the tree cells, which serve mainly to relay instructions and data to and from the leaf cells, may be simpler. The Mago machine uses a language similar to that proposed by Backus [Bac78].

In a *graph-reduction method* as proposed by several authors, the compiler removes all the variables in the expression to be evaluated and constructs a binary graphical structure, which is further processed by the reduction machine. This machine progressively transforms the structure by applying a succession of reduction rules.

3.2c Processor-Array Architectures

Introduction

There are many ways to build a parallel computer. Although general-purpose paralllel computers are highly desirable, there is a trend to build different machines, dedicated to a specific task. The focus on a specific application allows the user to develop high-performance processors. Generally, a processor-array architecture consists of three components:
a. *Processing elements* (PEs) or *cells*. In many cases, a set of identical, relatively simple, PEs are used.
b. An *interconnection network*, realizing the data path between the PEs.
c. *Interconnection-controlling circuits*.

Recent research in computer architectures has led to a plethora of parallel processing methods [Lei92]. Below, array architectures will be discussed.

Data Processing and Signal Processing

Supercomputers discussed in Subsection 3.2a are primarily meant for *general-purpose computing*. Advances in VLSI technology have made computing systems composed of thousands of processors economically feasible for special-purpose applications. The associated *processor-array architectures* are extremely suitable for implementing classical computational algorithms, particularly in digital signal processing and image processing. Whereas vector machines achieve their high performance by means of pipelining and vector handling, the processor-array architectures derive their power from the replication of identical cells. Such a cell, which combines arithmetic, logic and memory functions in a single unit, eliminates the need for primary-memory accesses. As a result, the system performance is improved and the number of input/output pins is reduced. Arrays of identical cells have several advantages, among which are cell-function flexibility, easy testability and adaptability to new IC technologies by restricting the design updating to a few types of identical cells. Some early applications of such cellular arrays are parallel-to-serial data conversion and sorting of data words according to some parameter.

By the end of the 1970s, new processor-array architectures for data processing, and somewhat later for signal processing, began to emerge. *Data processing* is concerned with sorting, transforming, computing and other operations on *data*, which is usually represented in binary form. The well-known paradigm is the digital computer. *Signal processing* deals with the processing of audible, visible and other analog forms of *signals* which, for the purpose of processing, must first be brought into digital form. The digital part is called *digital signal processing*. An important feature of signal processing is the necessity to subject the signals to a filtering process and transmit them through a suitable transmission medium. Though data and signal processing have many characteristics in common, some differences should be mentioned:

3.2 NON VON NEUMANN ARCHITECTURES

a. Signal processing (as well as image processing) requires "real-time" operations at high data rates.
b. Signal processing usually involves the handling of larger primitives (such as multipliers) than data processing, which uses simple operations, such as ADD and SHIFT.
c. Compared to data processing, signal processing involves very few data-dependent jump operations.

The principle of *processor arrays* has heralded a revolution in the design and hardware implementation of digital signal processing. To enhance the throughput, it is important to exploit and thus identify *parallelisms* and possible *pipelining* in signal-processing algorithms. Assignment and scheduling procedures in multi-processor systems have been discussed by several authors.

Systolic Arrays for Scientific Computations

Any implementation of a computational task involves arithmetic as well as I/O (input/output) operations. I/O operations are not only concerned with entering input data and delivering output data, but also with the operations related to storing and retrieving intermediate results. An algorithm in which arithmetic operations dominate over I/O operations will be referred to as a *computation-intensive algorithm*. A simple example is the multiplication of two $n \times n$ matrices **A** and **B** to obtain the matrix **C** = **A** * **B**. The complexity of the MULTIPLY-ADD steps is $O(n^3)$, while the I/O operations are of the order $O(n^2)$.

Parallel operation of a highly regular array of simple processing elements is instrumental in improving the operating speed of the computations. Computation-intensive algorithms can be implemented by processor arrays, which contain processing elements with a regular layout and interconnection pattern, depending on the class of computations to be performed. A *processing element* (PE) is a simple processor which can execute simple operations, such as multiplications and additions. A characteristic feature of this processor-array architecture is that a computational algorithm with many inherent parallelisms can be directly mapped onto a regular planar hardware layout, which thus avoids problems of placement and routing. The architecture is of the SIMD type, in which a global controller issues instructions to the individual PEs, eliminating the need for placing a complex control unit in every processing element. H.T.Kung *et al.* of Carnegie-Mellon University [Kun82] introduced a processor-array architecture which was given the name *systolic array*.

Systolic arrays can be applied to solving many problems, notably those related to matrix arithmetic in signal and image processing. The main task is to map cyclic loop algorithms into special-purpose VLSI processor arrays. When a computational algorithm is being designed, all inherent pipelining and multiprocessing possibilities must be exploited. The various forms of parallelism determine the most suitable interconnection network to be selected in a particular application. The communication between PEs is accomplished by a switching or routing network between the PEs. The basic principle of a systolic architecture will be explained below.

The systolic array consists of a regular arrangement of interconnected PEs within a bounded region with I/O cells at the boundaries of the region. The input data enter and leave the inner region via the I/O cells, while the information is passed in a pipelined fashion from PE to PE. Since the data are pulsed from PE to PE in synchronism with a system clock similar to the way in which the human heart pumps blood through the circulation vessels, the name *systolic array* becomes obvious. A characteristic feature of the systolic array is that I/O operations only take place at the boundaries, while the data within the array structure are transferred in a data-driven mode.

Example: A typical systolic array in which each PE has six connecting lines can effectively be used for the multiplication of two band matrices **A** and **B** to produce the resulting matrix **C** [Mea80]. When the PEs consist of suitable MULTIPLY-ADD processors (also called *inner-product-step processors*), the computation time is linearly proportional to the dimension n of the matrices. The elements of the matrices **A** and **B** enter as two diagonal data streams, while the elements of the resulting matrix **C** appear at the top of the vertical data stream.

The above example give rise to the following remarks:
a. The simplicity and regularity of the interconnection pattern obtained by direct mapping of the parallel algorithm onto a planar layout increase the reliability of very-large-scale networks for solving complex problems.
b. The repetitive arithmetic operations and PE-to-PE data transfers imply the desired property of *locality* of interprocessor communications which facilitates the routing problem. This array architecture reduces the loss of I/O time compared to the time required for truly arithmetic operations. Each PE communicates only with adjacent PEs.
c. The required PEs can be restricted to a few different simple types, each of which is used repetitively with simple interfaces in the array, making possible a cost-effective and reliable implementation using VLSI technology.
d. The concurrent use of many PEs through extensive pipelining and multiprocessing yields a major improvement in the computational speed, providing a high computational throughput.

The earliest applications of systolic arrays were concerned with matrix arithmetic, such as matrix-vector multiplication, matrix triangularization and QR decomposition [Mea80]. An application of a systolic array for string matching was described by Foster and Kung [Fos80]. The problem is to search all occurrences of a specific pattern string within a given text string. Another application is in database manipulation [Kun80]. An extensive list of

3.2 NON VON NEUMANN ARCHITECTURES

applications of systolic arrays (signal and image processing, matrix arithmetic and nonnumeric applications) was published by Fisher and Wing [Fis84]. Systolic arrays can be designed either for special purposes or as a general-purpose structure [Qui91].

A method for the verification of systolic algorithms was described by Hennessy [Hen86]. It employs an operator calculus, which contains a synchronous parallel operator × so that P_1 in the process $P_1 \times P_2$ may only perform an action if P_2 performs an action simultaneously.

System Timing

Three factors are essential to a processor array: the geometrical layout of the processor network, the functional capability of the network and the *system timing*. To perform the necessary functions, a PE usually consists of a small number of registers, an ALU and control logic. For the sake of cost-effectiveness, the number of PE types should be minimized. A system-timing mechanism sees to it that the right data are processed by the right PEs at the right times. The objective is to have the appropriate data streams processed repetitively by the PEs such that the proper results are obtained at the right place. The precedence relations of the computations which must be satisfied for correct functioning are directly related to the data and control dependencies in an algorithm. The absence of dependencies, which can be studied at several distinct levels, indicates that a number of computations can be carried out simultaneously.

A proper execution of the instructions requires a *proper sequence* of these instructions such that timing conditions are always met. In a *synchronous system*, a suitable global, system-wide clock distributes clocking signals over the entire chip such that the timing conditions are satisfied. In a *self-timed system*, the correct sequential operation is achieved by ensuring that local data flow satisfies the timing conditions.

The system operation is allowed to be *asynchronous* when data transfer from PE to PE takes place on a *handshake basis*, that is, whenever the required data is available, the transmitting PE informs the receiving PE of that fact and the latter accepts and manipulates the data as soon as it is convenient to do so. The systolic array discussed earlier has a centralized control and global synchronization mechanism. The *wavefront approach*, proposed by S.Y.Kung et al. [Hwa89], employs local clocking and a data-flow-based control mode.

Digital Signal Processing

The field of *digital signal processing* can be subdivided into two major areas: digital filtering and spectral estimation. An important field of research to employ digital signal processing techniques is *speech processing*. The problem of modeling speech behavior is effectively handled by the mathematical technique of *linear prediction* [Mar76]. The parameters of the speech-production model are obtained by using linear mathematics.

Speech processing falls into two areas: speech analysis and speech synthesis. *Speech analysis* usually amounts to determining the spectral characteristics of the filters that best whitens (usually in the mean-squares sense) the speech process, that is, a speech signal applied to the input of the filter produces white noise at the output. Obviously, the filter is called an *inverse filter*. Designing such a filter is usually based on linear *least-squares estimation*. Real-time speech analysis is now possible by deriving a system architecture which can implement the necessary equations.

Speech synthesis has come of age. Digital systems for speech synthesis on a single chip have become available for a multitude of applications. Many speech synthesizers are based on a simple model of human speech production. They use a stored set of control parameters to generate a limited number of phrases. A *vocoder* (voice coder) combines speech analysis and synthesis with the aim to transmit a speech signal efficiently with a low data rate of transmission. The techniques of speech synthesis and speech recognition have given the computer the ability to "talk" and "listen" respectively.

3.3 ARTIFICIAL INTELLIGENCE

3.3a Heuristic Methods of Solving

What is Intelligence?
Intelligence is hard to define. It has to do with the abilities to assimilate, comprehend, analyze and evaluate knowledge, to learn, to utilize and adapt appropriate pieces of knowledge for reasoning, problem solving and making the proper decisions in new situations, and to improve someone's skill in exploiting the acquired knowledge so that new insights can be gained, new problems can be formulated and new theories can be established. An intelligent person utilizes and adapts appropriate pieces of knowledge for making the proper decisions in new situations. For our purpose, intelligence is based on knowledge acquired by human beings who posses the capability of using this knowledge in a goal-directed manner.

Artificial Intelligence (AI) is concerned with the study of software and hardware information systems that are able to perform tasks which would be thought to require intelligence if done by human beings [Bar81, Nil82, Cha85, Ric85, Win85, Gen87, SCh87, Tan87, Sha90]. It is the branch of computer science that deals with the use of computers for performing intelligent tasks

3.3 ARTIFICIAL INTELLIGENCE

requiring cognitive abilities, such as perception, reasoning, learning and problem solving. Its goal is to understand the principles that make intelligence possible. AI investigates nonprocedural reasoning processes and their symbolic representations for use in machine-based inference. Extensive use is made of *heuristics* [Pea84], which imply intelligent search strategies, including educated guesses, intuitive judgments and plausible reasoning as means for implementing problem-solving methods.

Many useful AI applications are based on the availability of *knowledge*, which for our purpose is defined as consisting of facts, but also heuristic rules which enable one to derive new facts from known facts. The facts are clear-cut entities (pieces of information) and relations between entities which are undisputed. The totality of knowledge acquired from various sources is stored in a *knowledge base*. All the activities which have to do with the intelligent manipulation and utilization of knowledge for the purpose of problem solving are called *knowledge engineering*. Three major issues of knowledge engineering are knowledge acquisition, knowledge representation and knowledge inference. A system whose operation relies on a knowledge base is called a *knowledge-based system*.

AI requires special programming techniques. The programming strategy in AI is markedly different from that in traditional computer science. In AI, scientific results of the study of knowledge engineering, intelligence and reasoning are exploited for intelligent information processing with the computer as its principal tool. The first AI applications are programs for game-playing and theorem proving. At the present time, AI plays an important part in knowledge-based systems, such as expert systems and the fifth-generation computer.

Knowledge in a knowledge-based system is of two types: facts and rules. *Facts* correspond to the raw data in third-generation languages. Facts are easily acquired data. They can be written down, entered into a computer system and learned by rote. In expert systems, relationships between data form important facts of knowledge. *Rules* define conditional statements which allow the system to perform a reasoning process by appropriately operating on the facts.

The Concept of Inference

Third-generation languages, such as Pascal and C, are primarily meant for applications involving numerical computations. Accordingly, the data structures employed must represent numerical and symbolic data that are required in the process of solving numerical problems. In AI applications, it is assumed that rules, facts and relationships between these facts are stored in a knowledge base. As a consequence, the required data structures must allow the programmer to construct and match symbol structures so that conclusions can be drawn from knowledge statements in the database.

Deriving new facts from known rules, facts and their interrelationships is an

essential activity in AI, known as *inference*. Two aspects of programming in a fifth-generation language can be identified:
1. The specification of rules, facts about objects and their relationships.
2. The inference process of deriving new facts from the information given in 1.

In fact, fifth-generation languages provide the user with an interactive tool with which the user enters the facts, rules and questions at a keyboard, after which the system displays the answers on the graphics screen. The principle of operation is based on the use of logical deduction in addition to arithmetic calculations.

Applications of Artificial Intelligence
Though Artificial Intelligence is a many-faceted scientific discipline, two major fields of activity should be mentioned: computer science and cognitive science [Col88]. Artificial Intelligence is being applied in many branches of science and engineering [SRi86]. Typical applications of AI are robots and adult games (chess, draughts, bridge and backgammon). Subareas of AI include theorem proving, game playing, machine learning, problem solving, expert systems, natural-language understanding, pattern recognition, computer vision, and robotics.

3.3b Expert Systems, Powerful AI Applications

From Artificial Intelligence (AI) to Expert Systems
Early AI proponents were trying to apply artificial intelligence to solving all kinds of problems as human beings used to do [New72]. The present status is that expert systems, which can be considered as the most advanced application of AI, is based on expert knowledge in a restricted area of science or technology. The historical evolution from AI to expert systems can be illustrated by the thesis-antithesis paradigm.

Following several AI researchers [Fei63], there exists the following thesis: A universal AI system can be developed to replace human experts by computer programs which can solve all everyday problems, either engineering or nontechnical. It soon appeared that such a general-purpose problem-solving system is not feasible. The broader the scope of applicability of the system, the less efficient it is in solving special problems [Nil80, Win84]. A response to this observation is the antithesis: One must not aim at generality, but instead build expert systems for restricted areas of practical applications. As a realistic example, Feigenbaum *et al.* of Stanford University set up the expert-system project DENDRAL, which was implemented in Interlisp. DENDRAL infers the molecular structure of unknown compounds from the simple fragments which are identified after bombarding a small sample of the compound with high-energy electrons. The system enumerates all possible structures and uses

3.3 ARTIFICIAL INTELLIGENCE

chemical expertise to prune this list of possibilities to a manageable size.

Too much restriction would block the possibility of solving many-faceted design problems, such as encountered in chip design. As a consequence, current trends lead to the following thesis-antithesis synthesis: Try to build many-faceted expert systems based on an expert system shell which contains common attributes to which relevant knowledge are added to the knowledge base in special applications. It turns out that the following paradigm is useful in VLSI chip design. An appropriate set of expert systems, each dedicated to perform a specific task, is controlled by a general supervisor, which invokes one of the expert systems, whenever needed (see Fig. 1.4). The supervisor may be implemented as a blackboard system, as described in Subsection 4.3c.

General Architecture of an Expert System

An *expert system* is a knowledge-based system, since it relies on an appropriate amount of stored knowledge and experience imparted by human experts. It solves knowledge-based problems in a specific domain of application and offers intelligent advice or takes the proper decisions on demand, similar to the way in which human experts would do. The knowledge must be held in a representation such that intelligible inferences can be drawn by the system.

The earliest task of a computer is to store programs and execute complex computations. While a computer was originally a tool with a computational task, an expert system is primarily a knowledge-based problem-solving tool which aids human users in such tasks as decision making, planning, control, supervision, design and diagnosis.

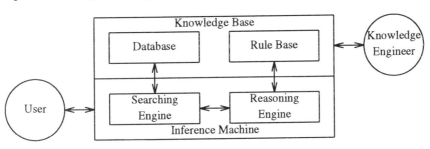

Fig. 3.3 General architecture of an expert system

The organization of an expert system depends on the characteristics of the specific domain of knowledge involved [Ste82]. Globally, an expert system consists of three parts: a knowledge base, a problem-solving or inference machine and an intelligent interface (Fig. 3.3). The *intelligent interface* provides a user-friendly communication between the expert system and its user [Boy91]. When a problem is posed or a question is stated, the system will respond in a conversational manner. The knowledge engineer creates, modifies, adds or deletes information from the knowledge base.

The *knowledge base* contains domain-specific knowledge of specialists in some field. It is a repository of the knowledge available in a system. In fact, the knowledge base consists of the *database*, which contains predefined *facts*, and a *rule base*, which contains predefined rules (including heuristics) and algorithms for solving specific problems. A fact is a fixed statement about some object. A rule is a conditional statement, such as an *if ... then* rule. Rules may contain constants (each representing a single value) or variables (symbols which may represent different values at different times). New facts can be derived from given facts by applying appropriate rules.

Example:
Fact: Aristotle is a human
Rule: *if* X is a human *then* X is mortal
New fact: Aristotle is mortal (The new fact is found by putting X = Aristotle)
In a programming language, facts and rules are represented in a compact form, e.g., human(Aristotle) may denote that Aristotle is a human.

The quality of the system responses depends on the quality and the amount of information stored in the knowledge base.

The *inference machine* controls the entire process that leads to the final solution of the problem presented to the system, that is, it decides when and which rules to fire and when the process is finished. It carries out the above-mentioned problem-solving algorithms and answers user queries either by retrieving facts from the database or by inferring new facts from these facts by using the rules in the rule base. The inference process may involve pattern matching (pattern-driven invocation) combined with heuristic search techniques.

Expert systems will be discussed in greater detail in Chapter 4. Below, some general properties, advantages and disadvantages of expert systems are enumerated.

Properties of Expert Systems

There are many reasons that may lead one to design and use an expert system for problem solving, amongst other things:
- Formal reasoning (inference, deduction) is required.
- An intelligent selection from many alternative solutions has to be made.
- The available knowledge is not complete, uncertain or subject to frequent changes.
- A prescribed algorithm for solving a specific problem is not known.

Expert systems are intelligent in the sense that they model, but not mimic the thinking of a human expert. Expert systems are suitable at applications that involve diagnostics, troubleshooting, symbolic processing and many other things. Medical diagnosis requires a very large computer. Many applications that

3.3 ARTIFICIAL INTELLIGENCE

currently require a large computer system can be run on a personal computer if the application domain is made smaller and more specific.

In order to indicate in what respects expert systems differ from conventional programs, the following properties of expert systems may be interesting.
- Expert systems are search-intensive rather than computation-intensive. Expert systems are pattern directed rather than control directed.
- Languages developed for programming expert systems are symbolic and declarative by nature (see Subsection 3.3c). Symbols are dealt with rather than numbers.
- An expert system starts a reasoning process using heuristics based on symbolic manipulation of facts and rules contained in a knowledge base, in contrast to a program in a third-generation language which prescribes a sequence of operations on data.
- There may be one, more than one or no solution at all to the posed problem.
- The reasoning process and the knowledge base are independent of each other.
- An expert system explains and justifies the procedure chosen for finding the solution.
- The knowledge base is restricted to a specific area of application. The reliability of the results depends on the extent to which the data and rules in the knowledge base are complete and reliable.

Advantages of Expert Systems
- Expert systems are cost-effective, because they replace scarcely available, and therefore expensive, human experts.
- Valuable expertise becomes accessible to many non-expert users.
- Even personal computers can be used to implement expert systems [Cha86, Tow86, Nay87].
- The knowledge base is consistent, easy to document and transfer, and it can be easily extended to an arbitrary size.
- Since the knowledge base is independent of the reasoning process, it can be modified and updated easily. This is of great importance as the knowledge is changing frequently.
- The expert system gives solutions to problems even when the data are only known with given statistical probabilities.
- Advice and support can be given in cases where the problems are incompletely understood and very unstructured.
- The expert system searches in a systematic way to a solution or advice which is ultimately the best possible from many alternatives.
- On demand, the expert system offers explanations and justifications of the underlying decisions and suggestions that it makes.

- An expert system is objective, has no prejudices to specific solutions and therefore judges unbiasedly. The expert system will not be affected by emotional and other factors which have nothing to do with the problem.
- Ever more attention is being paid to user-friendliness.

Drawbacks of Expert Systems
- For some applications which require rapid decision making, expert systems may be too slow. Question-reply dialogs often work slowly.
- A big problem is the acquisition of knowledge of a human expert and the translation of this knowledge into a codified form suited to the computer.
- Human experts generally make use of intuition and common sense to solve certain problems. Current expert systems lack common sense and are not capable of learning, i.e., improving knowledge by experience.
- An expert system is in principle only suitable for a relatively small knowledge domain.
- Conventional third-generation languages are far more suited to numerical processing using fixed algorithms.
- The system may give a wrong solution when the knowledge is not complete and lacks essential data or rules. The quality of the solution is directly related to the quality of knowledge base.
- For many complex applications, it is hard to fit the entire knowledge in the computer's memory, even with today's technology.
- The correctness of solutions to complex problems cannot be verified, since no adequate tools are available. As a consequence, we are not able to pronounce upon the reliability of an expert system.

3.3c Fifth-Generation Languages

The Fifth-Generation Computer
The proposals for the Japanese Fifth-Generation Computer System (FGCS) project were presented at an international conference in Tokyo in October 1981 [Mot82]. A report on this conference appeared in the January 1982 issue of *Datamation*. The FGCS project takes advantage of the advances in VLSI technology, supercomputers and artificial intelligence. A major requirement is that FGCSs support a high-level language and at the same time provide a user-friendly human-computer interface. FGCSs can be considered as sophisticated sets of expert systems.

There are at least three problem areas which must be expored:
1. Knowledge-base management: organization and maintenance of the knowledge base.

3.3 ARTIFICIAL INTELLIGENCE

2. Problem solving and techniques for inference from the knowledge base.
3. An intelligent human-computer dialog with user-friendly interfaces.

Some remarks on these points are in order.

Expert systems usually involve large-scale databases which require appropriate management and maintenance of the stored knowledge. The incorporation of new knowledge must not conflict with or contradict the existing knowledge in the knowledge base. A relational database machine with an adapted version of Prolog as the kernel language, extended with meta structures, is regarded to be suitable for providing the capability of inductive reasoning. Fifth-Generation Computer Systems are in fact knowledge-information processing systems, i.e., information can be inferred by processing knowledge-based data. The inference task should be performed in the way humans do it. To achieve this goal, basic AI functions, such as inference, association and learning, must be understood and implemented by suitable hardware in an efficient way. The processing machines exploit data flow, pipelining and highly parallel architectures. The supercomputer can be used in computational applications. More advances must be made in various branches of science, such as knowledge representation and acquisition, inference operation and efficient parallel execution mechanisms. It is appropriate to use dataflow architectures and parallel languages based on the functional or predicate-logic type. VLSI technology is ready to provide the high-performance hardware-implementation capability in the form of supercomputers and distributed databases.

A Fifth-Generation Computer System should have a genuine intelligent dialog capability. It should be able to answer questions, give summarized answers, prompt the human user, make suggestions etc. The system should be capable of adjusting its understanding in line with that of a human being. The human-machine interface must be designed so as to make the system easy to use. Therefore, future trends are toward friendly means to facilitate the human-machine dialog through the use of text, figures, images, sound and speech in some natural language.

Figure and image processing should handle thousands of pieces of figure and image information, which are stored together with information about their structure. Speech processing may involve a translation from one natural language to another which would facilitate the human interaction with the computer. The system must be able to handle a minimum number of basic words of a natural language and a fixed number of professional words and grammatical rules.

Languages of the Fifth Generation

For many years, a wealth of algorithms for solving all kinds of computational problems have been developed. The software tools required to implement these algorithms are procedural third-generation languages and newer languages adapted to parallel processing. Third-generation languages, such as Pascal and C, are oriented to numerical and consequently are extremely inefficient when used for data processing and reasoning problems.

The problems addressed by knowledge-based expert systems have a character different from the above-mentioned computational problems. Rather than computation-intensive operations, these problems involve a searching process in a knowledge base with the purpose of finding one or more solutions to the problem posed. The common use of heuristics and reasoning procedures require special, mainly nonprocedural, languages which are symbolic by nature. Because of the usually interactive character of the software methodology, it is better to speak of a knowledge-based *programming environment*.

The languages specifically dedicated to knowledge-based systems will be referred to as *fifth-generation languages*. Fifth-generation languages are the languages which are required in fifth-generation computers. These computers operate on knowledge rather than merely on data. Therefore, we have to deal with *knowledge programming*, which requires adequate tools for representing and manipulating knowledge in the computer.

Several languages can be used for knowledge programming. Since a broad spectrum of applications of the fifth-generation computer is aimed at, no single programming language is universal so as to be suitable for solving all kinds of problems. Just as numerical computations are best performed using a suitable procedural language, different kinds of knowledge-based problems are best handled by using specific knowledge-based programming languages. Such a language is generally not efficient for computational problems.

Depending on the way in which the expert knowledge is represented, different languages have been developed. In the course of years, four classes of fifth-generation languages have come to the fore:

1. *LISP and its dialects*. Though LISP, developed in the sixties, is one of the oldest languages known to the programming society, it can be viewed as one of the major languages for AI and expert systems (Section 3.4). In the United States, LISP is even the most widely used AI language.
2. *Rule-based languages*. Based on the concept of rules, which are the main ingredients of the knowledge base, rule-based languages have been developed for a class of expert systems, called production systems (see Subsection 4.3a). Representative of this class of languages is OPS5.
3. *Logic programming languages*. An important class of languages has been based on logic programming (Section 3.5). The prominent paradigm of this class is Prolog.
4. *Object-oriented languages*. The object-oriented approach to programming appears to provide a user-friendly environment for programmers (Section 3.6). The most representative language is Smalltalk. An intelligent integration of functional, logic and object-oriented programming provides a

powerful environment for handling complex knowledge-based tasks. Logic programming with its rule-based knowledge representation allows one to construct a knowledge base which can be queried for rapid answers.

Characteristic Features of Fifth-Generation Languages
The most important fifth-generation languages (LISP, Prolog, Smalltalk and OPS5) will be discussed in greater detail in later sections. The main purpose is to evaluate their relative merits, also in comparison to conventional languages.

Fifth-generation languages are in principle declarative, symbolic languages, that is, we are dealing with the manipulation of symbolic data as opposed to the third-generation data types, such as numbers and character strings.

Knowledge-based systems may handle large dynamic data or knowledge structures. Economic exploitation of the available memory resources is therefore important. Programming languages for AI applications, such as LISP and Prolog, should incorporate means for reusing memory space that is no longer needed by the program. This procedure of gathering available memory is called *garbage collection* [Coh81]. Algorithms for garbage collection can be implemented in software, hardware or a combination of both.

3.4 LISP

3.4a Basics of LISP

A List-Programming Language
LISP (a contraction of LISt Processor) was developed by John McCarthy *et al.* of the AI Group at the Massachusetts Institute of Technology as a tool for helping investigations into Artificial Intelligence [McC65]. This language has been widely used in the AI research community for more than two decades [Win88]. LISP is predominantly used as an interpreted language which allows easy program development. It is highly interactive and permits programming at a terminal with rapid response. In LISP programming, the distinction between programs and data is blurred. While FORTRAN is the oldest imperative language still in use, LISP has the honor of being the oldest functional language.

LISP has a simple syntax. Once a few basic concepts have been mastered, programming in LISP is very simple. LISP is capable of setting up data structures in the form of lists and manipulating the items on the list. It is therefore known as a list-processing language. A fact may be represented as a list, with the first element denoting the predicate and the remaining elements the argument values.

LISP has a rich set of data types, including symbols, lists and arrays, structures (record-like data types), floating-point numbers, integers of arbitrary size, complex numbers, true rational numbers, strings represented as vectors of characters and Boolean constants. It has also a complete set of operators, including square root, exponentiation, trigonometric, hyperbolic and logarithmic functions, symbol, list and Boolean operations, odd/even test, maximum and minimum, gcd and lcm and a random number generator. New data types can be defined based on those given by the language. The user-defined types can appear in declarations and be used freely throughout the program.

LISP is extremely suitable for processing lists and can be used in any application in which the data can appropriately be organized into and processed as a list of items. One major advantage of LISP is its simple list syntax. The basic primitive data types are *atoms* and *lists* of atoms. An *atom* is a string (sequence) of alphanumeric characters. It represents an identifier or elementary data. Symbolic atoms, such as *plus* and FIVE, are indivisible. The only operation which can be performed on them is *equality testing*. Numeric atoms, such as 348 and −1357, are sequences of digits, which may be preceded by a + or − sign.

An *S-expression* (symbolic expression) is the LISP way of defining lists and sublists of atoms by using an appropriate set of parentheses. A LISP program consists of S-expressions. An atom, but also a list of atoms, is an S-expression. A list whose elements are S-expressions is again an S-expression. The syntax of the language is defined recursively with atoms and lists of atoms.

Example: (*eq* A 5) defines a list of three atoms. Note that a blank is always used to separate neighboring atoms.

Symbols have associated property lists with the syntax: (property_name property_value).

Example: ((R1 1K)(R2 10K)) represents a list consisting of two sublists each of which contains two atoms, e.g., a symbol representing the resistor R1 and its value 1 k(ohms).

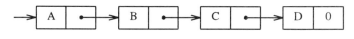

Fig. 3.4 Linked list

A linked-list structure is used to store a list of items. In such a data structure, the data representation of an item consists of a *data field* and a *pointer field*, each of which occupies a part of the total number of bits assigned to the item. The data field contains the data information of the item, while the pointer field contains the address of the next item in the list. The address 0 represents the end of the list. A simple example is given in Fig. 3.4, which shows the linked list for

3.4 LISP

the LISP expression (A B C D). More complex lists have a binary tree structure with two pointers which can refer to other lists or atoms. New nodes can be created by a special operator.

LISP Functions
LISP is based on a mathematical formalism, called the Lambda Calculus of Church. It is a functional programming language because functions (and not procedures) are used to manipulate the symbolic data structures. It defines a list of functions, each of which can be applied to given inputs to deliver the desired results as outputs. Operators or identifiers are represented by mnemonics: *plus*, *minus*, etc. Functions, which are the main structuring mechanism, are often executed as if they were subroutines. List can be considered as the sole data structure, while the main operations are function executions. LISP programs are essentially executed sequentially, though with an applicative style, that is, as a sequence of function invocations. A standard function is represented as a link whose first element is the funtion identifier which is followed by the operands.

Example: The addition A + B is expressed in LISP as (*plus* A B), where the first item of the list (placed in parentheses) is an identifier denoting the function type and the other items in the list are the operands or inputs for the function. An alternative notation is *sum*[A; B].

Lists can be used for expressing complex data structures, such as algebraic formulas composed of constants, variables and arithmetic or logic operators.

Example: The second-degree polynomial
$$2x^2 + 3x + 4$$
is written in LISP by the following S-expression:
(*plus* (*times* 2(*exp* X 2))(*plus* (*times* 3 X) 4))
The order of precedence for arithmetic operations must be determined by the programmer.

As seen above, the operators *plus*, *times* and *exp* are positioned in prefix form. Arbitrary deep levels of nesting of operators in association with their operands enables one to write very complex arithmetic or logic expressions.

The *setq* operator can be used to assign a value to an argument.

Example: (*setq* 'X 5) assigns the value 5 to the variable X. The quote symbol ' has the purpose of setting the variable X to the value 5 and not of substituting 5 for X. That is, X must always be considered as a variable, which can be called, whenever appropriate. Applied to the above S-expression, we obtain the result 69.

Nonnumerical Functions
LISP is particularly useful for nonnumerical applications. It is a flexible

symbol-manipulation language based on a small number of constructs [All78, Win88].

Three fundamental functions for processing nonnumerical lists are *car* (contents of *a*ddress *r*egister), *cdr* (contents of *d*ecrement *r*egister) and *cons* (*cons*truction). Applying *car* to a list delivers the first item of the list, while *cdr* delivers the remaining part, that is, the list without the first item.

Example: Define a list L of four items by
$$(set\ 'L\ '(W\ X\ Y\ Z)$$
Then, (*car* L) delivers W, and (*cdr* L) is the sublist (X Y Z). To obtain the second item in the list, we must specify (*car*(*cdr* L)).

In this way, the operators *car* and *cdr* allow one to extract any item, or a part, of the list.

The function *cons* allows one to construct a new list by adding an item preceding a list.

Example: (*cons* V L) or *cons*[(V) ; (W X Y Z)] constructs the list (V W X Y Z).

The function *append* adds an item following the list.

Example: *append* [L ; A] generates the list (W X Y Z A).

The *define* feature of LISP is the capability of users to define their own functions.

Example: Suppose we want to define a new function *sub* 1 which decreases the value of its argument X by 1, and mathematically is expressed as $X - 1$. The new function written in LISP is
$$(define\ (sub\ 1\ \ X)\ (minus\ X\ \ 1))$$
For example, (*sub* 1 10) gives the result 9.

Arithmetic operations are expressed as functions.

Example: *sum*[A; B] gives the sum of the atoms A and B.

Arithmetic relations are expressed as Boolean-valued functions or predicates.

Example: *lessp*[A; B] is TRUE if A is less than B.

The *conditional construction* is
$$[p1\ e1;\ p2\ e2; \ldots ;\ pn\ en],$$
where p1, p2, . . . are predicates and e1, e2, . . . are expressions. The construction is scanned sequentially, until a predicate which is TRUE is encountered, after which the expression is evaluated. An alternative construction takes the following form:
$$(cond\ (p1\ e1)\ (p2\ e2)\ \ldots\ (pn\ en)$$
A more conventional construction is the *if . . . then . . . else . . .* rule.

3.4 LISP

(*if* A B C) means that if A is TRUE, then the value of B is chosen, otherwise the value of C is selected.

A simple LISP function is *null*, which makes a test if a list does not contain any item.

Example: If L defines the above list, (*null* L) is FALSE, because L is not empty. (*null* ( )) is of course TRUE.

The following *cond* function gives the result 0 if the list M is null, 1 if M contains one item, and 2 otherwise.

(*cond* ((*null* M) 0) (*null* ((*cdr* M)) 1) (T 2))

Note that one of the tests will always succeed.

The *null* function can be used, for example, to define the sum of a list containing numeric atoms.

Example: (*define* (*sum* L)(*cond*((*null* L) 0)T(*plus*(*car* L) (*sum*(*cdr* L))))))
Note that the function is built up by using itself recursively.

A *recursive function* is a function which is built up by using itself as a building block. Such a function can be remarkably short and compact for the computation that it describes.

Example: Assuming positive integer values of N, the factorial function *fac* = N! is defined by the continuous computation N times (N − 1)! down to 0!, the latter being equal to 1. The LISP expression for N! is given by

(*define fac*(N) (*cond* ((*null* N) 1) (T (*times* N (*fac*(*sub* 1 N))))))

Properties of LISP

The main power of LISP is the easy handling of advanced data structures, such as lists, trees, queues and user-defined data types. This ability is important in applications that require the organization of large amounts of knowledge, such as in expert systems. Another strength is its suitability for interactively executing commands directly from the keyboard. Run-time structures used by the interpreter are accessible to the users. A special function call returns a list of all functions used in the system as well as the values of every variable.

Program modules developed and tested independently can easily be linked together as a system. Large-scale modularization of programs is facilitated by the concept of *packages* of software. These packages can be grouped together into modules, which may depend on each other in various ways. In addition to the list-processing speed, LISP performance is determined by the efficiency of garbage collection (i.e., reclaiming unused memory at run time).

Though LISP has the neccesary data types, it is less useful for numerical applications. The prefix notation and the extensive use of parentheses may be annoying to programmers. Distinguishing between constants and variables may also be awkward. LISP has no built-in pattern-matching capability.

Logo
A programming language based on LISP and originally created for introducing young children to computer use is Logo [Pap80]. A major feature of Logo is its friendly user interface by the availability of a graphic package (Turtle Graphics). The basic data elements in Logo are (integer or floating-point) numbers, words (strings of characters) and lists (of words delimited by square brackets). Recursion is used as a basic control structure. Logo is an easy-to-learn language for many applications, including artificial intelligence.

3.4b LISP Implementations and Machines

LISP Machines
Many sophisticated expert systems have been developed on specialized LISP machines because of the power and capacity of the basic hardware and the software development environments that run on them. All popular LISP machines have in common: high-speed LISP processing, a large physical memory, high-resolution, bit-mapped display, a mouse for pointing purposes, a communications link, and support for a powerful expert-system development environment. Many LISP machines have a tagged architecture. A tag, consisting of a small number of bits, is used to identify the data type of each data item and to guarantee that all operations performed on the location are appropriate.

Handling large dynamic data or knowledge structures requires support of algorithms and hardware to overcome memory limitations. *Garbage collection* is a technique of reclaiming memory space that is no longer used. It can be implemented in software, in hardware, or a combination of both. When high speed and efficiency are desired, dedicated hardware is preferred.

A LISP machine is a personal workstation designed to provide a high performance and economical implementation of the LISP programming language. Useful features of a LISP machine are a large address space, hardware data types, a microcode that can be compiled into pointer-manipulating instructions, a real-time garbage collector, a powerful editor, LISP used as a system implementation language, reasonable speed and a low price [Baw79, Gre84, Ple87].

The design of LISP machines was started at MIT's AI Laboratory. The first commercially available LISP machines were the Symbolics LM2, the Xerox 1100 Interlisp workstation and the Lisp Machine Inc. Series III CADR [Mye82, MAn83]. More recently developed LISP machines are based on additional hardware to support data tagging and garbage collection. Examples are LAMBDA, Symbolics 3600, Xerox 1100 series and ALPHA (see Subsection

3.4 LISP

4.4b). LISP-based systems come with their own Operating Environment which can access up to 16 Mbytes of working memory. LAMBDA is an open-ended, modular and expandable LISP machine with multiprocessor capabilities [MAn83]. It integrates expert system concepts with traditional numerical processing. It is based on the 68010/UNIX processor and a LISP processor. The NuBus architecture is designed around the LAMBDA processors to focus on a shared control system.

Symbolics 3600 (successor of the LM-2) provides support for LISP by compiling LISP programs into the 3600 instruction set [Moo87]. Instructions make use of a stack to store operands. A tagged architecture is used for hardware type checking, including a *cdr*-coded list representation and an intelligent instruction fetch unit. The 3600 provides special hardware for an incremental garbage collection.

A single-chip processor to support LISP has been implemented in the MIT SCHEME-79 [Sus81]. All LISP code in memory is represented uniformly as lists. SCHEME-79 used a tagged architecture, hardware support for virtual addressing and special hardware assistance for heap allocation.

LISP Implementations and Environments

A standard form of LISP is called Common LISP [Ste84, Tat87]. Versions of LISP available for personal computers include Golden Lisp, IQLISP, IQCLISP, PC Scheme and MuLISP. Interlisp, supported by Xerox LISP machines, and Zetalisp, designed for the Symbolics LISP machine, are AI environments which provide debugging, documentation and help facilities to assists LISP programming. MacLisp is an efficient dialect, developed at MIT [Moo74, Pit83]. Franz LISP [Fod81] is a descendant of MacLisp, designed for use under the UNIX operating system. Qlisp, an extension to Common LISP, is designed for parallel symbolic computation [GOl89].

The Interlisp environment [Tei81] contains a development system built to support LISP programming. Interlisp features include syntax extension, uniform error handling, automatic error correction, an integrated structure-based editor, a compiler, a sophisticated debugger and a filing system. Interlisp is easily extensible and customizable without losing compatibility. Interlisp has been used to develop and implement a wide variety of large application systems, including MYCIN and DENDRAL.

Another LISP-based environment is Explorer, developed by Texas Instruments. It contains 16k of 56-bit-word writable control storage, microprogrammed for LISP processing. It has 128 Mbytes of virtual address space and is centered around two high-speed 32-bit buses. Its detached mass storage subsystem has a disk capacity of 112 Mbytes. Explorer also supports Prolog. A VLSI custom-design application is Explorer CheckMate which runs on HP/Apollo and Sun workstations (see Subsection 6.3a).

3.5 LOGIC PROGRAMMING

3.5a Predicate Calculus

Introduction

Logic programming derives its name from *logic*, which studies the relationships of implication between assertions and conclusions [Kow79]. By expressing propositions and the relations between propositions, a formal procedure of logic can be devised so that one can infer some propositions from other propositions, as in theorem proving.

Most work on theorem proving is based on a particular form of logic, called *first-order predicate calculus* or *predicate logic*. Predicate calculus is a simplified language which consists of a set of statements or formulae used to express relationships. It provides a set of symbols to be used in expressing statements, and a set of inference rules for deriving new statements from the ones that have been given.

Propositional Calculus

A *proposition* is a legal statement in the form of a logic expression which has a value of either TRUE or FALSE. *Propositional calculus* uses propositions and relationships between them as a basis for a formal language for encoding and manipulating knowledge. It has rules of syntax and rules that derive new statements from existing ones. For example, a proposition is TRUE if all its premises are TRUE.

A	B	A $\wedge$ B	A $\vee$ B	$\neg$A	A $\equiv$ B
TRUE	TRUE	TRUE	TRUE	FALSE	TRUE
TRUE	FALSE	FALSE	TRUE	FALSE	FALSE
FALSE	TRUE	FALSE	TRUE	TRUE	FALSE
FALSE	FALSE	FALSE	FALSE	TRUE	TRUE

Fig. 3.5 Examples of logic connectives

Complex propositions can be expressed by *logic connectives*. These connectives and their symbols are given below:

AND	$\wedge$	or	$\cap$
OR	$\vee$	or	$\cup$
NOT	$\neg$	or	$-$
EQUIVALENT	$\equiv$	or	<->
IMPLIES	=>	or	$\supset$

3.5 LOGIC PROGRAMMING

Figure 3.5 applies to the above logic connectives. The statement A => B means that if A is TRUE, then so is B. However, the reverse statement (if B is TRUE, so is A) is not valid.

Example: *if* R1 is a resistor, *then* it is a circuit component. This statement does not imply that all circuit components are resistors.

The propositional calculus is based on the *modus ponens*. This rule states that *if* A => B and A is TRUE, *then* B is also TRUE. Bottom-up inference is a generalization of instantiation combined with the modus ponens rule. Instantiation is restricted to the minimum needed to match assertions with conditions so that modus ponens can be applied [Kow79].

Top-down inference is a generalization of instantiation combined with *modus tollens*. This rule states that *if* A => B and ¬B is TRUE, *then* ¬A is also TRUE. Here, instantiation is restricted to the minimum needed to apply to the modus tollens rule.

A useful property in propositional calculus is expressed by DeMorgan's Laws:

$$\neg(A \land B) \equiv \neg A \lor \neg B$$

$$\neg(A \lor B) \equiv \neg A \land \neg B$$

The process of *reasoning* can benefit from the following formal expression of the *reductio ad absurdum*:

$$(A => B) \equiv (\neg B => \neg A)$$

Suppose we have to prove that A => B, that is, assuming that A is TRUE, prove that B is TRUE. Assume first that B is FALSE and retract A is FALSE, which is in contradiction to the assumption that A is TRUE. This kind of reasoning is applicable in design situations.

Example: *if* APPLE => FRUIT *then* not_FRUIT => not_APPLE

Predicate Calculus

Suppose we want to state facts about objects in the real world and answer questions like the following: Is p a member of the set P? Do all members of the set Q possess a particular property? Such and many other statements cannot be expressed by propositional calculus. The tool we need is *predicate calculus*, which can be considered as an extension of propositional calculus. Predicate calculus provides a set of inference rules for deriving new statements from the ones that are given, and a set of symbols to be used in making statements. Predicate calculus allows the introduction of the concept of a *variable*.

Essential in predicate calculus is the *predicate*. It is a symbolic identifier that declares a statement about one or more objects. A consistent syntax for the predicate statements must be defined. For example, the syntax prescibes the predicate to be written first, followed by the objects (the arguments of the predicate) in parentheses.

Example: The predicate "computer(Macintosh)" states that Macintosh is a computer.

When variables are used, the value TRUE or FALSE is returned.

Example: The predicate computer(X), meaning "X is a computer", takes the value TRUE if X = Macintosh, but the value FALSE if X = CMOS.

It is common practice to use the last letters of the alphabet (e.g., X, Y, Z) for variables and the first letters (e.g., A, B, C) or symbolic identifiers, such as Macintosh or CMOS for constants.

A predicate can define a relationship or specify an action.

Example: The predicate gives(X,Y,Z) can mean "X gives Y to Z".

A *function* is a generalization of the concept of predicate, able to return a value of any type: Boolean, symbolic or numeral.

Example: The function datatype(X) returns the value INTEGER when X = 25, and the value SYMBOL when X = RX1.

Functions and predicates can be combined, but not without restrictions.

Example: The predicate "computer" and the function "datatype" as defined above can be combined in the order computer(datatype(X)), because if, say X = RX1, the predicate to be evaluated is computer(SYMBOL), which presumably has the value FALSE. The inverse order datatype(computer(X)), however, leads to the function datatype(FALSE), meaning "What is the datatype of FALSE?", which is clearly a nonsensical question.

Quantifiers may be *universal* ∇, meaning "for all . . ." or *existential* $\Leftarrow$, meaning "there is . . .".

Example: "Every dog is an animal" is expressed as
$$(\nabla X)(dog(X) => animal(X)$$
"Every student has a book" is expressed as
$$(\nabla X)(\Leftarrow Y)(student(X) => book(Y) \wedge owns(X,Y)$$
The second statement means that whoever is the student X, there is a book Y such that Y is owned by X.

"For all persons X, Y and Z, if X is a parent of Y, who in turn is a parent of Z, then Z is a grandparent of X" is expressed as
$$(\nabla X, \nabla Y, \nabla Z) \; parent(X, Y) \wedge parent(Y, Z) => grandparent(X, Z).$$

3.5 LOGIC PROGRAMMING

Logic Programming

Logic programming attempts to reach a goal, which may succeed or fail when solving a problem or answering a question. A set of statements is searched to find a statement that matches the goal. The matching process can be viewed as pattern matching. Logic programming provides a way of combining expert system devclopment and database access within a single framework.

A logic program consists of facts about certain objects expressed as statements that represent information which can be used to find the goal. Logic programs tend to blur the distinction between programs and databases, even that between programs and specifications. Logic can be used as a formalism that unifies AI, database management and programming.

There are many equivalent ways to express the same statement or formula in predicate calculus. To avoid redundancy and to unify the way in which a specific statement is expressed, a special form of predicate logic, called the *clausal form*, is used. The statements and formulas, called *clauses*, are sentences which contain m joint conditions implying n alternative conclusions, where m and n may be zero. Clauses which imply at most one conclusion are called *Horn clauses*. We give some examples of Horn clauses.

Examples: animal(horse). This clause with the meaning "A horse is an animal" is a Horn clause representing a fact, where $m = 0$ and $n = 0$.

$$A \; if \; B1 \; and \; ... \; Bm.$$

where A, B1, ... , Bm are statements and A is TRUE if B1, ... , Bm are TRUE. This Horn clause is a rule with m conditions and one conclusion ($n = 0$).

The general form of a Horn clause can be written as

head *if* body.

The Horn clause consists of a head and a body separated by the conditional identifier *if*. The head is a statement which is TRUE provided that all the conditions included in the body are fulfilled in such a way that the body is TRUE. When there are no conditions at all, the body is absent and the identifier *if* is dropped. In this special case, the Horn clause represents a fact. Horn clauses, which may represent facts and rules, constitute the basic elements of the logic programming language Prolog [Clo87, Bra86, Ste86].

3.5b Prolog

Introduction

Prolog (*Pro*gramming in *log*ic) was developed in 1972 by Alain Colmerauer and associates from the Artificial Intelligence Department of the University of Aix-Marseilles in France [Gia86]. Although it was initially restricted to such applications as automatic theorem proving, deductive reasoning and natural-language processing, Prolog has now gained a much wider applicability. Being related to mathematical logic, Prolog has a firm theoretical basis. The idea behind Prolog is that knowledge can be represented in terms of logic, from which appropriate inferences can be drawn. Each statement takes the form of a special logic formula, called a *Horn clause*, which can be interpreted operationally as a procedural declaration. The syntax of Prolog statements corresponds to that of Horn clauses (see Subsection 3.5a).

Prolog is a nonprocedural language whose program execution differs substantially from that of third-generation procedural languages, such as Pascal and C. It is an efficient language for nonnumerical problems which involve objects and relations rather than numerical problems (which are adequately handled by third-generation languages). For example, Prolog is very suitable for representing a hierarchy of components and subcomponents and answering questions about the hierarchical relationships between (sub)components. Prolog is suitable as a query language for relational databases and particularly suited as a language for expert systems.

A major issue in Prolog programming is how to formulate the problem to be solved with statements in formal logic and how to express all the known information relevant to the problem. All these statements will be called *clauses*. The syntax of Prolog rules corresponds to that of first-order predicate calculus, restricted to Horn clauses.

A Prolog program consists of three sets of clauses: *data clauses* to enter the assertions about facts and relationships between objects, *rule clauses* to enter the conditions to be satisfied in rules, and *goal clauses* to formulate a question or problem whose answer or solution is wanted. The process of problem solving differs radically from the way a problem in a third-generation language is solved. Rather than prescribing a sequence of steps taken by a computer according to a predefined algorithm written in an imperative third-generation language, the approach in a declarative language such as Prolog is based on inferencing methods, that is, conclusions are drawn from the facts and rules contained in a knowledge base.

A Prolog programmer specifies *what* the program is wanting to do rather than *how* it is executed. Once a program has been set up and some useful information has to be derived from the knowledge base, a goal clause simply specifies the question in the appropriate Prolog format. The internal matching mechanisms provide the searching process. The basic operation in Prolog is *unification*, that is, a pattern match which compares two identifiers to see if they are equal. Given a goal (the predicate to be verified), it allows the clause headers to be selected. A basic operation is that of matching variables with values. When an attempt to match the goal fails, a backtracking step is executed (see later).

The search for a solution to a problem is similar to the process of theorem proving. In the latter case, a set of axioms is entered as data clauses. The valid way by which deductions may be made from axioms are described as deduction rules and entered as ordinary clauses. Then, a hypothesis to be tested is entered as a query clause. When the hypothesis is shown to be true, the proof of the theorem is assumed to be established.

In the following, the standard Edinburgh Clocksin-Mellish (C & M Prolog) notation will be used [Clo87, CLo87]. Deviations from this notation will be indicated later. A Prolog program consists of clauses and operates on constants, variables and structures. Below, these clauses will be discussed and illustrated by examples. Comments or interpretations of the examples are enclosed by the symbols /* and */.

Facts

A *fact* is an assertion or axiom about objects. Facts in Prolog must be expressed in a consistent way as data clauses based on a predicate. A *predicate* in Prolog is an identifier that describes a property, activity or relationship of objects. A data clause, which consists of the *predicate name* and one or more *arguments*, is written in the syntax form:

> predicate(argument1, argument2, ... , argument$n$).

Each argument represents an object. The n arguments, separated by commas, are enclosed in parentheses, and the entire clause is terminated by a period. The period denotes the end of a clause and signals to the system to accept the clause. The predicate name preceding the list of arguments in parentheses is also referred to as the *functor*. Let us first consider *constants* as objects, which may be of different data types: symbolic (alphanumeric), numeric (integer or real) or Boolean (TRUE or FALSE).

The simplest predicate has only a single argument written as

> predicate(argument).

This clause has the meaning: "argument *is* or *is_a* predicate".

Examples:
male(bill).	/* Bill is a male */
female(susan).	/* Susan is a female */
horse(blacky).	/* Blacky is a horse */
capital(london).	/* London is a capital */
valuable(gold).	/* Gold is valuable */
personal_computer(macintosh).	/* Macintosh is a personal computer */

Note that the name of a predicate (functor) or a constant argument begins with a lowercase letter followed by any number of alphanumeric characters.

In the above examples, the predicate can be viewed as a class of objects, while the argument is an object which is an instance of this class. New arguments can be introduced to assign additional properties to this object.

Examples: pupil(john, 9).   /* John is a pupil of 9 years old */
person(barbara, 25, f, secretary).   /* Barbara is a 25-year old female

secretary */
car(chrysler, 13000, 3, red, 12000).   /* A car of make Chrysler, a mileage of 13000 miles, 3 years on road, red of color and with a price of $12000 */
Note that the subsequent arguments have specific meanings. When the predicates are built in, the meanings of the arguments are predefined. When a programmer introduces new predicates, specific meanings must be attached to these predicates.

A general class of predicates defines *relationships* between two or more arguments. Such a predicate can be expressed with the following syntax:
relationship(argument1, argument2, ... , argument*n*).
The order in which the arguments are positioned is directly related to the interpretations to be given to the predicate. Once this order has been predefined, it must be consistently maintained throughout the program.

Example: likes(john, mary).   /* John likes Mary */
Here, we assume that the first argument (john) likes the second argument (mary), and not conversely (Mary likes John). If the latter statement applies, we must write
likes(mary, john)   /* Mary likes John */
If John and Mary like each other, both clauses must be given.

Examples: parent(susan, gary).   /* Gary is a parent of Susan */
parents(mary, alice, bill).   /* Alice and Bill are Mary's mother and father */
Note that the first argument represents the child in the child-parent relationship. By definition, the reverse (i.e., the first argument is the parent of the second argument) may apply, provided that this interpretation holds throughout the program execution.

Examples: owns(john, horse).   /* John owns a horse */
owns(dick, book).   /* Dick owns a book */

The second argument can be augmented with new detailed information.

Examples: owns(john, horse(blacky)).   /* John owns a horse by the name Blacky */
owns(dick, book(from_here_to_eternity, james_jones)).   /* Dick owns the book "From Here to Eternity", which was written by James Jones */

Questions and Variables
Once a Prolog database has been constructed by entering data clauses as given above, questions can be asked whether or not a statement is true. The answer to such questions is a Boolean "yes" or "no". Let the following database KB1 be given.

3.5 LOGIC PROGRAMMING

likes(adam, mary).
likes(adam, john).
likes(mary, john).
likes(john, adam).

This database can be queried with a goal clause which begins with the Prolog prompt ?- .

Example: ?- likes(mary, john).
The reply is "yes", since the statement likes(mary, john). matches an entry in the database. The question

?- likes(mary, adam).

would be answered with "no", since no match is found in the database.

When questions as the above are entered, the database of facts is searched for a fact which matches the goal clause, that is, the data clause must be identical with the goal clause. If a match is found, the answer is affirmative. Otherwise, it is negative. From this it is clear that the arguments listed must always appear in a consistent order. Note that a data clause and a goal clause have the same syntax.

Horn clauses may include variables. In Prolog, a *variable* can be introduced to denote a thus far unknown object whose identity is sought. Predicates containing variables allow queries to be answered by performing a pattern-matching process. By convention, the name of a variable always begins with an uppercase letter or an underscore character _. Suppose a variable X is contained in a goal clause. The identity of X can be found by seeking clauses which are identical to the goal clause except for the argument in the same position in the clause as X. Let us assume that this argument is adam. Then, the solution is X = adam. The variable X is said to be *instantiated* (bound) to the constant "adam".

Example: Given the above database KB1, we can pose the question if there is a person X who likes John. The goal clause must be:
?- likes(X, john).
Searching the database leads to a match with the second entry in the database, leading to X = adam. A further search in the database yields the answer X = mary. The question
?- likes(adam, X).
would yield the answers
X = mary;
X = john;
no
When more than one answer is expected, the symbol ; can be entered following an answer. The reply "no" indicates that no further match is found in the database.

Two or more facts which are combined to a new fact is called a *conjunction*. Such a conjunction can be used to answer more complicated questions.

Example: ?- likes(adam, mary), likes(mary, adam).
The comma between the two predicates has the meaning of the logic connective AND. This question asks whether Adam and Mary like each other. The answer is "no".

Unlike the prespecified flow of instructions through programs written in a procedural language, Prolog procedures are executed as search processes using *backtracking*.

Example: ?- likes(adam, X), likes(X, adam).
The question is whether Adam likes a person X who in return likes Adam. To answer this question, two subgoals must be satisfied, namely, likes(adam, X). and likes(X, adam)., where X is of course identical in both subgoals. This question can be used to illustrate the important concept of backtracking in Prolog. Suppose that the above database KB1 is given to answer this question. In trying to satisfy the first subgoal, we find from the first entry in KB1 that X = mary. With this preliminary result substituted in the second subgoal, a match is to be found with likes(mary, adam). Since this action fails, the preliminary result X = mary is dropped and a new solution of X is sought. This is called backtracking. The new preliminary result is X = john, which substituted in the second subgoal matches the fourth entry in KB1, namely, likes(john, adam). Hence, the ultimate solution found is X = john.

Rules

A *rule* is expressed as a Horn clause involving one or more *conditions*. A Prolog rule consists of two parts: a *head* and a *body* separated by the symbol :- which has the meaning *if*.

head :- body.

The body may contain a number of subgoals (involving conditions) separated by commas with the meaning of logic AND. Each subgoal is associated with a condition and all the conditions in the body must be satisfied for the head to be TRUE.

Examples: grandparent(X, Z) :- parent(X, Y), parent(Y, Z). /* X is a grandparent of Z if X is a parent of a person Y who in turn is a parent of Z */

brother(X, Y) :- male(X), parents(X, Mother, Father), parents(Y, Mother, Father).
/* X is a brother of Y if X is a male and X and Y have the same mother and father */

3.5 LOGIC PROGRAMMING

Many problems boil down to solving for a variable X which is constrained to meet conditions in one or more Prolog rules. Although such a variable X is reminiscent of the unknown quantity X in algebra, the Prolog approach to finding X differs from that in algebra. In Prolog, the unknown variable X, when matched with a clause in an intermediate step is bound to a corresponding constant that can be viewed as a preliminary solution which satisfies some of the conditions or subgoals. This preliminary solution must further be checked if it satisfies the remaining subgoals. If an attempt to satisfy these subgoals fails, backtracking is needed to find an alternative preliminary solution. This match-and-backtrack procedure is continued until the system finds a solution which satisfies all conditions.

Example: Suppose that the following goal clause is given.
likes(beth, X) :- likes(mary, X), fruit(X), color(X, red).
/* Beth likes what Mary likes if it is a fruit which is red */
The body of this rule contains three subgoals each of which has to be satisfied, one by one, from left to right. Let the following relevant clauses be contained in the knowledge base. These clauses contain three different predicates corresponding to the three subgoals in the goal clause.

likes(mary, pears).	/* 1, begin of first subgoal clauses */
likes(mary, popcorn).	/* 2 */
likes(mary, apples).	/* 3, end of first subgoal clauses */
fruit(pears).	/* 4, begin of second subgoal clauses */
fruit(apples).	/* 5, end of second subgoal clauses */
color(pears, yellow).	/* 6, begin of third subgoal clauses */
color(oranges, orange).	/* 7 */
color(apples, red).	/* 8, end of third subgoal clauses */

The first subgoal likes(mary, X) matches the first clause in the first part of the database: likes(mary, pears) so that X is bound to the constant "pears". With X = pears, the predicate fruit(pears) matches the fourth entry in the knowledge base (first clause of the second part of the knowledge base). Examining the third subgoal reveals that a match of color(pears, red) cannot be found. Hence, the preliminary solution X = pears must be cancelled. A backtrack to the first subgoal leads to the preliminary solution X = popcorn. Attempts to satisfy the second subgoal will fail, since a clause fruit(popcorn) cannot be found. A second backtrack to the first subgoal yields X = apples. A search in the second subgoal succeeds since a match of fruit(apples) is found in the fifth entry of the knowledge base. The last clause in the third part of the knowledge base is color(apples, red). Hence, the third goal is also satisfied and the final solution is X = apples.

While the programming process is user-friendly, the internal mechanism may involve a very large number of ramifications and conclusions, before the desired information has been reached. In the simplest case, the search of an entry with certain properties would require an exhaustive searching procedure, which implies many redundant operations. To reduce this problem, Prolog implementations usually contain control structures which speed up the searching process. When the body of a rule contains *m* subgoals, the database is searched for successful matches for satisfying the subgoals one by one, fom left to right. Each time a subgoal cannot be satisfied by a variable which is bound to a name, backtracking is necessary to attempt a new match with a new name. In order to make the search for a match efficient, redundant operations should be avoided. When an attempt to match against a clause fails, it is not necessary to return to the very first clause in the knowledge base. The point from which to proceed in attempting alternative solutions must be the clause following the clause which matches the latest preliminary solution. By skipping clauses handled earlier, fewer clauses must be examined for successful matches. To save time, one can use the *cut* feature, which prevent unnecessary backtracks. Algorithms for optimizing the solution process in a Prolog-based system form a topic of intensive research. Prolog is inherently suited to parallel execution, leading to schemes for parallel computation based on predicate logic.

Let us consider the representation of a small family tree and the provision of the ability to answer questions on the relationships contained in the tree (see Fig. 3.6).

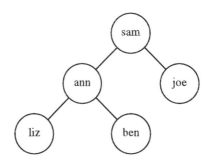

Fig.3.6 Tree structure

A database that defines relevant parent relationships follows below.
parent (sam, ann).
parent (sam, joe).
parent (ann, liz).
parent (ann, ben).
Note that the first argument is a parent of the second argument. The "parent" predicate can be used to define the "successor" predicate by two Horn clauses:
succ(X, Y) :- parent(Y, X).   /* X is a successor of Y if Y is a parent of X */

3.5 LOGIC PROGRAMMING

succ(X, Y) :- parent(Y, Z), succ(X, Z).   /* X is a successor of Y if there exists a Z such that Y is a parent of Z and X is a successor of Z */
These two clauses apply to all possible X and Y. The Prolog system can now be queried.

Example: ?- parent(ann, X).   /* Who are the children of Ann? */
X = liz;   /* One answer. Another answer will be given by typing a semicolon */
X = ben;   /* The second answer */
no   /* No further answers */

Example: ?- succ(ben, Y).   /* Who are the predecessors of Ben? */
Y = ann;   /* The first answer */
Y = sam;   /* The second answer */
no

Example: Let a two-NAND-gate circuit be given (Fig. 3.7).

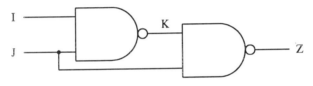

Fig. 3.7 NAND circuit

A 2-input-NAND gate is given by its truth table:
nand (0, 0, 1).
nand (0, 1, 1).
nand (1, 0, 1).
nand (1, 1, 0).
The circuit is defined as follows:
circuit (I, J, Z) :- nand (I, J, K), nand (K, J, Z).
Two queries along with their answers are given below.
?- circuit (0, 1, X).   ⇒   X = 0; no
?- circuit (X, Y, 1).   ⇒   X = 0, Y = 0; X = 1, Y = 0; X = 1, Y = 1; no

The List as a Data Structure

A *list* of elements enclosed in square brackets can be used as a data structure. The elements, separated by commas, represent objects which may be of type string, numeral of Boolean.

Example: [ ]   /* The empty list */
[1, 2, 3, 4]   /* The numerals 1, 2, 3, 4 */
[beethoven, chopin, grieg, mahler, mozart, schubert]/* Famos composers */

Like in LISP, the elements of a list can be divided into the *head* (the first element) and the *tail* (the remainder of the list). A list with the head H and the tail T is written as [H | T] in which a vertical bar is used to separate the head and tail. This division can be used to investigate if a given object is a member of the list. For this purpose, we introduce a predicate, member(name, [list]), which attempts to match a name with a list of names. The following two clauses can be used to answer questions about lists:

member(X, [X | _]).
/* X is a member of the list if X is the head of the list */
member(X, [_ | T]) :- member(X,T).
/* X is a member of the list if it is a member of the tail of the list */

Note that the underline character _ is used as an "anonymous" variable, that is, a variable whose name is irrelevant.

Example: Suppose we want to know which entries are in the list of famous composers as defined above. The goal clause is

?- member(X, [list]).

By applying the above-mentioned two clauses with predicate "member" recursively, the variable X will be identified successively with each of the elements in the list. In fact, the variable X is recursively bound to the first name of the list, where the list is successively deprived of the first name. In this way, the entries of the list will be presented one by one from left to right.

Suppose that we want to know if a given composer name, say, Chopin, is in the list. Since the first element of the list happens to be different from chopin, the tail of the list must be examined. In this example, a match is directly found and the answer to the posed question is "yes".

A function to append two lists is defined more easily in Prolog than in LISP. Also, the allocation of a free memory page is more easily done in Prolog.

Prolog Features

The above examples have shown that Prolog uses facts in conjunction with rules to make inferences in order to satisfy a goal. The specification of facts, relationships between facts and rules that govern specific relationships as the basis for solving a problem makes Prolog a *declarative* (or *descriptive*) language. Prolog can also be considered as a *database* and *query* language. Though Prolog determines the way of how the solution is obtained, its *procedural* aspects makes it possible to control the steps to be taken to find a solution quickly.

Prolog has the attractive feature that it has easy-to-understand semantics. Besides, a Prolog interpreter consists of a relatively small number of symbols. Prolog constructs differ in many respects from the standard constructs in third-generation languages. For example, the destructive assignment statement, such

3.5 LOGIC PROGRAMMING

as in X := 10, and DO or WHILE loops, are not used in Prolog. On the other hand, modification of a working memory is a central feature in logic programming. Consequently, logic programming tends to a more declarative style and Prolog is more suitable for implementation on parallel hardware.

In principle, any computable function can be computed by Prolog, which is a language with single assignments to variables. Prolog has a logic and a control component. The logic component is provided by the programmer who provides the facts and rules, which constitute the logic program. The control component is provided by the system, which contain mechanisms for several tasks, such as unification and backtracking [Kow79]. In fact, the programmer specifies *what* problem is to be solved, while the system's control mechanism decides *how* the problem is solved.

Prolog is an efficient tool for nonnumerical problems which involve objects and relations. It is, on the other hand, in its present form completely unsuitable for more traditional, numeric-type computations. It is rather obvious that Prolog is extremely well suited for the development of expert systems. It facilitates experimenting with alternative ways of representing and using knowledge in the system. Prolog can be naturally used as a query language for relational databases. It can also be used as a specification language for information system design. An important consideration in any Prolog system is a suitable procedure for garbage collection [App88].

Prolog has procedural and nonprocedural aspects. Though a Prolog program is written in a declarative style, Prolog rules are executed by an underlying *inference engine*, which imposes an order of execution. Because of that implicit ordering, Prolog rules also have a procedural interpretation. Unlike the linear flow through conventional procedures, Prolog is executed by *backtrack search*. This method is very powerful for many AI problems and it requires a new, declarative way of thinking.

Typical Prolog features include the following:
a. Applications in various areas, including mathematical logic, relational databases, natural-language understanding and VLSI chip design.
b. Efficient implementation of several programming aspects, such as non-determinism, parallelism and pattern-directed procedural call.
c. Predicates are defined by asserting facts, relationships between facts, and rules.
d. A query capability.
e. Data structures that can simulate Pascal-like records or LISP-like lists.
f. A backtrack-search procedure for evaluating goals.
g. A pattern matcher analyzing the data structure.
h. A set of built-in predicates for arithmetic, input/output, and system services.

3.5c Prolog Implementations

Edinburgh Prolog

Of the Prolog implementations released thus far the Edinburgh Prolog is the one that uses the Clocksin-Mellish notation as introduced in Subsection 3.5b. Edinburgh Prolog has been implemented on DEC-10 by David Warren [Per78].

A comment in Edinburgh Prolog is enclosed between /* and */ or as a line of text beginning with %. The user can define built-in operators by giving their representation, their evaluation, their arity (whether they are unary or binary) and their relative priority.

Prolog II

Prolog II is the current version of the Prolog implementation developed by the Artificial Intelligence Group at Marseille-Luminy [Gia86]. The notation differs substantially from the Edinburgh notation as shown by the examples below:

$val(add(x,y),z)$ /* $z = x + y$ */
$val(sub(x,y),z)$ /* $z = x - y$ */
$val(mul(x,y),z)$ /* $z = x * y$ */
$val(div(x,y),z)$ /* $z = x / y$ */
$val(inf(x,y),1)$ /* $x < y$ */

The meanings of the above clauses are given in the text between /* and */. Note that comments in Prolog II are given between a pair of double quotes, e.g.,
  "This is a Prolog II program.".

Examples: The Edinburgh rule
son(X, Y) :- child(X, Y), man(X).
is translated into the Prolog II clause:
son(x,y) → child(x,y) man(x);
The Edinburgh goal clause
?- brother(X, henry).   /* Find X who is Henry's brother */
is translated into the Prolog II goal clause: > brother(x, Henry);

A *variable* in Prolog II is written as a single letter, possibly followed by digits, a quote or a hyphen, followed by an arbitrary string of alphanumeric characters, e.g. X, x10, x', x-element-of-y. A *constant* can be an identifier, a string of characters, an integer or a signed real number, e.g., A, "John", 16, -2.5 .

The empty list in Prolog II is denoted by the symbol *nil*, which usually also serves to mark the end of a list. For example, the Edinburgh list [a b c d] is written in Prolog II as (a.b.c.d.nil) .

Micro-Prolog

Micro-Prolog is a Prolog implementation for personal computers developed in England by McCabe and Clarke [Cla84, Enn84, deS85]. The initial aim of Micro-Prolog is to provide a didactical tool for teaching Prolog to students

3.5 LOGIC PROGRAMMING

starting secondary school.

The goal clauses are different from the Edinburgh notation. Micro-Prolog starts goal clauses with Does or Which.

Examples: Does (Mary is-the-mother-of Ben)
Which (X is-a-parent-of Liz)

Built-in arithmetic clauses are specified in the form of lists.

Examples:
(*sum* x y z)   /* z = x + y   */
(*sum* y z x)   /* z = x − y   */
(*times* x y z)   /* z = x * y   */
(*times* y z x)   /* z = x / y   */

Lists are also used for Boolean operations:
(*and* x y)   /* x y   */
(*or* x y)   /* x + y   */
(*not* x)   /* $\bar{x}$ */

Turbo Prolog

Turbo Prolog has been developed by Borland International, Scotts Valley, CA, for the IBM PC and compatibles [Yin87, Tur88, Tef89]. Since Turbo-Prolog is widely available, it will be discussed in greater detail than the other implementations. Turbo-Prolog has the following features:

a. It is a *compiler*-based programming language and yet allows interactive program development. A programmer can test individual sections of a program and alter the goal of the program without having to modify the existing code of the program.

b. Tools are provided to build expert systems. The Turbo Prolog Toolbox includes 80 additional tools that can be incorporated Turbo Prolog

c. Programs can be run with full capabilities of *windows* which may contain mixed text and graphics. Turbo Prolog allows us to control such screen display characteristics as inverse video (black characters on a white background), underlining and colors. This information is passed to standard predicates via an attribute value which, among other things, determines the color of the characters (the foreground) and the color behind the characters (the background). It is possible to give attributes for single characters or for a whole screen area.

d. Program development with efficient memory usage is facilitated by Turbo Prolog's unique *type* system, defined as domains, objects and lists. Recursion allows us to process the elements of a list.

e. Turbo Prolog is extremely rich in *file*- and *string*-handling facilities. The standard predicates for reading and writing are elegant and efficient. With just a single command, output can, for instance, be routed to a file instead of being displayed on the screen. The standard predicate *findall* is used to collect the values of a variable that satisfy a given clause into a list. Random access files can be used.

f. *Mathematical* operations in integer and real arithmetic can be performed, as well as *bitwise* operations for control and robotic applications. Arithmetic expressions are written in infix notation. This includes relational operators, arithmetic functions and bracketed subexpressions.

g. Turbo Prolog provides unique debugging and tracing facilities for efficient program development. If unexpected behavior occurs, the *trace* compiler directive can be used to obtain a step-by-step trace of execution in the trace, edit and dialog windows of the screen.

h. Turbo Prolog allows *modular* program development. Modules written in Prolog or other languages (such as C and assembly language) can be linked into an executable unit.

A Turbo Prolog program consists of several program sections, each identified with a keyword: domains, global domains, database, predicates, global predicates, goal and clauses. The *domains* section is used for declaring the *data types* of objects (integer, real, symbol, character or string), *lists*, *compound objects* and *files*. Integers are whole numbers in the range from -32768 to +32767. *Real* numbers may range between $-1e^{-307}$ and $+1e^{+308}$. A *symbol* in Turbo Prolog is a sequence of alphanumeric characters, e.g., brand and color. A *char* is one of the ASCII characters enclosed in single quotes, e.g. 'A'. The *string* data type is represented as a sequence of alphanumeric characters enclosed within a pair of double quotes, e.g., "This is an example of a string.". A *list* is a predefined data structure consisting of an ordered sequence of objects enclosed in square brackets. A *compound object* contains itself other objects although it is treated as a single object. Turbo Prolog has built-in predicates for *file* management tasks.

The *predicates* section is used to define the structure of each relevant predicate by stating its name and the domains of its arguments declared in the *domains* section:

predicate_name(domain1,..., domain*N*)

A clause in the *clauses* section is either a fact or a rule corresponding to one of the predicates declared in the *predicates* section. A general rule consists of an atom followed by the keyword *if* (or :-) and a list of atoms separated by *and* (or commas) or *or* (or semicolons).

Example: /* Simple program */
domains
        brand, color = symbol
        age, price = integer
        mileage = real

3.5 LOGIC PROGRAMMING

predicates
    car(brand, mileage, age, color, price)
clauses
    car(chrysler, 13000, 3, red, 12000).
    car(ford, 9000, 4, gray, 25000).
    car(datsun, 8000, 1, red, 3000).
/* These clauses represent facts. Names of symbolic constants start with a lowercase letter */
goal
    car(Make, Odometer, Years_on_Road, Color, Cost) *and* Cost < 25000.
/* This is a compound goal looking for a car costing less than 25000 dollars. Names of variables start with an uppercase letter */

An important feature of the Turbo Prolog language is its *modularity*, which allows us to handle programs that are broken up into modules. Modules can be written, edited and compiled separately, and then linked together to create a single executable program. When a program must be changed, we need only edit and recompile one of the modules, not the entire program. This is useful in large programs. Modular programming also allows us to take advantage of the fact that, by default, all predicate and domain names are local. This means that different modules can use the same name in different ways.

Turbo Prolog uses two concepts to manage modular programming: *projects* and *global* declarations. Among other things, these features make it possible to keep a record of which modules make up a program (called a *project*), and to perform type checking across module boundaries. When a program is to be made up of several modules, Turbo Prolog requires a *project* definition specifying the names of the modules involved. A file (called the *librarian*) must be created containing the list of the module names.

Names are used to denote symbolic constants, domains, predicates and variables. Names may contain mixed uppercase and lowercase letters, digits and underscores, except that names of variables start with an uppercase letter or an underscore symbol. User-defined names must not be identical to built-in names: *if, and, or, domains, predicates, goal, clauses*, and 12 other reserved names; and *edit, exit, sin, cos, sound, time* and 100 other restricted names.

When the program consists of several modules, *global domains* and *global predicates* are used to define which domains and predicates are active in all modules. By default, all names used in a module are local. Turbo Prolog programs communicate across module boundaries using predicates defined in a *global predicates* section. The domains used in global predicates must be defined as global domains or else be domains of standard types. All the modules in a project need to know exactly the same global predicates and global domains. The easiest way to achieve this is by writing all global declarations in one single file, which can then be included in every relevant module via an *include* directive.

When Turbo Prolog is started, the screen displays four separate windows, labeled Editor, Message, Dialog and Trace. In addition, a menu line at the top of the screen shows seven fundamental commands: Run, Compile, Edit, Options, Files, Setup and Quit. Selecting the Edit option from this menu allows us to type in a program in the Editor window. Once the *domains*, *predicates* and *clauses* sections of a program have been finished (as in the simple program above), the Run option can be selected. This option first invokes the compilation process, after which the program will be run. Possible errors detected during the compilation are shown in the Message window. The appearance of a system prompt *Goal*: in the Dialog window is an indication that the Turbo Prolog system is in the interactive mode. When the *Goal* clause (as in the above simple program) has been entered, the system will respond with a True or False, depending on whether or not a match can be found in the database. Turbo Prolog looks for all possible solutions that satisfy the goal. We may continue entering arbitrary goals to get answers to our queries. At the beginning of the program, before the *domains* section, the compiler directive *trace* can be entered. Then, a step-by-step progress of the program execution is traced and intermediate results are displayed in the Trace window. The window shape, position and color can be changed during the course of a programming session.

Mathematical operations performed by Turbo Prolog include simple arithmetic, logarithmic, trigonometric and bitwise manipulations. *Arithmetic expressions* are written in prefix or infix notation.

Example: Prefix: −(a, *(b, c))
            Infix: a − b * c

The operators − and * in the prefix notation act as functors with the arguments enclosed in parentheses separated by commas. The precedence order of operator evaluation in the infix notation is: unary minus, mod (integer remainder) or div (integer division), multiply or divide (* or /), and add or subtract (+ or −).

The *sum* relation in combination with a *which* query can be used for adding and subtracting purposes.

Examples:
which (X: sum(3.5, −4.7, X)) − 1.2   /* Solution X = 3.5 − 4.7 = −1.2 */
which (X: sum(sum(X, −4.7, −1.2)) 3.5   /* Solution X = −4.7 −1.2 + 3.5 = −2.4 */

The *times* relation can be used for multiplying and division purposes.

Examples:
which (X: times (4, 3, X)) 12   /* Solution X = 4 * 3 = 12 */

3.5 LOGIC PROGRAMMING

which (X: *times* (X, 3, 12)) 4   /* Solution X = 12/3 = 4 */

The *relational* operators are: < (less than), <= (less than or equal to), > (greater than), >= (greater than or equal to), = (equal to), and <> or >< (not equal to). Below are some examples of arithmetic predicates:

7 *mod* 3 = 1   7 *div* 3 = 2
abs(−14) =14   *sqrt*(16.0) = 4.
round(2.35) = 2   *round*(2.67) = 3
random = a random real number between 0 and 1
$sin(\pi) = 1$ $cos(\pi) = 0$
$tan(\pi/4) = 1$ $exp(3) = e^3$
$log(100) = 2$ $ln(e^3) = 3$

A Prolog program can be viewed as a relational database of which the data (facts and rules) are organized in such a way that they can easily be retrieved and manipulated. The Turbo Prolog language can be utilized as a powerful query language for dynamic databases. A *dynamic database* is a database which can be altered during program execution, or fetched from a disk by a call to the standard predicate *consult*. Its unification algorithm automatically selects facts with the correct values for the known parameters and assigns values to any unknown parameters and its backtracking algorithm gives all the solutions to a given query. To increase its speed when processing large databases, facts belonging to dynamic databases, which must be frequently updated, are treated differently from normal predicates. Dynamic database predicates are distinguished from normal predicates by declaration in a separate *database* section.

Example:
domains
        name, address = string
        age = integer
        sex = male; female
database
        person(name, address, age, sex)
predicates
        male(name, address, age)
        female(name, address, age)
        child(name, age, sex)
clauses
        male(Name, Address, Age) *if* person(Name, Address, Age, Male).
        . . . .

The predicate *person* can be used in precisely the same way as the other predicates, the only difference being that it is possible to insert and remove facts for the *person* predicate during execution. Facts added in this way are stored in internal memory.

Built-in predicates for manipulating dynamic databases stored in memory are: *asserta* (insert a clause at the top of other stored clauses), *assertz* (insert a clause at the bottom of other stored clauses), *retract* (delete a fact), *save* (save all the facts to a disk file), *consult* (add a disk file to the current database), *readterm* (read specific clauses). A clause should be read as a fact or a rule. The clause to be inserted must be placed in parentheses.

Examples: *asserta*(likes(john, mary))
               assertz(likes(bill, joan))

Turbo Prolog also offers several built-in predicates for performing standard input/output and disk file management. During program execution, either the DOS operating system or ROM-BIOS services can be invoked for performing various I/O operations. In the latter case, the *bios* predicate can be used. One of the possibilities is to call various interrupt routines.

Other Prolog Implementations

There are several other implementations of Prolog. MS DOS versions of Prolog include Prolog1, version 2 from Expert Systems International Limited, Prolog86, version 1.2 from Micro-AI and Prolog V, version 1.0 from Chalcedony Software. Quintus Prolog, version 1.2 from Artificial Intelligence runs under UNIX and VMS operating systems of the VAX-11 and SUN-2 systems.

Prolog Compilers

The first Prolog implementation was an interpreter developed at the University of Marseille. The Prolog interpreter is based on the syntax of statements of the form:

         <left part> :- <right part>

with the righthand part implying a lefthand part. A statement with only a lefthand part is interpreted as asserting a fact. When the lefthand part is missing, the statement is taken to be a query to be proved. This query may contain a variable whose identity must be established. Several examples can be found in the above.

As is the case in conventional languages, compilation techniques can be used to increase the implementation efficiency of Prolog. Ever-recurring operations and decisions to be executed in an interpreter at run time can be simplified to a simple statement at compile time. Compiling also provides features for more efficient run-time performance (e.g., indexing), and memory usage (e.g., generalized tail recursion optimization). Warren's compiler translates Prolog to an intermediate form based on a set of operations fundamental to Prolog which can then be implemented on a target machine. The WAM code has properties of a procedural language, notably sequential storage and processing of the programming instructions. The highly interactive feature of Prolog makes the

3.5 LOGIC PROGRAMMING

compiling task a difficult one. Requirements imposed on compiler-based systems include dynamic compilation, loading and linking, dynamic modification of databases, and the retention of source clauses (e.g., for the purpose of debugging). These requirements would be easily satisfied in interpreter-based systems.

The execution of a Prolog program can be regarded as a process of traversing a search tree in the depth-first, left-to-right strategy (see Fig. 4.2), attempting to find a match. Traversing the tree and all necessary backtrackings can be realized by using tracks.

There are different schemes for designing a compiler-based Prolog system. Most important is the use of the Warren Abstraction Machine. Other schemes include direct compilation to the native machine code (e.g., RISC instructions) of a target processor, and compilation to native code via an intermediate code of a conventional high-level programming language.

Compiler systems mainly differ in the selection of the object code. Particularly for RISC processors, it is convenient to compile to the RISC instructions, whether or not via an intermediate code or a third-generation programming language. An innovative compiling method is due to Warren [War83]. The object code is Prolog oriented and is interpreted by the abstract machine which can be implemented with methods varying from software simulation to hardware machine.

The Warren Abstract Machine (WAM)

The WAM is an efficient execution model consisting of four parts: data types, memory space, machine state and an instruction set. The WAM specification calls for a tagged architecture, i.e., a few bits are appended to each location in memory to serve as a tag. There are four data types in the WAM model: references (variables), constants, compounds (structure: functor, an arity, n and n elements, each of which may be any Prolog item), and lists (linked collection of elements).

The memory space of the WAM is divided into two distinct areas: the Code Space (containing Prolog programs, procedures and clauses), and the Data Space (working memory containing Prolog data and computation state). The multiple-stack Data Space is further divided into three areas: Stack (primary working area, containing the machine state at some given moment, information about backtracking and recursive procedure invocations), Heap (which stores compound terms, i.e., structures and lists as well as globalized unsafe variables at execution time) and Trail (containing references to conditionally bound variables and used to keep track of variable bindings which must be unbound upon backtracking). In addition, a Push Down List is used as a scratch area during unification. The machine state is defined by a set of registers, including pointer registers which store pointers to various areas.

The instruction set of the WAM is divided into six classes: three for control (*procedure control*, *indexing* and *clause control*) and three for data manipulation (*get*, *put* and *unify*). WAM implements the AND/OR search tree of a Prolog program with the OR nodes containing the *procedure code* and the AND nodes containing the *clause code*. Fast compiler systems include BIM Prolog, Quintus Prolog and Aquarius Prolog [Van92].

Prolog Machines
While LISP leads to the development of LISP machines, Prolog forms the basis for the implementation of Prolog machines. Though several architectures (including uniprocessor CISC and RISC) have been proposed for implementing Prolog machines, a major archtecture is based on the WAM. Like in LISP machines, garbage collection of free memory space is an important feature in Prolog machines [App88].

Dobry [Dob90] described an approach to implementing the WAM instruction set in hardware resulting in a CISC style architecture. A variation on backtracking, called sidetracking, is used for more efficient implementation in many instances. Features providing special support for the Instruction Set Architecture (ISA) for Prolog include parallel internal data paths, support for tagged data, and memory buffers and caches to support operations specified in the ISA.

Parallel Logic Architectures
While most Prolog machines support the sequential execution of program instructions, it is important to exploit the possibilities of parallelism. Most implementations have concentrated on shared-memory machines [War88].

The large amount of parallelisms existing in logic programs has led researchers to design parallel architectures to implement Prolog programs. Two major types of parallelism in logic programs are the AND parallelism and the OR parallelism. The AND parallelism refers to the simultaneous execution of two or more subgoals in the body of a clause. An OR parallelism is the simultaneous unification of two or more clauses whose heads share the same calling goal. Variable binding conflicts may arise, when subgoals of a clause share variables. For that reason, the exploitation of AND parallelism is more difficult than that of OR parallelisms. Some execution models exploit both kinds of parallelism.

Tick [Tic89] compared the OR and AND parallel paradigms by analyzing Aurora (an OR-parallel Prolog system) and Kernel Language One Parallel System (KL1PS, an AND-parallel Prolog system). The two systems were subjected to executing seven different algorithms performing the N-Queens benchmark on the Sequent Symmetry multiprocessor.

3.6 THE OBJECT-ORIENTED ENVIRONMENT

3.6a Object-Oriented Programming

Introduction
In imperative languages of the third generation, there is a strict distinction between data and functions (procedures and subroutines). The functions are bound statically at compile time. This means that each function to be called in a program can be determined unambiguously at compile time. Data abstraction is usually not supported by these languages. As a consequence, data structures and functions are not explicitly related.

In *object-oriented languages*, we deal with objects, which contain data and functions as a unit, that is, data and functions are manipulated as associated entities [Pet88, Tel89]. An *object* is a unique, identifiable, self-contained unit, which is distinguishable from its surroundings. An object may be described by a set of attributes, which constitute an internal state, and a set of operations which defines its behavior. Data can only be accessed through the functions of the object.

Object-oriented programming is an interactive, incremental approach to software development in the sense that more or less autonomous software units are designed and combined to form a complex software system. An object-oriented (OO) language allows objects in the real world to be mapped directly into objects in a software implementation. The use of modular, self-contained objects makes OO languages more maintainable, reusable and extensible than imperative languages. OO languages have four basic features: (a) data abstraction, (b) class abstraction, (c) messaging, and (d) inheritance. Each of these features will be discussed below.

Data Abstraction
A major feature of object-oriented programming languages is *data abstraction*, which means that the programmer who uses a data structure or other software module does not have access to its complete implementation, but only to an abstraction or interface (in effect, a high-level, user-friendly representation of the language). This feature supports the creation of correct maintainable programs.

Different kinds of objects can be defined. To this end, the concept of class is introduced. A *class* is a set of related objects represented by the same general description. It defines the data fields and the functions of an object through which the data can be accessed. The data fields defined within a class are referred to as *instance variables*, and the functions as *methods*. Each type of manipulation associated with a specific object is described by a method, which is a sequence of actions to be performed by a processor. The name and arguments of a method must follow a predefined convention. Examples of the language Smalltalk are given in the next subsection. The packaging of data and procedures into a single programmatic structure is called *encapsulation*. In OO programming languages, encapsulation implies that an object's data structures are hidden from outside sources and are accessible only through the object's protocol.

An object always belongs to a specific class, which is defined by its *class variables*. An object is invoked as an *instance* of that class. Class variables are shared by all instances of a particular class. A class provides the set of available methods one of which is selected when an instance receives a message. The actions performed by a method may include sending other messages, assigning variables and returning a value to the original message. A programmer may create and modify classes that describe objects in an interactive way. A variable is not bound to a specific class. It can hold any object of any class. For example, an object of class *pair* can be used to hold any pair of objects that is needed.

Methods defined in a class can be divided into class methods for creating an object of a class and instance methods of accessing the data fields of an object. A class method returns the object that has been created by it. This makes it possible to assign newly created objects to variables. The object returned by an instance method depends on the specific kind of method.

Messaging

Communication with an object is accomplished by sending a message to it, for example, a command to an object to perform a computing operation. Messaging is the invocation of a method by means of a message. A *message* is a statement that consists of three parts: an object, a selector and arguments. The selector denotes the method which has to be invoked for the specified object in the message. The arguments, when present, are passed to the method. The association of a message with a method during run time (as opposed to compile time) is called *dynamic binding*. It means that a message can be sent to an object without prior knowledge of the object's class.

A variable can hold any object of any class. At compile time, it is not known to which class this object belongs. Messaging requires dynamic binding: the function to be invoked has to be determined at run time, depending on the object involved, while the class of the object is specified. The create a new object, a message is sent to a class and the result is assigned to a variable.

Inheritance

Classes in object-oriented languages are organized in a hierarchy. A hierarchy is an explicit ordering of objects in which each object is subordinate to the one above it, while it may have one or more direct subordinates and many indirect subordinates. Class B can inherit class A means that the instance variables and the methods for class A are automatically defined in class B. In addition, class B

3.6 THE OBJECT-ORIENTED ENVIRONMENT

may contain new instance variables and methods. In the above example, A is called the superclass and B the subclass. A subclass can have more than one superclass. A method defined in a superclass may be redefined in a subclass with the same name.

Object-Oriented Programming Languages

Data abstraction and the other features of OO languages facilitate the design of programming systems in two ways:

a. They simplify program design and debugging by allowing the program to be split into smaller, more manageable, and more easily separated units.
b. They help programmers reuse code, making them more productive and making it easier for users and other programmers to learn new systems built from existing familiar code.

Object-oriented programming languages are ideally suited for using a mouse with a graphic and a menu-based style of commands, as implemented in such personal computers as Macintosh (see Subsection 2.1a). The dominating paradigm of object-oriented languages is illustrated by Smalltalk [Sma81, Kae86, Pin88, Gol89, LaL90], which is discussed in greater detail in Subsection 3.6b. Smalltalk has borrowed some concepts from Simula67, notably data abstraction and the class concept. Unfortunately, the various OO languages employ a different terminology for the same entities. CLOS (Common LISP Object System) is an object-oriented language based on Common LISP [Kee89].

Stroustrup introduced an OO system for the C language, called *C++* [Str91]. This language implements most of the object concepts presented above. It provides inheritance and dynamic binding. In adition to C computational operators, control structures, pointers and records, C++ supports object-oriented programming features, including abstract data types, encapsulation, class hierarchy with inheritance and polymorphism. A typical C++ program consists of one or more header files and a set of program files. The header files contain the definition of classes and other declarations to be used in the program. The program files contain the bulk of executable code, i.e., the main program and the method definitions.

The C++ class allows the programmer to create new types and define operations on those types. The user-defined classes behave in exactly the same way as the system-defined classes, such as *int* anf *float*. C++ defines the class keyword to define new classes. *Names* are declared in three kinds of scope: *local* (in a block of code, local to that block), *class* (within a class, unknown outside the operations that belong to that class), and *file* (outside of any block or class but referenced in a file).

Borland's OO products include Turbo C++ for Windows, Borland C++ 3.0, and Applications and Framework. A real-time programming language, which is a derivative of C++, is *Flex* [Ken91].

Objective C from Stepstone is an illustrative example of the incorporation of OO programming features in a conventional language [Cox86]. Objective C includes the ordinary C language within an object-oriented language. The programmer always has the option of writing code in conventional C. In fact, Objective C is an extension of C by adding one new datatype (the object) and one new operation (message expressions). All objects in Objective C have to reside in the host computer's memory.

A number of commercial programming languages have facilities which include at least some of the OO concepts. These include languages as *ADA* [Buz85] and *Object Pascal* [Mac89]. *Turbo Pascal 6.0* from Borland International has a language extension, called Turbo Vision, which includes OO program functions. Mention should be made of the development of *Object-Oriented Prolog* and *Object-Oriented LISP* [Tel89]. For example, *Flavors* [Moo86] is a LISP version of an OO language.

Parallel Object-Oriented Programming

Some object-oriented programming languages have been developed for parallel processing. An example is POOL (Parallel Object-Oriented Language), which consists of a large number of objects which communicate by sending messages [Ann90]. Control of parallelism is provided. A parallel, general-purpose computer system DOOM (Decentralized Object-Oriented Machine) for the execution of POOL programs has been developed at Philips [Ann90]. The DOOM architecture contains many identical nodes with communication means which are connected in a packet-switching network. The prototype DOOM machine contains 100 nodes, each having a copy of the operating system kernel. DOOM is connected, as a satellite, to a host computer, where the programming enviroment resides. An application is a multi-level VLSI circuit simulator, designed at AEG [Loh90].

3.6b Smalltalk

Introduction

Smalltalk originates from the Xerox Palo Alto Research Center [Gol89, Gol84]. It is an object-oriented language and is therefore characterized by *data* and *class abstraction, messaging* and *inheritance*. *Objects* within the Smalltalk environment are functional modules which are independent of each other. The language descriptions to be given below pertain to Smalltalk/V [Sma86].

There are different kinds of objects, e.g.:

@A	the character A
324	the integer 324
#(1 2 3)	array of three integer objects

3.6 THE OBJECT-ORIENTED ENVIRONMENT

#(1 ('two' 'three') 4)   an object within an object

Objects are activated, when a message is received. A *message* is a request to an object to apply a *method*, that is, to perform a certain operation comparable to a function, a subroutine or a procedure in a third-generation language. A message contains three essential parts: a *receiver* object, a *selector* and zero or more *arguments*. The receiver is the object for which the message is meant and the selector indicates which type of operation is to be performed. The arguments serve to provide the message with data objects. Which operation an object can perform depends entirely on the *type* of object. For example, numerical objects can perform arithmetic operations. After performing the desired operation on an object, the message delivers a new object as the result of this operation.

Messages

The receiver and the arguments can be expressions, while the selector is a literal given by a keyword, such as *sin*, *sqrt* and *size*. A message can have 0, 1 or more arguments. By definition, we prescribe:
a. A keyword is always followed by a colon (:), except when there is no argument or when it is an arithmetic operation.
b. Arguments are separated by keywords. When the number of arguments is one or more, there must be the same number of keywords (selectors).

Unary messages are messages without arguments.

Examples: 20 *factorial* denotes the factorial 20*19*18*...*1. The integer 20 is the receiver, *factorial* is the message selector.
'how many characters?' *size* delivers the integer 20, that is, the number of characters equals 20.

Arithmetic messages include arithmetic expressions, e.g., 3 + 4, in which the integer 3 is the receiver, + is the message selector, 4 is the argument and the result 7 is delivered. Other examples of arithmetic messages are:

5 * 7   multiplication
5 // 2   integer part after division
4 \ 3   modulo, integer remainder after division
2 / 6   exact rational function

The operations are always carried out from left to right, unless priorities are indicated by parentheses.

3 + 4 * 2   delivers 14, not 11
3 + (4 * 2)   delivers 11

Keyword messages with one or more arguments:
#(1 3 5 7) *at*: 2 denotes 3
'hello' *at*: 1 *put*: @H   denotes 'Hello'
#(1 0 4 5) *at*: 2 *put*: (2 3)   denotes #(1 2 3 4 5)

Examples of *control structures* are:

```
3 < 4              TRUE
5 = (2 + 3)        TRUE
7 even             FALSE
```
Examples of *Boolean expressions* are:
```
(c < @0 or: [c > @9]        OR message
(c >= @0 and: [c <= @9])    AND message
```

Classes

Every object always belongs to a certain *class* which contains common attributes of the objects within the class. An object is said to be an *instance* of the class to which it belongs.

Examples: #(1 2 3) and #(sam joe) are instances of the class *Array*.
"north" and "south" are instances of the class *String*.

A class description contains the name of the class, a declaration of variables and a specification of methods. There are temporary and global variables. A *temporary variable* (beginning with a lowercase letter and the whole between vertical bars) exists locally for an object and disappears after its use. A *global variable* (beginning with an uppercase letter) can be called by all instances of all classes. *Class variables* are global variables which can be used by all instances of a given class.

The individual objects of the same class differ from each other only in the values of their instance variables and occasionally in additional instance variables. Objects may contain *named* as well as *indexed* instance variables. An example of a named instance variable is *numerator* in the class *Fraction*. Indexed instance variables are determined by an index beginning with 1. An example is:
'parts' *at*: 5 *put*: @y => 'party'

Hierarchy

There is a class hierarchy in which each class has a superclass above it and one or more subclasses below it. When an object receives a message, the class or one of its superclasses provides the corresponding method to be executed.

Example: *Boolean* is the superclass of *True* and *False*.

The inheritance property implies that instances of a class inherits the same instance variables as its superclass plus a number of new instance variables which belong to the class. An example of a class hierarchy is shown in Fig. 3.8.

The instance variables are placed in parentheses:
```
Animal (sort, habitat, feeding, reproduction, top_speed)
     Bird (wings)
          Eagle (bird_of_prey)
          Duck (swimmming_bird)
```

3.6 THE OBJECT-ORIENTED ENVIRONMENT

Mammal
    Dog (bark)
    Whale (sea)

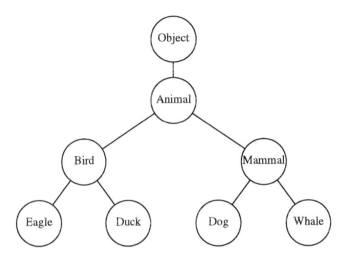

Fig. 3.8 Example of a class hierarchy

The class Animal has 5 class variables (in parentheses). In addition to these five variables, the class Bird has the additional class variable "wings", while Eagle has still an extra instance variable "bird_of_prey".

User-Friendly Interface with Smalltalk

Since its inception, Smalltalk has operated in a user-friendly environment with screen control of icons, menus, scroll bars and windows by manipulating a mouse. A window is divided into panes, each having a different function. Smalltalk offers the user the possibility to create complex object-oriented applications with a minimum of coding. The two major vendors of Smalltalk implementations for PCs are Objectworks/Smalltalk for Windows from Parcplace Systems, Mountain View, and Smalltalk/V Windows from Digitalk, Los Angeles. Smalltalk has been applied as the implementation language in solving VLSI CAD design problems [vAn87, Wal91].

ParcPlane has sprung from PARC (Palo Alto Research Center), where Smalltalk came into being. A distinguishing feature of Objectworks/Smalltalk is its system browser window with five panes (for class category, class, message category, message, and code editing).

Digitalk released commercial versions for DOS, OS/2 and Macintosh systems [Sma86]. It includes bitmapped graphics, a built-in Prolog compiler, object-swapping virtual memory and a debugger. The object-oriented user interface of Smalltalk/V can be used to integrate application components from a variety of sources written in different languages. The distinct feature of Smalltalk/V is that one can keep full advantage of the environment one is used to. The development environment has the familiar graphics user interface, including the availability of windows, e.g., the Class Hierarchy Browser window to enter programs into the system, and the Disk Browser window to browse and manipulate files. Application components from a variety of sources, whether they are written in Smalltalk or another language, can be integrated to create a consistent, object-oriented user interface.

The *DOS Shell* capability allows the user to exit Smalltalk/V to execute DOS commands and programs and then return to Smalltalk/V when finished. When *DOS Shell* is selected from the system menu, another menu is displayed with possible choices for DOS commands to be executed. The *go to DOS* command activates the processor COMMAND.COM from which the user can run several DOS commands and programs.

3.6c Object-Oriented Databases

VLSI-CAD Database Design

A VLSI chip can be described in an object-oriented environment. An object in VLSI-CAD has a functional description and an implementation, each of which is represented by a distinct set of heterogenous records [Bat85]. A *functional description* of an object (functional module) specifies the functional behavior with respect to the object's inputs and outputs. The *implementation* represents the internal structure of the object.

All *versions* of a design object share the same functional description, but differ in their implementation descriptions [McL83]. VLSI design methodology involves design representations at the various levels of abstraction from the system level moving down to the manufacturing level. At each level, several alternative implementations can be considered with different device technologies, design algorithms and architectures.

Given a functional description of an object, implementations can be given composed of lower-level components.

Example: Let an object represent a four-bit adder. The functional description states that two four-bit words and a carry-in bit at the inputs delivers the four-bit sum and a carry-out bit. An implementation may consist of a structural interconnection of four two-bit adders. Each of the two-bit adders can be implemented in terms of logic gates or as a MOSFET circuit.

In modeling databases, several concepts are relevant.
- *Hierarchical* levels of design database. This is illustrated by the above example of the adder.

3.6 THE OBJECT-ORIENTED ENVIRONMENT

- *Versions*. Given a functional description, alternative versions can be devised. An object type is an abstraction of the common features of its versions. All attributes of an object type are inherited by its versions.
- *Instantiation* is the invocation of a copy of an object. For example, suppose that a four-bit adder is to be composed of an interconnection of four two-bit adders. Then, each copy which is invoked for this purpose is called an *instance* of the two-bit adder object. Instances may have new attributes that are not inherited. For example, the pins of the input and output nodes of the four instances of two-bit adders for composing a four-bit adder have distinct pin numbers.
- When an instance of a two-bit adder is invoked, we have a *parameterized version*, when different implementations of the two-bit adder are available.
- A database system is called *active* if retrieval and update operations result in invocation of procedures. Such procedures, known as *triggers*, are associated with particular fields. When the field is accessed, the trigger is activated.

Object-Oriented Data Model

Objects have been introduced in Subsection 3.6a. To recapitulate:
a. Each real-world entity is modeled by an object, each identified by a unique identifier.
b. Each object has a set of attributes (instance variables) and methods. The value of an attribute may be another object or a set of objects.
c. The set of attribute values represents the object's status. This status is accessed or modified by sending messages to the object to invoke the corresponding methods.
d. Each object is an instance of some class. A class represents a template for a set of similar objects.
e. There is a class hierarchy in that a class inherits attributes and methods from its superclass, while the class itself may have subclasses.

In Subsection 2.3a, several data models have been introduced (hierarchical, network, relational, etc.). Object-oriented (OO) databases are based on the object-oriented data model, which has the following properties:

- *Object identity*. The elements with which they deal are typically records with unique addresses, just as in the network and hierarchical models.
- *Complex objects*. Typically, they allow construction of new types by record formation and set formation.
- *Type hierarchy*. They allow types to have subtypes with special properties.

The set of object structures definable in our model is very close to the set of possible schemes for database records in the hierarchical model. An OO data model is not limited to the notion of an object type. The basic notion is really the *class*, which is an object type for the underlying data structure, and a set of *methods*, which are operations to be performed on the objects with the object structure of that class. Another essential ingredient in the object model is the notion of subclasses, a formalization of "is_a" relationships.

Methods, being arbitrary procedures, can perform any operation on data whatsoever. However, in order to access data efficiently, it is useful to limit the operations that may be performed to something like what is possible in the hierarchical model. It is essential to allow navigation from an object A to the objects pointed to by fields of A. This operation corresponds to movement from parent to child, or along a pointer in a virtual field in the hierarchical model. It is also very useful to allow selection, as in the relational model, on fields that are sets of objects. Thus, we can navigate from an object A to a designated subset of the object found in some set-valued field of A.

Object-Oriented Database-Management Systems
Commercial Database Management Systems (DBMSs) based on relational concepts are being utilized in a variety of applications, particularly for business data processing. Recently, a new generation of database systems has emerged: the Object-Oriented Database Management Systems (OO-DBMSs). These systems prove to be very suitable for storing and manipulating VLSI design information [Hor91].

Object concepts have been employed in many areas of computer science, e.g., in the construction of object-oriented languages (Subsection 3.6a). These concepts have also been used as a basis for the development of OO-DBMSs. Such systems can store large quantities of information having complex structures, including data structures not supported by conventional database systems [BEr91, Gup91]. They are therefore useful in computer-aided chip design and expert systems. The object-oriented approach provides the ability to define complex data types that combine both data structure and procedure definition.

OO-DBMSs are DBMSs having the following features:
a. *Complex objects*, including the ability to define data types with a nested structure.
b. *Encapsulation*, which implies the ability to define procedures applying only to objects of a particular type and the ability to require that all access to those objects is via application of one of these procedures. For example, the operations PUSH and POP are defined to apply only to the data type "stack".
c. *Object identity*, which means that the system is able to distinguish two objects that "look" the same, in the sense that all their components of primitive type are the same. For example, two employees employed in the same department, having the same names in the relation *employee* (*name*, *dept*) can be distinguished in a relational database system by using two unique identification numbers.

3.6 THE OBJECT-ORIENTED ENVIRONMENT

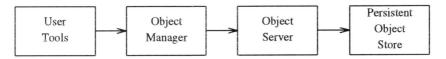

Fig. 3.9 Object-oriented DBMS

A typical OO-DBMS consists of three major components (see Fig. 3.9):
1. The *Object Manager*, which provides the interface between external processes and the OO-DBMS.
2. The *Object Server*, which is responsible for providing basic DBMS services, such as transaction management and object storage management.
3. The *Persistent Object Store*. A persistent object, created as one instance of a class, can persist even after the program that created it terminates. Then, programs can reference such a persistent object.

The query language SQL, widely used with relational databases, can be adapted for querying OO-DBMSs. The User Tools indicated in Fig. 3.9 include OOPL, DDL, DML and SQL-extended language processors and a browser.

The complete set of individual schema definitions which describe the logical structure of a database is called the *database schema*. In an OO database, the schema is expressed in the set of class definitions for a database. Classes define both the database objects and the retrieval and change procedures for the objects. The class interface defines for an object what kind of data is available. The implementation of the class determines how data is accessed. An *Object Server* is the software system which supports transaction management and storage management functions for objects.

Object-Oriented DBMSs

OO databases can be built on a suitable existing database technology, preferably a relational DBMS. The Object Management Extension of INGRES allows user defined data types, functions and operators to be used by the DBMS server in a client/server architecture. The opposite approach is to add DBMS capability to an OO programming language. *OPAL* is a database language, marketed by Servio Logic Development Corporation. Its data manipulation facilities are present in one language whose style borrows heavily from Smalltalk. In OPAL, even the most primitive operations must be declared for each class we define. The OO-DBMS *GemStone* [Bre89] is built on top of OPAL.

Several Object-Oriented Database Management Systems have been or are being developed. Various programming languages have been used for coding the systems. *POSTGRESS* [Sto88] is a project of the University of California, Berkeley [Sto88]. POSTGRESS is implemented in about 90000 lines of C code [Sto90]. *Iris* is a recent development of Hewlett-Packard [Fis89, Wil90]. *ORION* [Kim90] was implemented in Common LISP on the Symbolics 3600 LISP machine. *Vbase* [ANd87, Gup91] and O_2 [Deu90] use supersets of the C language. Other implemented OO-DBMSs include *Avance* [Bjo89] and *Encore* [Zdo90].

OO-DBMSs should provide a suitable object-oriented query language which can express query conditions involving inheritance relationships. The ANSI standard query language SQL should be enhanced to support object concepts.

An object-oriented database system is particularly suitable for handling the data-management problem in VLSI chip design (see also Subsection 5.2c). An example of such a CAD framework is *Cbase* [GUp91], which is built on top of Vbase. Its friendly, graphics-based user interface allows the user to access design data and invoke application tools, including a wide variety of VLSI-testing programs.

Chapter 4
EXPERT SYSTEMS

4.1 EXPERT SYSTEM FUNDAMENTALS

4.1a Expert System Architecture

What is an Expert System?
In Subsection 3.3b, general properties of expert systems have been discussed. An expert system is a sophisticated computer program that manipulates knowledge for efficient and effective problem solving in a restricted domain of application. Let us examine the essence of expert systems in greater detail. Solving a problem using an expert system requires at least two steps:
a. The choice of a useful *representation* of the knowledge and the problem to be solved.
b. The use of an effective *knowledge processing* method which is optimal in terms of processing speed and memory usage.

The techniques of knowledge representation and processing will be discussed in the next two sections.

An essential part of an expert system is the knowledge base. The construction of this knowledge base is of extremely importance, since it determines the reliability of the solutions. All the activities which have to do with the acquisition, the representation, and the intelligent manipulation of knowledge for the purpose of problem solving are called *knowledge engineering*.

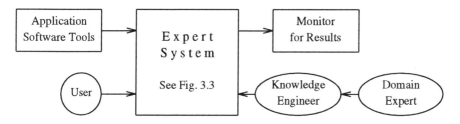

Fig. 4.1 Expert system environment

A *knowledge engineer* "extracts" from one or more domain experts the relevant facts, rules, procedures and strategies which are required to find an optimal solution. A *domain expert* is a knowledgeable person who through years of training and experience has become definitely proficient at solving problems in a particular field in a most effective way. The cooperation of a knowledge engineer and one or more domain experts leads to a knowledge base, which contains all required data and procedures. See Fig. 4.1 and Fig. 3.3.

Besides the knowledge base, a set of application software tools must be available for performing specific tasks. When a user enters a specific problem, the inference machine of the expert system employs the appropriate application tool provided with the required data and rules from the knowledge base to deliver the solution of the problem in an output device, e.g., a display monitor. When the user is familiar with the problem under consideration, common sense, creative and innovative features of humans can be exploited to improve the results.

An expert system represents knowledge symbolically, as sets of symbols that stand for some real-world concepts.

Examples: likes(mary, apples)   /* Prolog clause */
#(1 3 5 7) *at*: 2   /* Smalltalk keyword message */

The above symbols are combined in symbol structures such that a proper interpretation can be given to these structures. Solving a problem amounts to manipulating these symbol structures in a way that depends on the posed problem. Simple examples were given in Subsections 3.5b and 3.6b on Prolog and Smalltalk.

Problem solving requires a suitable control structure that determines the appropriate ways in which the knowledge must be processed. Many solving procedures can be considered as searching in a state space. Starting at an initial state, the search passes through a sequence of states, representing intermediate solutions, aiming at reaching the goal state, which represents the desired solution. As indicated in Fig. 3.3, it is expedient to distinguish two separate parts in a knowledge base: the *database* and the *rule base*. The database describes the widely shared facts agreed among practitioners, conditions, the initial state, the goal state, etc. It is also used as a working memory for describing the current states in an attempt to solve a problem. The rule base contains rules, heuristics and operators needed to solve a given problem. A specific control strategy indicates how the rules from the knowledge base are applied to solve the problem.

In the basic rule:
if antecedent *then* consequent
the antecedent and consequent may consist of a statement about an *object O*, its *attributes A* and the associated *values V*. An *OAV* triplet may have the form:
The <attribute> of the <object> is <value>

Example: The color (attribute) of the bird (object) is yellow (value)

4.1 EXPERT SYSTEM FUNDAMENTALS

When all the objects are identical, the object can be preassumed and an *AV* pair can be used in the form:

The <attribute> is <value>

Example: The color (attribute) is yellow (value)

An *AV* pair can be used to specify an "is_a" relation.

Example: The animal (attribute) is a bird (value)

A value may be an adjective, a noun or a verb.

Examples: The bird is yellow
　　　　　　　The animal is a bird
　　　　　　　The bird flies

An important feature of an expert system is that it has knowledge about its reasoning process. Suppose that the knowledge base is organized as a set of rules. Consider, for instance, the Prolog question in Subsection 3.5b whether Adam likes a person X who in return likes Adam. An inference chain of used clauses can be given to indicate the path that ultimately leads to the final solution. The knowledge that the system has about how it reasons is called *metaknowledge* which generally means knowledge about knowledge. A *metarule* guides the search for relevant rules, that is, a metarule is a rule about other rules.

Building Expert Systems

Expert systems are built for solving problems in a specific problem domain [HAy83, Nau83, Lug89, Rae90]. The appropriate domain knowledge is stored in the knowledge base.

There are two principal software alternatives for developing expert systems:
a. Using a suitable programming language for designing expert systems.
b. Building an expert system proceeding from a framework, called a *shell*.

A shell is a software package which contains basic features of expert systems, but lacks the knowledge base which determines the domain-specific application of the expert system. When an expert system for a specific application is required, the shell is provided with the knowledge of the relevant domain of application.

A most important consideration is the knowledge representation supported. We may distinguish three major categories:
a. Production rule systems based on production rules (see Subsection 4.2b).
b. Logic-programming based on Horn clauses.
c. Object-oriented representations based on semantic networks and frames (see Subsections 4.2c).

Although object-oriented systems generally have more expressive power, the majority of shells are oriented to production systems.

A typical expert system shell is marketed as a package of two components:
a. A compiler to translate the rule base, expressed in a language peculiar to the package, into an internal representation.
b. A run-time system to apply the compiled knowledge base in a consultation with the user. An interactive knowledge-base editor may be provided.

One of the major limitations of many shells is the lack of a powerful representation formalism. Many of them are equivalent to programming languages supporting only simple Boolean and numeric variables. For this reason, it may be desirable to consider using one of the AI programming languages, such as Prolog and LISP. These languages lack the run-time components of an expert system. Prolog has a built-in inference mechanism but if the rules of the knowledge base are encoded in "raw" Prolog, they are not available for inspection by the various run-time facilities. It is necessary, therefore, to build a rule interpreter and to code the rule as data. This means that Prolog and LISP are not as suitable for prototyping as an off-the-shell. However, their powerful data structures, built-in search mechanisms and interpretive nature makes them a better building tool than conventional Pascal-like languages.

Major tasks of expert systems include problem analysis, knowledge acquisition and project management. Problem analysis includes the identification of the problem and the system's technical and organizational feasability. Knowledge acquisition implies the selection of knowledge sources, but also the organization of the interaction with the expert and user friendliness.

Project management comprises the project definition and planning, the proper acquisition of expert knowledge, the construction of the knowledge base, the choice of the inference method, repetitive evaluation of the system during its development cycle and system maintenance, which includes iterative improvement of the knowledge base.

Expert-system reliability can be evaluated by applying appropriate techniques for validation and verification of expert systems [Aye91, GUP91]. An important verification tool in the project management of an expert system is the *prototype*, that is, a constructed version of the expert system that can be exercised. The objective of prototyping is to evaluate the quality of the system performance as achieved so far. Prototyping is a method of testing prototypes of increasing complexity in intermediate phases of the development cycle. Prototypes are milestones in the development process. A typical sequence of prototypes are: demonstration prototype, research prototype, field prototype and commercial product.

The prototyping approach allows the knowledge base to be refined in response to feedback and evaluation. It is also made possible by software that provides helpful tracing output. Software to implement an expert system must meet several criteria. It should support the knowledge representation formalism selected. It should support the chosen reasoning strategy with its own control

4.1 EXPERT SYSTEM FUNDAMENTALS

mechanism. It should have a good user interface and be able to integrate with other sources of data or other software. When commercial expert system shells do not match specific requirements, AI programming languages are the most flexible tool to invoke.

One important feature of expert systems that must not be underestimated is the provision of an intelligent interface for user-friendly communication facilities to state a question or pose a problem to the system.

4.1b Problem Solving Strategies

Approaches to Knowledge Processing

Once a problem has been formulated on some representation model, we must select a problem strategy which sets out a solution path from an initial state to the final solution. A widely used strategy is based on searching a knowledge base to find a solution. An exhaustive evaluation technique evaluating all possible states is usually time consuming and costly. For complex problems, the exhaustive technique may become unrealistic.

Most operations for knowledge processing are different from classical data processing operations of sorting, retrieving and performing calculations. We can identify two broad categories of knowledge processing operations:
1. Formulation and manipulation of rules.
    a. Deductive principle: New conclusions are drawn from established principles.
    b. Inductive principle: New general principles are inferred from specific evidence.
2. Searching for a matching pattern, or generate and test alternatives. Usually, search spaces and the number of alternatives to be generated and tested are very large.

Two serious problems must be overcome, when knowledge-based systems are to be designed.
a. Dealing with uncertain, incomplete and contradictory information.
b. Preventing combinatorial explosions during processing, in particular during search operations.

The first problem can be solved by the inclusion of probabilities or weighting factors in inference rules or by the application of fuzzy logic, based on sets which do not have definite boundaries (Subsection 4.1c).

Combinatorial explosion is a serious problem in knowledge-based processing. When an operation is undertaken by a simplistic approach, such as trying out all the possibilities, the number to be tested, in practical situations, soon exceeds the capacities of even the largest computers in existence. Fortunately, the memory capacities and processing speeds of computers are gradually being increased.

Moreover, knowledge processing techniques concentrate very strongly on intelligent searches, which are able to determine whether a particular line of investigation is going to produce a result or not.

Pattern Matching

A fundamental operation in the problem-solving process is pattern matching. The basic idea is that there is a *pattern* (e.g., a specific character string) to which some object is compared. The result of the comparison is a success or a failure.

Example: Let the following two Prolog clauses be given:
(1) color(pears, yellow)
(2) color(apples, red)
The following two clauses are presented to match the above clauses.
(3) color(X, yellow)
(4) color(Y, red)
The result is a match of (3) with (1) and (4) with (2), but a failure, when (3) and (4) are compared with (2) and (1) respectively. When the matching process is successful, a binding process associates a constant to a variable. In the current example, X = pears and Y = apples.

Pattern matching must conform to specific matching rules:
a. Two constants or two structures match when they are completely equal.
b. Within a structure, a constant matches a variable in corresponding places, making the variable take on the value of the constant.
c. Within a structure, two different variables in corresponding places match such that both are bound to the same entity.

Unlike Prolog, LISP has no built-in pattern-matching capability. The user has to provide a unification algorithm for LISP [Cha85].

State-Space Search

Starting from an initial solution, an optimal path through intermediate solutions should be traversed. Several searching techniques can be used differing in the order of intermediate steps or actions taken during the solution process. If a path is unsuccessful, alternative paths are tried. Control strategies are needed to guide the direction and execution at each intermediate step. Two major strategies for controlling the search mechanism are *state-space search* and *problem reduction*.

The problem-solving process in Artificial Intelligence is generally viewed as a search problem. The space in which the search takes place is known as the *state space* of the search. The state space represents the set of all attainable states for a given problem. The states may be the contents of a knowledge base or the set of possible intermediate solutions which can be generated. For example, in a game-playing application the state space embraces all the possible moves in the game. A state may represent a partial or a complete solution.

4.1 EXPERT SYSTEM FUNDAMENTALS

The objective of a problem-solving method is to find an effective path from some initial problem state to the state that represents the final solution. The search through the state space is carried out by utilizing problem-solving *operators*. Such an operator acts on a state to transform it into the next state. The sequence of states that are passed from the initial state to the solution state is called the solution path. The main objective in problem solving is to find a solution path that is as short as possible. An appropriate amount of domain knowledge in the form of rules may lead to the use of effective operators.

The most common structure of a state space is a tree, spreading from the root representing the start (or current) state of the system, and branching for each possibility at each subsequent stage. The final states of the system represent the goals towards which actions are directed.

Forward Chaining and Backward Chaining

Given the knowledge represented by production rules or otherwise, the inference engine has to use this knowledge to reason about the problem. An algorithm for pattern matching can be used to examine whether or not the conditions of a rule (the lefthand or *if* part) hold. By using an algorithm for pattern matching and consulting the database, the conditions of each rule (the lefthand part) can be examined to find whether or not they hold.

There are two major strategies for rule processing: forward chaining or backward chaining. *Forward chaining* is the process of examining the conditions (left-hand part) of each rule in turn and applying the rule whenever the conditions are found to hold. The process works from preconditions or subgoals toward the main goal by applying additional rules. It involves moving from the current state at the root of a tree towards a goal state which must satisfy specific conditions. An appropriate rule is applied to relevant (original or derived) facts, a variable may be substituted for a constant and a test is performed to see if a goal is found. If not, an alternative forward search is restarted. The process ends when it ceases to give any new facts. Forward chaining is in fact *fact-directed* or *data-driven reasoning*, since the procedure is guided by the data in the database.

In data-driven search, the problem solver begins with the given facts of the problem and applies the rules and legal moves to produce new facts, which are in turn used by the rules to generate new facts. This process continues until the path generated ends at a goal condition. The simple example in Subsection 3.3b leading to the conlusion that Aristotle is mortal represents a forward-chaining process.

In contrast, *backward chaining* is *goal-directed reasoning*, which attempts to reason back from a goal toward preconditions. A backward chaining system starts at the goal state and determines which path must be selected at each node to achieve this goal. The goal to be attained is given and the righthand parts of the rules are examined to find which of these include this goal. This sets up new goals by verification of the conditions in the lefthand parts of the relevant rules, which are subgoals for the original goal, and so on. The process continues recursively, until a known fact is reached or until either a necessary fact cannot be established or a dead end is reached. If either of the latter two events happens, additional hypotheses may be tried until some conclusion is reached or the process is terminated. Typically, such a system starts with a query in the form of a specification for a goal. Attempts are then made to derive an appropriate fact that satisfies the specification. The process of finding a solution involves a binding mechanism which assigns a constant to a variable parameter. In forward chaining and backward chaining, the constants come from predecessor and successor node respectively.

The choice between forward chaining and backward chaining depends on the specific problem under consideration. Each is best adapted to specific types of problem. When the main problem facing us is the attaining of a goal that can be stated precisely, it is desirable to use backward chaining. A depth-first, backward chaining approach was applied in the MYCIN expert system, while the corresponding EMYCIN shell is the goal-driven prototype of many commercially available expert-system shells. For some problems, the goal cannot be fixed in advance and therefore only forward chaining is feasible. Forward chaining is in order when the program has to follow the progress of a sequence of actions each of which are consequences of previous actions. Solving simulation problems usually require forward chaining. Many problems require forward chaining to be used in some stages of the reasoning process and backward chaining in others.

Both in forward chaining and backward chaining it is important to find the most relevant rule and associated facts to apply. The question arises of how to decide which strategy is best suited in a particular problem. A useful guideline is provided by the nature of the data and the goal. When the set of data is well defined, but the nature of the goal is vague, allowing the possibility of several alternative goals, the forward-chaining strategy would be obvious. On the other hand, when there is a single well-defined goal, the backward-chaining strategy may be the choice. In the latter case, a search via satisfying subgoals leading to the wanted solution is conducted. Forward chaining requires an intelligent procedure for identifying the relevancy of rules and associated facts. Rather than exhaustively tracing the entire rule and data base, an efficient and effective straightforward path to the goal must be followed. A bookkeeping process is used which avoids the use of irrelevant rules and facts. In a game-playing application, forward chaining is used to investigate alternative moves starting at the current state of the game, whereas backward chaining is used to determine what sequence of moves will lead to some specific goal state of the game.

Search Strategies

Forward and backward chaining constitute two basic control strategies. With a basic strategy, there are a number of substrategies which determine the order in which the states are examined in a tree or graph. A node in a tree or graph represents the current state of the search. There are two main strategies which can be used with a tree: depth-first and breadth-first search.

In *depth-first search*, all of its children and their descendents are examined before any of its siblings. The search goes ever deeper into the search space until no further descendents of a state can be found. Only then are its siblings considered. In the tree graph depicted in Fig. 4.2, the states are examined in the order 1 → 2 → 4 → 8 → 9 → 5 → 10 → 11 → 3 → 6 → 12 → 13 → 7 → 14 → 15.

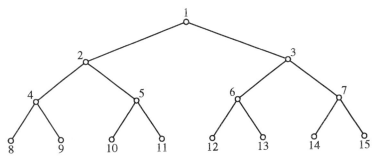

Fig. 4.2 Depth-first search or breadth-first search

In contrast, *breadth-first search* explores the space in a level-by-level fashion. Only when all siblings have been explored, the algorithm moves to the next level. In the tree graph depicted in Fig. 4.BF, the states are examined in the order 1 → 2 → 3 → .... → 13 → 14 → 15.

Much current research is aimed at finding increasingly intelligent control strategies for searches in order to avoid the dangers of a combinatorial explosion. Some measure of the value or benefit (or hazard) of each intermediate node of the state space is computed before any subsequent nodes are investigated. The processing time and effort spent on these intermediate evaluations is proving to be worth the effort in view of the reduction in the size of the space to be searched. Most contemporary approaches are a combination of depth-first and breadth-first searches.

A simple inferencing process may involve traversing a tree of possibilities, e.g., an AND/OR tree. Such a tree consists of AND and OR nodes. An AND node is TRUE if all the nodes below it are TRUE. An OR node is TRUE if any one of the nodes below it is TRUE.

Example: Suppose that a certain goal A has to be achieved by backward chaining, that is, the root A of the tree shown in Fig.4.3 must be TRUE. Since A is an OR node, either the leaf node B or the AND node C must be TRUE. If B is not TRUE, we need to prove that both D and E are TRUE.

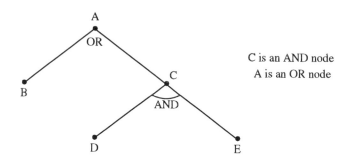

Fig. 4.3 AND/OR tree

Problem Reduction

Problem-solving methods have the objective to find a solution in the smallest possible time. One general method is to limit the search space of complex problems. The original problem may be partitioned into subproblems whose solutions combined appropriately give the solution of the original problem. This process can be repeated recursively, until only primitive problems remain for which standard solutions are available. Of course, it is possible to partition a problem in different ways.

All possibilities to partition a problem can be represented by an AND/OR graph. The nodes of the graph represent statements of the problem and the arcs represent the decompositions. The AND/OR graph distinguishes two types of branches. AND branches specify a certain decomposition of a problem in subproblems. The original problem is solved when all subproblems have been solved. OR branches represent alternative decompositions of a problem. In this case, when one of the alternatives leads to a solution, the problem is solved.

In *heuristic search* methods, the next node is selected on the basis of an estimate of the closeness to the goal. The most useful heuristic search methods include hill climbing, best-first search, branch-and-bound algorithm and the generate-and-test algorithm.

Generate-and-test or *hypothesize-and-test* is a search method through the solution space, consisting of the following steps:
a. Generate a specification criterion (e.g., a known symptom).
b. Try a path that satisfies the specification (test the hypothesis).
c. If the path is not plausible, prune that path removing a number of paths from consideration.

4.1 EXPERT SYSTEM FUNDAMENTALS

d. Try a new path.
e. When all specifications have been resolved, the process is complete. If not, re-iterate the above steps by returning to step a.

The value of this search method mainly depends on whether or not the pruning process effectively reduces the solution space.

A *demon* is an event-driven action which is executed whenever an event occurs, that is, when a certain condition is satisfied. This condition may be complex, involving ANDs, ORs and NOTs. The syntax of the demon clause is

when condition *then* action(s)

When the condition is TRUE, the action(s) following the *then* must be executed. After completion of the action(s), control is returned to the inference engine which resumes its procedure where it left off. Demons, which may be activated any time, are used to interrupt processing, identify unusual situations and specify remedial action.

Exploiting Parallelisms in Expert Systems

One of the most important attributes that expert systems must have is a high operating speed, particularly in real-time applications. As expert systems increase in complexity, the requirement of speed is harder to satisfy. A powerful way of solving this problem is to exploit the system's potential of parallel processing. Douglass [Dou85] indicated five levels of potential parallelism in expert systems: subrule, rule, search, language and system level, and several types within each level.

4.1c Dealing with Uncertainty

Introduction

Management of uncertainty is one of the key activities of intelligent persons. There are cases in which we are able to draw correct conclusions from poorly specified data. A well-known example is the establishment of the complete text from a text in which a small percentage of letters is illegible. In many cases, information is specified with a given degree of certainty. Many expert systems use techniques for calculating the degree of certainty of the conclusions gained in an inference process. The most widely used techniques are outlined below.

Probability Theory

Although certainty factors and fuzzy logic are the predominate theories used in expert systems, there are occasions when probabilities are an important part of the reasoning process [Zim87, Kli88, Pea88].

Let a production system have a rule of the form:

if E then H_i

where E represents an observable piece of evidence and H_i is the hypothesis to be tested. Suppose we are dealing with truly random events. Then, more complex probabilities can be calculated from known probabilities. When the probability of evidence E is $P(E)$, where $0 \le P(E) \le 1$, the following laws can be used:

$P(\text{NOT } E) = 1 - P(E)$
$P(E \text{ AND } F) = P(E) * P(F)$
$P(E \text{ OR } F) = P(E) + P(F) - P(E \text{ AND } F) = P(E) + P(F) - P(E) * P(F)$

Example: Given $P(E) = 0.6$ and $P(F) = 0.5$, we have
$P(\text{NOT } E) = 0.4; P(E \text{ AND } F) = 0.3; P(E \text{ OR } F) = 0.8$.

Bayes's Theorem

Bayes's Theorem provides a way of computing the probability of a hypothesis H being true, given some evidence E related to that hypothesis. For example, E might be a symptom and H a disease. The Bayesian approach to probability relies on the concept that one should incorporate the *a priori probability* of an event into the interpretation of a hypothesis, that is, it allows the calculation of probabilities from previously known results.

Let us define the following probabilities:

$P(H_i)$ is the *a priori* probability that H_i is true.
$P(E|H_i)$ is the probability that evidence E will be observed, given that hypothesis H_i is true.
$P(H_i|E)$ is the probability that hypothesis H_i is true, given evidence E.

Then, Bayes's Theorem can be written as

$$P(H_i|E) = \frac{P(E|H_i) \cdot P(H_i)}{P(E)}$$

$$P(E) = \sum_{k=1}^{n} P(E|H_k) * P(H_k)$$

and n is the number of possible hypotheses which display E. For example, n might be the number of diseases which display some symptom E.

The use of Bayes's Theorem implies two major assumptions:

a. All the statistical data on the relationships of the evidence with the various hypotheses are known.
b. All relationships between evidence and hypotheses, or $P(E|H_k)$, are independent.

Certainty Factors

Sometimes information is lacking and simply not available. In a more general case, information is known but not with certainty. Shortliffe introduced *certainty factors* in MYCIN to express the confidence of a given conclusion based on the evidence available at a given point [Sho76].

4.1 EXPERT SYSTEM FUNDAMENTALS

Example: In Prolog:
disease(milk_allergy) :- symptom(acute_stomach_pain),cause(just_drank_milk).

If more symptoms are available, you can obtain a growing confidence that the conclusion is true. Let us imagine a doctor examining a small baby who is crying and unable to express any symptoms verbally. The following rule can be stated:
if there is 20% confidence that there is an acute stomach pain
and there is an 80% certainty that the baby just drank milk
then there is a milk allergy problem.
In Prolog:
disease(milk_allergy) :-
symptom(acute_stomach_pain,.2),cause(just_drank_milk,.8).

In MYCIN and its derivatives, a certainty scale of -1 to +1 has been employed, where -1 and +1 means completely uncertain and completely certain respectively. A conclusion with a certainty factor < 0.2 is treated as false.

Fuzzy Logic
In some cases, questions allow only two discrete answers: TRUE of FALSE (or equivalently, Yes or No, 1 or 0). When only a partial truth can be given so that the answer is not a clear Yes or No, an extended form of conventional Boolean logic, called *fuzzy logic*, can be used [Zad89]. Fuzzy systems deal with the *likelihood* or *certainty* that a "proposition" is TRUE. The basic idea is that a proposition need not be simply TRUE or FALSE, but may be partly TRUE. In fuzzy logic, initiated by Zadeh in 1965, truth values can range from 0 to 1, with 0 representing absolute Falsity and 1 representing absolute Truth.

Suppose that conditions may appear in conjoined (AND) or in disjoined (OR) form:
 if cond1 AND cond2 AND ... AND cond$n$ *then* action
 if cond1 OR cond2 OR ... OR cond$n$ *then* action
Basic rules of fuzzy logic with truth values $T1$, $T2$ and $T3$ are:
 $T1$ AND $T2$ AND $T3$ = Min $\{T1, T2, T3\}$
 $T1$ OR $T2$ OR $T3$ = Max $\{T1, T2, T3\}$
 NOT $T1$ = 1 - $T1$

A major drawback of fuzzy logic is that it treats all the conditions as of equal weight. In practice, a specific condition may be much more significant than others and therefore it should affect the outcome of the rule adequately. This drawback is tackled by the introduction of certainty factors and the Bayesian theory.

Zadeh [Zad83] contends that probability theory is appropriate for measuring randomness of information, but it is inappropriate for measuring the vagueness of information. There is a need of having a measure of confidence in rules. Fuzzy logic is being applied to various applications that range from process control to medical diagnosis [Zad88, Sch92].

4.2 KNOWLEDGE REPRESENTATION

4.2a Knowledge Acquisition

Choosing a Domain Expert

The construction of the knowledge base for an expert system requires the acquisition and proper representation of the knowledge. *Knowledge acquisition* is the process of extracting knowledge about a particular domain of knowledge from an expert or group of experts. It is a critical task, since it is both people and time intensive. The proper selection of adequate and relevant knowledge as well as the way how to reason to solving the problems in the intended application field is assigned to the *knowledge engineer*. The biggest problem in building some expert systems is how to elicit the required body of knowledge from the human experts into a codified representation sufficiently complete to function as a reliable model of the expert knowledge. For this reason, knowledge acquisition is sometimes called the bottleneck in expert system design. Human experts apply a subjective kind of reasoning, experience, intuition and common sense in solving problems. Many of them have difficulty in explaining in descriptive language how they reason and arrive at a conclusion. The knowledge engineer must be able to communicate effectively with an expert whose understanding of his/her expertise may be difficult to verbalize.

In constructing the knowledge base, we should take into consideration what kinds of data are involved when using an expert system. A session with an expert system is a sort of human-computer communication. For example, in medical diagnosis, the user is asked to identify symptoms or observable characteristics so that the expert system can pose further relevant questions, until the ultimate conclusion is reached.

The knowledge engineer has the task to select the suitable objects and attributes used to represent the domain. The selection and order of the dialog questions (which, in turn, are driven by the rules in the system) are very important in preventing useless questions and the exploration of blind alleys. For example, when a patient gives an upset stomach as a symptom, there is no need to ask questions about a headache.

The knowledge engineer's task is to analyze the expert's approach to finding the solution to a problem and codifying this knowledge in a form suitable for expressing in an appropriate language so that it can be handled by the expert system. This is the phase of *knowledge representation* [Bra85]. The knowledge engineer often uses a trial-and-error method, beginning with small prototypes and gradually expanding the system as the confidence of both the expert and the knowledge engineer grows. The knowledge representation employed by the knowledge engineer is important since it determines the type of the eventual architecture of the system. Knowledge representations will be discussed in the

4.2 KNOWLEDGE REPRESENTATION

following subsections.

The accumulated knowledge acquired must be translated into a suitable code, tested and refined. The quality of the knowledge base affects the soundness of the solutions to the problems presented to the system. The problem of knowledge acquisition and representation resembles the problem of modeling a microelectronic circuit prior to the process of circuit simulation. The more accurate the circuit model, the more reliable the simulation result will be, though at the cost of more computer time. In knowledge acquisition, the task is how to reduce an exhaustive body of domain knowledge into a set of facts and rules that accurately represents the relevant knowledge. The knowledge engineer should have some mastery of the domain and be able to identify the knowledge that is required. Note that there exist several types of knowledge, including declarative knowledge (of facts and static heuristics) and procedural knowledge (of how to control the procedure that leads to the solution).

Knowledge Sources

A major source of knowledge is the human expert in the domain of interest. Great demands are placed on the domain expert who must be able to express his expertise adequately. An experienced expert who uses to rely on his or her intuition may have trouble to put design skills exactly into a wordly formulation. On the other hand, a designer not long in the job may have great skill in conveying the process of design. For that reason, the knowledge required for the circuit layout expert system developed by Joobbani [Joo86] is acquired by interviewing both an experienced and a young designer.

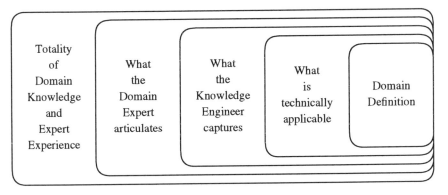

Fig. 4.4 Domain definition as a subset of the domain knowledge

In addition to expertise gained through interviewing human experts, knowledge (data, rules and heuristics) can be derived from textbooks and past experience with relevant problems. A textbook may contain suitable information or detailed algorithms for solving subproblems. Problems which have been solved and coded properly at an earlier date may be stored in the knowledge base and called whenever such a problem is encountered. As shown in Fig.4.4, the actual definition of the domain knowledge is only a subset of the totality of the expert knowledge and experience. It is crucial that the domain definition contains the essential information needed to have reliable solutions to posed problems.

A growing activity is the extension of existing knowledge by a learning process. Knowledge acquisition for expert systems may be carried out either manually or automatically by using an intelligent editing or induction program.

Knowledge Acquisition Stages
Rapid prototyping is the usual methodology involved in knowledge base construction. It entails the selection and rapid development of the knowledge base, testing, iterative refinement and further development. Buchanan *et al.* [HAy83] identify the following successive stages in knowledge base construction.
a. *Identification* of problem characteristics and selection of appropriate, available domain experts.
b. *Conceptualization* involving the specification of key concepts and their interrelationships.
c. *Formalization* of the recognized concepts and their relations into formal representation mechanisms.
d. *Implementation* or *prototyping* of the representational framework for the formalized knowledge.
e. *Testing* the efficacy of the prototype system.
If necessary, backtracking to earlier stages is carried out.

In addition to basic editing facilities, tools assisting in the generation of the knowledge base include syntax checkers to verify the grammar used in the rules, consistency checkers to ensure that the semantics of the rules or data are consistent with knowledge already present in the system, and integrity checkers to weed out contradictory facts, rules and redundant information or insert missing information.

Requirements of Knowledge Representation
The knowledge acquired from human experts must be rearranged into facts about the problem domain and rules which can be used for the reasoning process in problem solving. Any notation for knowledge representation must satisfy a number of requirements: the knowledge must be easy to read, to interpret and to change (including the addition of new knowledge). Manipulation of the knowledge, divided into declarative and procedural knowledge, must allow efficient problem solving. Programs with declarative knowledge imply what the problem to be solved is, not how the problems is solved. Blocks of procedural knowledge are intended to provide procedures for solving specific subproblems.

4.2 KNOWLEDGE REPRESENTATION

Object-oriented representations of knowledge make use of the notion of hierarchy, in which lower items are normally assumed to have some of the properties of higher items.

To be efficient, the domain-specific knowledge is represented in the form of *operators* or *pattern-invoked programs*. These programs have not the same function as procedures or subroutines in third-generation programs which can be called by the main program. Pattern-invoked programs, which may contain a great many statements, are activated by pattern recognition and checking if conditions are satisfied. Pattern-driven invocation is also applied in Prolog (see Subsection 3.5b).

In principle, rules in a knowledge base may appear in an arbitrary order. However, a judicious organization of the rules may speed up the process of searching for an appropriate rule. Some guidelines include: place the rules most frequently used first, group like rules together and organize rules hierarchically.

4.2b Production Rules

Production Systems

There are three common ways to represent knowledge: predicate calculus, production rules and semantic networks or frames. Predicate calculus was discussed in Subsection 3.5a. The rule-based system will be discussed below.

A *production rule* is an *if ... then ...* or an *if ... then ... else ...* formulation of a conditional rule relevant to a given problem. When the knowledge is organized on the basis of such production rules, an expert system can be said to be a *rule-based system* or a *production system*. The *if* clause is called the premise. A production rule has the format

if premise *then* action *else* action

which means that a specific action is executed depending on whether or not all the conditions in the premise are satisfied. The action part is not necessarily restricted to precepts, but may also represent inferences, assertions or probabilities. A widely used construct is

if cond_1 *and* cond_2 *and* ... cond_n *then* action

If all conditions cond_i, $i = 1 ... n$, are met, then some prescribed action must follow or some conclusion be drawn.

Rules may exist in a hierarchy in which a higher-level rule (a meta rule) may guide a search to lower-level rules. It is desirable for *if ... then ...* rules to have the properties of modularity (implying independent modules or pieces of knowledge), incrementability (allowing new rules to be added independently of existing rules) and modifiability (for replacing old rules by new ones). A *knowledge manager* should be able to maintain and update the knowledge base.

A production system (see also Subsection 4.3a) consists of three main parts:

a. A *rule base*, which contains the complete set of rules.
b. A *working memory*, which is a database holding data (represented by symbols) that are matched to the rules in the rule base.
c. A *rule interpreter*, which implements the matching process.

To be specific, let us look into OPS5 (Official Production System, version 5), which is a widely used rule-based tool [For77, Bro85].

OPS5

The working memory of a production system is the central medium in which the entities in the OP5 program are manipulated. It represents the state at a given point of time in the process of solving a problem. The basic entities in the working memory are called *working memory elements* (WMEs, pronounced "wimmies").

Each WME is an instance of an *element class*, identified by an *class name* and a set of *attributes* that refer to relevant features of the WME. It reminds us of the Pascal *record* type.

Example: Student   Name
                             Gender
                             Smoker?
                             Room

In this example, Student is the class name, while the list on the right side represents its attributes. Note that the attributes may represent different types of information (name, a property, the current state, a location, etc.).

Each WME attribute has a (symbolic or numeric) *value* associated with it. When no value is given, the default value is *nil*. A WME can be viewed as a sequence of fields, starting with the class name and followed by attribute names holding their values. A *literalize* declaration specifies the class name and the attributes associated with it.

Example: (*literalize*
            student   : Unique student identifier
            name   : Room number
            sex   : MALE or FEMALE
            smoker?)   : YES or NO

A comment in OPS5 is delimited by semicolon and line break.

OPS5 has a built-in inference engine. Though OPS5 in essence is a forward-chaining system, it can do backward chaining through the use of subgoals as attribute-value elements.

The rule base contains the OPS5 rules which are independent entities. A rule is executed when it is matched by a set of WMEs in the working memory, that is, when the rule's conditions are satisfied. The syntax of a rule is the

4.2 KNOWLEDGE REPRESENTATION

parenthesized expression

    (*p* rule-name <conditions> --> <actions>)

Each rule starts with the letter *p* for production, followed by a unique rule name. The arrow --> delimits the lefthand side (LHS) containing the conditions, and the righthand side (RHS), which represents the actions to be executed when all conditions in the LHS are met. In fact, a rule is an *if ... then ...* type of statement.

 The LHS of a rule consists of one or more *patterns*, which may use variables, predicate operators, disjunctions and conjunctions to specify the values in WMEs that can match the condition element. A *variable*, which may match anything, is enclosed within angle brackets, e.g., < x >. A *disjunction* refers to the case in which only one of several values is required to be equal to the matching WME. The set of possible values is enclosed in double angle brackets, e.g., resistance << 1 2 3 >>. A *conjunction* in a rule means that a specific WME value must match to each of the elements in a value set, which is enclosed in braces { }.

 The RHS of a rule may contain various types of action to perform. Specific identifiers are used to denote the desired actions: *make* (for creating new WMEs), *remove* (for deleting WMEs), *modify* (for changing WMEs), *bind* (for attaching values to variables), *call* (for invoking external subroutines), *write* (for output) and *halt* (for stopping program execution). The actions *openfile*, *closefile* and *default* indicate certain manipulations with files. The sequence of actions in the RHS must be executed in the specified order.

 Specific functions, contained in OPS5 rules, provide values to WMEs and manipulate inputs, outputs and vectors. Such functions include *accept* or *acceptline* (for reading the input), *crlf*, *tabto* and *rjust* (arguments of the *write* action for formatting the messages being printed), *substr* and *litval* (for manipulating vectors), and *compute* (for executing mathematical operations). For example, the function *crlf* causes the *write* to begin a new line.

Examples:
(*p* old-car (car ^year {<1950} ^owner <name>) → (write <name> | owns an old car | (crlf))
(*p* car (car ^year <<1990>> ^owner <name>) → (write <name> | owns an old car | (crlf))
The carat ^ is an operator that precedes an attribute which is followed by its value.

The order in which the rules appear in the rule base bears no relation to the order in which the rules are executed. To avoid evaluation of all the rules for matching purposes on every cycle, it is important to use a method that narrows the search space. By attaching indexes to the rules, the OPS5 interpreter remembers the previous searching steps which were taken so that this knowledge can be exploited to improve the searching efficiency.

OPS83

OPS83 includes the rule-based representation of OPS5 and features of procedural languages. In some respects, OPS83 [For84] has similar properties as Pascal (infix notation, type declaration, etc.). A variable in OPS83 is indicated by an alphanumeric word preceded by the symbol &. OPS83 allows declarations of complex structures: records of records, records of arrays, arrays of records, etc. OPS5, which was developed at Carnegie-Mellon University, follows LISP conventions closely in that all entities are expressed either as symbols or as parenthesized lists of symbols. OPS83 is a rule-based language, which includes most of Pascal.

4.2c Semantic Networks and Object-Oriented Representations

Introduction

Objects form an intuitively appealing way of representing knowledge in situations where there are many agents (say, people and computers) that work together on a task. This subsection deals with object-oriented knowledge representations. An object can be used as a template to generate other objects with similar properties. Objects are program structures that possess states (i.e., variables with updated values) and behavior (i.e., procedures). As indicated in Subsection 3.6a, object-oriented languages support the design and implementation of hierarchies of abstractions.

Semantic networks can be considered as an extension to the object hierarchy mentioned above. The hierarchy is replaced by a network that can define arbitrary relations among objects.

Semantic Networks

Knowledge about objects and their relations can be represented as a *semantic network* consisting of nodes connected by arcs. Usually, nodes and arcs are labeled. A *node* denotes an object, concept or event and an arc describes the relationship between two connected nodes. Semantic networks were first applied as psychological models of human memory [Qui68, Bra79]. Knowledge representation with semantic networks is considered to be the most general representation scheme.

Objects (represented by the nodes) are physical objects that can be seen or touched. They can also be events, acts, abstract categories (such as "carnivore"), or descriptors (such as "feathered" or "flier"). Descriptors, which are also nodes, represent additional information about a class of objects.

Arcs represent relationships between objects. Typical types of arc relationships include:

4.2 KNOWLEDGE REPRESENTATION

Is-a Indicates an object which is a member of a larger class (taxonomic relationship). Example: The albatross is-a bird.
Has-a Indicates the object which is a property of another node. Example: A giraffe has-a long neck.
Caused-by Used to represent causal relationships. Example: A low albumin is caused-by a liver disfunction.
Definitional Used to define a value for an object. A patient, for example, may have two brothers.

Knowledge that does not alter with time can be represented by a *static semantic network* in which the nodes denote objects (persons, things, events, concepts, etc.) and the arcs denote relations between connected nodes.

Example:

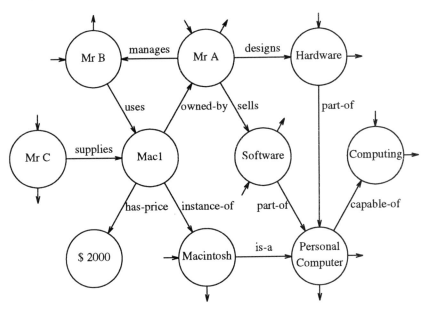

Fig. 4.5 A semantic network

Figure 4.5 illustrates (a part of) a simple semantic network, consisting of nodes (each indicated by a circle representing a semantic object) and directed arcs (each labeled with its particular relation between two nodes). For example, Mr A is the manager of Mr B who uses a Macintosh personal computer, called Mac1, owned by Mr A and supplied by Mr C. At the same time, Mr A designs hardware and sells software.

The advantage of semantic networks is their flexibility. They can be used for "pure" information in a declarativist context or include control arcs and nodes under a proceduralist regime. They are well suited to programs written in LISP or similar programs. Their structure can be tailored so that information which is required frequently is accessible via short paths and items needed less frequently are kept in the background. They have been used to investigate sentence structure, relations between medical symptoms, spatial relationships between objects and symptoms and causes of failure in mechanical devices. Semantic networks support hierarchical data and are suitable for VLSI applications [Che83].

The disadvantage of semantic networks is that their potential complexity runs counter to the need for a small number of underlying principles to govern the processing of a knowledge base. Because of their large size and consequent need to be updated in place, they do not lend themselves easily to processing by a data-flow or graph reduction computer.

Frames

A *frame* is a record-like representation structure, which has a name and a set of slots which can be filled with items or a reference to another frame [Min75]. Each frame represents an object and the slots contain attributes and their values associated with the object. Each slot can also contain a procedure for calculating the value (an algorithm), or one or more production rules for finding the value (heuristics). A slot may also contain multiple values. For example, a *brother* slot in a patient frame may contain multiple names. Some slots, defined as *facets*, are used to constrain the values used for frame attributes. Facets may be used to control the maximum number of values entered for an attribute (such as the maximum number of "brother" entries permitted) or a minimum and maximum value permitted for an attribute.

In a given system, the frames can be static or dynamic. In a static system, the frames cannot be changed or updated during the problem-solving process. In a dynamic system, it is possible to alter the frames.

Slots can be used to store values, procedures and rules. When a slot needs to be evaluated, the procedure or rules are activated. We could say that the slot procedures and productions are controlled by *demons*, which are activated when needed. For example, a frame may have attribute slots for a person's birthday and age. The birthday slot contains a value. The age slot contains a procedure that can calculate the age from the birthdate and the present date. A procedure attached to a slot in this way is said to be an attached procedure. The procedure is activated whenever the present date is changed.

Example: Figure 4.6 shows a simple frame structure that contains information about student years at a university. The University slot refers to another frame to specify the Faculty and other relevant information.

4.2 KNOWLEDGE REPRESENTATION

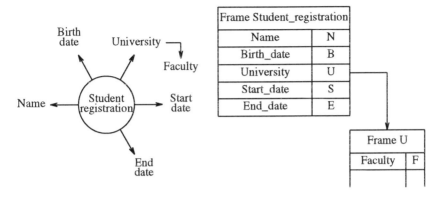

Fig. 4.6 A frame structure

The frame-based representation permits the hierarchical information about relationships to be stored in the knowledge base. The frames are organized in a hierarchical relationship, with the highest level frame containing information that applies to all frames below it. Any frame related to a frame of higher order is said to *inherit* the characteristics of the higher-order frame. A higher-level attribute could be used as a default value for frames below it in the hierarchy, and could be overridden if a lower-level frame contains an attribute of a different value. For this reason, the frame system can handle exceptions easily. In VLSI design, frames may contain all knowledge about a particular cell, chip or printed-circuit board. The frame system supports hierarchical logic design. The dynamic behavior of the hardware is built into frames as demons which are data-driven functions [Sai81]. Frames may also be filled with calls to processing procedures and the values of data items to be passed to and from these procedures.

Frames are generally used in a proceduralist knowledge processing context, since they can contain both data and control information, and they can include calls to processing procedures. They are more formally structured than semantic networks, and thus more suited to dataflow and graph reduction architectures. Frames are usually programmed in LISP, but are eminently suited to processing by functional and applicative languages. The limitations of frames is their fixed structure which makes the learning aspect of Intelligent Knowledge-Based Systems somewhat difficult to implement with them. Compared with rule-based representations, frames and semantic networks have the advantage that all information about an object is stored at a given place which is accessible. However, once the knowledge base has been established, it is difficult to change the knowledge-base hierarchy and to handle exceptions.

An early frame-based tool, implemented in Interlisp and developed by Xerox

PARC, is *KRL* [Bob77]. It is an attempt to integrate procedural knowledge in a frame-based system by associating procedures with the slots of a frame. The procedural attachment allows the subsequent steps for a particular task to be determined by characteristics of the specific entities envolved.

HPRL (Heuristic Programming and Representation Language) of Hewlett-Packard Research Laboratory integrates a frame-based representation and reasoning with rule-based knowledge [Ros83]. HPRL provides an initial set of tools for the interpretation of rules, including a capacity for forward chaining, backward chaining, meta rules and meta-interpretation.

Object-Oriented Representations

Object-oriented (OO) representations essentially involve grouping information in a more or less "natural" way as, e.g., in the form of frames. A usual representation is in the form of a graph with nodes interconnected by arcs. In semantic networks, the nodes represent concepts, while the arcs represent relationships between them. Concepts in the object-oriented representation have corresponding concepts in the frame representation: object → frame, class → parent, attribute → slot, value → value, method → attached predicate.

OO representations appear to be one of the most promising for database development and expert systems. There are also some drawbacks:
a. OO representations lack the distinctions that logic can make (AND, OR, for_all, etc.).
b. There is no knowledge in an OO system that tells us how to search for the knowledge that we want to find.

The most important OO systems will be enumerated in Subsection 4.3b Below, a brief account is given of LOOPS [Ste83], which is one of the languages used to implement Palladio. The Palladio system is one of the first CAD environments developed to study the applicability of expert systems in VLSI chip design (see Subsection 5.2d). Being an OO programming system, LOOPS provides for building hierarchies of classes and instances of those classes. LOOPS supports four main programming paradigms: the object-oriented, rule-based, access-oriented and the normal procedural paradigm. The access-oriented method uses demons, which in LOOPS make it easy to construct sophisticated visual displays for monitoring variables in a program. Developing applications in LOOPS involves a combination of writing codes in the editor and accessing a large number of convenient facilities in the mouse-oriented window and menu environment. A class can be accessed with a Class Browser. LOOPS offers six different categories of methods. The rules in LOOPS are organized into definite rule sets which may have various kinds of control structure to evaluate them. A rule set is always associated with some particular LOOPS object, which provides the workspace for the rules. These rule sets can be invoked in several ways [Tel89].

4.3 KNOWLEDGE PROCESSING

4.3a Production Systems

Introduction
This section discusses the various ways in which the knowledge in expert systems is processed. Let us first introduce the *production system*, which is a widely used type of expert system based on production rules for its reasoning process. Note that production systems are a subset of the broader class of *rule-based systems*, which also include logic programming systems. Production systems are successful when the number of rules is not too large.

What is a Production System?
A production system employs an unordered collection of data-sensitive rules, called *production rules*. These rules of the form

if antecedent *then* consequent

or

if conditions *then* action

are stored in the *rule base*. A database, called *working memory*, serves as a global database of symbols representing facts and assertions about the problem. It represents the current state during the search for a solution. The *inference engine* must determine which rules in the rule base are relevant to data in the working memory.

When forward chaining is used, the procedure is as follows. The inference engine seeks a suitable rule in the rule base. As soon as a match with a conditions part of a rule is found, the state in the working memory is changed according to the action part of this rule and the inference engine repeats the search procedure starting from the new state. These search cycles are iterated until the final state is reached, i.e., no rule applies any longer.

Control of the inference engine is performed in the *recognize-act cycle*, that is, rules with satisfied conditions are found (called *matching*), one of these rules is selected (called *conflict resolution*) and an action is executed. The match phase searches for matches between working memory elements and conditions, and then checks to see if the same variable names used in different condition elements are matched to the same values. Conflict resolution chooses one instantiation from the conflict set based on one of two strategies: LEX (LEXicographic-sort analysis) and MEA (Means-Ends Analysis). During the act phase, the actions on the right-hand side of the rule are executed in the order they are written.

The inference engine can be described as a finite-state machine with a cycle consisting of three states: *match*, *select* and *execute* rules (see Fig. 4.7). At the start, the working memory contains only the most essential data. During the dialog, this memory stores additional facts about the problem. In the first state (match), the machine finds all of the rules that are satisfied by the current contents of the data memory by a pattern-matching algorithm. Depending on the use of forward or backward chaining, the match may involve the conditions part or the action part respectively. The *rule interpreter* matches the data to the rules and chooses which rule to apply. It is this part of the inference engine that executes the rules by interpreting the conditions and actions of the rules.

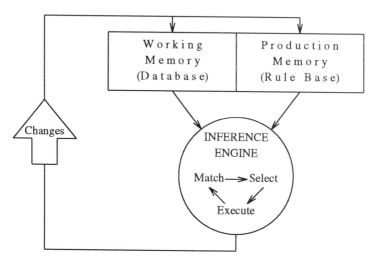

Fig. 4.7 Production system architecture

There is a clear distinction between a hypothesis and a conclusion. A *conclusion* is a fact reached by deduction from known facts that is accepted as true. Based on known facts, a *hypothesis* is a tentative conclusion that may or may not be true. Assuming the hypothesis is true, we can ask relevant questions. Conclusions obtained in examining rules are added to the database.

In a conventional third-generation program, instructions are executed sequentially. The control and execution sequence is fixed and well known ahead of execution. In a production system (like in all rule-based systems), control is based on re-evaluations of the data states, that is, examining rules involves the test whether or not a condition about data in the database is satisfied. Therefore, rule-based systems are data driven rather than instruction driven.

The pattern-directed modules to be executed are selected by the executive based on the interesting or important pattern of data in the database. This makes the sequence in which the modules are executed unpredictable and completely data dependent and suits applications in which the execution sequence is not known ahead of time or the number of possibles cases is too large to enumerate. Rule-based systems in which the executive follows the simple control structure

4.3 KNOWLEDGE PROCESSING

of recognize and act cycle are called *production systems*.
Production systems have the following advantages:
a. Modularity: The organization of knowledge in separate modular units facilitates the incremental expansion of the knowledge base. The knowledge base can be updated by simply adding, deleting or changing individual production rules in any modular unit without affecting other units.
b. Uniformity: The uniform representation of knowledge fosters easy understanding by the user.
c. Naturalness: Conclusions drawn by using production rules are reached in much the same way that a human expert would reason.
d. Flexible hierarchy: The hierarchy is controlled only by the relationship of the rules. By changing the rules we can change the hierarchy.

The disadvantages are
a. Inefficiency: Problem solving by production systems is relatively inefficient. Much computer time is wasted in unnecessary match-select-execute cycles.
b. Rigidity: It is difficult to follow the flow of control during the problem-solving process.
c. No hierarchy: The hierarchical relationships are difficult to visualize. There is no way to represent a hierarchy of rules.
d. Restricted conditions: Calculations in the (lefthand) condition part of a rule are not allowed.

Production systems have successfully been used in medical diagnosis, classification taxonomy and patient monitoring systems.

OPS Program Development
There exist several implementations of OPS5, including LISP-based OPS and VAX OPS5. Though OPS5 is inherently a forward-chaining problem-solving tool, backward chaining can also be used. OPS5 is primarily suited for solving problems that are difficult to formulate as an algorithm, and require symbolic expression of *if ... then ...* rules, while frequent changes in the rule base are expected.

Facts and rules relevant to a specific problem area can first be written in English and then translated in a suitable language. In the OPS5 environment, the necessary information in the working memory is entered as suitable class-attribute representations. An OPS5 rule base contains the rules and attribute declarations, with appropriate comments. It is important to introduce rules that produce meaningful results. A particular subtask can be accomplished by using a cluster of related rules. All element classes and attributes contained in the rule base must be declared prior to running the recognize-act process.

OPS5 uses two strategies for conflict resolution, when several rules match a given pattern:

a. Instantiations that have already fired are discarded and from the most recent instantiations the one is picked that contains the most tests for constants and variables.
b. When a rule has been added, the first strategy is used supplemented with instantiations with data from the working memory, which contains the expressions that the rules have to match.

During the execution of the recognize-act cycle, the user can interact with the program through a set of commands. These commands are read and executed by a *command interpreter*. Commands interpreted by the command interpreter include initializing the working memory, running the recognizing-act cycle and setting the conflict-resolution strategy. The VAX OPS5 *startup* statement contains commands and actions that are executed before the recognize-act cycle is started.

When OPS5 rules are compiled, they are transformed into a discrimination network that makes the program efficient to execute. In VAX OPS5, the rule base is compiled with the *ops5* command. In LISP-based OPS5, the rule base is loaded from rule files, or rules are entered individually at the LISP prompt.

Rete Algorithm

In the match-select-execute cycle, the matching phase consumes most of the execution time. The Rete algorithm was developed by Forgy [For82] to reduce this time. This algorithm is based on the fact that the firing of a rule causes only minor changes to the working memory. Only matches which are different from those in the previous cycle are stored within the Rete network.

The Rete algorithm proceeds from the individual elements of the left-hand sides of the rules and creates nodes that check
- whether the attributes that have a constant value are satisfied,
- whether the attributes that are related to a constant by a predicate are satisfied,
- whether two occurrences of the same variable within the condition are consistently bound.

Techniques for speeding up the match-select-execute cycle are being developed. Miranker and Lofaso [Mir91] describe a set of techniques that compile production systems to execute code, giving an increase in the execution speed of system programs by two orders of magnitude over the commonly used LISP-based OPS5 system. When the compiler has finished these tests for the individual elements, the consistency of variable bindings across the condition elements is tested.

Additional Comments

In the expert system shell EMYCIN, judgmental reasoning proceeds by chaining rules together to form deductions and by using a built-in algorithm to combine the numerical confidences at each step in the chain. So, there should be some way of assigning confidence numbers to rules so that the built-in algorithm will assign reasonable confidences to the answers that the system produces.

Experience has shown that maintaining and updating rule-based systems is not easy. Virtually, changing the knowledge base requires the help of the same knowledge engineer who was responsible for the original version of the knowledge base [Jac89].

OPS83 rules that execute the recognize-act cycle may contain Pascal-like expressions and subroutines in their condition and action parts [For84].

4.3b Logic Programming and Object-Oriented Systems

Predicate Logic

Logic programming systems are based on the use of predicate calculus, which is an extension of propositional logic (see Subsection 3.5a). Two major advantages of predicate calculus are its preciseness (logic is a precise method of determining the meaning of an expression) and its modularity (statements can be added, deleted or modified). A primary disadvantage is the exponential increase of inferences when the number of facts in the knowledge base increase.

Prolog

Even a cursory examination of the Prolog language as presented in Subsection 3.5b learns that Prolog allows easy prototyping. Positive features of Prolog in this respect are its symbolic notation and its extensibility as a programming language. Though a Prolog-based system as well as production systems use rules, we should bear in mind that Prolog rules are restricted to Horn clauses, while production systems allow more general structures for rules [Ste86, Fil88, Kim91]. Prolog's inference mechanism is basically depth-first backward chaining. On the one hand, Prolog takes place automatically, on the other hand control of the problem-solving process may become obscure.

Bratko [BRa86] distinguishes three approaches to implementing Prolog-based expert systems:

a. *Direct approach.* In this low-level approach, knowledge is stated directly as Prolog clauses, which are interpreted by a Prolog interpreter.
b. *Metaprogramming approach.* In this high-level approach, Prolog is used as a meta-language for implementing other languages.
c. *Meta-interpreter approach.* This approach lies in between the former two approaches. Knowledge is stated as Prolog clauses, but a meta-interpreter is used for exploiting the advantages of the metaprogramming approach.

Several Prologs have environments built on top of the language itself. For example, Arity Prolog has object-oriented programming features embedded in Prolog. MRS includes procedural code in a logic-oriented language.

Prolog-Based Circuit Specification and Design
Design of complex VLSI chips requires a hierarchical methodology with the modularity property of easy design modifiability. VLSI design implies an exploration of the space of possible choices. In order that logic programming be useful in VLSI chip design, the circuit must be described in terms of Horn formulae which are themselves amenable to formal manipulaton.

Clocksin [CLO87] applied Prolog methods to solve problems in gate assignment (mapping a circuit formulation onto a connected set of available modules), circuit rewriting (corresponding to Boolean simplification) and determination of signal flow (as a preliminary step prior to further circuit analysis, e.g., timing analysis). Prolog can be used for writing directly executable specifications of digital circuits, including the use of quantified variables, verification of hypothetical states and sequential simulation. Though a circuit can be represented in several ways, the method as exemplified in describing the NAND circuit shown in Fig. 3.7, Subsection 3.5b, proves to be effective.

Problem solving in a logic-programming language like Prolog involves the introduction of problem definitions rather than designing algorithms. By specifying the truth tables of the circuit components, the truth table of the entire circuit can be derived. Wos *et al.* [Wos84] used an automated reasoning program to design a circuit that meets a given input-output logic behavior, with special constraints imposed on the logic components. The opposite design problem of proving that the circuit in fact meets the specifications can also be handled.

Expert Systems in Prolog
As may be clear from the considerations in Subsection 3.5b, Prolog-based expert systems have certain characteristics. Prolog execution, founded on unification and automatic backtracking, is quite different from that of conventional third-generation languages. Data structures are handled using pattern matching, *if ... then ...* rules are represented using Horn clauses under procedural or declarative interpretation, and the inference engine is built on top of the resident Prolog theorem prover.

Prolog can be used to implement expert-system techniques on a personal computer [Tow86, McA87, Mer89]. Merritt [Mer89] described various Prolog-based techniques, including backward chaining with uncertainty, why and how explanations, forward chaining, defining and manipulation of frame structures, object-oriented interfacing, and prototyping.

Object-Oriented Systems
The emergence of object-oriented (OO) languages has led to the development of object-oriented expert systems. In OO expert systems, everything can be handled as an object, including the inference engine itself and the current state during a search procedure. The declarative representation makes it easy to implement and

4.3 KNOWLEDGE PROCESSING

modify modules. The class *rule* is defined by sending the adequate message to the root class *object*.

A main characteristic of OO expert systems is that objects contain their own procedures and data and communicate via messages, while inheritance plays a major part in the knowledge representation. OO languages have also provided greatly enhanced user interfaces.

Frame-Based Systems
A typical example of an object-oriented system is the *frame-based system*. An object is represented by a frame whose slots contain values associated with the attributes of the object. This system is extremely useful when static hierarchical relationships between objects exist.

Object-Oriented Programming
Whereas frame-based and object-oriented representations contain declarative information in their data structures, object-oriented programming contains most of its information in the code itself. An OO language, such as Smalltalk, has declarative semantics. OO programming allows the definition of component hierarchies, attribute relations and connectivity information. The data structures are descriptions of objects or structure descriptions of objects, usually enmeshed in a class/subclass taxonomy. A general inference engine of general reasoning algorithm operate on these data structures.

Object-Oriented Languages
The first full-fledged object-oriented language Smalltalk (see Subsection 3.6b) was developed at the Xerox Palo Alto Laboratories [Gol86]. The advent of LISP machines in the early 1980s gave impetus to the development of OO languages. Influenced heavily by Smalltalk, LISP-based OO programming systems were crucial to the design and implementation of LISP machines, operating systems and graphics software. Flavors [Kee85, Moo86] is an OO programming system, implemented on the Symbolics LISP machine. Other OO languages include LOOPS [Ste83] and Flavors [Can82]. LOOPS (LISP Object Oriented Programming System) integrates in one environment four programming techniques or paradigms for expert system development [Ste83]. As well as a rich implementation of the object-oriented paradigm, LOOPS offers procedure-oriented, access-oriented (demons) and rule-oriented programming.

An ANSI Common LISP standard is being developed for the LISP language which includes Common LISP Object System (CLOS) [Bob88].

Object-Oriented and Logic Programming
Object-oriented programming is particularly useful when a hierarchical structure in data objects is essential. On the other hand, logic-programming languages like Prolog have built-in facilities for deductive retrieval through backtracking and pattern matching. Several attempts have been made to exploit the best features of

each language in a common application.

One strategy to bridge the gap between the two programming paradigms is to implement objects in an existing logic environment. Another strategy is to implement a Prolog-like search facility in an object-oriented environment. A third approach is to use an interface between an existing object-based system and a logic-based system. As an illustrative example of the latter approach, Koschmann and Evens [Kos88] described an interface that was developed between LOOPS and Xerox Quintus Prolog.

4.3c Shells and Tools

Introduction
Building an expert system is considerably facilitated by the use of tools [Gev87]. The emerging expert-systems industry either serves as consultants or offers a wide variety of tools, which may assist in any stage of the development process, including knowledge acquisition, system construction, validation and testing. The various languages that can be used in the life cycle of an expert system can be considered as tools. Typical language tools are LISP, OPS and Prolog.

Expert System Shells
Tools may be useful in several problem domains. A typical tool is the *shell*, which forms the inference engine around which a knowledge base of a specific knowledge domain can be added. Expert system development involves two basic tasks: knowledge base construction and the design of an inference engine. By separating these tasks, the inference engine can be used in a number of different knowledge domains. Since the shell contains a particular reasoning strategy, the most appropriate shell must be employed when it comes to designing an expert system for a specific application. A shell may provide a knowledge-base development engine for constructing and editing the knowledge base.

A big advantage of a shell is that several expert systems for different applications can be built based on the same shell. Therefore, a shell contains many characteristics common to expert systems. A shell can be derived from a special-purpose expert system which was designed from the start for a specific application. Leaving out the special-purpose components, notably the domain-specific knowledge base, a shell is obtained which can be used for a different domain of application. An important example in the past is the shell EMYCIN, which was derived from MYCIN, the well-known expert system for medical diagnosis [Buc84].

Shell packages are the most straightforward means of rapidly prototyping an expert system. Run-time facilities include provisions as to points concerning the human-computer dialog. The importance of the concept of shell has led many

4.3 KNOWLEDGE PROCESSING

software firms to develop general-purpose expert system shells. It should be noted that a shell cannot be so general that every kind of expert system can be built upon it. A shell covers only a specific class of tasks efficiently.

Some development tools provide a hybrid combination of reasoning techniques to adapt the expert system more closely to the actual application. An example is KES of Software Architecture and Engineering, which provides both an IQLISP and a C-based microcomputer expert-system development tool. KES supports three reasoning strategies: hypothesize-and-test, Bayesian statistics and production rules.

Expert system shells available on the market are differentiated by the problems they address. The majority address the run-time capabilities, whereas others address the problems of knowledge-base development and refinement. Individual marketed products vary according to the power of the knowledge representation formalism they support, the inference strategies employed (and whether a choice or a mixture of strategies can be used), their human factoring, their external interfaces, as well as such pragmatic considerations as knowledge-base size and performance limitations, machine ability and cost.

Expert-System Tools

The design and application of expert systems can be facilitated by using special development tools providing various interface utilities or user-friendly programming environments. The armory of the knowledge engineer consists of a set of tools, which is usually referred to as a *tool kit*. Some of the well known tool kits are described below.

UNITS is a combination of an AI language and a frame-based representation language [Ste79].

LOOPS is a LISP-based object-oriented programming system, developed at the Xerox Palo Alto Research Center [Bob83, Ste83].

ARBY, developed at Yale University, supports a rule-based representation which uses predicate-calculus notation for expressing rules, a backward-chaining facility for hypothesis generation and an interface subsystem that manages a set of interaction frames [MCD82].

KEE, developed by IntelliCorp, Menlo Park, supports a frame-based or other representation and reasoning schemes (forward and backward), and provides a ready-made rule interpreter [Int84]. It has an excellent graphics interface.

ART (Automated Reasoning Tool) of Inference Corporation provides a rule-based perspective based upon a blackboard architecture [CLa84, Wil84]. Declarative knowledge is stored in schemata or contexts and relationships between schemata are defined. Hierarchies are created via rules. During knowledge-base construction, ART permits the user to browse through the rules and modify them. ART is a general-purpose tool in contrast to KEE, which serves special purposes.

Another general-purpose tool is KL-TWO from Bolt Beranek and Newman Inc. It permits the system developer to switch to other languages or reasoning paradigms without the need to reassemble the whole language base. The above-mentioned tools are based on LISP.

Knowledge Craft was developed at Carnegie-Mellon University [Kno87]. It combines the deductive power of Prolog with the representational power of object-oriented programming. It allows complex inheritance paths to be defined through user-defined relations and grammars of inheritance paths.

Mettrey [Met91] made a comparative study of five tool kits: ART-IM (ART for Information Management), CLIPS (C Language Integrated Production System), KES (Knowledge Engineering System), Level5, and VAX OPS5.

TEIRESIAS

In large knowledge bases with a large number of rules acquired from human experts through the years, a rule may contradict another rule. TEIRESIAS, developed at Stanford University, is a knowledge acquisition aid that can handle such problems [DAv82]. When new rules are to be incorporated, it makes sure that the new data fit into the structure without introducing contradictions.

TEIRESIAS ensures completeness, consistency, type checking and syntactic integrity. In fact, it is a knowledge-base management system with a backward-chaining inference mechanism that produces an exhaustive search of a goal tree accomodating both AND and OR structures.

The Blackboard Architecture

Many problem-solving processes may be considered as sequences of state transitions, starting with the initial state (the first guess of a solution) through intermediate states (partial solutions) and ultimately leading to the final state (the solution of the problem). Each state transition, initiated by some action, must bring the current state to a new state, which is "nearer" to the final state. The choice of appropriate actions determines the efficiency of the problem solver.

Solving complex problems may need the help of a number of human experts, each one being proficient in a specific part of the problem domain. Imagine a group of such experts sitting in a room collaborating to solve a problem. A blackboard is available to record the state of the problem-solving process. Each expert may jump onto any point in the process to make a suggestion in the direction of the final solution. A coordinating expert is needed to control the solving problem, that is, to determine which actions to perform in what order. This expert must take a decision as to what the next action must be, when two or more experts may intervene at the same time or when simultaneous suggestions are contradictory.

The above design approach in which a problem is partitioned into loosely coupled subtasks performed by specialized experts, has led designers to devise the *blackboard architecture* of expert systems [Nii86, Mor88]. This architecture

4.3 KNOWLEDGE PROCESSING

consists of three components: a *blackboard* (a working database, corresponding to the blackboard in the above case), a set of *knowledge sources* (containing pieces of subdomain knowledge, representing human experts mentioned above) and a *coordinating expert*, which controls the operation of the knowledge sources. See Fig. 4.8.

The blackboard is a working database that stores input data, hypotheses, (sub)goals, partial solutions, alternatives, activities to be performed, etc. In any case, information recorded on the blackboard constitutes the current state of the problem-solving process. The coordinating expert has the scheduling task of controlling which knowledge sources are bound to be active. Depending on the information on the blackboard, one or more knowledge sources may contribute to the solution. Usually, this results in a modification of the information on the blackboard. When some information proves to be false, all information based on it must be cancelled and a new action should be introduced. To be able to explain why certain decisions were made, the system must keep a record of all rules used and failures during the session. Some problems require some form of probabilistic reasoning. Consistency of information must always be preserved any time new information is added to the blackboard.

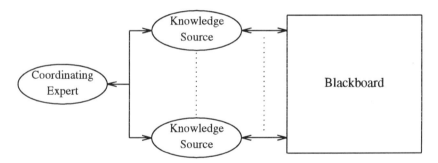

Fig. 4.8 Blackboard system

Usually, the knowledge sources can be considered as task-specific expert systems. Specific tasks in solving a complex problem can be relegated to such expert systems. Intelligent coordination of these expert systems can resolve the potential conflicts in decision making that may occur.

The knowledge sources play the role of independent expert systems for solving subtasks in the problem-solving process. The coordinating expert controls the interaction between the knowledge sources. It monitors the blackboard and determines what actions to take at each stage, while any type of reasoning (backward, forward, etc.) can be applied. It directs the current focus of attention of the problem solver, which may be a particular knowledge source to apply or a set of blackboard objects to process. Given a new state of the blackboard, any knowledge source may contribute to the solution by stating hypotheses, providing a confirmation or denial of current information and improving the current state of the process.

The blackboard model was first applied to the HEARSAY speech-understanding system which responded to spoken commands and queries about computer science abstracts in a database [Erm80]. HEARSAY evolves to HEARSAY-III, a knowledge engineering language for controlling multiple knowledge sources [Bal80].

4.4 PROGRAMMING ENVIRONMENTS

4.4a Knowledge-Based Management

Database versus Knowledge Base

Expert systems are developed and delivered on many different types of hardware, including mainframes, minicomputers, personal computers, LISP machines and general-purpose engineering workstations. Highly valuable expert systems are those that integrate with existing software and database systems [Jar84, Fro86]. It is interesting to identify the differences between a database and a knowledge base, the latter being an essential part of an expert system.

Comparing databases with knowledge bases leads us to the distinction between data and knowledge. The following distinctive features can be given:

a. *Data* are verifiable, factual statements and correspond to the fixed facts in a knowledge base. In addition to facts, *knowledge* implies rules and all kinds of heuristic information as acquired from experts in a specialized field.

b. The distinctions between the concepts of data and knowledge stem from the different environments in which they are used. Data are often used for computational purposes, whereas knowledge is connected with interpretations of objects and their interrelations.

c. Though all programs embody knowledge and reasoning of some sort, the knowledge in knowledge-based programs is made explicit in that it is manipulated in declarative form [Fre86]. Problems involving knowledge are solved by an inference engine.

d. A database is designed to query and update a large volume of data. The data in a database are a permanent resource, shared by different application programs.

e. A knowledge base is an essential part in expert systems. The inferencing process inherent in expert systems is a manipulation of knowledge in the

4.4 PROGRAMMING ENVIRONMENTS

knowledge base.

f. Though search is fundamental both in database-management systems (DBMSs) and in knowledge-based systems, the search objectives are different. An essential task of database systems is query evaluation, that is, a query initiates a search for finding the required information. Characteristic of knowledge-based systems is the inferencing process. Note that various methods of inference exist. In deductive inferencing, there is a set of axioms and the response to a query is to find one or more solutions by using a reasoning process.

g. The data in databases have a simple format, e.g., as records of predetermined length with a fixed number of fields. The query language enables only the specification of a limited class of queries (retrieving and updating).

h. Facts in knowledge bases may contain elements of different nature. Moreover, the facts may be imprecise or given with some probability measure. Domain-specific knowledge is stored via one of the known representation languages (see Section 4.2).

i. Data in a database usually exist in such large volumes that storage in secondary memory is inevitable. Efficient access methods are required to retrieve the data efficiently.

Hypertext

At present, we experience an overwhelming emergence of textual, numeric and visual data. Modern electronic information systems have been developed to assist both writers and readers in taking full advantage of the dissemination of vast amounts of updated knowledge. These systems provide mechanisms for compact storage and rapid retrieval of this knowledge.

One typical example of such an information system is *Hypertext*. When non-textual information, such as graphics, images, audio and full-motion video is considered in addition to textual information, the term *Hypermedia* is sometimes used. In what follows, we use the term Hypertext and Hypermedia interchangeably, where text is to mean all kinds of knowledge (visual, audio or otherwise). Hypertext is a paradigm of electronic writing and reading that goes beyond mere editing and display [Con87]. An essential characteristic of Hypertext is the existence of a link from a text fragment to other text fragments in a complex network in such a way that users can traverse the network quickly. For example, when a word or phrase is highlighted, the computer would retrieve and display other relevant information (e.g., a definition, a reference to a book or even the page numbers in the book).

For authors, Hypertext systems provide facilities which are not available in conventional word-processing systems, including linking information together, creating paths through a body of related material, and annotating existing texts. For readers, Hypertext provides a sophisticated form of database management which allows the user to traverse complex networks of information to find sources of quotations, references to articles or books, and related passages.

Intelligent Databases
The information glut that comes over us calls upon the use of intelligent databases which allows one to have easy access to all kinds of information, including tools helpful in making decisions. According to Parsaye *et al.* [Par89], intelligent databases have evolved as a result of the integration of traditional approaches to databases with more recent fields, such as text management, object-oriented programming, expert systems, hypermedia and machine learning.

While database systems were initially developed in the early mainframe environments, typically applied to numeric and record-based data, later on advanced text management permits a user-friendly way of on-line information processing. Gradually, the object-oriented approach to programming has proven to be powerful in bringing database technology to a more intelligent level (see Subsection 3.6c). The emergence of expert systems has shown the importance of knowledge bases, which can be considered to be at a higher intelligent level than the traditional database. In the past decade, the quality of databases was further heaved up by the introduction of hypermedia systems, which deal with information in a variety of forms, such as text, images and sound. A learning capability further improves the intelligence level of database systems.

The top-level architecture of an intelligent database system comprises three components: high-level tools, and advanced user interface and an intelligent database engine [Par89]. High-level tools provide the user with a number of facilities, such as intelligent search, data integrity and quality control, hypermedia management and knowledge discovery. The user interface should enable the user to create new object types, browsing, searching and asking questions. The intelligent database engine has the capability of query processing and deductive reasoning. The engine includes forward and backward chaining inference procedures as well as drivers for the external media devices, version handlers and optimizing compilers.

Chip-design (CADCAS) management is more complex than pure database management in several respects [Kat86]:
a. CADCAS deals with objects that are more complex than the data in DBMSs. Objects in CADCAS may have multiple representations: logic schematics, transistor circuitry, geometric layout, etc.).
b. Modern chip design involves the use of a wide diversity of design tools.
c. Several alternative CADCAS designs with different performance characteristics may be available.
d. As improved design techniques and technologies become available, new versions of a design may be developed and released.

4.4 PROGRAMMING ENVIRONMENTS

e. Chip design is an incremental process proceeding from a high-level specification passing several intermediate stages ending at the final chip.
f. Interactive design requires a suitable user interface to evaluate intermediate design results.
g. Advanced expert systems in CADCAS require the combined use of a database and a knowledge base.

Knowledge-Based Management Systems

Knowledge-based systems imply the use of some sort of database [Ull88]. Therefore, they may benefit from the efficiency of advanced database management systems, particularly when the relevant domain of knowledge require large volumes of stored data. On the other hand, databases systems may exploit the reasoning methodology of expert systems to enhance their capabilities. No wonder, there is currently a great interest in interactions between knowledge-based systems and database-management systems.

Though Fig. 4.9 suggests a possible merger of both database and expert systems, it is important to bear in mind their essential differences [Fre90]. A system that combines both knowledge base and database concepts is called a *knowledge-based management system* (KBMS) or an *expert database system*. Such a system provides a user-friendly environment for the manipulation of shared data as well as knowledge in large distributed systems.

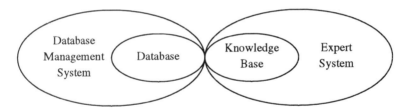

Fig. 4.9 Interactions between database and expert systems

While the necessity of integrating knowledge base and database technologies is widely recognized, the question arises to which extent this integration should be carried. Many solutions would be conceivable, ranging from a loose coupling of an existing expert system and a database management system to a complete merger of an expert system and a database management system [Jar84]. Such a complete merger is not necessarily the ideal solution for all applications. Its usefulness depends largely on the kind of application. When a straightforward query search in a database is needed, a pure database-management system is cheaper and more efficient than a sophisticated KBMS. The most appropriate fifth-generation language for programming the problem and for representing the knowledge must be used.

The object-oriented approach represents a successful unifying paradigm in a wide variety of applications, including databases, knowledge representation and computer-aided design [Zan86]. Object-oriented languages are particularly suited to applications needing graphical, highly interactive user interfaces with complex data-modeling requirements and ready modifiability.

In order to move toward an open system architecture, standard interfaces should be defined for the user interface and interfaces with other processes, such as Hypertext and Cobol systems. There is a move toward using graphic and window-based interfaces provided with the hardware platform, for presentation of the user interface. A standard interface to this software is usually specified. The user interface, as generated by the inference engine, would need to be compatible with this interface.

The SQL standard has been developed for interfacing a DBMS and application software, and a similar standard for interfacing a KBMS application with other software is needed. A recent trend is the *distributed computer network*, where individual computers are connected across networks to one another and to large mainframes. The mainframes act as file servers, which are central repositories for information, data and programs.

4.4b AI Architectures

AI Machines

To overcome the inefficiency and difficulty in encoding AI processes by using conventional procedural languages, such as C, declarative languages have been developed. Lambda-based (LISP-like) and logic-based (Prolog-like) languages are two popular classes of declarative languages. Specialized AI architectures providing support for the unique features of declarative languages are emerging [Wah86, Kow87, HWa87, Uhr87, Alb92]]. A recent trend is to integrate lambda and logic languages with procedural languages. Besides that, object-oriented programming is attracting a growing interest.

Most AI architectures are software-oriented. Many architectures (data flow, reduction, direct execution of high-level languages, data/knowledge base, logic) form the basis for the design of hardware-implemented AI machines [Wah86]. Such machines are meant for intelligent purposes: theorem proving, automated reasoning, automatic programming, planning and problem solving. Hwang *et al.* [Hwa89] identify four major classes of AI machines: based on a particular language, knowledge, connectionism and intelligent interfacing.

Language-Based AI Machines

Language-based AI machines are aimed at an effective use of Prolog, LISP or another functional language. Particularly LISP and Prolog machines capitalize on the wide availability of software written in these languages. These machines

4.4 PROGRAMMING ENVIRONMENTS

support a wide variety of knowledge representations and heuristic search procedures, which must be writtten in the supported language. The latter leads to a rather slow execution time.

Commercial LISP machines include the following:
a. Lisp Machines Inc.'s LAMBDA, supporting Zetalisp and LMLisp [Mye82, Man83].
b. Xerox 1100 series, supporting Interlisp-D and Smalltalk [Man83].
c. Fujitsu ALPHA machine, a back-end LISP processor for a general-purpose computer, supporting Maclisp [Hay83].
d. Texas Instr. Explorer, supporting a Common LISP dialect.
e. Berkeley's SPUR (Symbolic Processing Using RISCs) [Hil86].
f. Symbolics 3600, supporting Zetalisp, Flavors and Fortran 77 [Moo87].

A concurrent LISP machine design requires a definition of a suitable concurrent LISP language, and a set of representative benchmark programs. Though the designer community is divided on several issues, the use of the shared-memory model of access to data is generally accepted (see Subsection 3.4b). Prolog machines developed include: PIE [Mur85] and PIM [Tan86], which were developed in the Japanese Fifth-Generation Computer System project. ASCA is a Prolog machine based on a content-addressable-memory machine architecture with a hierarchical pipelining scheme [Nag88]. The Xenologix X-1 [Dob87] is a coprocessor board for the Sun workstation that implements Prolog with extensions for LISP.

Language-based AI machines based on a functional language include ALICE [Pou85], supporting the functional language HOPE using graph reduction on a shared-memory multiprocessor.

Knowledge-Based AI Machines

Knowledge-based AI machines and their languages are related to expert systems. Three types of knowledge-based AI systems can be identified, based on rules, semantic networks, and objects [Hwa89].

Rule-based AI machines use a forward-chaining sequence: match - select - fire (see Subsection 4.3a). A prototype of a 1023-processor rule-based machine with a binary-tree interconnection network is DADO2, which exploits the inherent parallelism in the match phase, where many rules must be matched against the facts in the working memory [Sto87]. The working memory is stored in a tree architecture, and matches are carried out in parallel.

Knowledge-based AI machines processing semantic networks include the massively parallel fine-grain machines NETL [Fah83] and the Connection Machine [Hil85, Tuc88]. FAIM-1 [And87] is an AI machine which implements logic programming and procedural programming within an object-oriented framework. Smalltalk-80 is the basic language in Dorado [Deu83], which is a microcoded single-user workstation, and in SOAR [Pen86, Ung87], which is a

RISC-based microprocessor with the same functionality and speed as the Dorado.

Connectionist Systems

The connectionist architecture is inspired by biological systems, such as the human brain, where knowledge is stored and processed by a huge number of interconnected neurons (see Subsection 9.3b). Compared to knowledge-based systems which use coarse-grained knowledge representations (rules, frames or objects), a connectionist system uses a fine-grained knowledge representation. Unlike the other architectures, which are based on symbolic processing, the connectionist architecture is characterized by the interconnections rather than the cells (processing elements) which are interconnected [Fah87]. Each connection has a weight associated with it. The knowledge is represented by the pattern of weights of all the connections. All cells are concurrently active and the state of each cell is given by the weighted sum of the states of the connected cells. The final solution to a problem is the result of a complex process of interactions between the processing elements in the connectionist machine.

The significance of connectionist architecture becomes manifest in the growing field of *neural networks*. A characteristic feature is the self-learning ability of connectionist machines. Connectionist architectures may be hierarchical as in the Hypernet [Hwa87], or nonhierarchical as in the Boltzmann machine [Fah83].

Intelligent Interface Machines

In order to enhance the human-computer communication, intelligent interfaces require such disciplines as speech analysis, natural-language understanding, image processing and computer vision. Several hardware systems that perform these tasks have been developed [Hwa89].

Examples of speech-recognition systems are Harpy, Hearsay-II, Dialog System 1800, NEC DP-100 and IBM Nat. Task [Tor85]. Examples of interface machines for pattern recognition and image processing are Cytocomputer, Tospics and Pumps [Hwa83]. An example of an interface machine for computer vision is Butterfly [Hwa89], a MIMD machine consisting of up to 256 processor/memory nodes interconnected through a butterfly switching network.

Content-Addressable Memory

In *content-addressable memory* CAM (or *associative memory*), data items are searched on the basis of (a part of) their contents rather than their addresses or location. Accessing data in a CAM requires a comparison of an external search argument with the data content [Chi89].

A CAM allows parallel access of multiple memory words and hence has an inherent parallelism. Parallel search and matching make CAMs very useful for pattern matching in production systems and logic programming. Building static

4.4 PROGRAMMING ENVIRONMENTS

and dynamic CAMs at a relatively low cost has been possible by advances in MOS technology.

4.4c Linking Declarative and Procedural Languages

Introduction
Declarative languages, such as LISP and Prolog, are suitable for AI applications and hence are most obvious to implement the AI software in expert systems. At the same time, computation-intensive tasks, which are well established in various chip design steps, can more effectively be performed by using algorithms implemented by procedural languages, such as C. Because procedural programs run faster than AI-oriented programs, development tools are written in C, Pascal or Ada. In many AI applications, LISP and Prolog are used to implement the original version of a program. Once this version has been tested and reached its final form, it is translated to a traditional language, such as C.

Chip design incorporates both algorithmic procedures and intelligent decision steps. The question arises of how to combine procedural with declarative languages in one design system [Klt86]. Coding intelligent decisions and implementing search algorithms in procedural languages is at least arduous. On the other hand, using AI languages for implementing computational algorithms is cumbersome. The best solution is achieved when AI modules implemented in an AI-oriented language are combined with procedural modules written in a conventional language, such as C. In this way, the good qualities of both declarative and procedural languages can be exploited. Improvements in LISP have made it possible to run LISP almost equally well in a conventional operating system as on a LISP machine.

Linking declarative and procedural languages is usually similar to linking symbolic and numeric methods. This linking activity can be realized at three levels:

a. *Control level.* This level involves several decision processes, particularly the decision of which is the most appropriate step to be followed next, or the selection of the proper simulation or testing method to be used. Suitable selection criteria must be defined, an appropriate linkage between the control structures of the symbolic and numeric/procedural parts of the intelligent control process must be designed.

b. *Method level.* Linking at this level occurs when a numeric procedure is augmented with symbolic methods, e.g., expressing circuit equations in symbolic form allows one to study the effect of parameter variations.

c. *Data level.* A mapping between symbolic and algorithmic variables is required, e.g., to transfer data from the numeric part to the symbolic part.

Coupling Architectures
The simplest coupling architecture is based on loose coupling of algorithmic and symbolic modules with mutual communication taking place via data files. Several hybrid coupling architectures have been proposed of which some important ones are outlined below.
  a. *Object-oriented modules.* Algorithmic procedures can be treated as objects, in which all information pertinent to the selection, construction, the use and interpretation of the procedure is encapsulated within an object.
  b. *Message passing in object-oriented methods.* Communication with algorithmic modules is performed by message passing in an object-oriented language, such as Smalltalk. The modules may be programmed in different languages and have different operating features, including loading subprograms and use of verification tools. The attributes of procedures and their state variables are often represented in frame-based structures.
  c. *Blackboard architecture.* Knowledge sources with different tasks react state changes maintained on the blackboard (see Subsection 4.3c).

Commercial Software
Several software houses have developed application tools in which AI-oriented languages can be called from a procedural language or vice versa [Fal88]. A procedural language, notably C, can be linked to Quintus Prolog from Quintus Computer Systems (Mountain View, CA), to Arity Prolog from Arity (Concord, MA), to Prolog-2 from Expert Systems International and to Lucid Common Lisp from Lucid (Menlo Park, CA).

4.5 APPLICATIONS OF EXPERT SYSTEMS

4.5a Types of Expert System

Classification of Expert Systems
Expert systems have been developed for a wide variety of applications. Waterman [WAt86] presented a comprehensive Guide to Expert Systems which contains over 200 descriptions of expert systems and system building tools. This guide allows one to devise classification schemes based on different criteria. Firstly, classification can be based on the application areas in which expert systems are used. Waterman discussed the following application areas (in alphabetical order): agriculture, chemistry, computer systems, electronics,

4.5 APPLICATIONS OF EXPERT SYSTEMS

engineering, geology, information management, law, manufacturing, mathematics, medicine, meteorology, military science, physics, process control and space technology.

Expert systems can also be classified on the basis of the tasks that they perform: analysis, classification, interpretation, instruction, decision making, prediction, planning, design, monitoring, control and diagnosis/debugging. Other classification criteria are the language used, the control type (forward or backward) and the knowledge model (shallow or deep knowledge).

Forward chaining and backward chaining can be thought of as a bottom-up approach and bottom-up approach respectively to problem solving. Backward chaining, which starts from a distinct goal, is commonly used when values of some key facts eliminate a number of other facts. Backward chaining is appropriate in such applications as diagnosis (of medical problems, fault diagnosis), selection systems (choosing a suitable method) and advisors (as in law and regulation interpreters).

Diagnosis

In medicine, diagnosis is the identification of a disease from its signs and symptoms. Diagnosis in the broadest sense of the word (the analysis of cause-effect relationships in cases of malfunctioning system behavior) is among the dominant applications of expert systems.

Diagnosis involves two sets of concepts: hypotheses, which may be possible causes of the malfunction, and a set of deviant observations, which indicate the presence of the malfunction. Typically, a diagnostic problem starts with the observation of a malfunction, which is a deviation of the expected or desirable behavior. One strategy is to invoke one or more hypotheses which may represent the cause of the observed malfunction.

Precompiled pieces of knowledge relating malfunction observations to possible causes can be stored in a rule base. This cause-effect mapping can be used to find the causes when the malfunction observations are specified. Going from behavioral observations to the identification of the associated causes is a backward matching process as indicated in Fig. 4.10.

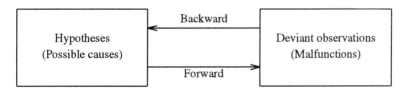

Fig. 4.10 Diagnostic system

The opposite process, forward matching, starts from a set of hypotheses and gives the resultant effects. For example, a set of blood tests may reveal a quiescent or manifest form of a specific disease. This is comparable with the case of fault simulation (see Subsection 7.2b) in which a fault in a circuit may or may not be visible as a deviant signal at the output of the circuit.

Shallow and Deep Knowledge Models
The above simple knowledge model involving stored relationships between effects and their causes does not require the detailed knowledge of the internal structure of the system being considered. Such a model is referred to as a *shallow knowledge* model. It is usually based on earlier results of diagnostic experiments or on past human experience about system behavior that deviates from the normal behavior. Shallow knowledge uses high-level models in a black-box manner relating specific faulty behaviors and their causes. In diagnostic systems based on shallow knowledge, frames and predicate calculus are normally used as knowledge representations of symptoms and faults. Production rules and clauses are used as representations of relations between symptoms and faults.

When the knowledge of the system implies detailed low-level information on the structure and behavior of the system, the model is referred to as a *deep knowledge* model. A deep model assumes a structural decomposition of the system in an appropriate number of subsystems. This means that malfunctions can be explored down to the lowest level of knowledge. Whereas a shallow model is concerned with a high-level black-box behavior, a deep model is a low-level representation, often expressed by mathematical relationships. Deep knowledge refers to knowledge about the structure and behavior of the system under test in addition to knowledge of diagnostic methodologies. Reasoning directly from deep knowledge is often called *first principles reasoning* or *model-based reasoning*. Deep knowledge penetrates to the low levels of abstraction, i.e., to low-level components of the system. First principles include such laws as Ohm's Law and the Kirchhoff Laws.

The type of knowledge model used determines the required strategy for diagnosis. Usually, either the shallow model or the deep model is used. There are also diagnostic systems which use both shallow and deep knowledge. Usually, the shallow knowledge is first used to identify the faults by relating symptoms to possible causes. If this process fails, the system then uses deep knowledge, which contains structural details of the unit under test.

By introducing realistic fault modes into a circuit, the effects can be observed by using a fault simulator. In this way, systematic symptom-to-cause associations can be compiled to create a test-knowledge base. The knowledge can be used for developing a knowledge-based expert system to perform fault diagnosis [Noo87]. Kramer [Kra87] argued that the computational efficiency of various reasoning tasks in digital circuit design is improved when reasoning about design is reformulated to include mathematical knowledge.

4.5 APPLICATIONS OF EXPERT SYSTEMS

4.5b An Anthology of Expert Systems

Classification Based on Knowledge Representation

The diversity of expert systems developed and used is phenomenal. The present subsection takes samples from the extensive range of expert-system applications known thus far. Subsection 4.5c deals with the special domain of electronics and computer systems. Chapters 5 through 8 discuss the major phases in VLSI chip design (system design, verification, testing and layout design). These chapters also contain numerous descriptions of expert systems dedicated to be applied in each of these phases. These descriptions give an overview of VLSI applications of expert systems as published in the open literature.

The classification given below is based on the major knowledge-representation schemes used in expert systems. A small selected group in each of the three major ways in which the knowledge is represented is collected in the tables given in Fig. 4.11 (production systems), Fig. 4.12 (logic programming systems) and Fig. 4.13 (object-oriented systems). A short description of each of the expert systems is given in the subsection indicated in the last column of each table.

Name	Application	Subsection
DENDRAL	Mass spectroscopy	4.5b
MYCIN	Medical diagnosis	4.5b
PROSPECTOR	Mineral exploration	4.5b
DART	Fault diagnosis	7.1c
IDT	Fault diagnosis	4.5c
ISA	Scheduler	4.5c
XCON	Computer configurer	4.5c
XSEL	Component selector	4.5c
DAA	VLSI synthesis	5.2d
EL	Circuit analysis	4.5, 6.3b
SYN	Circuit synthesis	6.3b
TALIB	Cell layout	8.2c

Fig. 4.11 Production systems

DENDRAL

DENDRAL (DENDRitic ALgorithm), implemented in Interlisp and developed at Stanford University, was one of the first expert systems implemented for a practical application [Buc78, Alt84]. This system infers the molecular structure of unknown organic compounds from mass spectral and nuclear magnetic response data obtained from measurements with a mass spectrometer. Given a large number of possible molecular structures, DENDRAL uses fragmentation rules, chemical expertise and the data from the mass spectrometer to eliminate the vast majority of these structures, leaving a small candidate set to be considered by the human chemist.

Meta-DENDRAL employs a learning algorithm to refine and extend DENDRAL's rule set.

MYCIN

Medical expert systems put the collected expertise of hundreds of specialists at the disposal of any doctor faced with a difficult medical problem. MYCIN, developed at Stanford University, is the first major expert system for medical diagnosis [Sho76, Buc84]. It is a rule-based system meant to select an appropriate antimicrobial therapy for hospital patients with bacteremia (infections that involve bacteria in the blood), meningitis (infections that involve inflammation of the membranes that envelop the brain and spinal cord) and cystitis infections (inflammation of the urinary bladder). The system collects all the relevant information about a patient and examines this information to diagnose the cause of the infection (e.g, the identity of the infecting organism is pseudomonas) using knowledge relating infecting organisms to patient history, symptoms and blood test results. To accomodate uncertainty, all information given to MYCIN can be accompanied by a certainty factor. As soon as a reasonable diagnosis has been made, the system recommends drug treatment (type and dosage) according to procedures followed by physicians experienced in infectious disease therapy.

MYCIN is a production system that incudes mechanisms for reasoning with uncertain and incomplete information, and providing explanations of the system's reasoning process. Over 500 rules are linked together through shared antecedents and consequents. For example, the system links the consequent of rule R concerning the identity of the organism with the antecedents of therapy rules that need to know the identity of the organism. Similarly, any rules that concluded that the type of the infection is bacterial gets linked to the corresponding clause in the antecedent of rule R. A rule can be thought of as a discrete function which determines the value of a consequent variable (such as the identity of the organism) from the values of other variables in the network. MYCIN, written in LISP, is goal-driven, using backward chaining to find the values of variables.

The cumbersome interface, particularly the lengthy and error-prone question-and-answer process, is a reason why the original MYCIN system has been hardly used by medical doctors. Anyhow, MYCIN has a strong influence on the research into reasoning processes in medical diagnosis and even other applications.

PROSPECTOR

At SRI International, geological exploration expertise was captured in a

4.5 APPLICATIONS OF EXPERT SYSTEMS

consultation program, called PROSPECTOR [Har78, Dud79, Gas82]. The knowledge is based on geological rules which form models of ore deposits (including massive sulfide, carbonate lead/zinc, porphyry copper, nickle sulfide, sandstone uranium and porphyry molybdenum) and a taxonomy of rocks and minerals. Information about rock types, minerals, etc., provided by the user, is matched against a set of models of geological structures containing different minerals and metal compounds. At any stage of the process, the user may provide new data, change existing information or request an evaluation from the system.

PROSPECTOR's inferences are based on the use of certainty factors and the propagation of probabilities associated with the data. In its interactive mode, PROSPECTOR gathers evidence by backward chaining, looking for the data-level node that has the greatest potential for changing the probability of a top-level node. As the user provides information, the system will update the probabilities for each proposition according to Bayes' rule. This results in a data-driven propagation of information through the inference network.

PROSPECTOR, implemented in Interlisp, uses a combination of rule-based and semantic network formalisms to encode the knowledge.

Name	Application	Subsection
MECHO	Mechanics	4.5b
PEACE	Circuit design	4.5c
DFT	Circuit testability	7.3c
FOREST	Fault diagnosis	4.5c

Fig. 4.12 Logic programming systems

MECHO
MECHO, developed by Bundy *et al.* of Edinburgh University is an expert system for solving mechanics problems [Bun79, Lug81]. The problems which deal with point masses, inclined planes, strings and pulleys are stated in English. MECHO statements are based on predicate calculus and Prolog. MECHO can control its search procedure by developing meta rules. For solving a mechanics problem, MECHO first invokes sets of schemata that provides semantic knowledge for the particular problem domain. Predicate-calculus statements represents the information content of the sets of schemata. After creation of the knowledge base, an algorithm using means-ends analysis is invoked to produce sets of simultaneous equations sufficient for solving the problem.

CASNET
CASNET (Causal ASsociated NETwork), developed at Rutgers University, is a system for diagnosing glaucoma, an eye disease [Wei84]. The knowledge is

represented by a dynamic semantic network that models causally linked pathophysiological states. The nodes represent system states and the arcs the causal relations. The diagnosis strategy is to find a pattern of causal pathways between pathophysiological states observed in the patient. The pathways are related to classification tables containing diagnostic categories so that the pattern can be identified with a disease category. The causal model facilitates prediction of the development of the disease in a range of treatment circumstances.

Name	Application	Subsection
CRIB	Fault diagnosis	4.5c
CASNET	Glaucome diagnosis	4.5b
SPEX	Molecular biology	4.5b
WHEEZE	Pneunomia diagnosis	4.5b
MOLGEN	Molecular genetics	4.5b
RABBIT	Database queries	4.5b
PALLADIO	VLSI design	5.2d

Fig. 4.13 Object-oriented expert systems

MOLGEN
MOLGEN, developed at Stanford University, is a prototype expert system for experiments in molecular genetics [Ste81]. It assists the genetist in planning gene cloning by using the concept of constraint propagation. Knowledge about genetics is used to create a plan for specific laboratory steps. MOLGEN determines which task decompositions from a set of alternatives are still consistent with constraints generated in other steps of the plan.

MOLGEN provides different control strategies within a single system. The "least-commitment" approach takes precedence, but other heuristic strategies are available to take over when it fails. Least commitment means that decisions should not be made prematurely but postponed until there is enough information. MOLGEN treats interactions between subproblems as constraints. The generation of new constraints from old ones is called constraint propagation. The objective is to reach a situation in which the constraints are satisfied. A meta-planner reasons with constraints and alternates between least-commitment and heuristic strategies to solve the problem. MOLGEN uses an object-oriented and frame-based representation and control scheme and it is implemented in LISP and UNITS (an object-oriented representation).

SPEX
SPEX (Skeletal Planner of EXperiments), developed at Stanford University, assists scientists in developing a plan for complex laboratory experiments [Iwa82]. It is implemented in the frame-based language UNITS.

4.5 APPLICATIONS OF EXPERT SYSTEMS

WHEEZE
WHEEZE, developed at Stanford University, diagnoses lung diseases by interpreting results acquired from pulmonary function tests [Smi84].

RABBIT
RABBIT, developed at XEROX PARC, assists users of a database in formulating appropriate queries [Tou82]. It is implemented in Smalltalk and operates on databases represented in KL-ONE.

4.5c Expert Systems in Electronic Engineering

Expert Systems in Microelectronic Chip Design
CADCAS has abundantly been using algorithmic tools. However, a completely algorithmic approach to flexible chip design from design specification to the lowest chip level is not likely to become the dominating design approach in the future for the following reasons:

a. A silicon compiler (see Subsection 5.3b), which is supposed to automate the whole design process by a sequence of algorithmic procedures has the restriction that a fixed design methodology and a predefined architecture are used. This is a serious constraint if flexibility in design is a desirable feature.
b. When system complexity increases, algorithmic tools can be used at specific levels for accomplishing restricted subtasks. A human designer is needed to connect results at one level to the next level. The designer usually takes the freedom to select one appropriate solution out of a number of alternatives.

It is this choice out of many possibilities that inhibits a complete design process to be totally algorihmic. By definition, an algorithm involves a fixed sequence of instructions leading to one solution (if there is one). In contrast, a characteristic feature of a real-world chip design is that virtually an infinite number of solutions exist in the solution space. From these solutions, which satisfy given criteria within certain constraints, a number of trade-offs are made leading to an appropriate solution, which is acceptable in terms of operating speed, dissipation and cost.

Armed with an appropriate amount of design experience, a human designer uses heuristics to handle the non-algorithmic subtasks in the design process. As indicated in Fig. 1.4b, expert systems are very well suited to take over these human subtasks. Expert systems are being developed and used in virtually all phases of the chip-design process. In the subsequent chapters, the most important developments of expert-system applications in CADCAS will be discussed.

In the present subsection, an anthology of expert systems developed for applications in various aspects of electronic and computer engineering will be given below. The main attributes (knowledge representation, tool or language, given below. The main attributes (knowledge representation, tool or language, origin) of each of these systems are collected in tables. A short description of each system is given to present its main features. The expert systems to be considered are divided into three classes: diagnostic systems, design systems and miscellaneous.

Diagnostic Systems

The table in Fig. 4.14 gives a list of diagnostic systems which help diagnose and locate faulty components in electronic equipment, computers and telecommunication networks.

Name	Knowledge Representation	Tool/Language	Origin
FOREST	Frames/Clauses	Prolog	Univ.Pennsylvania/RCA
CRIB	Semantic nets	CORAL66	ICL/RADC/Brunel Univ.
IDT	Rules (forward)	OPS5/Franz LISP	DEC
ET	Rules	OPS5	DEC
IN-ATE	Rules	Franz LISP	Aut. Reasoning Corp.
FG502-TASP	Objects	Smalltalk-80	Tektronics
FIS	Rules	Franz LISP	US Naval Res. Lab.
TEMPIC	Rules	KEE	Temple University
BDS	Rules (backward)	LES/PL1	Lockheed (Palo Alto)
ACE	Rules (forward)	OPS5/Franz LISP	Bell Labs (Whippany)
SHOOTX	Objects	PEACE	NEC
NDS	Rules	ARBY/Franz LISP	Smart Sys. Tech./Shell
ART-FUL	Rules	ART/LISP	RCA (Burlington)
IPT	Frames/Rules	HP-RL	Hewlett-Packard
AESOP	Objects	Hyperclass	Stanford University

Fig. 4.14 Diagnostic systems

When the knowledge is represented by production rules, the way of reasoning (backward or forward chaining) is specified. The table also contains the tool or language used and the institution from which the expert system originates.

FOREST is an expert system for the detection and isolation of faults in electronic equipment [Fin84]. Particularly, it handles the faults that cannot be detected by current Automatic Test Equipment diagnostic software. The knowledge base contains three categories: heuristic *if* ... *then* ... rules based on experience, information obtained from circuit diagrams, and general electronic troubleshooting meta-rules. FOREST addresses problems involving multiple faults and problems caused by components or systems which gradually drift out of calibration. It combines an object-oriented (frame-based) representation

4.5 APPLICATIONS OF EXPERT SYSTEMS

reasoning with certainty factors and an explanation-generation system.

CRIB (Computer Retrieval Incidence Bank) is a computer engineer's diagnostic aid, described by Hartley [Har84] of Kansas State University. Its highly flexible, user-friendly and pattern-directed inference system is adequate for both hardware and software fault diagnosis. The input to CRIB is a quasi-English description of the symptoms of the fault. CRIB matches these symptoms against the known fault conditions stored in the database. The (sub)system is split into faulty and non-faulty parts. The testing procedure is performed down the hierarchy of subunits until the location of the fault is sufficiently pinpointed so that the faulty module can be repaired or replaced. A new, more flexible version of CRIB, called NEOCRIB, is able to adequately explain its reasoning [Ker87]. NEOCRIB is implemented in Franz LISP on a VAX 750 under UNIX.

IDT (Intelligent Diagnostic Tool) helps technicians locate defective components inside a PDP computer system [Shu82]. It uses knowledge about the unit under test, such as the functions of its components and the interrelations between the components. IDT can be run in either interactive or automatic mode. A choice from a number of operating modes can be made depending on the skill of the technician. In automatic mode, IDT performs diagnostic tests and interprets the results. The selection of the tests is based on the outcome of previous results. Figure 4.15 shows the block diagram of the entire IDT testing system. The IDT expert system is housed within a remote computer VAX11/780. Tests are applied to the unit under test via a local computer PDP11/03, which provides the interface with the user. This computer receives commands from the user and test results are displayed on a screen.

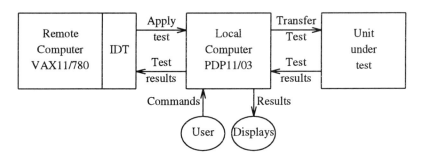

Fig. 4.15 The IDT system

IDT has a mechanism for interpreting the results, (by "reasoning" from the interpreted results which subfunctions may be faulty) and a strategy for choosing the next test to be applied. The reasoning capability is achieved through the use of symbolic formulae which are used to represent the relationship between the test results and the faulty system. Test procedures are run under the single fault assumption. The IDT system has been successfully applied by Digital Equipment

Corporation to diagnose faults in the floppy disk subsystem in PDP11/03 computers.

ET (Expert's Toolkit) assists test technicians to diagnose problems in computer modules [Che89]. It contains the Knowledge Acquisition Tool (KAT), which enables human experts to capture problem-solving strategies in a knowledge base, and the Intelligent Decision Engine (IDE), which can be used to pinpoint defects in failing computer modules. See Fig. 4.16.

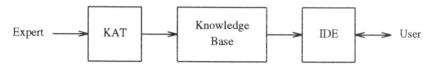

Fig. 4.16 The ET system

A characteristic feature of ET is that the knowledge base can be easily updated whenever appropriate, while a user with little test experience can handle a diagnosis problem. ET reasons by manipulating shallow knowledge. The expert knowledge is represented in a decision tree with a question at the root and a repair recommendation at a leaf node.

IN-ATE, recently renamed I-CAT, has been developed for diagnosing faulty circuit boards [Can83]. Rules are derived from hierarchical representations of the circuit schematic and of component and block functions of the circuit. I-CAT uses reliability information to propagate probabilistic measures of belief for each component in a unidirectional path across the circuit. It offers the prospect of constructing a diagnostic system based on the knowledge of the circuit schematic, with the ability to incorporate repair information as it becomes available. Experience with I-CAT emphasizes the importance of a close interaction between the developers and the user community [Ken89]. Micro IN-ATE is an Apple Macintosh version of IN-ATE [Mil89].

FG502-TASP (FG502 function generator Troubleshooting Assistant System Prototype) assists technicians in the diagnosis of malfunctioning Tektronix FG502 function generators [Ale84].

FIS (Fault Isolation System) assists technicians in locating faults in a piece of analog electronic equipment [Pip86]. Given adequate knowledge of the electronic unit to be tested, a knowledge engineer uses FIS to create a computer model of the unit. The model is then used to recommend tests which can be used for fault diagnosis. The knowledge in the model includes such information as causal rules for each module, relative *a priori* probabilities of failure of modules, connectivity and function descriptions. Test results are analyzed by a technician until faulty replaceable modules are identified. Using a probabilistic reasoning method, faults are isolated to the level of amplifiers, power supplies and larger modules. FIS is able to reason qualitatively from a functional model of the

4.5 APPLICATIONS OF EXPERT SYSTEMS

system, with recourse to numerical simulation.

TEMPIC (TEMPle Integrated Circuit diagnosis system) is an expert system, which assists process and device engineers in determining the origin of fabrication faults in VLSI CMOS chips [Sul90]. Two classes of data are used:
a. parametric data (electrical parameters measured from test chips), and
b. process control data, used to monitor the processing steps.
A Data Interpretation Module, which includes an expert system, resides within the TEMPIC system with the function to interpret the test results. Analysis of parametric data may indicate potential problems of different weights in various parts of the fabrication process. The part with the highest weight is considered as the origin of a fabrication fault. The process control information is used to verify if this fault indeed is most likely to occur.

BDS (Baseband Distribution Subsystem) helps locate faulty modules large signal-switching network [Laf84]. The diagnosis is based on the strategies of the expert diagnostician and knowledge about the structure, function and causal relations of the components in the electronic device. BDS is implemented in the LES (Lockheed Expert System) language.

ACE (Automated Cable Expertise) is an expert system for troubleshooting telephone cables in a certain area [Ves83]. Fault diagnosis is based on theoretical knowledge and accumulated experience about cable faults. ACE runs automatically on AT&T 3B-2 microcomputers, which are activated by the mainframe's UNIX operating system. Every night, a cable repair administration computer program CRAS generates trouble reports, listing where and at what time cable failures occurred. ACE analyzes the day's trouble reports and draws up a maintenance plan for repairing the faulty cables. It may also suggest preventive maintenance and the type of maintenance most likely to be effective. A database stores its recommmendations and network analysis strategies to facilitate decisions to be made in the future.

SHOOTX is a troubleshooting expert system for the NEC NEAX61-series telephone switching system [Wad88]. It assists inexperienced maintenance technicians in performing diagnosis tasks. The system enables location and repair of a fault which cannot be detected by built-in diagnosis functions. Suspect components are figured out, based on an abstract signal-flow model of the target switching system. Structure, symptom and test knowledge are represented in a unified network based on PEACE (Prolog-based Expert AppliCations Environment).

NDS (Network Diagnosis System) is an ARBY-based expert system for fault isolation and correction in a nationwide communications network [Wil83]. The system handles multiple component failures (either dependent or independent) and intermittent failures. The knowledge base contains different types of knowledge, including the topological structure of the communications network, geographical organization, and frequency of failure information. Fault isolation is performed as a heuristic search through a space of successively more refined hypotheses. The available diagnostic tests impose a refinement hierarchy on the space of hypotheses, enabling the exploitation of hierarchical search. Back links to more general hypotheses at higher levels in the refinement hierarchy are introduced to ensure the isolation and repair of multiple and intermittent failures.

ART^{TM}-FUL is a rule-based diagnosis system for multiple-link, multiple-resource communication systems [Bra86]. Implemented using Inference Corporation's ART software on a Symbolics 3670, it is capable of forward as well as backward reasoning. Both single and multiple faults can be detected. The objective of designing ART^{TM}-FUL is to overcome the weaknesses inherent in the go-chain method. Go-chain involves a predefined sequence of pass-or-fail tests, which cause the program to terminate as soon as a fault is detected. The diagnostic subsystem is part of a larger expert system used in mobile military communications shelters. Its job is to verify lack of functionality, isolate problems to specific pieces of equipment or transmission media and recommend corrective action.

IPT (Intelligent Peripheral Troubleshooter) is an expert system that diagnoses malfunctions in Hewlett-Packard disk drives and other peripherals [Got86]. Its aim is to predict failures before they happen, based on aspects of the current state of the peripherals. The knowledge base contains both forward and backward chaining rules. IPT uses a peripheral-independent inference engine. Once the diagnosis has been completed, IPT instructs the user how to proceed with the repair.

AESOP [Dis89] is a simulation-based knowledge system for diagnosing CMOS process problems using parametric test data from end-of-line test structures. The source of knowledge which supports AESOP process diagnosis is founded directly on fundamental principles of semiconductor device physics and processing theory instead of human experiential knowledge. This was accomplished through extensive use of process and device simulation. Th diagnosis stage invokes the diagnostic reasoning mechanism of HyperPIES. It derives a set of failure candidates which can account for deviations observed in the results of electrical tests. This project was supported by Intel, Schlumberger, SRC and DARPA.

Pan et al. [Pan89] described a framework for knowledge-based computer-integrated manufacturing. Its goal is to demonstrate the ability to capture manufacturing knowledge and use it to design, simulate, monitor, control, diagnose, and schedule semiconductor maufacturing operations. The testbed for this framework is the semiconductor fabrication line at Stanford's Center for Integrated Systems. The SMART (Semiconductor Manufacturing Analysis and Reduction Tools) system is another application of AI to semiconductor manufacturing, described by Murphy Hoye [Mur86].

4.5 APPLICATIONS OF EXPERT SYSTEMS

Design Systems

The algorithmic approach to system design is unsuccessful when many constraints must be taken into account. Usually, a design problem has a large number of possible solutions from which the optimal one must be selected. Heuristics, as contained in expert systems, can speed up the search for the solution. Figure 4.17 lists a number of expert systems developed for design purposes in electronic and computer engineering. The knowledge representation, the tool or language used and the place of origin are also given.

Name	Knowledge Representation	Tool/Language	Origin
EL	Rules (forward)	ARS/MACLISP	MIT
XCON	Rules (forward)	OPS5	CMU/DEC
MICON	Rules	OPS5	Carnegie-Mellon Univ.
EURISKO	Objects/Rules	Interlisp	Stanford Univ.
PEACE	Pred. Calculus	Prolog	Univ. of Manchester
SADD	Frames	LISP	Univ. of Maryland
TSS	Frames/Objects	LOOPS/Interlisp	Stanford University
IBIS	Rules	FORTRAN77/Franz LISP	Univ. of Genoa

Fig. 4.17 Design systems

A short description of the expert systems mentioned in Fig. 4.17 is given below.

EL employs the principle of constraint propagation to analyze the steady-state behavior of electrical circuits containing resistors, transistors and diodes [Sta77]. EL's knowledge base exploits familiar electrical laws, notably Ohm's law and the Kirchhoff laws. Given the circuit schematic, EL determines the values of node voltages and branch currents in the circuit. A rule-based representation is used with forward chaining. The system is written in ARS (Antecedent Reasoning System), which allows the use of a rule-based representation scheme with forward chaining. See also Subsection 6.3b.

XCON (eXpert CONfigurer for computer systems), a cooperative development of Carnegie-Mellon University and Digital Equipment Corporation, is one of the widely used expert systems implemented in the OPS5 language (using on the order of 5000 rules) and currently operating on a commercial basis [MCD82]. It configures all VAX computer systems as well as the full range of PDP-11 computers for DEC, that is, it selects computer components and interconnects them to meet a customer's purchase order. A floor plan for the CPU cabinet and components is made to facilitate cabling connections. The configuration problem deals with a relatively small number of components (e.g., boards and cabinets) and highly structured ways of assembling them. Using a simple control structure, a small fraction of the knowledge rule base is active at any one time with limited backtracking. Given a customer order, XCON produces diagrams showing the spatial and logical relationships between the many components that comprise the complete system. XCON is a member of a family of expert systems for configuration at Digital Equipment Corporation [Bar89].

MICON (MIcroprocessor CONfigurer) is an expert system that supports the design and configuration of circuit boards, specifically single-board computers [Bir84, Bir88]. Given the specification of requirements, MICON derives the best implementation, including the CPU. One of the processor types stored in the system library (e.g., iAPX286, Z80, TI9900, 8086) must initially be selected, followed by a selection of memories, controllers and I/O devices. When the configuration is known, the components are connected according to the specification. The configuration requires no backtracking because of the hierarchical procedure, e.g., a selection of a processor defines the type of memory or the I/O devices which may be used. The design is completed by examining and tuning the performance, characteristics, including cost, power consumption and timing constraints. The synthesis stage is followed by placement and wiring of the components on printed circuit boards. See also Subsection 5.2d.

AM is an expert system for exploring the domain of elementary mathematics [Dav82]. EURISKO is an improved version of AM with extended applicability to other domains [Len83]. One particular application is the design and fabrication of new kinds of three-dimensional microelectronic devices using recrystallization techniques. One of the objectives of EURISKO is to invent improved semiconductor devices by exploring alternative device configurations and comparing these with existing devices. An interesting feature of EURISKO is its learning ability. New concepts in a problem domain are created by applying heuristic rules to existing concepts and even new heuristics can be learned. This was made possible by using the object-oriented language RLL-1. The EURISKO system has been implemented for the Xerox 1100 series workstations.

PEACE is a knowledge-based design tool that performs analysis and synthesis of analog and digital circuits [Din80, Ste83]. The input is a functional description for an analog circuit or a logic expression for a logic circuit. The knowledge base contains information on basic circuit components, rules for circuit transformation (e.g., delta <-> star) and heuristics for anticipating failures.

SADD (Semi-Automatic Digital Designer) is an interactive knowledge-based system for designing digital electronic circuits [Gri80]. The knowledge base contains generic information for each high-level digital module (e.g., counter, clock) and the methods needed to transform functional descriptions into

4.5 APPLICATIONS OF EXPERT SYSTEMS

realizable circuits. From the description in English, a circuit model is constructed and a structured, modular circuit is designed via an interactive user interface. The design of a TV video display circuit is shown as an example.

Transistor sizing, e.g., the selection of the proper length/width ratios of pullup and pulldown transistors in NMOS technology, determines performance characteristics, such as switching delay, power consumption, etc. It provides requirements for the generation of the fabrication masks. Since switching delay is inversely related to power consumption and capacitance, an expert system can be used to find a suitable compromise among these parameters. Foyster [Foy84] of Stanford University developed TSS (Transistor Sizing System), which selects a suitable trade-off between power and delay in an NMOS circuit. This system is embedded in the PALLADIO environment for logic circuit design (see Subsection 5.2d). Another system which provides transistor sizing is EXCIRSIZE, developed by Dawson [Daw86] of DEC, Hudson.

Miscellaneous

Expert systems for a variety of applications in electronic and computer engineering are collected in Fig. 4.18.

Name	Knowledge Representation	Tool/Language	Origin
SOPHIE	Rules	Interlisp/FORTRAN	UCI/Xerox
CADHELP	Rules	Franz LISP	Univ. of Connecticut
MTA	Rules	Prolog	Univ. Waterloo
TIMM/Tuner	Rules	FORTRAN	General Research
YES/MVS	Rules (forward)	OPS5	IBM
XSEL	Rules (forward)	OPS5	CMU/DEC
PTRANS	Rules	OPS5	DEC
ISA	Rules (forward)	OPS5	DEC
MIXER	Clauses	Prolog	Tokyo Univ.
PEX	Frames	Prolog/C	Lehigh Univ.
IBIS	Rules	FORTRAN77/Franz LISP	Univ. of Genoa
PEDX	Rules	LISP/Explorer	Texas Instruments

Fig. 4.18 Miscellaneous systems

These expert systems will be described briefly.

SOPHIE (SOPHisticated Instructional Environment) arises from the desire to have an interactive learning environment in which the implications of hypotheses during a problem-solving process are evaluated and the user receives immediate detailed response to ideas communicated to the system [Bro82]. Electronic troubleshooting in analog circuits is recognized as a particularly good application domain. SOPHIE enables one to insert arbitrary faults into the circuit and watch the system perform its troubleshooting task. An interesting feature is that the system explains its strategies for attacking the problem. SOPHIE has evolved through three versions. SOPHIE III consists of three major expert modules: the electronics expert (containing the local constraint propagator LOCAL and the submodule CIRCUIT for handling the circuit-specific knowledge), the troubleshooter (which evaluates hypothetical measurements for choosing the most informative one), and the coaching expert (which determines whether or not to interrupt or advise the user). SOPHIE encourages the user to make suggestions as to what should be the next step to take in the fault-finding process, while the system acts as a tutor monitoring and critiquing the user's decisions and giving advices and explanations whenever asked for.

CADHELP (Computer-Aided Design HELP) simulates an expert demonstrating the operation of the graphical features of a computer-aided design (CAD) system [Cul82]. It consults a knowledge base of feature scripts to explain a feature, generate prompts as the feature is being operated and to give certain types of help when a feature is misused. Research has been conducted into the integration of natural language with graphical capabilities within the framework of a CAD system developed for the design of digital logic circuits. CADHELP consists of two major subsystems: the CAD tool which interacts with the user during design and an explanation mechanism which can describe the operation of the CAD tool to users in varying levels of detail. A data tablet is used as the input medium, while outputs are displayed on a graphics screen. CADHELP runs on a DEC VAX 11/780 under UNIX.

MTA (Message Trace Analyzer) helps debug real-time systems such as large telecommunication switching machines containing hundreds of processors [Gup84]. The system examines interprocess message traces, identifying illegal message sequences to localize the fault to within a process. The system considers the sender process ID, the receiver process ID, the message type, and the time stamp fields of messages in the trace. General debugging heuristics and facts about the specific system being debugged are represented as rules and applied using both forward and backward chaining. The system contains a limited explanation facility that allows it to answer questions about its reasoning.

TIMM/TUNER (The Intelligent Machine Model TUNER) assists in tuning VAX/VMS computers to yield maximum performance [Kor84]. It was built using TIMM (The Intelligent Machine Model), an expert system development tool, which is notable for automated question-and-answer knowledge acquisition and flexible inferencing. Iterative refinement and improvements of the expert knowledge have been made with little effort.

YES/MVS (Yorktown Expert System for MVS operators) is a continuous, real-time expert system that helps computer operators monitor and control the MVS (Multiple Virtual Storage), IBM's operating systems for large mainframes

4.5 APPLICATIONS OF EXPERT SYSTEMS

[Gri84, SCh86]. The following six subdomains of operator activities are implemented: JES (Job Entry System) queue space management, monitoring and correcting problems in channel-to-channel links, scheduling large batch jobs off prime shift, detecting MVS hardware error, monitoring software subsystems, and performance monitoring. YES/MVS routinely schedules the queue of large batch jobs. Network link problems are sent immediately to the console operator.

XSEL (eXpert SELling assistant) is a salesperson's assistant that permits a potential customer to specify what components of a computer system are needed, considering the constraints due to space [McD82]. XSEL is an interactive system, which uses information from XCON. It checks the user requirements for consistency, selects a central processing unit, primary memory and peripheral devices and the software, and then asks XCON for the physical configuration, including a floor plan. Much of XSEL's knowledge base is used to assist the buyer in the selection process, providing the salesperson with explanations of the buyer's proposals.

PTRANS supports the manager in the production of DEC computer systems as ordered by customers [McD84]. It must be able to perform two different tasks simultaneously: (1) assist in determining when, where and how each system should be built, and (2) recognize events which make implementing such a build-plan infeasible and make appropriate modifications to that plan. The management task involves quite diverse subtasks, which are performed by PTRANS' cooperating assistants. PTRANS is designed to work with XSEL in order to keep the delivery date in conformity with the order's specifications.

ISA (Intelligent Scheduling Assistant) is an expert system that schedules all of DEC's customer orders. The database of ISA's prototype system [Orc84] contained information on orders, material availability and the customers. The input to the system were customer orders, change orders, and cancellations. The output was a schedule date for each order, corrections when necessary for certain administrative errors, reasons for schedules not in the customer request month, and in some cases alternative schedules. Later versions of ISA contain substantial expert knowledge and effective scheduling strategies.

MIXER helps programmers write microprograms for the Texas Instruments' TI990 VLSI chip.

PEX (Package EXpert) is an expert system for designing suitable packages for chips [Vor89]. Existing databases contain design rules which are used to reason about properties of chips and potential packages, including dimensions, possible materials and packaging requirements. PEX uses a frame-based hierarchical knowledge-management approach to databases, where frames serve as the interface between rule-based knowledge and the databases.

IBIS (Interpretation of Biomedical Images of the Slice type) is a knowledge-based system for the recognition of tomographic images of the head obtained by means of nuclear magnetic resonance (NMR) [Ver87]. Starting from conventional image processing, IBIS can recognize slices and locate automatically their principal anatomical organs in the NMR images. The knowledge, with 3-D and 2-D aspects for anatomical description, is represented by an extended semantic network. The control structure consults production rules and, according to their content, activates the various processing procedures in order to attain the recognition goal.

PEDX (Plasma Etch Diagnosis eXpert system) is a rule-based system which detects and diagnoses problems during a plasma etch process in semiconductor manufacturing [Dol88]. The material removed from a silicon wafer contains an amount of particular chemicals (called endpoint traces) which can be measured by emission spectroscopy. PEDX interprets the traces to detect problems due to defective wafers and determines their causes during the etch process. Several other rule-based systems have been developed for the diagnosis of failures during the chip-manufacturing process [Mit89, Ars91].

Expert systems have been and are being used in a diversity of environments, including capacity allocation in a multiservice application environment [Erf91].

Chapter 5

VLSI SYSTEM DESIGN AND EXPERT SYSTEMS

5.1 LOGIC-CIRCUIT DESIGN

5.1a Logic Modules

Logic Circuits and Systems

Many logic system designs are based on standard MSI functions, such as gates, flip-flops, latches, registers, multiplexers, barrel shifters, parity generators, comparators and counters. Also LSI module functions, like 8-bit RALU slices, 8-bit sequencer slices, adaptable-size PLAs and 256 × 9-bit static RAMs are included, as well as various MOS/TTL drivers.

Logic circuit or system design is discussed in a wide variety of books [McC86, Puc88]. MOS technology is playing a dominant role in VLSI circuit and system design [Mea80, Gla85, Wes85]. Within certain limitations, the scaling property, as device feature sizes shrink to submicron dimensions, is extremely useful in designing complex systems.

Principles of computer hardware are treated elsewhere [Cle91]. In this subsection, an incomplete selection of logic modules is given.

Decoders

An N-to-2^N *decoder* transforms an N-bit binary code at its input to a 2^N-bit singular code at the output. Whereas a binary code may contain any number of 1's and 0's, a singular code consists solely of 0's with the exception of a single bit which is 1. One of the 2^N output lines is forced to accept the logic-1 state depending on the address code present at N INPUT lines.

Example: The binary inputs and singular outputs of a 1-to-2 decoder and its possible implementation are shown in Fig. 5.1.

Input	Outputs	
a	b_1	b_0
0	0	1
1	1	0

(a)

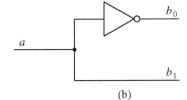

(b)

Fig. 5.1 1-to-2 decoder

Example: The binary inputs and singular outputs of a 2-to-4 decoder and its possible implementation are shown in Fig. 5.2.

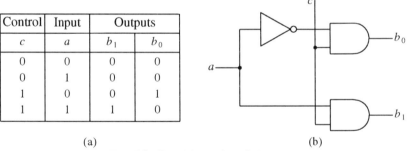

Inputs		Outputs			
a_1	a_0	b_3	b_2	b_1	b_0
0	0	0	0	0	1
0	1	0	0	1	0
1	0	0	1	0	0
1	1	1	0	0	0

(a)   (b)

Fig. 5.2 2-to-4 decoder

A conditional decoder can be enabled and disabled by additional control inputs.

Example: A conditional 1-to-2 decoder is shown in Fig. 5.3. When $c = 0$, the decoder is disabled and the output consists of 0's. When $c = 1$, the decoder is activated so that it functions equivalently as the implementation of Fig. 5.1b.

Control	Input	Outputs	
c	a	b_1	b_0
0	0	0	0
0	1	0	0
1	0	0	1
1	1	1	0

(a)   (b)

Fig. 5.3 Conditional 1-to-2 decoder

The 2-to-4 decoder can be disabled by introducing third inputs to all gates which, connected together, can be used as the disabling control input.

Multiplexers

A *multiplexer* (sometimes called *data selector*) is a combinational logic subsystem that routes one of its $M = 2^N$ input signals (DATA SOURCES) S_1 ... S_M to only one output (DESTINATION) D. The selection of the wanted input source is accomplished by N DATA SELECT lines C_1 ... C_N. Typical multiplexers are 4-to-1, 8-to-1 and 16-to-1 multiplexers.

5.1 LOGIC-CIRCUIT DESIGN

Example: Figure 5.4a shows a circuit which realizes a 2-to-1 multiplexer. Figure 5.4b shows the situation when $C = 0$ and $C = 1$ respectively. The selection of which input signal must be routed to the output is accomplished by a 1-to-2 decoder, which is shown in Fig. 5.4a within broken lines.

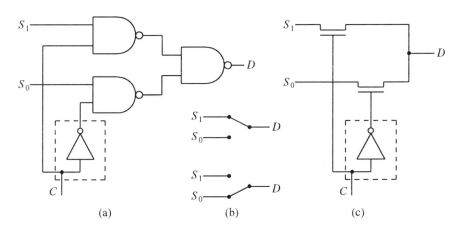

Fig. 5.4 2-to-1 multiplexer

Figure 5.4c gives the NMOS implementation of the 2-to-1 multiplexer with one comtrol input C. In the case of an N-to-1 multiplexer, a singular address code $C_1 \ldots C_N$ can be used to route the wanted input to the output.

Example: A 4-to-1 multiplexer is shown in Fig. 5.5. The position of the 1 in the singular code $C_3 \ldots C_0$ determines which input S_i, $i = 0 \ldots 3$, is connected to the output D.

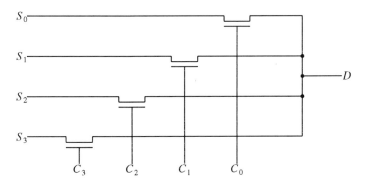

Fig. 5.5 4-to-1 multiplexer circuit

Demultiplexers

A *demultiplexer* has a function opposite to a multiplexer. It connects a single source to one of 2^N destinations. An N-bit control variable C is needed to select the required destination.

Read-Only Memories

Applications of Read-Only Memories (ROMs) are: code conversion, generation of alphanumeric characters, looking-up in data tables, microprogramming and the synthesis of combinational logic functions. For the last application, the hardware implementation consists of a decoder and the factual ROM. The decoder converts N input variables into 2^N possible minterms of these variables. In the ROM, the available minterms are used to implement given logic functions. The decoder as well as the ROM consists of an array of transistors (mostly MOSFETs) which are or are not connected to a common line.

Programmable Logic Arrays

Like ROMs, Programmable Logic Arrays (PLAs) are suitable for the synthesis of combinational functions of logic variables. PLAs give a more efficient layout, since not all possible 2^N minterms are needed, but only those minterms which appear in the logic functions. When a PLA is used for the synthesis of combinational functions, these functions are first put in the form of logic sums-of-products. The product terms are implemented in an AND array, after which the summation of these product terms is performed by an OR array, yielding the desired outputs. By placing feedback registers between outputs and inputs, sequential functions (e.g., finite state machines) can be realized.

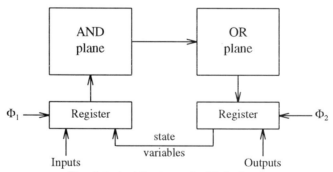

Fig. 5.6 Architecture of a PLA circuit

Like in ROMs, a PLA consists of rows (and columns) of MOS transistors which are arranged at regular distances. In this way, a very compact layout can be obtained. For realizing given logic functions, specific transistors must be connected to given common lines. The positions of the transistors and their interconnections are fixed so that placement and routing algorithms are not needed. As indicated in Fig. 5.6, a PLA circuit can be used to implement a synchronous finite-state machine with a two-phase clock (Fig. 5.26). Registers at the PLA input and output are clocked by the clock signals Φ_1 and Φ_2, while appropriate outputs (state variables) are fed back to the input.

Suppose the number of inputs N of the AND array is equal to the number of different logic variables. The number of outputs U of the OR array is equal to the number of sum functions. Hence, the number of columns of transistors in the entire PLA is $\gamma = N + U$. The number of transistor rows ρ is equal to the number of different product terms in all sum functions. Hence, the area needed for the PLA realization is proportional to $\rho \times \gamma$. In general, a large part of the transistors is not connected and hence inactive, while they occupy an amount of chip area.

There two classes of methods for minimizing the chip area required in PLA implementation:
a. Minimizing the number of rows ρ,
b. PLA folding techniques.

A method according to a. is applied in the program MINI [Hon74]. This is a heuristic program that attempts to minimize, in an number of steps, the number of terms (implicants) required in the logic sum functions, and hence ρ. The combinational-function minimization problem can be formalized as a heuristic state-space search problem [Kab88].

The term *PLA folding* [Hac82] refers to the operation by which two rows (or columns) are implemented in a single row (or column) with inputs and outputs on both sides of the row (or column). With the objective being to minimize the chip area, we must find a permutation of rows (or columns) of the PLA which allows a maximum number of column (or row) pairs to be implemented in shared columns (or rows).

5.1b Synthesis of Sequential Circuits

Design Techniques
Design of sequential digital circuits may be based on different approaches:
a. Use of programming languages, particularly hardware-description languages.
b. Finite state machine techniques, including the algorithmic state machine (see Subsection 5.1c).
c. Graph-theoretic methods, e.g., using flow-graphs and state-graphs (see Subsection 5.2b).

Finite State Machines
Synthesis is the formal approach to system or circuit design in which a well-defined sequence of steps is executed to arrive at the hardware implementation of a given design specification in a straightforward manner. This is in contrast

with *design* which is the conventional procedure of iterative improvements of the design based on heuristics and the designer's experience.

The *sequential function* that describes a sequential circuit specifies the output as a function of both the present input and previous inputs of the circuit. In other words, a sequential function defines a mapping between input sequences and output sequences. The effect of all previous signal values of the inputs is represented by a finite number of *internal states* of the circuit which will be denoted by the state vector **Q**. If the input specification is for a finite state machine, equivalent states are collapsed by successive optimizations in order to find a state encoding that minimizes the combinational logic to be generated.

When the circuit contains binary memory elements, each of which may be represented by a state variable, which can take on the value 0 or 1. The number of internal states is 2^m, where m maximally equals the number of memory elements. When the number of inputs if p, the output Z depends on $p + m$ variables. These variables define 2^{m+p} states, which sometimes are referred to as *total states*. Since 2^{p+m} is finite, the mathematical model that describes a sequential function (or the sequential circuit that realizes the function) is referred to as a *finite state machine*.

Let A_n, Q_n and Z_n denote the vectors of the inputs, present states and the outputs respectively, at time t_n, and let X_{n+1} represent the next states at time t_{n+1}. Following ideas in Huffman's work [Huf54], a finite state machine is represented as a combinational network provided with memory or delays (see Fig. 5.7).

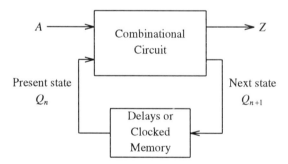

Fig. 5.7 Finite-state machine

For clocked circuits, the memory elements are clocked by the clock generator. It is assumed that the clock period is greater than the worst-case summation of delays in the combinational part (see Subsection 6.2b). Transient conditions occurring between two successive clock pulses do not affect the correct operation. In asynchronous circuits with feedback connections and memory elements, such as latches, finite delays τ_i are assumed in the feedback part. Then,

5.1 LOGIC-CIRCUIT DESIGN

an input change causes a sequence of state changes in the feedback loop until the state is stabilized.

Two models of sequential machines have been proposed by Mealy [Mea55] and Moore [Moo56], and are referred to as the Mealy machine and the Moore machine respectively. The output Z_n in the *Mealy machine* at any time t is uniquely determined by the present input A_n and the present state Q_n. The output Z_n in the Mealy machine is

$$Z_n = Z(Q_n, A_n)$$

In the *Moore machine*, the output is a function only of the present state Q_n:

$$Z_n = Z'(Q_n)$$

When an input signal A is applied to the circuit, the *present state* Q_n of the circuit will change into the next state Q_{n+1}, which is a function of A_n and Q_n. In both models, we have

$$Q_{n+1} = Y(Q_n, A_n)$$

Z or Z' and Y are referred to as the output function and the state-transition function respectively. In the case of clocked circuits, the time instants t_n and t_{n+1} refer to subsequent clock times.

Since sequential functions depend on previous inputs as well as on the present input, the truth table representation as used for combinational functions is inadequate for describing sequential functions.

State Table for Describing Sequential Functions

The functional relationships between the input, present state, next state and the output of a sequential machine are represented by a *state table* or *state-transition table* of a finite state machine. It is a general means for describing the system requirements prior to realization and displays the results of the next-state and output functions in tabular form. See Fig. 5.8.

Present state	Inputs A_j			
↓ Q_n	A_1	A_2	.....	A_p
$Q_{n,1}$	$Q_{n+1,11}/Z_{11}$	$Q_{n+1,12}/Z_{12}$	.....	$Q_{n+1,1p}/Z_{1p}$
$Q_{n,2}$	$Q_{n+1,21}/Z_{21}$	$Q_{n+1,22}/Z_{22}$	.....	$Q_{n+1,2p}/Z_{2p}$
...	...	...	.....	...
...	...	...	.....	...
$Q_{n,m}$	$Q_{n+1,m1}/Z_{m1}$	$Q_{n+1,m2}/Z_{m2}$	.....	$Q_{n+1,mp}/Z_{mp}$

Fig. 5.8 State table

Its rows correspond to the possible present states X_i and the columns correspond to all possible inputs A_j. The entry in row i and column j represents the next state $Y(Q_i, A_j)$ produced if A_j is applied while the finite state machine is in state Q_i. The entry is represented as Q_{n+1}/Z. See Fig. 5.8. The state table specifies all next states and outputs for all possible combinations of present inputs and present states. The finite state machine is a mapping of the pairs $\{Q_n, A_j\}$ into $\{$next state Q_{n+1}, output $Z\}$.

Example:

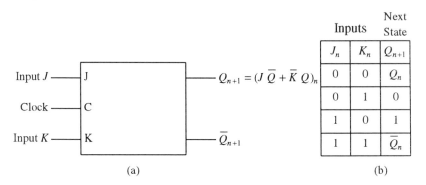

Fig. 5.9 JK flip-flop

A JK flip-flop (Fig. 5.9a) is a clocked flip-flop with inputs J and K and outputs Q and $\overline{Q}$. The output Q defines the internal state of the circuit. Given the present state Q_n and inputs J and K, the next state Q_{n+1} is given as shown in Fig. 5.9b. The state table is shown in Fig. 5.10. The present states A and B are assigned the values 0 and 1 respectively.

Q_n	JK			
↓	0 0	0 1	1 1	1 0
$A = 0$	A/0	A/0	B/1	B/1
$B = 1$	B/1	A/0	A/0	B/1

Fig. 5.10 Description of a JK flip-flop

The entries in the state table can be interpreted easily. For example, if the present state is A and the input is $JK = 10$ (the set condition), a transition from state A to state B takes place producing an output and next state of 1.

5.1 LOGIC-CIRCUIT DESIGN

State Diagram

A *state diagram* or *state-transition diagram* provides a graphical representation of the operation of a finite-state machine. It consists of nodes (depicted as circles) and directed branches. Each node corresponds to a possible present state, while each directed branch represents a state transition from Q_i to the next state with the input A_j and output $Z_k = Z(Q_i, A_j)$, indicated by a branch label of the form A_j/Z_k. For each ordered pair of (not necessarily distinct) states Q_i and Q_r, a directed branch connects node Q_i to Q_r if and only if there exists a value A_j of the input signal such that $Q_r = Y(Q_i, A_j)$. A state diagram contains exactly the same information as a state table.

There are two types of state diagrams: the Mealy model and the Moore model. In the Mealy model, each node (circle) contains the symbol of a possible state (Fig. 5.11a). Given a state Q_n and a specific input, the output can be evaluated. The branch emanating from the node with state Q_n and labeled with A_n/Z_n, i.e., Input/Output, is directed to the node for the next state Q_{n+1}. In the Moore model, the output Z is directly related to the state Q and both quantities are contained in each node (Fig. 5.11b). The branch is labeled with the input A_n, which together with Q_n leads to the next state Q_{n+1} and hence to the next output Z_{n+1}.

Fig. 5.11 Mealy and Moore models

Example:

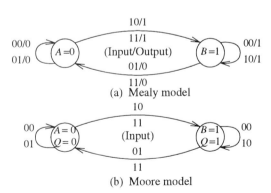

Fig. 5.12 State diagram for a JK flip-flop

The Mealy model of a JK flip-flop, which can be derived from the state table (Fig. 5.10), is given in Fig. 5.12a. Note that the output is identical to the next state. This is more expounded in the Moore model given in Fig. 5.12b.

Though in general both models are equivalent and can be translated into one another, the Mealy model is usually easier to handle. The state table given in Fig. 5.10 corresponds to the Mealy model. In the case of the Moore model, the outputs in any row would be the same since they are independent of primary inputs.

Example: Figure 5.13 shows the model of a serial binary adder which adds together two positive binary numbers represented in serial form.

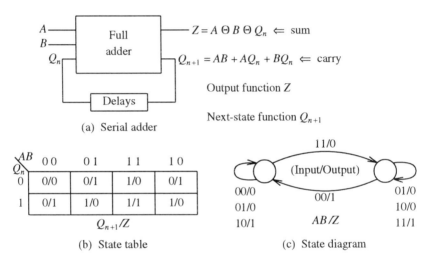

Fig. 5.13 Serial adder, its state table and state diagram

The logic functions for the sum Z and the output carry Y are included in the figure. The input carry Q_n and the output carry Q_{n+1} may take one of two possible states 0 and 1. The state table and the state diagram of the serial adder are shown in Figs 5.13b and 5.13c respectively.

Excitation Table

In analyzing sequential circuits, the next state is specified for a given input and the present state. In logic synthesis, the required transition from the present state to the next state is known, and we wish to find the input conditions that will effect the required transition. An *excitation table* lists the required input combinations for all possible changes of states $Q_n \rightarrow X_{(n+1)} = Y$.

5.1 LOGIC-CIRCUIT DESIGN

Example: Consider a JK flip-flop. This is a simple sequential circuit whose excitation table is shown in Fig. 5.14. Given the columns Q_n and Q_{n+1}, the required transition is achieved by the J and K inputs given in the third and fourth columns. A symbol X denotes a don't-care condition, i.e., it does not matter whether the input to the flip-flop is 0 or 1.

Q_n	$\to$	Q_{n+1}	J	K
0	$\to$	0	0	X
0	$\to$	1	1	X
1	$\to$	0	X	1
1	$\to$	1	X	0

Fig. 5.14 Excitation table of a JK flip-flop

Sequential-Circuit Synthesis

The problem of deriving a sequential circuit which realizes a finite-state machine specified by a state-table (or a flow table) is discussed in many textbooks [Lew77, Bol90]. The solution to this synthesis problem depends on the type of sequential circuit and the number of internal states. As an example, let us briefly touch on the synthesis of a simple clock-mode sequential circuit using a JK flip-flop.

Apart from the clock generator, a general clock-mode sequential circuit consists of flip-flops, which are memory elements, allowing the circuit to assume a finite number of states, and a combinational circuit, which generates the flip-flop excitations and outputs (Fig. 5.7). The circuit inputs A and the states Q determine the outputs Z of the circuit.

The first step in the synthesis procedure is *state assignment*, i.e., a state is assigned to each flip-flop. State assignment is the important problem of how to find coding assignments to the state variables that will yield the minimum amount of circuitry. Once the state assignment has been specified, the sequential circuit design is completed by deriving the combinational logic which implements the next-state functions $Y(Q,A)$ and the output function $Z(Q,A)$. When the number of inputs and states is small, we can derive an excitation table in the form of a Karnaugh map whose entries denote the necessary JK excitations for specific present states (corresponding to the rows) and given input values (corresponding to the columns).

Example: Consider the synthesis of a binary adder using a JK flip-flop. We are given the state table which is shown in Fig. 5.10 and redrawn in Fig. 5.15 in the form of a Karnaugh map. It suffices to use a single JK flip-flop with the state assignment $Q_1 = 0$ and $Q_2 = 1$. Using the excitation table for the JK flip-flop (Fig. 5.14), the required excitations can be entered into the table as given in Fig. 5.15.

From this table, separate maps for the required J and K excitations can be derived (Fig. 5.15c and d). Finally, the following minimized functions can be obtained. From Fig. 5.15c and d the next-state functions are

$$J = A\,B, \quad K = \bar{A}\,\bar{B}$$

From Fig. 5.15b, the output function is

$$Z = \bar{A}\,\bar{B}\,Q + \bar{A}\,B\,\bar{Q} + A\,B\,Q + A\,\bar{B}\,\bar{Q} = A \odot B \odot Q$$

AB \\ Q_n	0 0	0 1	1 1	1 0
0	0/0	0/1	1/0	0/1
1	0/1	1/0	1/1	1/0

Q_{n+1}/Z
(a) State table of serial adder

AB \\ Q_n	0 0	0 1	1 1	1 0
0	0X/0	0X/1	1X/0	0X/1
1	X1/1	X0/0	X0/1	X0/0

JK/Z
(b) Excitation table of serial adder

AB \\ Q_n	0 0	0 1	1 1	1 0
0	0	0	1	0
1	X	X	X	X

J
(c) Map for JK excitation

AB \\ Q_n	0 0	0 1	1 1	1 0
0	X	X	X	X
1	1	0	0	0

K
(d) Map for K excitation

Fig. 5.15 Synthesis of a serial adder

The general problem of synthesis of a synchronous sequential circuit with finite memory can be divided into two parts:

1. The construction of a finite-state machine which models the input-output behavior of the sequential circuit. The finite-state machine is described by its state table or its state diagram.
2. The realization of this state table by a sequential circuit made up of logic gates and flip-flops.

5.1c Data Processor and Controller Design

Introduction

For the sake of conceptual clarity, it is useful to identify a data processor (or *data section*, for short) and a controller (or *control section*). While the data section of a logic system (e.g., an ALU with its gating circuitry) has the function of performing the data processing proper, the control section has an organizing task of generating control signals so that the appropriate hardware units of the system operate in the proper sequence, as specified by a program. In a *stored-program computer*, the sequence of instructions to be executed is contained in a

5.1 LOGIC-CIRCUIT DESIGN

computer program which is stored in the computer's memory. The data section consists of registers and combinational circuitry with links to the memory such that the computer program can be executed. When the appropriate signals are sent, data transfers from register to register will take place. At times, a control signal is fed back from the data section to the control section. When necessary, an I/O peripheral unit is commanded to perform an I/O function.

One possible design approach is directed toward the hardware implementation of the data transfers between registers. Gating circuitry must be connected between pairs of registers. When a large number of registers is involved, the use of buses will simplify the design problem. While data transfers via a bus may imply large numbers of bits (e.g., 32 bits), a control-signal transfer requires only one bit or a small number of bits.

Since the transfer of data between the registers in the data section is essential, the data section is often called the *data path*. Likewise, the controlled section, which provides the time-dependent control signals that cause the data transfers among the registers to take place in a proper sequence is referred to as the *control path*. A careful design of the data path and the control path is essential for the quality of the total system.

There have been several attempts to formalize the logic-system-design process into a straightforward procedure. Clare [Cla73] introduced the use of an ASM (Algorithmic State Machine) chart as an intermediate phase between the conceptual phase of a design and the actual circuit implementation. An ASM is a model for describing the input-output behavior of any module of a subsystem. An ASM chart is a diagrammatic description of the output function and the next-state function of an ASM. When a satisfactory ASM chart for the design problem at hand has been found, the chart must be transformed into a hardware implementation in some logic family or a microprogrammed logic machine.

Following Davio *et al.* [Dav83], a logic system can be viewed as a machine which can carry out an algorithm. Since an algorithm implies a general sequence of data-processing steps, the designer of the logic system must decide which hardware and software are needed to execute these steps. Once the basic system organization has been established, we must decide on how to control the sequence of data transfers and processing operations prescribed by the algorithm.

Two components of the algorithm are realized on the data part of the design:
a. Assignment statements concerned with arithmetic and logic operations.
b. Relational operations used by conditional statements.

Boolean assignments to external or internal units are performed directly by the controller. External or internal relational variables that are already Boolean are connected directly. Thus,

                count := count + 1

and   count < 10

are data-section tasks. The objective is to realize the assignments so that single control lines (instructions) may invoke the assignment from the control section and the relational operations so that single Boolean variables (qualifiers) may represent any complex relationship. The design is clearly a simple logic representation taking no account of implementation constraints. Intuition is still the key design skill although it should be recognized that a complex data part subfunction may well be realized as a combined data/control section pair which interprets instructions from the higher level. The control section of the design implements the control sequence of the algorithm using the well-established model of a finite state machine.

Data processor and controller design can be accomplished by several methods:
a. The ASM (Algorithmic State Machine) method.
b. The use of PLAs or PALs.
c. The use of ROMs (Read-Only Memories).

Algorithmic State Machine

For complex synthesis problems, the construction of the state diagram becomes cumbersome and we need an algorithmic tool that specifies which state transitions and outputs are produced for all possible inputs at each possible inputs at each possible present state. A useful tool for this purpose is provided by the *Algorithmic State Machine* (ASM), which gives an algorithmic specification with a flowchart notation for the required operations of the finite state machine [Cla73]. The ASM chart not only provides a sequence of control operations bearing a strong resemblance to the conventional software flowchart, but it also provides precise values of the time durations involved in the sequence of events.

Given the present state of the system, the ASM should unambiguously determine the next state at the proper time and produce the required output signals for any values of the input variables. A state is described within a rectangular box with the symbolic state name enclosed in a small circle adjacent to the upper left corner of the box. As in software development, a conditional branch is represented by a diamond symbol with a TRUE and FALSE output. Outputs may either follow the Moore model or the Mealy model. In the former case, the output is directly associated with the state and indicated in the state box. In the Mealy case, the output is generated as a result of input and present-state conditions and indicated in a separated oval-shaped symbol.

Example: Figure 5.16 shows an ASM chart in which the above-mentioned symbols are employed. The output SP in state *P* is independent of the input *A*. The output SQ is determined by both the input *A* and the state *P*.

Since both the Moore and Mealy model outputs can be incorporated in the same ASM model, a corresponding pure Mealy or Moore model can in general not be

5.1 LOGIC-CIRCUIT DESIGN

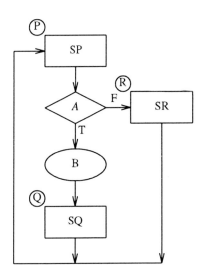

Fig. 5.16 ASM chart

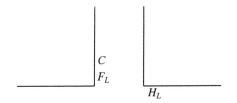

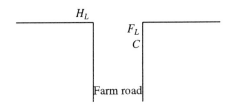

Fig. 5.17 Highway intersection

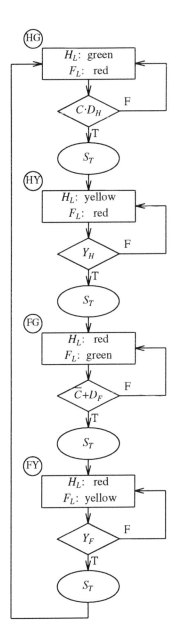

Fig. 5.18 ASM chart of the traffic controller

derived from the ASM model. The hardware implementation can be directly realized from the ASM chart.

The Traffic-Light Controller

The traffic-light controller, as discussed earlier by Mead and Conway [Mea80], is chosen below for illustrating the use of the ASM chart.

A busy highway is crossed by a little-used farm road, as shown in Fig. 5.17. A traffic-light controller controls the traffic lights installed at the places indicated by HL and FL. The highway traffic is given priority in that its light is set to green as long as there are no cars on the farm road. The presence of cars on the farm road is detected by a sensor S located at the places indicated by C in Fig. 5.17. When a car is detected by S, a binary variable C changes value from 0 to 1. At this moment, a timer driven by a high-speed master clock is initiated and the highway light switches to yellow and remains yellow during the time interval TYH, after which it turns to red, while at the same time the farm-road light turns to green. The green farm-road light is to cycle through yellow during TYF to red and the highway light to green after a time interval TDH or earlier when no more cars are detected by the sensors. The highway light is to remain green during a minimum interval TDH. Only when after a time interval TDH the presence of cars on the farm road is detected, will the highway light cycle through yellow to red and the farm-road light to green.

The following hardware is needed: four traffic light installations (indicated in Fig. 5.17 by H_L and F_L) to display any of the three colors (red, yellow, green), a car detector (S) for producing a signal when the presence of a car is detected, and a timer (initiated by the logic variable S_T) which, when started, produces an output signal after any of the intervals TDH, TYH, TDF and TYF. At each state transition, a timer is reset and started. After an appropriate time interval TDH, TYH, TDF and TYF corresponding to the next state, the timer delivers a signal 1 which is assigned to the logic variable D_H, Y_H, D_F and Y_F respectively. In clocked logic, the timer can be implemented as a counter.

Four different states corresponding to specific color combinations of the traffic lights are initiated by four different events:

State HG: The highway lights are green and the farm-road lights are red. This is the default state which applies when no cars are present on the farm road. A change to the next state HY is initiated when cars are detected on the farm road on the condition that state HG has been green for at least TDH. A transition from state HG to state HY takes place when $C \cdot D_H = 1$.

State HY: The highway lights are yellow and the farm-road lights still red. The transition to the next state is initiated when the time TYH has elapsed.

State FG: Farm-road lights are green and the highway lights are red. The transition to the next state FY is initiated either at the end of the time interval TYF or when within this interval no more cars are detected on the farm road. A

5.1 LOGIC-CIRCUIT DESIGN

transition from state FG to state FY takes place when $\overline{C} + D_F = 1$.
State FY: Farm-road lights are yellow and the highway lights are still red. At the end of the yellow-light interval TYF, the state HG will be resumed.

The variable C is logic 1 if cars are present on the farm road; otherwise C is logic 0. The inputs of the finite state machine are $C \cdot C$, Y_H, $\overline{C} + D_L$ and Y_L. Together with the present state, these inputs determine the next input and the outputs. The outputs are the three colors: green, yellow and red. Rather than constructing a state diagram, an ASM chart is a natural tool for describing the finite state machine. The ASM chart for the traffic-light controller is given in Fig. 5.18.

One of the possible implementations of the traffic-light controller is the PLA circuit shown in Fig. 5.19.

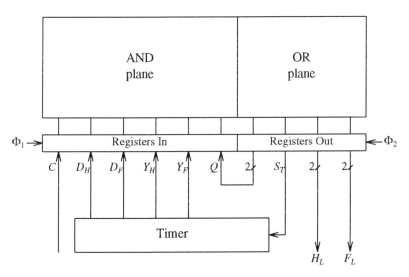

Fig. 5.19 PLA implementation of the traffic-light controller

The inputs to the AND plane are C, D_H, D_F, Y_H, Y_F and the present state Q_n. The outputs consist of the next state Q_{n+1}, the start-timer signal S_T, and H_L and F_L. The inputs are stored in the In-Register during Φ_1, while the outputs are stored in the Out-Register during Φ_2. This means that the next state, which is stored as output in Φ_2 will be the present state stored as input in the next active cycle of Φ_1. The two registers can be implemented as a parallel array of transmission gates which alternately are made conducting. Two bits are needed for encoding the four possible states (HG, HY, FG, FY) and the three possible colors (green, yellow, red) of H_L and F_L.

The state-transition table in text form is given in Fig. 5.20. The entries in this figure have the form next-state/output. The outputs have the syntax S_T H_L F_L.

The colors of H_L and F_L are coded G for green, Y for yellow and R for red. The timer signal S_T is coded 0 or 1 if it is OFF or ON respectively.

Present State	Inputs							
	$C \cdot D_H$		Y_H		$\overline{C} + D_F$		Y_F	
	0	1	0	1	0	1	0	1
HG	HG/0-G-R	HY/1-G-R						
HY			HY/0-Y-R	FG/1-Y-R				
FG					FG/0-R-G	FY/1-R-G		
FY							FY/0-R-Y	HG/1-R-Y

Fig. 5.20 State table of the traffic-light controller

Note that the empty entries represent don't cares X. For the sake of clarity, the symbol X has been left out of the table. The inputs $C \cdot D_H$ and $\overline{C} + D_F$ can be generated from C, D_H and D_F by using an AND gate and OR gate respectively.

Present State	Inputs									
	C		D_H		Y_H		D_F		Y_F	
	0	1	0	1	0	1	0	1	0	1
0 0	00/ 0-00-10		00/ 0-00-10	01/ 1-00-10	01/ 1-00-10					
0 1					01/ 0-01-10	11/ 1-01-10				
1 1		10/ 1-10-00	11/ 0-10-00				11/ 0-10-00			
1 0								10/ 0-10-00	10/ 0-10-01	00/ 1-10-01

Fig. 5.21 Encoded state table of traffic-light controller

5.1 LOGIC-CIRCUIT DESIGN

If we use the three individual variables C, D_H and D_F as the basis, as indicated in Fig. 5.21, we need three possible inputs for $C \cdot D_H$ and $\overline{C} + D_F$ instead of two as follows:
$C \cdot D_H = 0$ or 1 is replaced by $C\, D_H = 0X, X0, 11$,
while $\overline{C} + D_F = 0$ or 1 is replaced by $\overline{C}\, D_F = 00, 1X$ or $X1$,
which is identical to $C\, D_F = 10, 0X$ or $X1$.

Product terms ↓	Stored in Register In during Φ_1									Stored in Register Out during Φ_2				
	Inputs					Present State		Next State		Outputs				
	C	D_H	D_F	Y_H	Y_F	Q_{p0}	Q_{p1}	Q_{n0}	Q_{n1}	S_T	H_{L0}	H_{L1}	F_{L0}	F_{L1}
R_1	0					0	0	0	0	0	0	0	1	0
R_2		0				0	0	0	0	0	0	0	1	0
R_3	1	1				0	0	0	1	1	0	0	1	0
R_4				0		0	1	0	1	0	0	1	1	0
R_5				1		0	1	1	1	1	0	1	1	0
R_6	1		0			1	1	1	1	0	1	0	0	0
R_7	0					1	1	1	0	1	1	0	0	0
R_8			1			1	1	1	0	1	1	0	0	0
R_9					0	1	0	1	0	0	1	0	0	1
R_{10}					1	1	0	0	0	1	1	0	0	1

Fig. 5.22 Rearranged state table

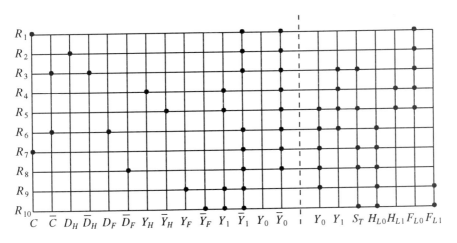

Fig. 5.23 Symbolic representation of the PLA layout

Introducing the coding: HG = 00, HY = 01, FG = 11, FY = 10, and G = 00, Y = 01, R = 10, the encoded state-transition table is shown in Fig. 5.21. Note that the inputs in the above state-transition diagram are given in terms of the five variables (C, D_H, Y_H, D_F and Y_F) being assigned the value 0 or 1 instead of the $2^5 = 32$ possible bit patterns according to Fig. 5.21.

In order to determine the necessary row-column intersections in the PLA implementation of Fig. 5.19, Fig. 5.21 is rearranged in accordance with the normal layout as illustrated in Fig. 5.6. The result is shown in Fig. 5.22. The symbolic representation of the necessary row-column intersections in the AND and OR planes of the PLA implementation is given in Fig. 5.23.

Karnaugh-Map Methods

These methods cover conventional methods using TTL and CMOS integrated circuits as well as gate-level methods of semicustom VLSI design. The starting point is the logic-level description of a state table and the basic method is the classical approach of using Karnaugh maps followed by Boolean Algebra.

For each output variable, a Karnaugh map is produced. The problem is coping with the large number of input variables. The solution is to produce maps with the state variables as the map variables and then to enter the qualifiers on the Karnaugh maps as map-entered variables. Thus, the mapping process breaks down into solving each subtable for each state in the state table and entering the result in the square on the map for the state concerned.

Multiplexers

The implementation with multiplexers recognizes the fact that a multiplexer is a direct implementation of a Karnaugh map. That is, each input represents one map square if the state variables are connected to the select inputs of the multiplexer. Therefore, having determined the entries for the Karnaugh map as above, they become the functions that are connected to the multiplexer inputs.

Multiplexer-based solutions suit standard MSI integrated circuit methods and semicustom VLSI methods where a regular structure is desirable but PLAs or ROMs are not appropriate. The advantage of a multiplexer approach is that the design is quite flexible, permitting easier changes once built and the hardware structure is simple. The circuit can be fast since only a single chip delay is involved in the feedback path.

For VLSI designs, a small multiplexer can be configured from pass transistors and is quite compact. It can offer an alternative to a full PLA implementation. Multiplexers would not normally be used for the output function since they are usually much simpler.

Programmable Logic Arrays

The Programmable Logic Array (PLA) was discussed in Subsection 5.1a. It is ideally suited to the implementation of ASM designs as shown earlier in this

5.1 LOGIC-CIRCUIT DESIGN

subsection. The table is entered directly into the PLA. Lines in the table with no active outputs are omitted.

Example: A typical PLA has 96 product terms, 16 inputs and 8 outputs in a 28 pin package.

For some years, Field-Programmable Logic Arrays (FPLAs), which can be electrically programmed, have been available from a few manufacturers. The Field-Programmable Logic Sequencer (FPLS) includes a state register on the same chip, thereby increasing the logic power without requiring many more pin connections.

The Programmable Array Logic (PAL) is an optimized variant of a PLA, providing more inputs or outputs or functionality by reducing the width of the OR array.

For large designs, it may be necessary to expand the capability of the single PLA component either by using more of the same, or a mixture of PLAs or even using MSI and SSI components. There are three ways of expanding the component PLA: outputs, product terms and inputs.

ROM Based Machines

The increasing size and speed of ROMs and Programmable ROMs make it more attractive to use them as logic components in a finite state machine. The ROM implements a truth table directly with each output bit representing each entry in the output side of the table. Therefore, the state table of a design problem is mapped into a ROM if enough inputs and outputs are available. Direct mapping in this way is only economical if the table has a limited number of inputs, at least less than the available number on the ROM.

To reduce the number of inputs to the memory, it is necessary to remove the more redundant inputs, i.e., the qualifiers.

5.2 HIGH-LEVEL SYSTEM DESIGN

5.2a VLSI Design Aspects

Hierarchy of Abstraction Levels

When a complex digital system is to be designed, it is convenient to consider the hierarchical levels of abstraction which are relevant in the top-down approach of design. Figure 1.1 in Subsection 1.1b shows a schematic hierarchical structure indicating the various activities of design verification, simulation, testing and layout in the subsequent stages of the design cycle.

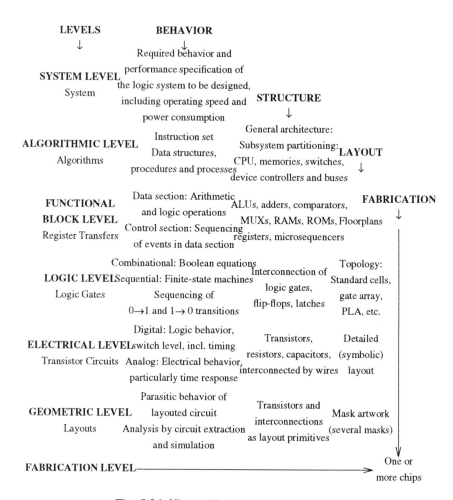

Fig. 5.24 Hierarchical approach to design

Figure 5.24 shows a more detailed scheme of the various activities which are essential in the process of creating a hardware implementation, given a high-level design specification. The hierarchical levels indicated are: the system level, the algorithmic level, the register-transfer level, the logic level, the circuit level, the geometric level and the fabrication level. It is expedient to distinguish three aspects of a complex design: the behavior, the structure and the layout.

Behavior refers to the functional circuit or system performance as specified in the design requirements. A behavioral aspect can be associated to all hierarchical levels from the system level down to the fabrication level. In general, the starting point of a design process is the behavioral specification of

5.2 HIGH-LEVEL SYSTEM DESIGN

the logic system to be designed. This is a specification at the system level. The given required behavior must be satisfied in all following levels down to the fabrication level, which yields the ultimate hardware implementation of the system in the form of one or more silicon chips satisfying the given behavior. Thus, the logic behavior at the logic level and the electrical behavior at the circuit level must be compatible with the system specification given at the system level.

Figure 5.24 suggests that three aspects (behavior, structure and layout) play a role in a fixed order. First, the system *behavior* is considered. The original design specification is primarily concerned with the system behavior and the performance requirements. The second aspect to be considered is the *structure* of the intended implementation. Structure has to do with the set of hardware modules which must be interconnected to build up the system. Usually, many different structures (i.e., hardware implementations) satisfy the design constraints. A goal in synthesis is to select an implementation that is optimal with respect to given performance criteria. As lower levels are considered, the system components get an ever more structural refinement. A module at a specific level is decomposed into smaller submodules at the next lower level. Figure 5.25 shows the hierarchical tree structure of a computer system.

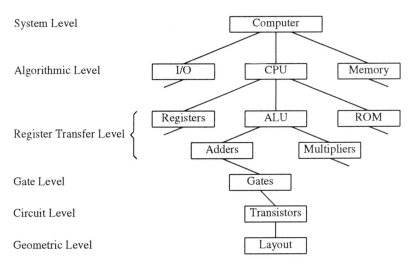

Fig. 5.25 Hierarchy of a computer system

Before the manufacturing process can be considered, the *layout* aspect has to be dealt with. Hardware modules, which together form the implemented system must be placed on a geometric plane in such a way that the chip fabrication can be started. An efficient placement procedure aims at using the smallest possible area on the chip. Floorplanning is the efficient placement of high-level modules.

When the circuit-level description is available in the form of transistors and their interconnections, the layout design stage ultimately produces the mask-artwork description. For a transistor circuit to be implemented on a silicon wafer, a number of masks are needed. The reason is that a transistor has a structure that extends into the body of the wafer, while the interconnections between transistors are implemented in aluminum, silicon or polysilicon in different layers which are isolated from each other by silicon dioxide layers.

Referring to Fig. 5.24, a system design is a mapping of the system behavior to the system structure and subsequently a mapping of the structure to the layout given by the complete description of the required masks. Let us study the synthesis at the higher levels of the hierarchy.

High-Level Synthesis

Synthesis is the process of mapping an input specification of the required behavior and performance of a logic system into a hardware implementation of the system which satisfies the given specification. Synthesis is usually performed with a number of objectives in mind: a *minimum* design time, manufacturing cost, power consumption, occupied chip area, pin count, and a *maximum* operating speed, reliability and testability. Each objective implies a constraint, e.g., although a maximum operating speed is aimed at, the speed must exceed a specified minimal value. A minimized overall delay in the circuit is conducive to attaining a high operating speed [Lew84].

A major problem is that synthesis is a one-to-many mapping. In fact, the solution space which contains all possible design solutions is very large. Though optimizations can be used to meet the objectives as closely as possible, trade-offs must be made between conflicting goals. Referring to Fig. 5.24, design involves a transformation from the behavioral specification to a structural specification, which in later stages is transformed into a layout and eventually to the fabrication step. Let us restrict ourselves to the higher levels of synthesis.

Algorithmic Synthesis

System-level synthesis includes partitioning an algorithm into multiple processes that can be pipelined or can run in parallel. A system description at the algorithmic level resembles a program, i.e., a formulation of an algorithm in a procedural high-level language, such as Pascal, C or Ada. Such a program contains declarations of variables and operations performed on variables and constants, together with conditional statements. The algorithm may be that of a data processor, a peripheral controller or a signal processor, and, depending on the data dependence of the variables in the program statements, concurrent processes can be detected.

The input language may be a nonprocedural hardware description language or a declarative language, such as Prolog. Generally, an input description contains bindings of variables to storage elements (registers and memories), operations to

5.2 HIGH-LEVEL SYSTEM DESIGN

functional units (ALUs, multipliers, shifters, etc.) and register transfers to control states.

Register-Transfer Synthesis
Algorithm synthesis is followed by synthesis at the register-transfer level which consists of two parts:
1. *Data-path synthesis*, dealing with the functional units for storing and operating on data, and
2. *Control synthesis*, for ensuring the proper sequences of events to occur in the data path.

Generally, data-path synthesis is performed prior to control synthesis, although there is some interdependence. For example, a series processing design with processor sharing needs less hardware but more complex control sequences and hence a worse operating speed than parallel processing design. The behavior of a logic system is usually represented by a data flow-graph and a control flow-graph.

Data-Path Synthesis
The data-path specification reveals how many registers are in principle needed, and which register transfers are required.

Example: The register-transfer statement
$$IR \leftarrow M[PC]$$
indicates that a program counter PC and an instruction register IR are needed, while the contents of the memory M pointed to by the PC must be moved to IR.

A major part of data-path synthesis concerns the allocation of functions to specific hardware as the implementation of the mapping from behavior to structure. Variables or values are allocated to registers and operations to operators (e.g., an addition is allocated to an adder). Where necessary, registers and operators are interconnected by using multiplexers. A data flow-graph may be useful for representing the data path. Control may be integrated into the data path by specifying a finite state machine.

Control Synthesis
When the register-transfer data-path structure is known, along with a partially or completely specified control flow, the register-transfer control synthesis is performed. Control synthesis consists of two parts:
1. The selection of the clocking scheme, and
2. The design of the controller.

The first part deals with the selection of the number and lengths of clock phases and the allocation of events, which must be synchronized to clock phases. A frequently employed scheme is the two-phase clock with the two phases Φ_1 and Φ_2, as sketched in Fig. 5.26.

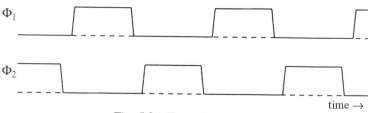

Fig. 5.26 Two-phase clock

The controlling task can be performed by several design styles, including random logic, microcode and a PLA-based design. Usually, one of these styles is preselected for control synthesis. With the knowledge of the actual control signals and a decision on bit lengths and memory sizes, control synthesis can be completed. Potential parallelisms and data-precedence relationships between operations can be derived from the control flow-graphs.

Logic-Level Synthesis

The outcome of a register-transfer level synthesis given in the form of a high-level logic schematic or a hardware description language may be the basis for logic-level synthesis. Data-path logic synthesis is followed by control-path logic synthesis. Data-path logic hardware is the implementation of registers and operators in gates, flip-flops and latches. A complex module may be composed from simpler modules. For example, an eight-bit adder can be composed from two-bit adders. Control logic synthesis involves the gate-level implementation of the control path, synthesized at the register-transfer level. This may consist of the generation of microcode or PLA programming.

Data Flow-Graphs and Control Flow-Graphs

The *data flow-graph* can be generated from a high-level program [Ora86]. The statements of this program are examined successively. The righthand side of a statement, which may contain operators acting on values to obtain a new value can be represented by a subgraph. Arithmetic or logic operations are represented by operation nodes, which have at most two inputs. The new value obtained is assigned to the lefthand-side variable of the statement and corresponds to the outgoing arc of the subgraph. By connecting outgoing arcs to incoming arcs of new subgraphs, a program can be converted to the complete flow-graph. Conditional statements require a special data-flow element, a choose-value node. This node represents an *if* or *if . . . then* and *case* statements, the latter allowing a selection from a number of possible cases. A conditional statement generates a separate subgraph.

Conditional statements can also be represented in a *control flow-graph*. The control flow-graph shows the sequence of executions between different data-flow blocks. Control-flow nodes may be related to corresponding data-flow

5.2 HIGH-LEVEL SYSTEM DESIGN

representations.

Data dependencies in a program affects the required structure of the circuit implementation. For example, consider the statements

(1) $e := a + b$
(2) $g := e - 1$
(3) $f := c + d$
(4) if $g >= f$ then $h := f$ else $h := g$

The statements (1) and (2) are data dependent and must be executed successively. Statement (3) can be performed independently of (1) and (2). Statement (4), however, can only be performed if the values g and f are available. Hence, statement (3) and (4) must precede statement (4).

This simple example shows that the search for data dependencies in program statements is an essential issue in implementing optimal implementations of algorithms. Any parallel program execution must take these data dependencies into account.

5.2b Intelligent High-Level Synthesis

Introduction

High-level synthesis includes several design steps in which a choice from different alternatives must be made. Automating the whole high-level synthesis task boils down to using an algorithm which inevitably implies a predefined sequence of steps each of which is a preselected procedure. When a great variety of synthesis problems are to be solved, it is not very likely that such a predefined algorithm would lead to an optimal solution.

A satisfactory degree of optimization can be achieved by using an intelligent decision approach, which involves intelligent choices from alternatives in subsequent steps of the synthesis procedure. In such a step, alternative design options are analyzed so that the best one can be selected. Brewer and Gajski [Bre86] have referred to this procedure as the "knobs and gauges" method. An *evaluator* examines several measures of design quality ("gauges") of various design alternatives. A *planner* uses these measures to select the appropriate design style and strategies ("knobs"). The "knobs-and-gauges method" is an iterative procedure embracing the steps: planning (envisioning possible design styles and strategies), refining (the structure), optimization (improving the quality of the structure) and evaluation (of the design quality), as depicted in Fig. 5.27. Envisioning is a forward running tree process of pruning unacceptable branches.

Measures of *design quality* include such design criteria as performance, cost, power consumption, operating speed, chip area, the required operators and their frequency of use. A *design style* includes such elements as the use of pipelining and parallel processing, clocking and control styles (data bus, multiplexer, etc.), the number of addresses in microinstructions, the number and types of components, and the choice of implementation (CMOS, NMOS, etc.) and layout structure. A *design strategy* deals with such factors as ranking trade-offs, the optimization order and the use of fast or slow components. Below, a general approach to intelligent high-level synthesis is outlined in the way as proposed by Gajski *et al.* [Bre86, Ora86, PAn87].

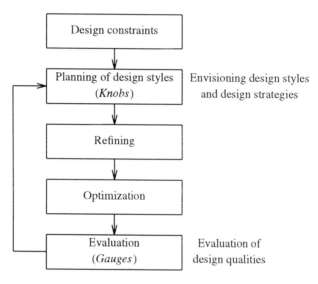

Fig. 5.27 Knobs-and-gauges method

The purpose of high-level synthesis is to translate a behavioral input description into an output description in terms of an interconnected network of components. The input description, which may take the form of a high-level program, contains variables and operations on these variables. An operation, e.g., an addition, is assigned to a component which can perform this operation (e.g., an adder or an ALU). Variables are used to transfer the data between operations and are assigned to data buses and registers. The order of execution is governed by the synthesis problem to be solved.

An intelligent choice from alternative solutions in intermediate steps can be made by an intelligent controller, called Design Critic by Pangrle and Gajski [PAn87]. The Design Critic can be considered as an expert system whose knowledge base contains rules which allow responsible decisions to be made. A scheme for an intelligent high-level synthesis system is given in Fig. 5.28.

The whole synthesis procedure consists of three main steps:

5.2 HIGH-LEVEL SYSTEM DESIGN

1. From input specification to flow-graph,
2. From flow-graph to state-graph, and
3. From state-graph to microarchitecture.

These steps will be discussed in detail.

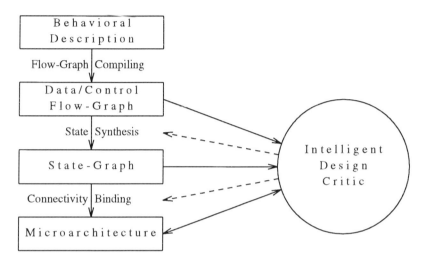

Fig. 5.28 Pangrle-Gajski intelligent high-level synthesis

From Input to Flow-Graph

The input specification includes a functional description of the required system behavior, design requirements on the speed of operation, power consumption, chip area and information as to which conficting requirements can be traded against each other. The input language consists of a type definition (e.g., bit length of a component), register declarations (for explicit specification of registers which must hold values of variables), read/write statements and a set of procedure statements. The first step in high-level synthesis is the compilation of the input language into a suitable graph representation. The purpose is to extract from the input description the appropriate information as to which functional units must be assigned to each operation and the control steps which define the sequence of operations to be performed.

Based on the functional input description, a flow-graph is generated. Operations contained in the input description are mapped into operators which realize the required behavior. Three types of flow-graph can be distinguished: data flow-graph, control flow-graph and data/control flow-graph. In a *data flow-graph*, an operator is represented by a node, while variables are indicated by arcs connecting two nodes. A data flow-graph representation can be used to handle area/performance bounds and trade-offs. In a *control flow-graph*, pieces of a program are put into time blocks which correspond to subsequent control states.

When data dependencies are insufficient to preserve the semantics of the description, control dependencies will be added, leading to a mixed *data/control flow-graph*. Since a flow-graph may contain structures for data flow and control flow, a choice between these two must be made for each operation. Often, this choice is suggested by the functional description. The following statements can be distinguished: operations (such as $x + 1$), loops (*for, while, until*), conditions (e.g., *if* A *then* B) and memory (e.g., $x[1] := 3$).

The data flow-graph can be generated from a high-level program [Ora86]. The statements of this program are examined successively. The righthand side of a statement, which may contain operators acting on values to obtain a new value can be represented by a subgraph. Arithmetic or logic operations are represented by operation nodes, which have at most two inputs. The new value obtained is assigned to the nodes, which have at most two inputs. The new value obtained is assigned to the lefthand-side variable of the statement and corresponds to the outgoing arc of the subgraph. By connecting outgoing arcs to incoming arcs of new subgraphs, a program can be converted into the complete flow-graph. Conditional statements require a special data-flow element, a choose-value node. This node represents an *if* or *if . . . then* or *case* statement, the latter allowing a selection from a number of possible cases. A conditional statement generates a separate subgraph. Conditional statements can also be represented in a control flow-graph. The control flow-graph shows the sequence of executions between different data-flow blocks. Control flow nodes may be related to corresponding data-flow representations.

Data dependencies in a program affects the required structure of the circuit implementation. For example, the statements
$$x := x + 3$$
$$y(1) := x + 1$$
are data dependent and must be performed sequentially. On the other hand, the statements
$$b := a + 1$$
$$d := c + 1$$
are not related and can therefore be performed simultaneously by using a parallel architecture. The data flow-graph and its data dependencies can be used to determine how to partition the program instructions into control steps and which functional units must be assigned to the required operations in such a way that performance and area goals are met. The control flow-graph is used to generate the control unit. Orailoglu and Gajski [Ora86] developed the Array Algorithm which adequately integrates memory references in the flow-graph representation. This algorithm has two purposes: minimizing memory elements in the flow-graph and minimizing critical paths resulting from control dependencies. A *critical path* is the longest path in the flow-graph consisting of the successive operations to be performed in a given procedure. Other relevant algorithms in

5.2 HIGH-LEVEL SYSTEM DESIGN

this stage have the objective of reducing array elements and control dependencies.

From Flow-Graph to State-Graph

Given the flow-graph, a state-graph can be generated by partitioning the flow-graph into clock periods representing subsequent states. State synthesis involves the allocation of operation modes in the flow-graph to machine states in the state-graph. A state-graph synthesizer known as *Slicer* [Pan87] will be discussed here.

State-graph generation embraces three subsequent steps: envisioning, refinement and evaluation. *Envisioning* deals with the optimal exploitation of the clock period. Parallel operation gives a higher speed but requires more components than a serial architecture. On the other hand, a series architecture requires fewer components paid for with a lower operation speed. An intelligent design controller would make a decision according to the available components, the number of permitted clock cycles and the clock period.

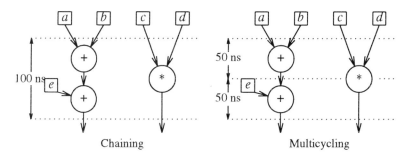

Fig. 5.29 Chaining and multicycling

The possibility of *chaining* and *multicycling* gives the system more freedom to fit the available operations in the given clock period. Figure 5.29a shows a chaining example in which a multiplier operates in parallel with two sequentially operating adders. In Fig. 5.29b, the clock period is halved, forcing the multiplier to operate in two clock periods. Note that we need only one adder which performs different additions in two subsequent clock periods.

Important in state-graph generation is the concept of *mobility* of an operation, which is the range in which an operation can be shifted between ASAP (As Soon As Possible) and ALAP (As Late As Possible) partitioning. The ASAP partitioning performs an operation as soon as the data and control data are available. The ALAP partitioning performs an operation as soon as the data and control information are available. The ALAP partitioning performs an operation when its successor needs the result of the operation. The node mobilities can be used to give different priorities to state bindings. Nodes with zero mobilities may form a critical path, that is, the operations in this path are performed successively.

Starting from an initial design, suitable rules are used to improve this design. When the design does not satisfy the requirements, the strategy can be changed, e.g., by using faster components in a series architecture. *Refinement* binds operations to clock cycles. When a component needed for performing an operation is available, the operation is bound to the current state, otherwise the operation is shifted a clock period. After the nodes in a partition (future state) have been handled, the partitioning list is updated, as a consequence of the shifted nodes. In order to perform this procedure, Slicer contains a number of routines which iterate as long as there are nodes in the partitioning list. *Evaluation* in Slicer amounts to estimating execution times of the fastest available components to determine the total execution time.

From State-Graph to Microarchitecture
The microarchitecture generator binds operation elements to hardware components. The objective is to minimize the hardware costs at given timing specifications (e.g., from Slicer). As in state-graph generation, this process follows the three steps: envisioning, refinement and evaluation. The operation-component bindings made by Slicer are temporary. Only the timing specifications derived from it are used for the microarchitecture. When a component needs n clock periods in the state-graph, this will also be the case in the microarchitecture.

Envisioning takes into consideration that the microarchitecture has different options to meet the design requirements and possible trade-offs. The choice of components and design structure determines the behavior of the implementation. The choice is governed by cost functions, the number of clock periods and the design time. A cost function generated depends on the design style and strategy. It limits the components to be used and determines the design structure. When a cost function declares the use of multiplexers to be too expensive, another less expensive structure will be suggested.

By considering several clock periods of the state-graph simultaneously (look-ahead), the global yield of a design can be evaluated. Considering many states simultaneously improves the design, though it takes much time. By considering fewer states at a time, the critical parts of the design in the remaining design time can be improved locally. The implementation of the design is obtained by assigning operations to components and data to registers, while the interconnections are made. These steps are controlled by the cost function, where cheap steps are preferred to expensive steps.

The same cost function that controls the above refinement procedure is used to decide whether an implementation is satisfactory. Depending on the style and strategy, certain implementation steps are expensive. This may be caused by a

5.2 HIGH-LEVEL SYSTEM DESIGN

parameter in the cost function which makes this function increase excessively when the step is taken. The cost function also suggests which alternative steps can be taken to make the design cheaper.

Splicer is a microarchitecture generator developed by Pangrle and Gajski [Pan88]. It distinguishes three types of design structure: the multiplexer structure, the data-bus structure, and the point-to-point connection. The three types of design structure can be used simultaneously. The cost function determines which type will dominate. In a multiplexer/data-bus structure, several components are interconnected via a shared data bus. The multiplexer sees to it that the appropriate components are connected via the data bus. When many components make use of a data bus, the multiplexer will be large and hence expensive in which case we deal with a multiplexer structure. When several data buses are used, we deal with a data-bus structure. When only one input and one output are interconnected via a one-line databus, we have a point-to-point connection. In a databus structure, all data pass through a bus, whereas in a multiplexer structure as many connections as possible are point-to-point.

The cost function in the design requirements determines the design structure to be selected. Splicer uses a Branch-and-Bound search method in each step of which the design costs are compared with the cost bounds to determine if a component satisfies requirements. For example, if a bus structure is attempted, the costs of the data buses can be reduced. This can be achieved by multiplying the cost component by a parameter that reduces this component. By quantifying design costs during the design and search into statistical data, a new cost function can be defined based on an earlier search action, which fails to find a solution. Splicer distinguishes two cost levels: connections to a data bus and connections from a data bus. For each level, the costs of the various components (adders, multipliers, multiplexers, data buses) are updated in terms of numbers. For every new state, Splicer carries out the microarchitecture in four steps. Use is made of a connectivity model consisting of four levels, each representing a generation step Op (see Fig. 5.30).

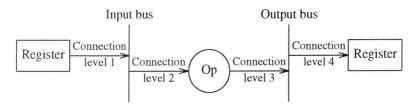

Fig. 5.30 Four microarchitectural levels

Illustrative Design Example

Let us now illustrate the above synthesis steps by considering a very simple

design example. Let the following procedure statements be given:
begin
>$a := a1$
>$b := b1$
>$c := c1$
>$d := d1$
>$e := a + b$
>$g := e - 1$
>$f := c + d$
>if $g >= f$ then $h := f$ else $h := g$

end

Based on this functional input description, the flow-graph of Fig. 5.31 can be generated.

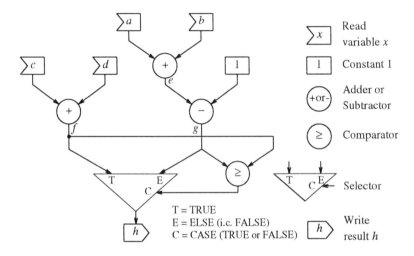

Fig. 5.31 Data flow-graph of a simple synthesis example

In an attempt to minimize the chip area by using a minimal number of components, let us select a serial architecture. This implies that for each type of operation only one component is used and that the clock period is adapted to the fastest component.

The state-graph follows from the flow-graph by drawing appropriate cut lines across the flow-graph such that two neighboring lines bound a specific clock period corresponding to a specific machine state. See Fig. 5.32, which for each operation gives the required component, the ASAP and ALAP partitionings and the mobility. The critical path consists of the adder, the subtractor, the comparator and the selector so that the sequence of these operators is predetermined. The critical path is formed by nodes with zero mobility.

5.2 HIGH-LEVEL SYSTEM DESIGN

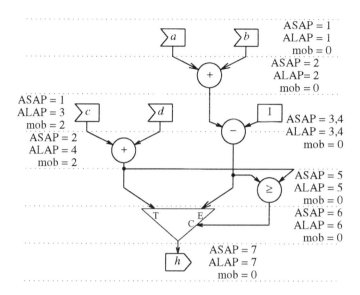

Fig. 5.32 State-graph of the synthesis example

Let us assume that the adder, the comparator and the selector has the same delay τ, which fits into one clock period. Assume further that the subtractor has twice this delay, then multicycling must be applied to it. The second position of the adder which does not lie on the critical path is chosen on the basis of the ALAP partitioning. Reading the a, b, c and d values takes place just before its use in order to avoid the generation of unnecessary state registers. When the state-graph is evaluated, it appears that the clock period is 0.3 ns and the number of clock periods is 7, which gives an execution time of $P = 2.1$ ns. Suppose that this result does not satisfy the original design requirements.

An alternative state-graph is shown in Fig. 5.33. By chaining the registers a, b and the adder as well as the comparator and selector, the critical path (a, b registers → adder → subtractor → comparator → selector → register h) can be reduced. By setting the clock period to the subtractor delay of 0.6 ns, this chaining is possible. Addition of c and d takes place according to the ALAP partitioning to avoid unnecessary use of state registers. Evaluation reveals that the clock period $τ = 0.6$ ns and the number of clock periods is 4, resulting in a total execution time of $P = 2.4$ ns.

Suppose that this execution time is considered to be too large. T can be reduced by choosing faster components, e.g., a clock period of 0.4 ns, which gives an execution time of 1.6 ns, which is acceptable.

Based on the last state-graph, a microarchitecture is generated. For every state or for multiple states (look-ahead =2), the following connections are made: (1)

input to data bus, (2) data bus to functional element, (3) functional element to data bus, and (4) data bus to output. Several alternative designs are possible.

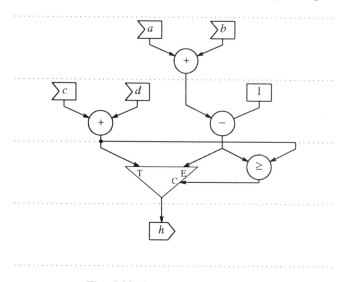

Fig. 5.33 Alternative state-graph

Let us first initiate a slow and expensive design. Envisioning prescribes a multiplexer or databus structure with look-ahead 1 style and a strategy with fast components. Let us examine a multiplexer structure. The result is shown in Fig. 5.34. A 5-line bus functions as the interconnection medium between operators. In Fig. 5.34, the sequence of operations takes place from left to right, starting from the addition $c + d$ and ending with presenting the result h. The multiplexers see to it that at each subsequent clock period the correct operators are activated. Evaluation yields as a result: 4 clock periods of 0.2 ns and hence a total execution time of $T = 0.8$ ns.

The cost of the design is evaluated on the basis of the chip area occupied by the components and the number of data-bus lines. The cost amounts to 2.1 μm^2 (seven registers) + 0.3 μm^2 (one adder) + 0.6 μm^2 (one subtractor) + 0.3 μm^2 (one comparator) + 0.3 μm^2 (one selector). This sum multiplied by the factor 10 (five data-bus lines) gives a total of 36 μm^2. Multiplied by the total execution time, the performance-cost product is 28.8 10^{-15}m^{2}s. The Design Critic decides that this design is too expensive so that another design must be generated.

In order to make the design cheaper, the strategy decision rules suggest to use slower components and/or mixed components. For this example, slower components do not produce the desired trade-off. For example, the performance decreases faster than the cost. The use of mixed components appears to be not feasible without changing the state-graph. Since the strategy does not offer

5.2 HIGH-LEVEL SYSTEM DESIGN

further profits, another style must be chosen in order to make the design cheaper. There are two possibilities: a data-bus style can be used instead of a multiplexer style, or a larger look-ahead value (look-ahead 2) can be taken.

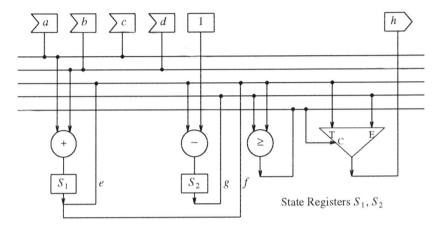

Fig. 5.34 Initial microarchitecture

Since the use of a data bus instead of a multiplexer structure does not entail additional costs and a larger look-ahead is more expensive, the Design Critic chooses a data bus + look-ahead 1 style. Furthermore, the use of slow components is again selected. In this design style, the fewest possible data buses are employed. For a data transfer from element a to b, a direct connection is laid. When more components must be connected to a component, a multiplexer is added to the existing connection. On the basis of these data, Splicer can generate a stepwise design according to the first initial design.

The following operations are performed in four clock periods.
1. The a and b registers are connected to the adder.
2. The c and d registers are connected via multiplexers to the adder, and at the same time the subtractor and state register 2 are generated and connected.
3. The comparator and the selector are generated and connected. Via a multiplexer the comparator and the selector are connected to the state registers.
4. The result is written in register h.

The number of clock periods (four) multiplied by the clock time $\tau = 0.2$ ns gives a total execution time of $T = 0.8$ ns. The area cost is 2.1 µm² (seven registers) + 0.3 µm² (one adder) + 0.6 µm² (one subtractor) + 0.3 µm² (one comparator) + 0.3 µm² (one selector) = 3.6 µm². The four multiplexers are evaluated as a factor $m = 8$, which gives a total performance-cost product of 28.8 10^{-15}m²s. This result is accepted. When no satisfactory design has been generated, a fault report is given and on this basis another state-graph must be generated.

Hwang et al. [Hwa91] developed an integer-linear-programming model for the scheduling problem in high-level synthesis.

5.2c Automated Systems for High-Level Logic Synthesis

Introduction

High-level synthesis aims at achieving the logic circuit or system description in an intermediate form, e.g., as Boolean functions of a combinational circuit or a suitable representation of a sequential machine [Lew77]. There exist several methods which convert the intermediate form into a form ready for hardware implementation. A survey of automated systems for logic synthesis, starting from a specification in a hardware-description language can be found in [Sch87].

There is a strong motivation to automate the high-level synthesis process. The purpose of automated synthesis aids is to allow a system designer to specify an appropriate behavioral description of the system, which is then used to generate a structural description suitable for further hardware implementation. In what follows, an anthology is given of special automated methods for high-level logic synthesis. Silicon compilers, which constitute a special class of automatic design tools for translating a high-level specification into a mask description, will be surveyed in Subsection 5.3b.

Special Approaches to Logic Synthesis

Synthesis of logic circuits may be performed with ROM implementations [Sho75] or PLAs in mind. One major consideration in designing logic circuits is which *design style* should be selected in a specific logic-design project. The problem of selecting the proper design methodology was considered in Subsection 1.1c. A design style introduced in Subsection 5.1b is the ASM (Algorithmic State Machine) technique, which allows one to formalize the logic-system design. First, an algorithmic description of the control structure and data-path requirements is established. From an ASM chart, which specifies the behavior of an ASM module, a state table is desired. A design representation is used which is suitable for direct implementation in hardware, either in gate logic or a regular array structure. Attempts to utilize the ASM technique in automated logic synthesis have been made by several authors [Edw83, For83].

The *flowchart method* as proposed by Tredenick [Tre81] appears to be a powerful approach to designing the Central Processing Unit of a computer. Flowcharts are graphical notations which are manipulated so that the design can be pursued from the architectural level down to the circuit implementation. The major feature is the use of a simple, natural and readable notation to express the flow of machine actions. To facilitate the design procedure, programs may be

5.2 HIGH-LEVEL SYSTEM DESIGN

used to translate the input in textual form into block diagrams, Boolean equivalents or lists of states. This flowchart method has been successfully applied to design the controller in the MC68000 microprocessor.

Hardware synthesis from high-level specifications involves both behavioral and structural aspects. Structural data may be visualized by block diagrams, while behavioral information requires a behavioral description language. Odawara *et al.* [Oda84] developed a diagrammatic description language, called SFDL (Symbolic Functional Description Language), which enables the designer to enter both structural and behavioral data into a computer by means of a digitizer. SFDL has been implemented on LDSS (Logic Design Supporting System). This logic-design system consists of two parts: CIP (Circuit Information Processor), which interprets and processes the designer's hand-drawn input diagrams, and LOVE (Logic Verifier), which is a mixed-level simulator to be used at different levels from the functional level down to the gate level. Below, some special synthesis methods will be discussed in greater detail.

The Use of Local Transformations

An obvious approach to logic synthesis is to start from an initial nonoptimal design, which can easily be obtained. From this point on, local or global transformations can be performed on the design in order to achieve an optimized design which satisfies the design specification. The objective of *local transformations* is to optimize small circuit portions of a design one at a time so that the total design will be improved incrementally. *Global transformations* are concerned with the system as a whole, e.g., in the top-down decomposition of a system into smaller subsystems.

A design approach in which local transformations form the chief ingredient is attributed to Darringer *et al.* of IBM [Dar81]. Their approach, called LSS (Logic Synthesis System), is intended for designing synchronous logic systems, in particular control logic. LSS is aimed at producing a system which has been optimized in terms of operating speed, power consumption, chip area and design time. LSS-based designs consist of the following steps:

a. A functional specification in a register-transfer language, such as BDL/CS, is automatically translated into an initial design.
b. By means of local transformations, this design is converted into an implementation with interconnected primitive elements, such as ANDs, ORs, inverters, and higher-order modules, such as decoders and adders.
c. Simplifying transformations reduce the number of elements in this implementation.
d. The AND/OR implementation is converted into a NAND (or NOR) implementation, in which redundant NANDs (or NORs) are eliminated.
e. The NAND (or NOR) gates and higher-level modules are implemented in a specific technology, taking into acount the technology-specific constraints,

such as I/O-pin and timing constraints.

This procedure is a top-down approach which is completely controlled by the human designer. Each transformation is evaluated with respect to some performance criteria so that design improvements are obtained in an interactive manner. The major problem in this approach is to identify where and when transformations can be applied successfully so that the necessary optimizations have been performed before the technology-specific implementation is considered. As applications, the design of three gate-array chips has been reported as well as a transformation of a TTL gate array to an ECL gate array. The technique used to guarantee testability of the logic takes into account any redundancy used in the logic hardware. Note that no tests exists for the redundant part in the circuit.

A similar example of transforming existing implementations from one specific technology to another was described by Nakamura *et al.* [Nak78]. Such a system can help a designer translate an existing SSI or MSI implementation into large-scale integration. In the design system LTS (Logic Transformation Subsystem), described by Bendas [Ben83], a user-programming interface was provided so that human design experts can write their own transformation programs. LTS is particularly suitable for delivering input data to gate-array layouts.

The CMU-DA System

CMU-DA (CMU Design Automation) is a major research project of the Carnegie-Mellon University, Pittsburgh, aimed at providing a complete methodology for designing complex digital systems, starting from a behavioral description and going down to the circuit-layout level [Dir81]. The basic input-description language for the system to be designed is ISPS.

In the higher-level steps, an optimal solution from alternative design representations at the behavioral level is selected without bothering about the structural and technological aspects of the design. When the data-path design style (centralized or distributed, bus-oriented, use of pipelining) and control-path style (random logic, PLA, ROM-based style or a busing system) have been selected, a module binder specifies the hardware implementation of the data path and the controller by evoking an interconnection of hardware modules from the module database.

A complete design cycle in using the CMU-DA system consists of the following stages:

a. *System-Input Entry and High-Level Optimization.* An ISPS behavioral description is compiled into a Global Database, which can be used to deliver the input to a behavior-level simulator. The ISPS description is transformed into an abstract design representation, called *Value Trace* (VT) [Sno78]. The Value Trace is a directed acyclic graph whose nodes are represented by abstract functional blocks, called *VT bodies*. Each VT body consists of a list

5.2 HIGH-LEVEL SYSTEM DESIGN

of inputs and a data-flow graph which represents the operations performed by the functional blocks [Dir81]. The arcs connecting two nodes of the graph represent the data flow from one operator to another. A *Global Optimizer* applied to the Value Trace enables one to explore alternative design representations at the behavioral level without bothering about the internal structure and technological aspects of the design. High-level transformations applied to the Value Trace have the purpose of finding a realization which minimizes the global cost and improves the system performance.

b. *Selection of the Design Style.* Given the behavioral description, this step selects an appropriate design style based on the design specification and the design facilities provided by the CMU-DA system [Tho81]. The problem of logic synthesis is to establish the transition from the behavioral level to the structural level. The synthesis task is concerned with the data path of the system and the control section which will evoke the operations in the data path. Possible data-path design styles include the centralized and distributed design styles, a bus-oriented style and extensive use of pipelining. The design style which is selected determines the specific algorithms and heuristics needed in further design stages. Control-path synthesis may be based on random logic, a PLA or ROM based style or a busing system.

c. *Data-Path Synthesis.* This stage consists of two steps: data-path allocation and module binding [Hit83]. The *data/memory allocator* [Haf82] has the function of generating the appropriate set of funtional modules which implements the required behavior. It provides a register-transfer description of the data path in the form of a *data-path graph*. This graph depends on the design style selected in b. Although the Value Trace is usually taken as the input to the data/memory allocator, Hafer and Parker [Haf83] developed an alternative data-path allocator which takes the original ISPS description of the system directly as input. The *module binder* allows the data path to be implemented by specifying an interconnection of physical functional modules which are evoked from a *module database* [Tho83]. Each candidate set of functional modules, which corresponds to a point in the design space, is evaluated and the one that meets the designer's requirements most closely is selected. This is the first step in which the implementation technology plays a role. In addition to specifying or synthesizing a logic structure, the module binder specifies the control requirements for the data path, including a description of the control sequence. Hitchcock and Thomas [Hit83] developed a minimum-cost data-path synthesis method, given the Value Trace with its associated control flow.

d. *Control Synthesis.* Like in data-path synthesis, control synthesis consists of two steps; control allocation and module binding. Based on control information provided in c., the *control allocator* produces a description of the controlling procedure which evokes the data-path modules in an order consistent with the desired behavior [Nag82]. The *module binder* specifies the hardware implementation of the controller by using physical modules contained in the module database. The microcode, PLA style or any other appropriate design style may be used. The program BLINK (Behavioral Linking) implements a multilevel representation for the CMU-DA system [THo83]. From the detailed circuit implementation, the program can extract timing information which can be used in the ISPS behavioral simulator.

The implementation technology appears on the scene not earlier than in stage c. of the design cycle. Based on the designer's specification of requirements (in terms of optimization criteria and constraints imposed on the final design), a number of candidate functional structures composed of data modules is generated. By means of an appropriate evaluation procedure, depending on the particular technology and design style used to implement the target system, the most satisfactory candidate structure is selected for hardware implementation. In the current version of the module database, at least TTL and CMOS module sets are available. The module database allows one to define and store new functional modules in any desired technology. Each modular unit is described at different levels: the *functional level* (for specifying the register-transfer operations), the *logic level* (in terms of gates and flip-flops), the *circuit level* (in terms of transistors and capacitors) and finally, the geometric *layout level* (for specifying the mask geometry). New building blocks to be incorporated in the module database have to be synthesized and specified in terms of the above-mentioned four levels. The output of the module binder in c. and d. is a net list for the geometric layout of the complete circuit.

Summarizing, the CMU-DA system encompasses all design levels of abstraction from the system level down to the mask level, in which the design can be optimized in several stages of the design cycle to conform to the designer's objectives by exploring a large number of possible design alternatives, while new building blocks and design constraints can be incorporated as technology evolves. The complete system may take an ISPS specification as input to generate the required mask descriptions as output. This methodology allows one to quickly redesign a complex system in a new technology. The CMU-DA system has passed through several phases of development. Some of the newer development will be outlined below.

Tseng and Siewiorek [Tse81] studied the modeling and synthesis of bus-oriented digital systems. With a Value Trace as input, the allocator produces a bus-styled data path. It is essential to minimize the number of buses and the total number of drivers and receivers. Tseng and Siewiorek [Tse86] also developed an automated program for the allocation of the data path at the register-transfer level, called FACET. The program formulates the data/memory allocation as a clique/partitioning problem. Such problems are concerned with partitioning a connected graph of nodes into a minimum number of disjoint clusters of nodes

5.2 HIGH-LEVEL SYSTEM DESIGN

(each forming a connected subgraph, called a *clique*) such that each node appears in one and only one cluster. A minimal number of clusters corresponds to a minimal number of physical operators. To facilitate extensive experiments with FACET, a design generator, based on FACET and called EMERALD, was developed. EMERALD takes the Value Trace as input and generates functional-level structures as output. Nestor and Thomas [Nes98] described a technique for behavioral synthesis of synchronous digital systems with interfaces that tie structure, behavior and timing specifications together. More recent developments of the CMU-DA system concern knowledge-based systems (see Subsection 5.2d).

Hafer and Parker [Haf83] reduced the high-level synthesis task to a strictly mathematical problem. By formulating behavioral specifications and design rules as a set of relations, this set can be solved as a mixed-integer linear programming problem. The advantage is that a global solution is obtained, taking into account all constraints relevant to the problem. Moreover, a straightforward verification is implied by the preassumption that all constraints are met. Apparently, this method is only suitable for small circuits.

Another optimization-oriented approach to high-level synthesis is EMUCS, developed by Hitchcock and Thomas [Hit83]. The objective function to be minimized is related to the cost which represents a quantitative parameter, such as dissipation power or chip area or the ease of binding values and operators in the Value Trace to registers and hardware operators. Estimates are made of the cost of incremental design decisions. The objective function to be minimized is the largest difference between the costs before and after a choice of an alternative has been made. A minimax strategy is used to find the minimum.

A recent high-level synthesis system, developed at Carnegie-Mellon University, is the System Architect's Workbench [Tho89]. An abstract behavioral description of a piece of hardware is converted into a set of register-transfer components and a control sequence table. The system supports two approaches to synthesis, one specifically tuned to design microprocessors and the other being suitable for a more general design style.

CAD Frameworks
The complexity of many current VLSI designs requires a large number of design tools. A VLSI chip designer would take advantage of a design environment in which the design process can be managed from a single point, while an appropriate tool can be called from a tightly integrated set of tools, whenever needed. A problem is, however, that there are many advanced CAD tools, which do not fit easily in an integrated set of tools.

A modern trend is the use of a *design framework*, which can be defined as an open, loosely coupled set of procedures and tools which share data and resources [Har90]. The design process is then an intelligent exploitation of the use of the

CAD tools which are available. An advantage of a design framework is that its modular architecture allows each individual design tool to be replaced by improved tools without affecting the other tools. In this way, the set of tools can be easily updated to the latest developments.

Design-automation vendors, computer semiconductor suppliers, end users of CAD tools, and government, research and academic institutions started the CAD Framework Initiative (CFI) in 1988 [Gra91]. Their belief is that developing and using standards-based frameworks will raise productivity among users by providing interchangeable, interoperable tools. CFI's goal is to make available a set of standards for portable, high-performance framework technology.

The procedures and tools are usually based on the object-oriented approach for several reasons, including the capability of data abstraction, inheritance and run-time method determination [Wol89]. An example of an object-oriented CAD framework is MACLOG, a family of layout generators, which generate parameterized, custom layouts for many of the cells in the AT&T Cell Library [Bow91].

Cadweld Design Framework

Cadweld is an object-oriented design framework that treats CAD tools as objects [Dan91]. It is implemented as a loose collection of heterogeneous modules, allowing the user to easily add or delete modules as new or improved subsystems become available. Cadweld incorporates the concept of database as well as a design system.

Each Cadweld tool is represented as an object-oriented entity with associated specialized functions. A hierarchical structure of CADCAS tools is established to exploit the use of the inheritance concept. The designer can choose from a list of possible tools to perform a given task. The CAD Object Manager attempts to match the perceived needs at a given time in the design process with the abilities of the CAD tools which are available. A blackboard is used as the basic medium through which the various modules interact. The implementation language of Cadweld is Lucid Common Lisp with Xerox Portable Common Lisp included.

HILDA

The HILDA framework [Hsu87] supports the integration of design tools and the management of design data. The interaction between HILDA and its user is provided by the interface DEBBIE [Yoo90], which offers a variety of services to create, update or evaluate designs by activating various design tools. To ensure consistency, every time the user performs such an action, DEBBIE calls an inference server in OPS83, which checks its rule set to determine whether the action is allowed.

VLSI Design with OO Knowledge Bases

A design framework may be based on object-oriented knowledge bases in which

5.2 HIGH-LEVEL SYSTEM DESIGN

the design knowledge is abstracted and organized as classes [SHe88]. A descriptive specification of the desirable behavior of a computing system is matched against the existing design knowledge. If a match can be found, the abstract knowledge is instantiated and reused. If not, a heuristic synthesis process is performed, based on the design knowledge stored in the knowledge base. If necessary, a structural design procedure at lower levels is started. A refinement process must lead to a satisfactory design.

ESPRIT CAD Frameworks

A project for the construction of a framework for computer-aided design of complex integrated circuits was started under the name ICD (Integrated Circuit Design) [Dew86, Dew88]. It was funded by ESPRIT (European Strategic Programme of Research and development in Information Technology). The framework includes a structured storage of design data, a database and programs which control the data flows, such as a database manager and a tool manager. It is assumed that several users work at the same VLSI design. A good cooperation between the users is essential. A project manager should look into the problems and the time schedule.

A new project, Common Frame, is being carried out under the JESSI (Joint European Submicron Silicon Initiative) program. The overall objective of the JESSI Common Framework project is to provide an open, integrated CAD/CAE environment for IC designers. Though domain-neutral services are supported, the JESSI project focuses on CAD facilities applied to microelectronic circuits and systems. Prototypes of this CAD framework are SIGMA (Philips, Siemens and SGS-Thomson), NMD-CAD (Nmp-Cad and Cadlab) and NELSIS (Delft Univ. Tech., DIMES).

DIMES and DDTC

DIMES (Delft Institute of MicroElectronics and Submicron technology) is a national Dutch institute that stimulates research in microelectronics and submicron technology. It provides facilities for bipolar and CMOS production processes. DIMES participates in research projects in cooperation with the faculties of Electrical Engineering and Applied Physics of the Delft University of Technology and with other institutes in Europe.

DDTC (DIMES Design and Test Center) provides software and hardware support for VLSI design and test within the Delft university environment. DDTC supports and develops the NELSIS software package (see below). It offers an assortment of advanced VLSI design and verification tools based on standards, such as X Window version 11, UNIX, CIF, GDSII, SPICE and an in-house software framework.

NELSIS

NELSIS (NEtherLands System In Silicon) is a framework for designing VLSI circuits and systems resulting from a cooperative research and development project. It was partially funded by the Commission of the EEC, the Dutch Government under the NELSIS program and the participating partners: Delft University of Technology, Eindhoven University of Technology, Twente University, British Telecom Research Labs (Ipswich), PCS (Munich), INESC (Lisbon) and ICS (Utrecht).

NELSIS is an open design system to be operated from designers' workstations. The NELSIS framework allows a chip designer to retrieve relevant information on the design process, to select design objects and activate design tools in a uniform and integrated fashion [Dew86, van88, van89, van90, vAn90, vAn91, van93]. Its kernel is a configurable meta data storage module, based on the semantic data model OTO-D (Object Type Oriented Data model) [ter86, vAn88, ter91]. The NELSIS architecture is shown in Fig. 5.35.

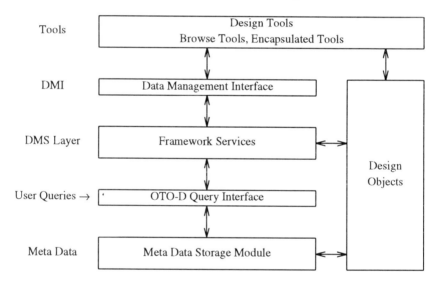

Fig. 5.35 NELSIS system architecture

Meta data are concerned with global system aspects related to design objects rather than detailed design aspects. A query interface allows the user to access the meta data. An OTO-D Data Schema permits information to be retrieved and stored. The DMS (Data Management System) layer contains framework services, such as versioning and design management, developing tools for hierarchical verification and keeping track of the design process by framework browsers, which make high-level information about the structure and status of the design available to the designer.

A design object may contain detailed descriptions, e.g., a net list, a circuit diagram or a mask description. An object type is defined in terms of its

5.2 HIGH-LEVEL SYSTEM DESIGN

attributes: module, version number and status. Design objects may be related by either hierarchical or equivalence relationships. Meta design data contain information about the hierarchical decomposition of the design, the status of the design objects, their version history, etc. The meta data are collected by the framework, while the design tools communicate with it to obtain access to the actual design descriptions.

Design tools, browse tools and encapsulated tools may be invoked via the DMI (Design Management Interface) [van87]. The DMI hides the tools from the other framework parts of the NELSIS system, allowing the DMS and the design tools to evolve separately. Whenever appropriate, a design tool (e.g., a simulator) is applied to a design object. Using a transaction schema, the tools obtain access to the design data.

The Design System User Interface (DSUI) provides a user-friendly working environment for designers [Bin90]. It exploits the facilities offered by modern workstations to enable convenient and effective user interaction based on the X Window System. DSUI is the central platform from which all design activities can be supervised. Depending on the structure and the status of the design, activities can be initiated. The DSUI combines a meta design browser for information retrieval from the database with a CAD shell for tool activation.

NELSIS design tools support the synthesis, verification and testing of VLSI circuits and systems. HIFI allows the designer to synthesize VLSI processing circuits (see below). The schematic entry ESCHER (Eindhoven SCHematic EntRy) is an interactive graphical program for creating and modifying a hierarchical circuit diagram. ESCHER has dedicated options, such as commands for editing buses and signal types, and a built-in checker for network consistency and electrical rules.

Design verification of MOS circuits can be performed by the switch-level timing simulator SLS (see Subsection 6.2c). Circuits to be simulated using SLS can be described at different levels, including the transistor level, the gate level and the functional level. Using piecewise-linear voltage waveform approximations, SLS computes min-max voltage waveforms to account for circuit-parameter inaccuracies. SLS is in fact a switch-level simulator which provides reliable delay computations [van91].

An automatic placement and routing system was developed to place and route designed circuit blocks in a hierarchical manner. A fast and powerful gridless channel router generates compact wiring patterns, particularly in a sea-of-gates structure [Gro91]. The Delft Placement and Routing System allows the efficient implementation of a full-custom general-cell VLSI chip. A slicing channel structure in combination with the channel router guarantees a 100% complete routing result.

SPACE is a layout-to-circuit extractor that finds parasitics in submicron integrated circuits, including resistances and capacitances computed by finite-element methods [van88]. By employing a new approximate matrix-inversion technique, SPACE is capable of rigorous finite-element modeling of VLSI circuits even on small workstations.

HIFI

HIFI (Hierarchical Interactive Flow-graph Integration), developed at the Delft University of Technology, is a design methodology for implementing signal and image processing algorithms on VLSI processor arrays [Ann88, deL89, deL91, van92]. The high-level design process consists of the translation of a numerical problem to a parallel algorithm and the subsequent mapping of this algorithm on either dedicated VLSI hardware or a programmable multiprocessor system.

HIFI supports the generic specification of both the behavior and the structure of a design, including the use of functions and cells. The system uses a special form of a signal flow-graph (i.e., nodes connected by directed arcs). A HIFI node is an abstraction of a processor element that can perform any one of a set of functions. The HIFI process of designing a multiprocessor system consists of several steps:

1. The starting point is a technology-independent specification at the functional level. The system behavior is specified as an Applicative State Transition (AST), which is an input-output map.
2. An algorithm may be given in the form of imperative nested loop programs, e.g., *do* loops in FORTRAN or *for* loops in Pascal-like programming languages [Bu90]. These programs can be converted into regular algorithms.
3. Construction of an algorithm that satisfies the AST specification. Its graphical representation is a dependence graph, which explicitly contains all the parallelisms present in the algorithm. The dependence graph is a static representation of the program, i.e., independent of time with stateless nodes and no loops.
4. Space-time partitioning of the dependence graph, i.e., mapping it onto a data-flow graph which represents a processor architecture. The data-flow graph can be repeatedly refined by mapping the input-output functions of nodes into refined structures.
5. Synchronization of the data-flow graph by assigning clocks to nodes in the data-flow graph and by partitioning the network in ripple zones to obtain reduced processor arrays. Execution of the delay timing model is based on using Petri Net Theory [Wal83].
6. Clustering of nodes to obtain fixed-size processor arrays.
7. A partitioning and clustering process generate controllers, memories and bus structures in a correct-by-construction fashion. Parameterized arithmetic building blocks are instantiated from a data-path library.
8. The complete architectural specification, including the net lists and the clock strategy, is transferred to the NELSIS VLSI silicon compiler that includes

5.2 HIGH-LEVEL SYSTEM DESIGN

floorplanning, placement and routing.

Tools supporting the design process in the HIFI environment include: the Graphics Design Entry HiEntry, the HIFI simulator HiSim, and interactive graphics tools for the synthesis of massively parallel pipelined processor arrays. An illustrative application of the HIFI design system is the design of a systolic CORDIC processor for massively parallel operations in computational algebra and signal processing [deL91].

5.2d Expert Systems for Logic Circuit and System Design

Introduction

Due to the fact that in general a great many hardware solutions meet a given design specification, the designer has to choose between various acceptable alternatives. Decisions must be made as to the implementation technology, the design style and trade-offs among conflicting requirements. Optimizations in terms of operating speed, chip area and design time force the designer to make decisions on the initial goals to be met.

A design often involves an iterative cycle of: design specification → synthesis → evaluation of the outcome of the synthesis → modification of the design specification or circuit → synthesis (possibly using a different design style and implementation technology) → etc. This sequence of decisions to be made in several phases of the design cycle lends itself very well to a knowledge-based approach of VLSI circuit design.

To give an idea of how rules can be used in the high-level synthesis stage of the design cycle, a number of implemented design systems will be discussed below.

Local Transformations

In Subsection 5.2c, a short account is given of the local transformations performed in the LSS design system. Some activities in the design process has the aim to transform an intermediary circuit solution to either a more optimal one or a solution which is more suitable for circuit implementation in the desired implementation technology. For example, suppose it is desirable to restrict the type of gates to only NANDs or NORs. In the fourth stage of the LSS procedure, the AND/OR circuit is converted into a NAND/NOR implementation. LSS uses nine rules to perform these conversions [Dar84].

Logic Reorganization System LORES-2

The rule-based system LORES-2 transforms an existing logic circuit into another circuit of a different implementation technology, or into a circuit with improved performance [Eno85]. The rule base consists of different reorganization rules: for assigning macrocells available in the target technology, for circuit optimization (such as deletions of redundant elements), for satisfying design constraints (such as load adjustments), and for standardizing a source circuit according to a certain regulation.

DDL/SX
This expert system, developed by Fujitsu Laboratories [Kaw82, Kaw85. Ueh85], converts technology-independent function diagrams into technology-dependent logic diagrams. The objective is to achieve an optimal design, which satisfies the design constraints, e.g., with respect to fanout and fanin of gates. Functional modules are replaced by appropriate gates or IC packages in the desired technology, while redundant gates are removed. The knowledge about IC packages is stored hierarchically using frames, which can be readily updated.

Another version of DDL/SX uses macro-expansion and local transformation to convert high-level function diagrams into technology-dependent circuits, particularly CMOS gate arrays [Sai86]. A function diagram includes arbitrary bit-width decoders and registers as well as ANDs, ORs and EXORs. DDL/SX transforms function diagrams into circuits by local transformations.

Logic Synthesis
Yoshimura and Goto [Yos86] described a logic-synthesis system based on rule-based and algorithmic methods. The knowledge base contains rules for local transformations and logic minimization. Physical constraints, such as the longest path between registers, fanin and fanout, and polarity propagation are checked whenever a rule is applied.

Kurosawa *et al.* [Kur90] of Toshiba Corporation described another expert system for logic synthesis. Given the register-transfer description of a logic circuit, the expert system automatically generates the logic circuit composed of functional elements registered in a cell library. The synthesis procedure consists of three phases:

a. The input description is translated into a data structure of technlology-independent functional elements.

b. Circuit transformations: expansion of abstract functional elements into AND/OR gates, elimination of redundant logic, and the assignment of library cells to functional elements.

c. The logic circuit is transformed into a net list, representing the connection relations among cells.

The knowledge base contains knowledge about the transformation between abstract levels, optimization at an abstract level, transformation from abstract level to cell level, and optimization at the cell level. In addition, meta knowledge controls the application of basic knowledge and its ordering.

5.2 HIGH-LEVEL SYSTEM DESIGN

VEXED

VEXED (VLSI EXpert EDitor), developed at Rutgers University, New Brunswick, is an interactive knowledge-based consultant for VLSI design [MIt85]. VEXED works with three categories of domain-specific knowledge: knowledge of implementation methods, control knowledge and causal knowledge. A set of functional specifications is transformed into a network of submodules, each of which is iteratively refined to lower-level submodules, until eventually the detailed circuit is formed. During transformations to lower levels, special requirements are imposed on the interfaces between submodules.

VEXED propagates constraints forward as well as backward through the circuit in order to assure that all interactions among subproblems are considered. An interesting feature of VEXED is the possibility of expanding the rule base by learning from previous situations. To that end, a user's solution to a design subtask provides a training instance from which a general rule might be inferred.

DTAS

One of the key issues to be addressed in chip design is technology adaptation, i.e., the task of maintaining the integrity of a design system against technology changes. DTAS (Design and Technology Adaptation System) is a rule-based design system that incorporates technology adaptation in its knowledge-acquisition process [Kip91]. DTAS synthesizes generic logic components from a library of technology-specific cells. A learning tool LOLA (LOgic Learning Assistant) uses knowledge of fundamental principles for logic design to generate design rules that map generic components into library cells. As technology develops, new library cells become available so that DTAS can be used in updated form.

SYNAPSE

SYNAPSE (SYNthesis Aid for Parallel System Environment) is a means to explore language and expert system issues in VLSI design [Sub86]. A main goal is to exploit the common basis for different levels of abstraction in VLSI chip design, avoiding the problem of disjointed concepts at each level. This requires a translation process consisting of repeatedly transforming expressions that represent system descriptions at lower levels. Other features include the cohesive formal mathematical framework with algebra-based synthesis, verification and analysis tools, the use of expert-system tools and accomodation of machine-learning techniques.

Design Automation Assistant (DAA)

DAA [Kow83, Kow85] is a knowledge-based high-level synthesis program which has been developed within the framework of the CMU-DA project (see Subsection 5.2c). An algorithmic description of a VLSI circuit is transformed into a technology-independent representation consisting of operators, registers, data paths and control signals. DAA is implemented as a production system written in OPS5 code running on a LISP-based system. A newer version based on the BLISS language is about ten times faster.

The knowledge required for DAA has been acquired partly from established knowledge taken from textbooks, partly from previously developed algorithms [Haf82], and partly by interviews with human design experts. The information was processed into a knowledge base, then applied to practical design cases, discussed with the experts and modified, where necessary. The knowledge base can be divided into three main parts:

a. *Working memory*. This part of the knowledge base describes the current situation in the problem-solving process and contains the attributes with their values as needed in using the expert system. It is comparable to a data structure in a conventional programming language:

$$\text{structure} \quad \{ \quad < \text{attribute1} = \text{value1} > $$
$$< \text{attribute2} = \text{value2} > $$
$$\ldots\ldots\ldots\ldots\ldots\ldots $$
$$< \text{attribute}n = \text{value}n > \quad \}$$

b. *Rule memory*. This component contains the knowledge in the form of conditional statements:

if $\quad < \text{antecedent1} >$
$\quad\quad < \text{antecedent2} >$
$\quad\quad \ldots\ldots\ldots\ldots\ldots$
$\quad\quad < \text{antecedent}k >$
then $\quad < \text{consequent} >$

c. *Rule interpreter*. This part interprets the relations between the rules and the facts in order to determine in a given situation which rules apply to which facts. This is accomplished by matching working-memory elements against the rule memory. The rule memory is searched for a rule whose conditions are all true. A distinction should be made between data driven and goal driven selection.

The rule memory is the heart of the expert system in that it contains the rules as formulated by the expert whose knowledge has been exploited for the creation of the system. Unfortunately, the expert is not able to put his or her knowledge directly in the knowledge base. It takes much time and insight to translate what the expert has to say into rules. The value of the expert system depends directly on the correctness of the rules in the rule memory. If some rule is not correct, the results obtained are unreliable.

The biggest problem in setting up an expert system is the acquisition of the information. The coding of this information afterwards is only a matter of finding a suitable representation and scrutinously transforming this information into code. The problem in interrogating specialists is such that the specialists often take big steps in their reasoning, because many things appear natural to

5.2 HIGH-LEVEL SYSTEM DESIGN

them. A complete set of rules must be used like a chain without a missing link. In the case of DAA, four human designers with different experiences were interrogated, including one with little experience. The reason is the hope that such a designer also tells design steps which have been so natural to an experienced designer that the latter does not talk about these steps.

After having acquired the knowledge from the literature, the first round of interviews and using knowledge stored in an existing allocation program, the knowledge must be brought in coded form. Use is made of the OPS5 KBES writing system. OPS5 is a tool which, based on the human way of pattern recognition, divides a problem in working memory, rule memory and rule interpreter (see Subsection 4.2b). To verify the adequacy of the acquired rules, these rules are applied to a number of real problems. The prototype DAA system which was used to design MCS6502, a microprocessor of MOS Technology Inc., contained 70 rules. Based on critiques of human experts, rules were modified and new rules were added. Eventually, the rule base contains more than 300 rules. The redesign of the MC6520 required more computation time, but the quality of the result was improved.

The DAA system is organized by the stages shown in Fig. 5.36.

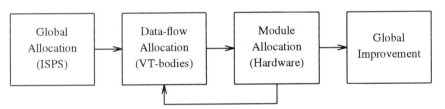

Fig. 5.36 DAA system

With a target system in mind, the design follows a certain pattern, starting at a high level and going down to lower levels. The input is a functional description written in ISPS. Global allocation includes the assignment of storage modules (such as registers and memories) to values declared in the ISPS description. The input is converted into a Value Trace (VT), which consists of VT bodies, each corresponding to an ISPS procedure, loop or block. This data-flow allocation step is followed by the module allocation step.

The Value Trace is converted into a network of functional blocks. A technology-independent data-path synthesis results in an interconnection of ALUs, registers, buses, etc. A controller synthesis includes the allocation of the clocking cycles. The eventual hardware implementation is performed by a module binder. Area, dissipation and operating speed with a given implementation technology (e.g., NMOS or CMOS) are criteria which govern the decisions to be made. A functional block which has been designed earlier and stored in a module database can be used, when needed.

An optimizing loop leads to a network consisting of hardware modules, connecting links and multiplexers, where necessary. Superfluous registers are eliminated or modules are combined. For example, several function operators are combined into ALUs or increment, decrement and shift operations are performed by the same register. The local optimization is followed by a global one, in which unnecessary ports, multiplexers or buses may be eliminated. For example, duplicated multiplexer input ports are combined. Where appropriate, registers, memories and ALUs are assigned to buses. DAA seeks to find a parallel design. Only when strictly necessary, for example, when too many hardware modules are needed to support the proposed parallelism, a serial design is carried out.

The result of the synthesis process is a technology-independent representation of memories, operators, registers, data paths and timing signals. The data structure is divided into the algorithmic representation Value Trace, the technology database (which sets technological constraints on the final design) and the technology independent hardware description (given in the Structure and Control Specification language). The control structure is divided into four control functions.

According to Kowalski [Kow85], two categories of rules can be distinguished by the knowledge type:
a. Domain specific knowledge, which is specific to the IC design task. Rules in this category include the way in which registers can be shared.
b. Domain independent knowledge. Rules include overhead tasks, such as counting fanouts or cleanup of working memory elements.

A new version of the DAA system combines top-down and bottom-up analysis, global and local optimization, and procedural and rule-based methods [McF87]. The main extension to the DAA system is the use of the program BUD (Bottom-Up Design), which acts as a preprocessor to DAA [McF90]. The improved global analysis and bottom-up evaluation provided by the BUD lead to more effective scheduling and resource allocation.

ULYSSES
An offspring of the CMU-DA project of Carnegie-Mellon University is ULYSSES (Unified LaYout Specification and Simulation Environment for Silicon) [Bus88]. To facilitate the integration of various design tools available to ULYSSES, a common blackboard stores three types of information: design data, assertions about tool failures and scheduling parameters for each knowledge source (or design tool). Each CAD tool is mapped to a knowledge source. One knowledge source, the scheduler, decides which CAD tools will be activated when there is a conflict. The decision is based on conflict resolution parameters associated with each knowledge source.

5.2 HIGH-LEVEL SYSTEM DESIGN

SOCRATES
Given a functional specification in the form of a Boolean expression, a rule-based system developed by Gregory *et al.* [GRe84] generates an optimal implementation of combinational logic. The Boolean expression is mapped into a network of multiplexers and inverters by using binary-decision diagrams [Ake78]. Incremental optimizations are achieved by successive replacements of local configurations.

SOCRATES (Synthesis and Optimization of Combinatorics using a Rule-based And Technology-independent Expert System) optimizes logic using Boolean and algebraic minimization techniques. It uses a rule-based system to optimize circuits derived from this logic in a user-defined technology [Gar84, deG85, Gre86]. The circuit optimizer improves measurable circuit characteristics by iteratively replacing and rearranging small portions of a circuit.

SOCRATES consists of four modules, each with a specific function:
(1) mathematical reduction and synthesis, (2) circuit-level optimization, (3) extraction and comparison, and (4) rule generation.

MICON
Subsection 4.5c reports on the original version of MICON (MIcroprocessor COnfigurer) [Bir84]. The extended version of MICON is an integrated collection of programs which automatically synthesizes small computer systems with rapid prototyping [Bir88, Bir89]. Starting from high-level specifications, it addresses multiple levels of the design down to the logic and physical level. MICON relies on two major knowledge-based tools: the synthesis tool M1 and an automated knowledge-acquisition tool CGEN [BIr89], which is used to teach M1 how to design.

M1 performs a hierarchical, stepwise-refinement synthesis process leading to the selection of existing physical parts (processor, memory, peripheral). CGEN (Code GENerator), which is tightly coupled to M1, contains knowledge about various microprocessor families, brought in by domain experts during a training process.

CHIPPE
Chippe is a design system for constrained behavioral architecture synthesis [BRe90]. It uses algorithmic as well as a rule-based expert system. In a closed iteration loop, the design is evaluated with respect to the goals and then modified for the next iteration. The objective is to meet the global constraints imposed by the designer in a way as indicated by the "knobs and gauges" approach, illustrated in Fig. 5.27. A design evaluator examines the present state of the design and reports its findings to the expert system. Driven by the global constraints and the present state of the design, the expert system makes design trade-offs.

PALLADIO

Palladio is a flexible CAD environment, developed at Stanford University with the purpose of experimenting with methodologies and rule-based tools for the hierarchical design of VLSI circuits [Bro83]. Within Palladio, CAD tools as well as description languages are integrated in a universal concept.

Palladio is implemented in the programming environments Interlisp, LOOPS and MRS, which offer various semantic concepts: rule based, data oriented, object oriented and logical reasoning. Several tools make use of knowledge of human experts stored in a knowledge base. All kinds of heuristic knowledge can be supported by the system, such as relations between certain decisions at the higher levels of abstraction (e.g., the architecture) and their consequences at lower levels. A level of abstraction is called *perspective* in Palladio terminology.

The design methodology of Palladio is a process of stepwise refinement of the system or circuit specification, alternated by a periodic validation of these specifications by means of simulation. It is possible to specify the behavior as well as the structure of a design, or parts of it, simultaneously and to refine these specifications step by step.

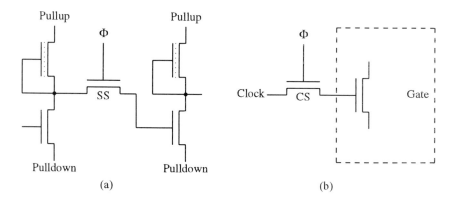

Fig. 5.37 CSG component types of Palladio

An example of such a methodology is the CSG (Clocked Switches and Gates) perspective, specially directed to the class of NMOS circuits, which use two-phase clocks. To this end, the CSG perpective contains a number of primitive building blocks (see Fig. 5.37) with the associated composition rules (constraints). The most important composition rules are:

a. The input of a "steered switch" (SS) can only be connected with the output of a logic pullup-pulldown gate. This avoids devaluation of the logic level by a sequence of threshold drops.
b. The input of a "clocked switch" (CS) can only be connected with one of the basic clocks.

5.2 HIGH-LEVEL SYSTEM DESIGN

c. The output of a clocked switch can only be connected to the input of a logic gate. This avoids certain cases of charge sharing.

In addition, there exist alternative composition rules related with the fanout, the maximum length of a chain of pass transistors, etc.

Within Palladio, system behavior is represented as "rules" which can be activated by changes in the internal state of a circuit (or parts of it). This approach has a number of advantages:

a. The rule concept is surveyable and comprehensible for the user, notably because of the universal character of the concept.
b. For the different levels of abstraction, it is relatively simple to define rules, which appropriately describe the behavior of the primitives at these levels. An example of a rule for a pass transistor at the CSG level is:

if Signal(Port CLT) = HIGH at time t
$then$ Signal(Port OUT) = Signal(Port IN) at time t+1

c. There is a possibility to carry out mixed-mode simulations, in which the different levels of abstraction and the hierarchical character of the design are utilized. In this way, parts of a circuit can be simulated considerably faster.

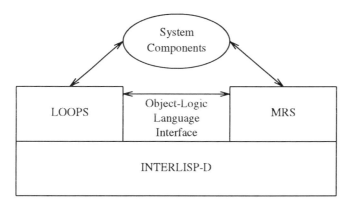

Fig. 5.38 Palladio system architecture

The properties of a rule-based, event-driven simulator are mainly determined by the environment in which it is implemented. The Palladio environment, called Multilevel Reasoning System (MRS) contains data in the form of assertions which describe the internal state of the circuit and a number of rules which on the basis of the existing assertions can derive new assertions. In addition, the system provides the facility, though in a rather primitive way, to explain the way in which and with which rules a certain result was reached. Finally, the results of a simulation can be displayed on a graphical screen.

The structure of the Palladio system is implemented in the object-oriented environment LOOPS. A schematic of the system architecture is given in Fig.

5.38. The basic entities in the Palladio environment are the object and the message. In this environment, the important system components (editors, simulators, etc.) as well as the units which are manipilated by the user (wires, circuits, prototype components, etc.) are viewed as objects. Since large groups of these objects have common properties, it is expedient to describe these as a class. Individual objects which differ only in specific parameter values are the instances of such a class. The larger system components, such as a simulator, can be started by sending the message ACTIVATE. The data structure of a LOOPS object is a frame, consisting of a number of attribute-value pairs. The classes within LOOPS are organized in class inheritance network.

Palladio provides the designer the ability to decide on optimization issues. In the layout phase, a desired point on the power/speed curve can be selected, after which all transistors are resized. A companion system, Helix, renders data paths in a range of colors that depend on the intensity of data flow. Palladio's simulator, MARS (Multiple Abstraction, Rule-based Simulator) uses inference rules to determine how and when new assertions are added.

HAL

HAL (Hardware ALlocator) is an object-oriented data-path synthesizer, developed on a Xerox LISP machine using LOOPS [Pau86]. The goal of the performed design space exploration is mainly the optimization of the sharing of data path resources like ALUs by the minimization of concurrency. The necessary decisions are rule based, whereas the design objects are represented by hierarchical frame-based classes.

The HAL system is constructed from three hardware-allocation modules (INHAL, MIDHAL and EXHAL), each with its own heuristic and algorithmic method to carry out its task. INHAL converts a data flow-graph into a state-graph on the basis of simple rules. MIDHAL performs operation-component bindings on the basis of the state-graph, timing constraints and a component table. From these bindings, cost/performance estimations are made. Using heuristic rules leads to alternative state-graphs and component bindings from which an optimal design can be selected. EXHAL converts the generated state-graph into a microarchitecture by performing the definitive operation-component bindings and using detailed component data.

FRED

Fred, described by Wolf [Wol89], is another object-oriented approach to support logic synthesis. Similar to HAL, operations and operators or components are related by so-called functional and generic relations. The design space exploration is supported by the message passing mechanisms of object-oriented programming.

5.2 HIGH-LEVEL SYSTEM DESIGN

Rule-Based Design Assistant

Drongowski [Dro85] described a rule-based system that assistss designers making control-graph and data-path trade-offs. The system is writted in CProlog and runs on Apollo workstations. A design goal is to increase the execution speed of a control graph and to reduce the component count. A user-friendly graphical interface is used to improve interaction between the designer and the rule-based system. When design flaws are detected, suggestions are made to improve the design.

ProLogic

ProLogic, developed at Hitachi Limited, is a programming methodology for logic synthesis which unifies the behavioral and structural specification of the hierarchical architecture of the design [Ham90]. Logic programming, object-oriented frames and a rule base have been implemented in Prolog. New knowledge of a functional module, e.g., a microprogram control module, can be easily described.

Rule-Based Object Clustering

Traditionally, separate data structures are used to support analysis (node and net lists) and synthesis (e.g., CIF). Larsen [Lar86] introduced the concept of object clustering as a data structure to support interactive analysis as well as incremental synthesis of custom VLSI devices. A top-down symbolic design methodology is assumed that embodies the management of design complexity through the principles of hierarchy, modularity, regularity and correctness by construction. Object clustering refers to the collection of symbolic objects to form a group which possesses some common property, in this case, the electrical continuity. The clustering of symbolic objects to form an efficient VLSI design data structure is controlled by a set of rules residing in the technology file which represent the present state of the given process.

MBESDSD

MBESDSD (Model-Based Expert System for automated Digital System Design) is an expert system that generates digital system designs from high-level specifications [WU90]. This system is model-based which means that the reasoning process is based on a model of the objects in the application and the specific operations that act on those objects. In MBESDSD, the objects in the behavioral specification are various electrical signals acted upon by five primitive operations: AND, OR, NOT, STORE and TRANSMIT.

The design process consists of three phases:
1. The high-level behavioral specifications are translated into a sequence of primitive operations.
2. These operations are grouped to form intermediate-level behavioral functions.

3. Structural function modules are selected to implement these functions.

The model-based reasoning, based on primitive operations, allows the designer to refine the set of given specifications with a great deal of flexibility to optimize the design.

Intelligent Librarian

Many design systems take advantage of the use of libraries containing predefined modules, which can be retrieved when needed. As various modules become complex and their variety grows, the selection of the appropriate module requires some intelligence.

An Intelligent Librarian [Ho85] has been developed to assist the designer in choosing the most appropriate module from the library. Heuristics direct the search process to the desired module. If an exact match with the desired module cannot be found, a module with relaxed requirements is chosen. On the other hand, if several matches exist, the requirements in the specification are made tighter.

VLSI System Planning

VLSI System Planning [Dew90] is a design tool meant for aiding designers during the conceptual stage of a VLSI design. A conceptual design may result in either an efficient implementation or the conclusion that the initial specifications are not realizable.

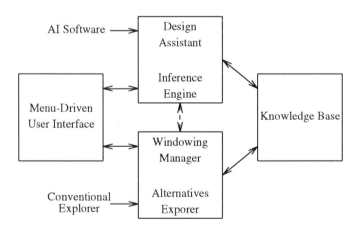

Fig. 5.39 VLSI System Planner

Proper decisions lead to a particular design plan which is followed by the detailed design phase. The final design is obtained through iterative improvements. The idea behind this approach is that the number of redesigns is minimized by selecting the proper design plan in an early design phase. This selection process is performed by the VLSI System Planner. Figure 5.39

5.2 HIGH-LEVEL SYSTEM DESIGN

illustrates the system architecture of the VLSI System Planner.

The User Interface is based on windows (for diplaying different pieces of information concurrently) and mouse-activated pulldown menus (for invoking the System Planner's commands). The Knowledge Base supports the procedure-driven Alternatives Explorer and the Intelligent Design Assistant. The AI Design Assistance subsystem has a blackboard architecture with advice and prediction "experts" as knowledge sources. An advice may be a preferable choice from available options for a particular design issue. If the selected option is quantitative by nature, the associated prediction must be computed.

The VLSI System Planner has been used for designing digital filters. Such a design involves decisions as to choosing the right filter algorithm (finite or infinite impulse response), filter type (Butterworth, Chebyshev, elliptic), semiconductor technology (e.g., CMOS or BiCMOS), etc.

Automatic Chip-Partitioning System
APS (Automatic Partitioning System) is a problem solver for performing chip apportionment, particularly of CMOS gate-array chips [Mou88]. Given a complete digital-system specification, the output of APS is a set of suitable partitions each of which can be implemented in a chip. Several alternative chip apportionments are presented for each design. Objectives include the minimization of the total number of chips and the interconnections between chips. APS is implemented in the LOOPS environment on a Xerox LISP Workstation.

CHARM
CHARM (CHip ARchitecture Maker) is a chip-architecture planner [TEm88]. The knowledge base consists of chip-architecture schemes containing architecture principles, refinement strategies and evaluation knowledge. A given formal algorithmic behavioral description is transformed into a structural chip-architecture plan at the register-transfer level. In a stepwise manner, well-suited schemes from the knowledge base are instantiated until the best one is selected.

CHARM has been implemented in a blackboard architecture. The knowledge sources represent four experts: for process mapping, memory allocation, data transfer and control transfer. These experts are supported by the regularity assistant, the timing assistant and the tactician.

USC ADAM Design System
The ADAM (Advanced Design AutoMation) system, developed by the University of Southern California, unifies a number of chip design tools into a single framework [GRa85, Afs86]. Its goal is to produce correct, testable implementations as a result of trade-offs between various characteristic design features. ADAM combines hard-coded and knowledge-based techniques. It uses a mixture of three basic kinds of representation: rules, frames and semantic

networks.

The design process traverses a planning phase and an execution phase. Planning is a sort of stepwise refinement process. When a satisfactory design emerges in the planning phase, the execution phase can be started. As soon as violations of the design constraints are detected, control is returned to the planner.

The ADAM Design Planning Engine [Kna91] is an expert system that uses a set of planning rules adapted to the needs of a particular set of specifications and constraints. Expert systems dedicated to specific tasks include the fault-diagnosis system TDES (see Subsction 7.3c) and a logic synthesis system. A database provides a common representational scheme for designs, programs and constraints.

Analogical Reasoning in Logic Synthesis
Human designers exploit their learning capacity to build new designs upon their experience with previous design efforts. Relatively few expert systems have been provided with some learning mechanism which can store valuable information on previous design results. An attempt to use a learning tool in an expert system is LEAP [Mit85], which is incorporated in REDESIGN (see Subsection 6.2d).

Acosta *et al.* [Aco86] proposed an analogical reasoning approach to digital system design in which old design efforts can aid in solving new problems, even when there is only a partial match between old and new problems. A behavioral description written in VHDL is translated through a refinement process into a structural specification. A sequence of rules transforms and decomposes a design problem by building a hierarchical design tree. Circuit-design rules are applied deductively to transform and decompose VHDL specifications of a design problem. Because a hierarchy of independent subentities is created by this process, rule sets leading to final designs are partially ordered when they are saved as plans. These plans can be followed to design circuits for new problem specifications. In order to make use of the previous problem-solving experience embedded in a plan, portions of it can be used to solve, at least partially, the new problem. Abstracted plans have more general preconditions and are applicable to a wide class of problems. These plans are analogically transformed to suit specific problems by appropriately filling in missing steps. The system employs rules for the circuit synthesis written in Common LISP.

High-Level Synthesis Learning Apprentice
Herrmann and Witthaut [Her92] described LEDA (LEarning Design Assistant), which is a learning apprentice system that acquires design plans for high-level synthesis of integrated circuits. Given a behavioral algorithm and a specification of the design goals (minimal chip area, maximal speed), LEDA selects a design plan, i.e., a sequence of allocation operators that creates register-transfer

5.2 HIGH-LEVEL SYSTEM DESIGN

components according to the statements of the algorithm. Every time a synthesis problem has been solved, a learning apprentice system memorizes and generates the situation by using an inductive learning mechanism. Should a known situation occur later on, LEDA proposes to execute the corresponding design plan.

Intelligent Signal Processing
Signal processing involves the use of numerical algorithms for quantification, filtering, frequency transformations, etc. [Bat89]. More and more, knowledge-based techniques penetrate into the domain of signal processing. Intelligent signal processing combines AI capabilities (symbol manipulation and knowledge representation) with the numerical and mathematical tools of signal processing.

Past approaches attempted to break down the problem into separate signal-processing and AI components. An early application is the Hearsay-II speech-understanding system developed at the Carnegie-Mellon University [Erm80].

Essential differences exist between AI and signal processing. Signal-processing algorithms are typically based on mathematical models. They prescribe a sequence of instructions to be executed, whereas AI programs specify a strategy rather than a sequential procedure. Knowledge in AI systems usually has a heuristic basis with a symbolic representation. A functional paradigm for the signal-symbol combination is a signal-processing front end, which extracts features of an input signal, followed by a symbolic inference unit for further processing of the front-end results.

Researchers have come to realize that a close and effective interaction between AI and signal processing may be instrumental to successful results. There is no consensus as to which form this interaction should take. Heuristics and AI appear to be useful in various aspects of signal processing, ranging from the selection task of what to do, what type of algorithm to employ and how to tune the selected algorithm [Bro88].

An expert system can be used for digital-filter design [MAr90]. A curve that approximates the desired magnitude plot is sketched by the designer by using a mouse. The expert system derives a set of transfer-function coefficients from a set of points which defines the above curve. The system determines design parameters, computes the transfer function and finally produces a magnitude plot. The current filter possibilities (IIR Butterworth, Chebyshev I, Chebyshev II) are designed as cascaded second-order sections with real coefficients.

Glover *et al.* [Glo88] described a coupled system for knowledge-based signal understanding. In their view, signal understanding requires expertise in the use of signal-processing algorithms but also knowledge of how to select the appropriate algorithm for the task at hand. Their system maintains two separate, collaborating knowledge bases, one for the signal-analysis knowledge and one for the domain-specific knowledge. An object-oriented representation is used for both fields of expertise. A vocabulary (or language) is provided to facilitate interaction between the two experts and the two knowledge bases.

Gass et al. [Gas87] described an expandable, multiple digital signal processor architecture with a symbolic processing host. This host is Texas Instruments' Explorer, a LISP machine workstation, which provides an environment to perform many intelligent signal-processing tasks by associating meaningful relationships between quantitative (signal processing) and qualitative (symbolic processing) entities to develop inferences using expert system technology. A multiple processor board, called Odyssey, has been developed to operate with Explorer to facilitate these tasks. Implemented applications include the Connected Word Recognizer, a text-to-speech system using natural language and digital signal processing, neural-network simulations and an EEG analysis.

Sarkar et al. [Sar90] described a knowledge-based system for designing high-performance VLSI chips for digital signal processing. The high-speed requirements have been met by using a unique systolic architecture based on a special-purpose cell which utilizes the properties of finite ring arithmetic. The knowledge base was designed to be an extensible system, where tutorial help is available through elaborate explanation facilities and also to serve as an input specification tool.

Design Frameworks

IDEAS (Integrated DEsign Automation System) is a CAD framework developed at AT&T Bell Laboratories [Meh87]. IDEAS/Executive provides a common user interface to all the tools through a linkable interface library.

TACOS (TAsk COnfiguration System) is an interactive expert system that assists chip designers in configuring and controlling design tasks according to their intent [Cha89]. This object-oriented system allows designers, who are familiar with VLSI design methodologies to effectively manage their design tasks without having in-depth knowledge about CAD tools.

Design frameworks (see Subsection 5.2c) have been developed with a focus on design data management and tool integration. An important part of a design process using a framework is the management of the design flow. Bretschneider et el. [Bre90] described a knowledge-based method of design-flow management. The knowledge regarding design-flow management is modeled by Predicate-Transition Petri nets and production rules. Both static and dynamic behavior of a design flow are supported. Different forms of knowledge are implemented by OPS83 and integrated in the CAD framework HILDA to guide the users through the design process.

In the NELSIS CAD framework (see Subsection 5.2c), design flow management is performed by using a flowmap [ten91]. A flowmap is built of functional units and channels.

5.3 ASIC DESIGN

5.3a ASIC Design Methodologies

Introduction

For a long time, standard modules and microprocessors have been major implementations of integrated circuits produced in high volumes. Standard modules, such as the TTL Series of Texas Instruments, were used as basic components for constructing complex systems. The microprocessor was and still is meant as a general-purpose component capable of performing a wide variety of functions. Microprocessors can solve a wide variety of problems. Unfortunately, the one-instruction-at-a-time execution of the instruction sequence requires a long time to solve even moderately complex problems.

The increasing need for chips designed for specific applications has led to the development of ASICs (Application-Specific Integrated Circuits). This has been possible by the availability of automatic design tools for nearly all stages of the design cycle [New87, Ein91]. The design methodology and tools used should match the semiconductor technology and fabrication facilities provided by the silicon foundry. When compactness of logic circuits (e.g., control circuits of cameras) is mandatory, we usually need to use full-custom chips [Har91].

An ASIC is a chip designed for a specific application dedicated to a single function or a small set of functions. Unlike general-purpose mass products (such as microprocessors and memories), which usually are fabricated in very large quantities, the ASICs may be required in small batches, although large amounts may be wanted in some applications. In any case, savings in cost and reduced development time are at a premium. Though general-purpose microprocessors and memories share a large part of the market, ASICs are gaining in importance.

Four important design criteria have to be considered, when ASIC-based systems are to be implemented:

a. A low total cost, including nonrecurring engineering cost as well as design, production and manufacturing costs.
b. Fast design turnaround time with the aim to come to market earlier than the competition.
c. Ability of the available tools to support the design approach. Suitable design tools can help shorten the turnaround time.
d. A high reliability and a performance which agrees with the requirements in the design specification.

The behavior of the ASIC to be designed may be in terms of Boolean expressions, state transitions or conventional flow-of-control statements. An important task in ASIC design is the intelligent use of automatic tools and the judicious choice of the architecture, design style, layout structure, implementation technology and chip production. The use of parallelism and pipelining has made it possible to design special-purpose hardware dedicated to solve specific problems in an efficient manner.

The ASIC designer should be aware of some pitfalls which may occur in the handling of clocks, clock signal skews, bidirectional and tri-state I/O signals, and multiple test pattern sets.

Layout Structures

There are five major *approaches to automated design* based on the *layout structure*:

a. *Standard-cell approach*, which involves the use of single or double rows of standard cells. A standard cell represents a simple logic function at the logic-gate level. The logic system to be designed must be made up of standard cells whose layouts are stored in a standard-cell library. Placement and routing can be accomplished in a fully automatic way. The basis is a library of optimal layouts within rectangles of a constant width and variable lengths. The chip manufacture includes the whole range of manufacturing steps.

b. *Gate-array approach*, which involves the use of prefabricated gate-array chips containing regular arrays of identical cells of one or more semiconductor components in fixed positions. A circuit design using a gate-array chip consists of the assignment of circuit components to gate-array cells, after which the routing phase is performed. Customization of the desired logic circuit amounts to implementing the proper connections of the gate-array elements.

c. *Design of Programmable Logic Devices*. With reasonable complexity of the circuit, the possibility of using regular structures (ROM, PLA, etc.) must be exploited. It speeds up the design time at the cost of a large layout area. Whereas the design approaches a. and b. require the logic circuit to be described as an interconnection of hardware components in the form of a parts and net list, the ROM or PLA design starts from the logic functions to be realized. By using a suitable regular layout structure of transistors, an automatic synthesis procedure can be called upon to generate the ultimate layout without the need of any placement or routing procedure. The circuit design is customized by connecting the appropriate transistors according to the input specifications.

d. *Programmable Technologies*. Programmable technologies have a short turnaround time, though at the cost of the system performance. When getting to market early is a first consideration, we may use a programmable logic device or a gate-array style, and possibly a standard-cell or macrocell style. When the product has come to stay, attention is focused on improving the performance. For low-volume applications, costly and time-consuming redesign must be avoided. Field-Programmable Logic Devices include

5.3 ASIC DESIGN

Field-Programmable Logic Arrays (FPLAs), Erasable Programmable Read-Only Memories (EPROMs), Electrically Erasable ROMs (EEPROMs) and Field-Programmable Gate Arrays (FPGAs). With FPGAs, the designer is no longer constrained by a traditional programmable logic device's architecture. Programmable gate arrays are not only used for rapid prototyping mechanism for gate arrays, but also changing system characteristics can be quickly implemented. Effective use of FPGAs requires appropriate design methods [Hil91, Pel91].

e. *Hierarchical Block Design Approach.* A very complex system is generally composed of subsystems which can be designed separately. The system may be viewed as a hierarchical structure, as indicated in Fig. 8.6. Suppose that the layouts of the subsystems are given as prespecified rectangular blocks, while the interconnections of the blocks are also specified. Then, the system design amounts to placing the blocks on a plane in such a way that some optimizing criterion is satisfied. The placement problem is usually referred to as *floorplanning*.

Below, some additional remarks relevant to ASIC design are given.

Handcrafted Layout
For very small (sub)circuits, a handcrafted layout design leads to the smallest possible layout. When the size of the circuits increases, the design time and the probability of errors will increase.

Gate Arrays
A gate array consists of a matrix of logic gates arranged in rows and columns. Usually, a fixed number of (four or more) unconnected transistors are pregrouped into a gate or a more complex cell. Logic modules predefined by templates that specify interconnections among a fixed group of transistors can be stored in a module library. Various technologies have been used (CMOS, ECL, BiCMOS, GaAs). Advanced gate arrays contain millions of transistors. By using cheap prefabricated chips containing an appropriate number of transistors, only the necessary interconnections between the transistors in one, two or three layers must be designed. As a consequence, the fabrication can be restricted to the implementation of these interconnections. This means a shorter design time and the use of a simpler production facility. Mass production of gate-array chips and the need for a small number of masks reduce the costs. When an inappropriate gate array chip is used, there is a large possibility that many transistors contained in the chip are not used. Because of the fixed wiring channel width, it is difficult to achieve optimal routing.

Sea-of-Gates Structure
A large wiring connection affects the circuit performance. The CMOS or BiCMOS *sea-of-gates* structure is an attempt to avoid this drawback by allowing

interconnecting wires to be laid over the (preferably unused) transistors. A major feature is the high density of transistors and interconnections which is achieved. When complex systems are to be designed, a mixed design style may be the solution. Assembling hand-crafted cells and macrocells, preferably in regular structures, combines a small area with a fast turaround time. The SPARC microprocessor core uses sea-of-gates mixed with prediffused, highly optimized macros.

Complex Building Blocks
As an extension to the standard-cell library, a library of predefined layout blocks of functional modules can be used. The chip design amounts to a proper placement of the blocks in such a way that the interconnections are laid in a minimal layout area. Instead of fixed layout with predefined dimensions and pin positions, use can be made of parameterized blocks, i.e., blocks which can easily be modified as to the external dimensions and pin positions.

Silicon Compilers
This automatic design style should be exploited, whenever possible. See Subsection 5.3b.

Clocking Schemes
Digital operations may be synchronous or asynchronous. A design is synchronous if the clock input to every edge-sensitive storage element is directly controlled by the primary clock generator, and there are no level-sensitive storage elements in the system. Timing requirements must be met, e.g., clock skew is a major concern (see Subsection 6.2b). The use of synchronous techniques for ASIC design is highly preferable for many reasons.

Three main clocking schemes can be distinguished for synchronous design:
a. *Single-Phase Static Design.* All storage elements respond to the rising edge of a common clock signal. When the clock supply is switched off, the system stays in its current. This explains the classification as static.
b. *Multi-phase Static Design.* This is static design with two or more phases. Two-phase clock waveforms are depicted in Fig. 5.26.
c. *Multi-phase Dynamic Design.* This is level-sensitive dynamic design with two or more phases. Dynamic logic can be easily implemented by special MOSFET circuitry which relies on the use of clock signals and temporary charges stored on capacitive circuit nodes. In order to retain the charge storage on a capacitive node during every clock period, a continuing supply of clock pulses is required. The clock rate is constrained by the RC constants associated with charging and discharging a capacitor.

Example: A one-bit section of a two-phase shift register is shown in Fig. 5.40.

5.3 ASIC DESIGN

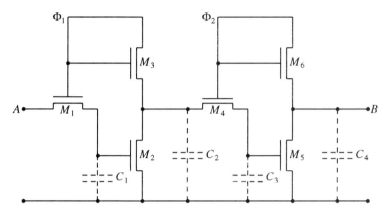

Fig. 5.40 Two-phase ratioless shift register section

Suppose that the input A is at logic 1. When Φ_1 goes to logic 1, the MOSFETs M_1 and M_3 will turn on and the capacitor C_1 will be charged to logic 1. As a result, M_2 will also become conductive. When Φ_1 returns to logic 0, M_2 remains conductive and C_2 is discharged to ground via M_2. When Φ_2 goes to logic 1, M_4 and M_6 will turn on and M_5 will be cut off so that C_4 is charged to logic 1. This completes one bit cycle.

When an asynchronous interface must be implemented, the data communication takes place on a "handshake basis", i.e., whenever the data are available on the transmitting side, the receiving side is informed of that fact and the latter accepts and manipulates the data as soon as it is convenient to do so. The implementation of the interface is different for mutually synchronous systems with a common external clock, and a synchronous system which deal with an isolated asynchronous input. The data integrity in transmitting data can be improved by inroducing more protection around the handshake mechanism, but at the cost of the system performance (e.g., a worse bandwidth).

Clock Buffering
The loading of clocks in synchronous systems may lead to unacceptable circuit delays and clock skew. There are mainly two methods for overcoming this problem: *geometric* and *tree buffering* [Nai88]. Geometric buffering uses a chain of buffers (e.g., inverters) of increasing fanout. In Fig. 5.41a, the relative fanout (i.e., the ratio of a fanout of a buffer and that of the preceding buffer) is 3. The theoretical optimum relative fanout is e = 2.71828 [Mea80].

In the case of tree buffering, the clock distribution splits into a number of branches and each branch then increases its fanout geometrically. All inverters are equivalent.

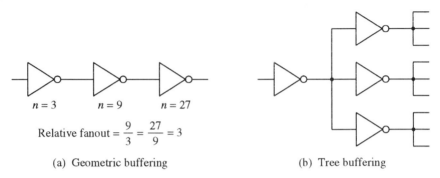

(a) Geometric buffering   (b) Tree buffering

Fig. 5.41 Clock buffering schemes

Standardization

A great variety of CAD tools have been developed in the past. This is the result of continuing efforts made more or less independently by a large number of companies and universities all over the world. An important task to be attacked is the standardization of operations, representation schemes and data formats for CAD applications. Particularly in the interface between the intelligent machine and the user, there is a need for industry standards. Some promising results have already been achieved in the area of two-dimensional graphics with GKS. Nevertheless, considerable efforts will be required to avoid incompatibilities between standards.

ANSI (American National Standards Institute), IOS (International Organization for Standardization), and IEC (International Electrotechnical Commission) support efforts of VLSI designers and manufacturers to integrate product data standards, based on the existing standards: VHDL (Very high-speed integrated circuit Hardware Description Language), Verilog, EDIF (Electronic Design Interchange Format), IPC (Institute for Interconnecting and Packaging Electronic Circuits Series 350), IGES (Initial Graphics Exchange Specification), STEP (Standard for the Exchange of Product Model Data) and PDES (Product Data Exchange Using STEP). Digital bus standards are discussed in [DiG90].

VHDL, adopted by the IEEE as the Standard 1076-87, will have a significant impact on VLSI architecture design [Wax86]. VHDL is a powerful language for describing concurrent behavior of a digital system, and covers a wide range of description levels, including structure, data flow and algorithmic-level description. VHDL features a powerful construct for describing simulator behavior. Another popular hardware-description language is Verilog of Cadence Design Systems [Tho91].

A standard format for communicating circuit and layout data is EDIF, which emerged from a joint effort of at least six semiconductor vendors [Cra85, Kah92]. EDIF, version 1.0, provides a foundry-designer interchange format for

5.3 ASIC DESIGN

gate-array and other semiconductor chip designs in a multivendor environment.

Dedicated Hardware as Design Tools

Most problems of simulation, testing and layout are solved by appropriate software programs running on conventional general-purpose computers. To enhance operating speed, special hardware has been developed to handle specific design problems [Bla84, Amb87]. Examples are hardware simulators [Kit86], and wire-routing machines [Hon83].

Usually, a hardware system is dedicated to perform one specific task (e.g., simulation or wire routing). MARS (Microprogrammable Accelerator for Rapid Simulations) is a multiprocessor-based hardware accelerator that can address a wide range of problems [Agr87]. This flexibility is achieved by using microprogrammable and reconfigurable processors. When programmed as a logic simulator, MARS is able to perform 1 million gate evaluations per second.

5.3b Silicon Compilers

Silicon Compilers and Silicon Assemblers

A *silicon compiler* is a design tool that translates a given high-level system description into the mask-artwork description of the hardware implementation. The term "compiler" is reminiscent of the software compiler that translates a program written in a high-level language into a machine-interpretable code. A silicon compiler encompasses all levels in the logic-system top-down hierarchy given in Fig. 5.42, from the algorithmic level via the logic and circuit levels to the mask-artwork level. Most published design tools presented as silicon compilers do not cover the whole range of levels and hence do not deserve the name "compiler".

It is sensible to consider a silicon compiler consisting of at least two tools which are used successively: a *high-level synthesizer*, which translates the algorithmic level to an intermediate level (e.g., the logic level) and a *silicon assembler*, which takes the intermediate level description as input to deliver the mask-artwork level description as output. The name "silicon assembler" corresponds with the software practice, where an assembler translates a program in an intermediate-level language (i.e., an assembling language) to a machine-interpretable language. A silicon compiler may then consist of a logic synthesizer followed by a silicon assembler. Figure 5.42 shows a schematic representation of the silicon compilation process. This figure indicates the actual function of the proposed tools which have been published under the name silicon compiler.

The first part of a silicon compilation process, high-level synthesis, has been discussed in Subsection 5.2c. Silicon compilers and expert systems provide two different ways to automate the VLSI design process [Par87]. Silicon compilers, which are inherently automatic design tools, presuppose a specific standard target architecture whose layout structure is governed by the particular class of applications intended. A typical example of a standard layout structure is that of the data path of a microprocessor. Expert systems allow the user to select one of several alternative architectures or design methods.

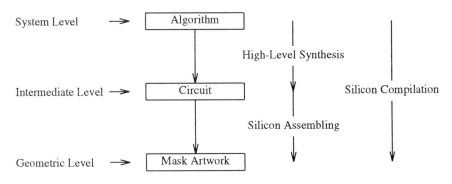

Fig. 5.42 Silicon compilation

Below, an anthology of silicon compilers is given.

Survey of Silicon Compilers

Numerous methods of logic synthesis and silicon compilation have been proposed [DeM87]. Below, a selected number of silicon compilers is described briefly.

MacPitts (a contraction of the names of two researchers, McCulloch and Pitts), a silicon compiler developed at MIT's Lincoln Laboratory [Sou83], accepts a behavioral system description and implements the data path and the (random-logic) control section of the chip, according to the declarations contained in the source program. The data section is composed of a set of bottom-level primitives, called *organelles*. In programming the design, the user has to know which organelles are available. However, the library of organelles can easily be updated.

The *ICEWATER compiler*, developed at the University of Waterloo [Pow85], is a CMOS methodology with an integrated set of VLSI design tools. The ICEWATER language allows the topology and geometry of IC layouts to be described in a hierarchical manner.

The *PLEX project* of Bell Laboratories, Murray Hill, is intended for the automatic generation of NMOS microcomputer layouts [Bur83]. The user specifies the program (either in assembly or C language) that the microcomputer is to execute. The required instructions, which are derived by the software

5.3 ASIC DESIGN

compiler of the language, determines the hardware specifications, such as the number of registers in the data path and the size and contents of the ROM.

The *Scheme-81 methodology* [Sus82] is used for the construction of large-scale single-chip microprocessors. An abstract microprogram, written in LISP, is required for deriving the specifications for the data path and the actual microcode of the microprocessor. Use can be made of module generators, which derive special layout modules for the data and control paths, and routing programs for determining the interconnections between the major blocks. Gajski [Gaj84] developed the *ARSENIC silicon compiler* which translates an instruction-set functional description in a Pascal-like language into the circuit layout.

The *CAPRI design methodology*, adopted by the IMAG Computer Architecture Group in Grenoble, France, is built upon an extensive study of the architectures of existing microprocessors [ANc83]. A major feature of CAPRI is that it can be used in a wide range of applications, such as automata, microprocessors and peripheral controllers. *APOLLON* [Jam85] is a data-path compiler incorporated in the CAPRI methodology. Given a list of steps to be executed to realize an algorithm, it produces the layout of the data-processing section of a VLSI system. *LUBRICK* [Sch83] is a silicon assembler developed for the data-path design for FISC (Familiar Instruction-Set Computer), which is an application of the CAPRI Silicon-Compiler project.

The silicon-compilation method adopted by Lattice Logic, Edinburgh, generates an intermediate-level description which can be used as a basis for simulation, test generation and further physical design [Gra82]. The compiler is primarily intended for gate-array implementations. A hierarchically structural language MODEL [Gra83] is used to create and interconnect the appropriate gate-array cells in such a way that a near-optimal solution in terms of chip area is obtained. Lattice Logic also allows the use of prefabricated CMOS gate arrays for which the metal personalization must be added.

Silicon Assemblers and Module Compilers

According to the definition of a *silicon assembler*, as given at the beginning of this subsection, PLA generators and standard-cell based design systems, which automatically provide all necessary mask-level information required for IC production, can be viewed as silicon assemblers. Below, let us outline a number of typical solutions which are meant as back-end stages of silicon compilers.

When a high-level design specification is used for designing circuits or modules which can be used as building blocks, the design program is called a *module compiler*. Module compilers (or cell compilers) have been developed to automatically generate layouts of specific functional modules ranging from simple gates to complex building blocks, such as PLA modules. Small module sizes induce a high degree of flexibility in the circuit implementation and performance of the module. In a *parameterized module*, one or more adjustable parameters can be specified by the designer to change certain electrical or geometrical attributes of the module. Matheson *et al.* [Mat83] developed cell generators that take into account both electrical and physical constraints. The distinction between a silicon assembler and a (parameterized) module compiler is rather vague. A module compiler is often referred to as a design tool that integrates a legal composition of parameterized modules into a system [Eli83].

SLAP (Silicon Layout Program), developed by Reiss and Savage [Rei82], accepts Boolean equations, which are parsed. A characteristic feature of SLAP is the representation of the circuit by a directed acyclic graph, which is called *level graph*. In such a graph, each edge connects two nodes whose levels differ by exactly one. The level graph is mapped into the chip layout, while efficient heuristics are used for placement, routing and control of the layout.

DUMBO of Stanford University [Wol83] accepts cells described by its structural primitives, such as geometric aspect ratios, and electrical primitives, such as circuit connectivity in terms of Boolean and transmission gates. By subjecting the logic circuit to a series of layout procedures (placement of functional blocks, expanding into transistor circuits and wire routing), we obtain a stick diagram which is first compacted before the circuit is implemented in NMOS technology.

The *cell compiler* of Lursinsap and Gajski [Lur84] translates a given cell described by Boolean equations and pass transistors into CIF. Constraints which are taken into account in this compiler include the height and width of the cell and the positions of the I/O pins on the boundary of the cell, while the size and the power consumption may also be specified. The target architecture is the PLA structure.

Parameterization has been applied in the design of array multipliers [BEn83]. Chu and Sharma [Chu84] developed a parameterized multiplier generator that can produce either NMOS or CMOS layouts from the same source software.

LiB is a compiler for automatically generating the layout of CMOS cells [HSI91]. The layout style is a variant of that proposed in [Ueh81], where a static MOS gate is converted into two graphs, one for the P transistor network and the other for the N transistor network. The detailed routing is performed by a general-purpose channel/switchbox router, based on a rip-up and reroute approach and the simulated evolution technique.

Silicon Compilers for Signal Processing

The *FIRST* (Fast Implementation of Real-time Signal Transforms) silicon compiler, developed at the University of Edinburgh [Den85] demonstrated the superiority of bit-serial architectures over bit-parallel schemes in terms of clock rate, flexibility and extensibility. For that reason, the bit-serial architecture is used in programming FIRST. The circuit is constructed by mapping the specified

5.3 ASIC DESIGN

function into an interconnected network of bit-serial operators. FIRST maintains a library of parameterized operators (add, multiply, etc.) and more complex procedural definitions so that architectures for different applications can be configured. The FIRST silicon compiler is equipped with a high-level simulator and a circuit simulator which can be provided with data extracted from the chip layout.

SYCO
SYCO is a silicon compiler which transforms an algorithmic description into a microprocessor-like circuit that realizes the algorithm [Jer86]. A large interpreter of a given command language is split into a set of interpreter levels. Commands at a given level breaks down into subcommands at a lower level. Each interpreter level is implemented by a layout block. A chip contains one data path slice and several control-section slices

CATHEDRAL
CATHEDRAL-II, developed at IMEC, Leuven (Belgium), is a silicon compiler, which starts from a high-level behavioral description at the algorithmic level and produces a chip layout [DeM86]. The architecture is targeted at a synchronous multiprocessor chip for digital signal-processing applications in the audio-frequency range up to a 1 Mbit/s sampling rate. From the input description written in the behavioral language SILAGE, a rule-based synthesis of the data path of the processors is performed. A heuristic scheduling program then generates the microcode for the processor controllers and the interprocessor communication network.

The layout of the processors is synthesized in terms of modules called from automated re-usable module generators created by silicon designers and adaptable to new layout rules. A floorplanner produces the chip layout. An expert subsystem verifies correctness during the design process and generates functional and timing models for verification at the module levels.

Silc Silicon Compiler
Silc [Bla85], developed at GTE Labs, produces fabrication masks for VLSI chips, particularly in telecommunication engineering. The input to Silc is a specification of the chip's behavior written in a LISP-like language defining a set of finite-state machines, which include logic, control as well as memory elements. The input description is parsed into abstract objects using Flavors, which is provided on a Symbolics LISP machine.

The logic design is implemented in standard-cell, gate-array or full-custom architectures. Parameterized module generators creates PLAs, ALUs, multiplexers, counters, etc. The Silc silicon compiler includes a functional simulator to verify the encoded chip algorithm. Also, design-for-testability aspects are taken into account.

Regular-Expression Compilers

Some silicon compilers accept regular-expression inputs. A *regular expression* can be used to specify sequential processes and identify certain bit patterns. The corresponding nondeterministic automata can easily be defined. Algorithms for hardware implementattions of pattern recognition were described by Mukhopadhyay [Muk83]. The compilation of regular expressions into reasonably compact layouts has been described by several authors [Flo82, Tri83].

LAGER

LAGER is an integrated CAD system for designing algorithm-based ASICs for signal-processing applications, such as speech processing, image processing, telecommunications and robot control [Shu91]. LAGER, initially used in a research project at the University of California, Berkeley, is now being applied at several academic and industrial institutions.

LAGER takes both a behavioral and a structural input with user interfaces at behavioral, structural and physical levels. It is a form of silicon compiler consisting of a behavioral mapper and a silicon assembler (Fig. 5.43).

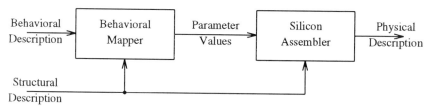

Fig. 5.43 LAGER system

The Behavioral Mapper can be retargeted to different architectures with, among others, the word length of the data path as a parameter value. It delivers as output a parameterized description of the processor architecture (a control unit and a number of data paths). The mapper employs three independent input languages: Silage (a data-flow language suitable for describing a signal-processing algorithm), RL (a C-like procedural language whose compiler generates a control program for a processor of specified structure), and Sass (an assembly language for defining the control unit).

The Silicon Assembler generates the chip layout, given a structural description of the chip circuitry. Designing new cells with minimum effort is achieved by re-using parameterized modules or leaf cells. The Silicon Assembler is implemented using an object-oriented database that makes the integration of new cells and CAD tools easy.

Intelligent Silicon Compilers

The short turn-around time in silicon compilation enables the designer to

5.3 ASIC DESIGN

evaluate the design at an early time. Silicon compilers can be distinguished by the type of input description and the prespecified target architecture.

Normal silicon compilers are limited by using only one predefined design methodology and prefixed target architecture. *Intelligent silicon compilers* allow trade-offs among chip area, operating speed and power during the design procedure. Tuning of intermediate designs can be performed as long as user constraints are not satisfied. Intelligent compilation involves optimization procedures in several stages.

In intelligent compilation, the designer submits the traditional interaction during the design process to the design system. Such a system is given a flexibility in design, providing features such as multicycle, chained or pipeline units and selection out of options, e.g., choosing between bus- and multiplexer-based solutions. Such an intelligent procedure is outlined in Subsection 5.2b. The combination of automatic and heuristic subprocedures can be regarded as an expert system for CADCAS.

5.3c Expert Systems for ASIC Design

Knowledge-Based ASIC Design

One of the characteristics of intelligent circuit design is coping with design trade-offs. Suppose that a new design must satisfy a number of requirements. For example,
- the speed, in terms of gate delay, setup time, clear time or load time, must be smaller than a ns,
- the operating frequency must be greater than b MHz,
- the dissipated power must be smaller than c mW.

A problem arises when we deal with contradictory requirements. Generally, a smaller delay is accompanied by a larger power dissipation. An intelligent trade-off can be achieved by assigning different weights to the different requirements. The choice of a suitable semiconductor technology may reduce the power-delay product of the circuit.

When knowledge-based techniques are introduced in ASIC design, the knowledge base may contain knowledge about options with respect to the layout structures and the available design tools and methods. The availability of a wide variety of design alternatives is essential since ASIC design turnaround time is at a premium. Expert systems mentioned in other subsections of this book may be useful in ASIC design. Some other examples of expert systems are outlined below.

ASIC Design Specification

ASIC design requires computer assistance during the specification phase as well as in the intermediate and final design stages. A knowledge base for VLSI design may maintain numerous frames that represent all artifacts or objects of a specification: either design objects, specification parameters or relationships [Ado86].

A knowledge-based approach, developed at the University of Erlangen/Nürnberg, involves the acquisition, checking and processing of electrical, timing, environmental and test-oriented specification data using object-oriented data representation and consistency checking based on Prolog rules [Mue89].

When a complex system is to be implemented with multiple ASIC chips, a partitioning process is required. In addition, there is a need to estimate the chip size, performance, power and cost early in the design cycle. Bournazel and Piednoir [Bou90] described a Prolog-based expert system which addresses these issues. This system offers a top-down design strategy from the high-level description to the final implementation.

SCHEMA

SCHEMA is a highly interactive CADCAS system, developed at MIT [Zip83, Cla85]. The underlying object-oriented model is based on Flavors, allowing objects to be generated, stored and retrieved, whenever needed. Modules are represented in terms of schematics, layouts and waveforms. The automatic tools are mainly meant for the lower abstraction levels. Analog as well as digital circuit behavior can be handled. Decisions above the module level are to be made by the user. SCHEMA synthesizes a description at transistor level from the functional specification of a circuit. The successive design steps are indicated in Fig. 5.44.

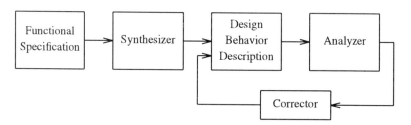

Fig. 5.44 SCHEMA ES system

In the first step, the synthesizer module generates a provisional design, which is then analyzed by the analyzer module. The generated data describe the behavior of the current design and are used by a corrector module to improve the performance of the circuit. This procedure is iterated until the given specification is met. If improvements are to be made, the corrector module is informed of the circuit behavior. This information is stored in the form of current and voltage

5.3 ASIC DESIGN   299

values for the various circuit nodes, the signals being described in terms of levels and ramps.

PAMS
PAMS (PArameterized Module Synthesis) generates CMOS modules, parameterized in terms of area, speed, and power dissipation [Tsa88]. This expert system permits human interaction at any level in the design process to customize the design.

PAMS requires two input files: a file describing technological contraints and data, and a description of the transistor connectivity of the circuit. Placement heuristics are used to produce an optimal placement. The PAMS routing step is a rule-based maze run. A frame-based route planning mechanism has been implemented to improve the quality of the interconnection wiring by providing a more global view of the routing area. Layout compaction, circuit extraction, performance evaluation and scaling form a loop for performance enhancement.

Designing ASIC Architectures
Given a set of specifications and requirements concerning an ASIC's behavior and performance, a large number of possible solutions in the design space must be examined. Various trade-offs must be evaluated in order to obtain a satisfactory architecture for the desired ASIC. An initial circuit synthesized to realize an algorithmic expression of the procedure can be modified according to the user's requirements using a rule-based technique [Shi89]. The knowledge base of the rule-based system contains deep knowledge and experience about the generation of register-transfer-level hardware designs, given a set of high-level specifications.

DESCART
DESCART [Gaj86] is an expert silicon compiler, which accepts as input a high-level functional specification of a circuit and generates a design using standard CMOS cells. The idea behind DESCART is that the process of translating a functional specification into a structural representation is not straightforward and many heuristics are needed to obtain an optimal design.

Module Generator KD2
The VLSI Group of the University of Utah developed knowledge-based tools for VLSI chip design. For example, KD2 is an intelligent circuit-module generator [Gu86]. This is part of the Intelligent Silicon Compiler, which in addition includes: the Architecture and Task Planner (KD1), the VLSI System Configurer (KD3), the Circuit Module and System Compactor (KD4), and the Simulator and Explanator (KD5).

The input to the KD2 system is a set of circuit specifications, including module functional, behavioral, performance and geometrical requirements. KD2 produces a set of implementation strategies for each circuit module with the assistance provided by the circuit Performance Evaluator, the Cell Knowledge Base and the Module Knowledge Base. The Circuit Module Synthesizer generates the circuit module. A Circuit Module Characterizer is a postprocessor which extracts important module parameters of the circuit module. A special silicon assembly language PPL is used which allows SLA (Storage Logic Array) methodology with CMOS transistors to be used.

NEPTUNE

NEPTUNE is a frame-based system for selecting VLSI functional modules [Foo86]. The knowledge base contains implementation alternatives and design constraints. The selection of modules depends on the design trade-offs (speed, area, power consumption) associated with particular implementation alternatives (random logic, PLA, etc.). An evaluator invokes a backtracking routine which explores different implementation alternatives to select an optimized solution. NEPTUNE, coded in C, runs on a VAX 11/780 computer and SUN workstations.

Rule-Based Controller Synthesis ASYL

Data-path design can be generated automatically using commercially available silicon compilers. One approach was described by Saucier and Hanrat [Sau86]. This approach was illustrated by two examples: a Boolean function compiler and a controller design. ASYL (Aide à la Synthèse Logique) is a rule-based VLSI design environment, developed at the University of Grenoble, particularly for determining the state assignment of controllers and performing logic minimization [Sau87, SAU87]. The controller implements the set of dedicated functions or algorithms of the ASIC. A flexible environment for controller design is achieved by an integrated structure of six modules, each accomplishing one of the design tasks: control-graph verification, synchronization, state encoding, automatic synthesis, Boolean minimization, floorplanning and topological optimization.

ASYL recognizes typical patterns in a control flow-graph, deduces constraints on the state encoding and looks for an optimal state assignment satisfying these constraints. The controller synthesis is facilitated by using several intelligent design assistants: a Prolog rule-based logic minimizer, a synchronization assistant, a PLA partitioner, a floorplanning assistant and a test expert.

KINDEN

KINDEN (Knowledge-based INtelligent ASIC Design ENvironment) provides a VLSI design environment for rapid prototyping of VLSI design and heuristic optimization [Hek90]. KINDEN combines object-oriented modeling and model-based reasoning to capture, integrate and manage VLSI design process attributes and hierarchies. The methodology is based on a new hierarchical modeling architecture that can represent the deep knowledge characteristics of VLSI

5.3 ASIC DESIGN

expert's conceptualized design models [Hek91]. The knowledge, encoded in an object-oriented frame formalism, includes well-established CAD tools as well as knowledge-based tools. Using KINDEN allows novice designers to design VLSI ASICs that satisfy specified requirements. It is argued that the object-oriented approach provides a natural and efficient means for representing and better understanding of the relationship between the physical description of the design and the designer's mental conceptualization of the solution space of the problem.

KINTESS (ASIC Process Technology Selection System) [Hek89] and KINFIDA (KINDEN Digital Filter Design Assistance) [Hek90] are two knowledge-based tools that have already been successfully implemented in this environment.

KMDS

KMDS is an expert system for integrated hardware/software design of microprocessor-based digital systems [Kuo89]. It performs the following functions on a personal computer:

a. Automatic synthesis of the circuit of microprocessor-based digital systems at a high architectural level.
b. Automatic generation of control programs for the implemented microprocessor-based digital system.
c. Automatic ASIC design to realize special-purpose circuit functions at algorithmic level.
d. Automatic Printed-Circuit-Board layout.
e. High-level knowledge-base editing and debugging.

Besides single-board microcomputers, KMDS supports the design of any intelligent I/O controller. It synthesizes digital systems at system-specification level and combines AI techniques, such as knowledge structuring, frames and demons, and algorithmic approaches to perform the design. A knowledge-base editor and debugger facilitates the extension and modification of the knowledge base. KMDS is implemented in OPS5.

VLSI Design Correct by Synthesis

Subrahmanyam [Sub86] discusses a formal algebraic framework underlying an expert system for VLSI design. An abstract behavioral description is transformed into a detailed architectural description, which can be used as input to a silicon assembler. The major objective of the system is to guarantee logical correctness of the designs generated. The formal basis supports a broad spectrum of abstractions from axiomatic behavioral specifications to switch-level MOS networks. If some performance criteria are not met, plausible causes for the failure are identified and remedies will be suggested. The expert system combines heuristic and algorithmic procedures. SYNAPSE [SUb86] is an expert system based on the above principles of conceptual integrity in VLSI design.

Silicon Compiler in Prolog
ASP (Advanced Silicon compiler in Prolog) is a Prolog-based hierarchical silicon compiler developed by Bush *et al.* [Bus87, Bus89] of the University of California, Berkeley. It is particularly suited for rapid generation of microprocessor designs. Starting from instruction-set architectural specifications, the design process traverses the behavioral, logic-functional and geometric level, ultimately producing VLSI masks.

Chapter 6
DESIGN VERIFICATION AND EXPERT SYSTEMS

6.1 SYSTEM SIMULATION

6.1a High-Level Simulation and Design

The System Level

Any system design starts with a *design specification* which specifies the complete set of requirements that the target system must satisfy. At the system level, the design specification is examined to see if the requirements are consistent and practically realizable. Proposals are made as to which feasible solutions may be considered. It is appropriate to introduce two distinguishable, though related, system levels which emerge when the system behavior and its architecture is emphasized.

a. *Behavioral Level.* As the name suggests, this level is concerned with the system *behavior*. The aim at this level is to optimize the design costs, the system performance and throughput. Important system-performance measures at this level include resource utilization, queue lengths, congestion and other potential bottlenecks. Relevant information, obtained by system simulators, will facilitate the evaluation of potential high-level architectures.

b. *Structural Level.* This level is concerned with the *structure* of the system. The system is partitioned into functional subsystems, each having a well-defined interface with its neighbors. After having considered hardware-software trade-offs, functions are allocated to hardware or firmware. Parameters to be evaluated are memory capacities, information flow rates, power dissipation, global costs, etc.

Languages for System Simulation

Let us discuss some important points about *simulation languages* at the system level. In the mid-1960s, several different system-simulation languages were developed. Widely known system simulators, which are currently in use, are outlined below. For references see [Sch87]. The choice of a simulation language is heavily dependent on the problem to be simulated and solved.

Digital systems are discrete-time systems. For the purpose of system simulation, we have to consider *entities* (comparable to subscripted variables in a program, corresponding to the states of a logic system) with their *attributes* (e.g., the discrete values in multiple-valued logic), which are determined by *activities* (e.g., logic operations). Signal changes in digital systems, called *events*, occur at discrete time instants.

The most important system-simulation languages are GPSS, which is entity-oriented, and SIMSCRIPT, which is event-oriented. The language GPSS (General-Purpose Simulation System), which is based on a flowchart or a block diagram, is easy to learn and is meant for users who have little or no programming experience. It is particularly suited to simulating traffic and queueing problems. Programming is also possible by using more familiar general-purpose languages, such as FORTRAN. Event-oriented simulation languages which are based on FORTRAN are GASP (General Activity Simulation Program) and SMPL (SiMPle Language). SMPL is especially suited to the analysis of computer systems and is applicable in various situations. More advanced simulation languages are the process-oriented languages SIMULA and ASPOL (A Simulation Process-Oriented Language). The latter is particularly suitable for the computer-aided design of computer systems.

Multilevel Simulation

Since every VLSI system design has both a behavioral and a structural aspect, it would be convenient if both the *structural* and *behavioral* modes of system simulation could be unified. Several methods of multilevel system simulation have been proposed. Below, a short survey will be presented.

The SCALD System

SCALD (Structured Computer-Aided Logic Design), developed at Stanford University and the Lawrence Livermore National Laboratory, provides the user with a means to design a complex logic system in a hierarchical manner, as exemplified by the design of the S-1 (Stanford-1) multiprocessor with at least ten times the computational power of the Cray-1 supercomputer [McW78]. A graphics-based, high-level input description of the target system is manipulated in a number of hierarchical design steps so as to generate low-level documentation which can be used to implement the hardware.

SARA

SARA (Systems Architect's Apprentice) is a system-design methodology developed at the University of California, Los Angeles [Est77]. It provides facilities for describing and simulating both system structure and behavior by means of the modeling tools STRUCTURES and BEHAVIORS respectively. Although system structure and behavior are related, they are considered separately. STRUCTURES uses the language SL1 (Structure Language 1) which defines a hierarchy of the structured primitives: modules, their interconnections and sockets (such as the interface between modules and interconnections). It allows the designer to interactively construct multilevel structural models by employing fully nested modules. Successive refinement of a module can be

6.1 SYSTEM SIMULATION

accomplished by describing the internal structure of a module in terms of modules, sockets and interconnections.

SABLE
SABLE (Structure And Behavior Linking Environment) is a simulation tool developed at Stanford University to support system design in which the system behavior and structure are linked at multiple levels of a structures hierarchy [Hil79]. To this end, SABLE uses the languages SDL and ADLIB for describing the interconnection structure and the system behavior respectively.

SDL (Structural Design Language) describes the structural properties of the system at all levels from the system level down to the circuit-element level. Some structural requirements may be part of the original design specification. SDL is able to expand a system description into a lower-level description that suits a particular purpose. For example, when logic simulation at the logic-gate level is desired, the system has to be expanded into a network of logic gates and flip-flops. For convenience, the user may generate the SDL description automatically through the use of the interactive graphical structure editor SUD2.

ADLIB (A Design Language for Indicating Behavior) describes the system behavior of synchronous or asynchronous systems in terms of Pascal-like constructs for describing control and data flow at different levels of abstraction [Hil86]. It provides facilities for multilevel description, hardware/software description and type specification at the behavioral, register-transfer, gate and circuit-element levels. It allows users to define the "data level" at which each component operates and to specify mechanisms for translating information between these levels. For example, an integer variable (denoting a logic level) is at a higher data level than a real number that represents the instantaneous voltage values. ADLIB supports a range of applications, including fault simulation, timing verification, software simulation and symbolic simulation.

MIMOLA Design System
Given a behavioral or functional specification of a digital processor, the MIMOLA (Machine Independent Microprogammable Language) Design System generates hardware-structural descriptions [Zim79]. The main ingredients of this system are a high-level description language, a synthesis procedure, a set of available hardware components and an analysis tool. MIMOLA requires as input either an ordered list of instructions at the architectural level or a control sequence at the behavioral level. An architectural-level input is translated directly into hardware functional modules and microcode. When the input is a behavioral description, it is first iteratively transformed into an ordered control list at the functional structured level, after which the above-mentioned translation procedure can be started. The output is a register-transfer-level description.

VISTA

VISTA (VLSI Simulation Test and Artwork) of Hughes Aircraft combines a related set of four CAD tools (SHIELDS, SCAMPS, ASCAP and MAGIC) into an integrated system for VLSI design [Das82]. SHIELD (Simulator for Hierarchical Integrated Error-free LSI Design) is a multilevel simulator based on high-level and low-level simulation models. A global multiplexer schedules and synchronizes the incremental execution of the individual simulations of different subcircuits performed by independent simulators. SCAMPS (Semiconductor, Characterization, Analysis, and Modeling Parameter System) is a computer-controlled measuring system which is used for extracting device parameters. ASCAP (Artwork and Schematic Analysis Program) incorporates a library database organized in a pseudo-hierarchical manner. MAGIC (Multilevel Artwork Generation for Integrated Circuits) employs symbolic and stick schemes for describing modules at different levels of abstraction.

Other Integrated CAD Tools

The CMU-DA System, a major research design project of Carnegie-Mellon University, was described earlier in Subsection 5.2c. For more detailed descriptions of the above CAD tools and other multilevel and mixed-mode tools, see [Fic87] and [Sch87].

6.1b Hardware Verification and Expert Systems

Logic Verification by Comparison Methods

The final result of a logic design is a hardware implementation which satisfies the design specification. It is therefore natural that several authors have proposed verification methods based on checking the functional equivalence of a synthesized result and the designer's hardware implementation. Some verification methods which rely on comparing the actual hardware result with adequate information derived from the design specification are outlined below.

Several methods of verification are based on comparing a synthesized result and the designer's hardware implementation to check for functional equivalence. The *cyclic simulator* [Par76] assumes synchronous logic and considers the system state in one clock cycle at a time, ignoring the effects of gate delays. That is, the effective time delays within the combinational part of the circuit are assumed to be less than the period of the clock which controls the flip-flops. The purpose of a cyclic simulator is to determine the next stable state as a function of the system inputs and the present state. The cyclic simulator performs the technique of *Boolean comparison*. i.e., control and data-path logic in flowchart form is compared to the logic pertaining to the actual hardware implementation. Boolean comparison has been applied to the IBM 3081 machine project [Mon82,

6.1 SYSTEM SIMULATION

Smi82] for design verification. A synthesized flowchart is EXCLUSIVE-ORed with the hardware logic. A result $f = 1$ indicates the presence of an error. Another logic verifier based on Boolean comparison was proposed by Odawara *et al.* [Oda86].

In addition to cyclic simulation, other verification tools, such as timing verifiers, can be used. DAV (Design and Verification) is a logic-design subsystem of IBM's Engineering Design System [Dun84]. One main feature of DAV is its graphics entry facility via a workstation equipped with two display screens and a number of graphics input/output devices. DAV includes three types of simulator: a cyclic simulator, a functional simulator and a hardware-implemented logic simulator. In the functional simulator, called IFS (Interactive Functional Simulator), circuit modifications can be incorporated incrementally and thus the effect can be examined by comparing the simulation results, which can be shown as waveform diagrams or in some other graphical form on the display screen.

Temporal Logic

To describe and prove properties of concurrent programs, we have to consider complete bit sequences. A language for specifying such sequences may be based on regular expressions. Another language which is adequate for expressing a wide variety of properties of the execution sequence of concurrent programs, such as partial correctness, termination, mutual exclusion and accessibility, is *temporal logic* [Mos86]. Temporal logic allows a notation suitable for specifying and reasoning about digital circuits and programs with consideration of timing. Verification of both sequential and parallel programs by using temporal logic is based on temporal reasoning in which the time dependence of events is the basic concept.

Temporal logic is an extension to traditional logic and predicate calculus. Instead of using conventional timing charts and specifying properties of the system states as functions of time, an expression of temporal logic is assumed to specify properties of all possible temporal sequences that may evolve from the present system state. Propositional temporal logic is propositional logic extended with four operators to define timing relations: H (henceforth), E (eventually), N (next) and U (until). These operators are usually denoted by special symbols, but, for simplicity, they are denoted here by H, E, N and U. Let F and G be temporal sequences. Then H F is TRUE is F is TRUE from now on forever; E F is TRUE if F is TRUE now or at some time in the future; N F is TRUE if F is TRUE in the next state in the sequence; F U G is TRUE if F is TRUE, until G is TRUE for the first time.

Simulation programs verify the presence of good properties. When temporal logic is used, such as in the DDL verifier of Uehara *et al* [UEh83], the program can in addition verify the absence of bad properties. Fujita *et al.* [Fuj84] applied temporal logic to specify hardware and synthesize state diagrams. Pattern matching and automatic backtracking mechanisms inherent to Prolog can be exploited to decrease the design time. Moszkowski [Mos85] introduced a temporal-logic formalism for specifying and proving several properties of a wide variety of logic circuits. Fujita *et al.* [Fuj90] applied temporal logic for the automatic and semi-automatic verification of switch-level circuits.

Formal Methods of Design

The usual approach to system design consists of a design procedure followed by verifications at several levels [Cor80]. The traditional technique of design verification is to use a simulator with a given set of input patterns. This form of verification is incomplete since all that it proves is that the circuit functions correctly for the set of applied inputs.

A desirable approach would be to adopt a formal, algorithmic method of synthesis. This input-independent method is automated to avoid human errors. Once a formal synthesis algorithm has been established, a synthesis result is deemed to be correct. This correct-by-synthesis approach can be applied in automated logic synthesis. The definition language used may be a subset of LISP. The LISP interpreter can be used as a simulator and hence as a verification tool.

Formal methods give the designer a firm foundation for developing integrated circuits. The correct-by-synthesis property has led designers to study formal methods of VLSI design [MIl86, Mil88, BIr88, BIR89, Har89, Cla89, Cla90, STa90]. Kabat and Wojcik [Kab85] described an automated technique for synthesizing combinational logic circuits using theorem-proving techniques. The synthesis procedure involves the execution of subtasks to represent the elements of the design as a set of axioms of a formal system or as a theory, state the problem of realizability of the target function as a theorem, and prove it in the context of the theory. Once the proof of the theorem has been found, an automatic procedure for the recovery of the logic circuit is to be executed to complete the design.

The formal approach to design and verification can be extended to other aspects of the design process. For example, Kljaich *et al.* [Klj89] described a formal verification system, which applies automated reasoning techniques to validate fault tolerance. Extended Petri nets have been used for modeling and analyzing digital systems. A Petri net is a directed bipartite graph used for modeling as well as control and data flow in systems. The Stark Draper Laboratory Fault-Tolerant Processor [SMi84] was taken as an example to illustrate the verification of its fault tolerance.

Formal Hardware Verification

In the above, the importance of formal methods of logic-circuit verification was emphasized. VERIFY [Bar84] is a functional verification system for proving the design correctness and for the straighforward but very detailed logic-circuit design involving many thousands of transistors (without consideration of the timing problem). It is implemented in Prolog. Other formal design-verification methods based on Prolog have been described by Wojcik *et al.* [Woj84] and Woo [Woo85].

Maruyama and Fujita [Mar85] developed a verification technique for hardware logic designs based on temporal logic and Prolog. Suzuki [Suz85] discussed the use of Concurrent Prolog as an efficient VLSI design language.

PROVE

PROVE (PROlog-based VErifier), written in Prolog, verifies the functional correctness of digital circuits in a hierarchical manner [Sri88]. The circuit is viewed as an interconnection of functional modules at different hierarchical levels: module level (multipliers, etc.), lower module level (registers, adders, etc.) or gates (Boolean gates). PROVE performs hybrid simulation, which implies that specific signals are specified numerically and others in symbolic form. The primary verification mechanism is a comparison of two symbolic expressions, one generated from the implementation being verified and the other from the specification. Pattern matching is incorporated to allow formal verification of large circuits.

Formal Verifier PRIAM

PRIAM, developed by Bull Research, is a tool that automatically proves the functional correctness of synchronous circuits [Mad89]. It handles circuits whose specification and implementation are both described in Bull's hardware description language LDS. Functional verification boils down to comparing LDS programs for equivalence or implication.

Verifying Silicon Compilers

Many methodologies for realizing logic functions by means of silicon compilers provide facilities for verifying the correctness of the designs. The nature of the available verification tools depends on the particular design methodologies chosen for the silicon-compilation process. Some silicon compilers use the intermediate level between the front-end and the back-end stages as a basis for simulation or testing purposes. The shift of attention in VLSI designs to higher levels of abstraction imposes the necessity of new verification tools. Cheng [Che84] reported on a silicon-compiler project in which verification can be restricted to high-level functional simulation (for verifying the architectural design of the chip in terms of interconnected functional blocks) and timing analysis at the layout level (for verifying the timing performance).

A desirable design approach is to proceed in such a way that design correctness is automatically established. This *correctness by construction* principle obviates the need for exhaustive simulations or other verification techniques. Subrahmanyam [Sub84] described the technique of transforming high-level behavioral specifications into lower-level architectural descriptions that are amenable to VLSI implementation. One realized system [Org84] involves the transformation of an ADA program code into a specification and NMOS implementation of an Internet Protocol. Evans *et al.* [Eva84] developed a C-based language ADL (Algorithmic Design Language) for describing the functional, circuit, schematic and mask aspects of integrated circuits. An attempt to develop a formal procedure for proving the correctness of a design was made by Milne [Mil83].

6.2 LOGIC SIMULATION

6.2a Gate-Level and Functional Simulation

Introduction
Let us first discuss the simulation of a *logic circuit at the gate level*, i.e., the network consists of logic gates (NOT, AND, OR, NAND, NOR) and their interconnections. For simplicity, assume that each gate has a single primary output, whereas multiple primary inputs are allowed. *Primary inputs* are the externally accessible inputs pins of the circuit through which logic values can be injected into the circuit. *Primary outputs* are the externally accessible output pins of the circuit through which logic values can be observed from the circuit. In contrast to analog simulation (see Section 6.3), *logic simulation* implies the analysis of the behavior of logic signals with a finite number of discrete values, in the simplest case, the two logic values: logic 0 and logic 1. A transition of a logic signal from logic 0 to logic 1, or vice versa, is called an *event*. Actually, a logic gate does not respond directly on an event at an input of the gate, but only after a finite propagation delay, called *delay* in short. Gate delays may cause unwanted signals which lead to an erroneous operation of the logic circuit under consideration. To investigate the effect of gate delays on the logic behavior of a network, let us consider the logic simulation of a combinational network.

Simulation of a network always involves two steps: (a) selection of a model of the network, and (b) analysis of the network. The logic signals in the network are analyzed at discrete time instants. Let us first consider the modeling problem. Since the network consists of gates, we must use suitable models of the gates.

Delay Models of Logic Gates
A logic gate model contains two essential attributes: the description of its ideal

6.2 LOGIC SIMULATION

logic function, and the deviations of this ideal situation. These deviations have primarily to do with the delay properties of the gate. Gate models differ in the way the gate delay is modeled. A *zero-delay model* describes ideal gates without delays and can be used for *logic verification*, i.e., verification of the logic function without taking delays into account.

Unit-Delay Model. The simplest way to take gate delays into account is the assignment of an equal, constant delay τ to each gate in the network. The value of τ depends on the device technology used. Unit-delay simulation involves the analysis of the logic signals occurring at discrete time instants with fixed intervals of τ. This means that an event at an input of a gate causes a response at its output at the next time instant. Simulation with unit-delay models is suitable for networks consisting of one type of components (e.g., NANDs) or for a first rough analysis of the network.

A gate with a unit delay can be modeled either by a delay element at the output or a delay element at each input of the gate.

Example: The simulation of an AND gate with two inputs is illustrated in Fig. 6.1.

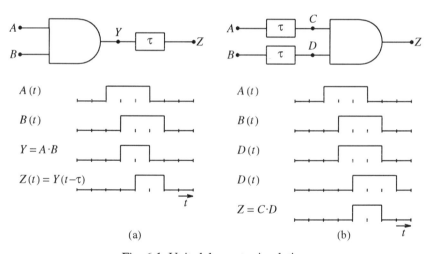

Fig. 6.1 Unit-delay gate simulation

Usually, the simple model of Fig. 6.1a is used. In case the delays from A to Z and from B to Z are different, the model of Fig. 6.1b must be used.

Assignable-Delay Model. The simplest form is the assignment of different types of gate, e.g., each NAND gate has a delay of 2 ns, each NOR gate has a delay of 3 ns, etc. The *rise delay* (i.e., the delay associated with a $0 \rightarrow 1$ transition at the output) and the *fall delay* (delay associated with a $1 \rightarrow 0$

transition at the output) have usually different values. When the pulse duration is too short, erroneous results may be obtained.

Example: Figure 6.2 shows that the simulation produces a faulty signal if the pulse duration is smaller than $\tau_R - \tau_F$.

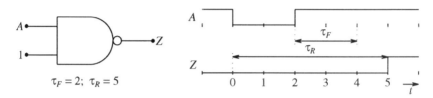

Fig. 6.2 Erroneous simulation result

Min-Max Model. Usually, exact nominal values of gate delays in an actual circuit are not available. For that reason, one can use a *min-max model*, where it is assumed that the gate delay lies somewhere between a minimum value τ_m and a maximum value τ_M. The range $\tau_M - \tau_m$ is called the *ambiguity region*.

Example: Figure 6.3 shows time diagrams for a simple network with $\tau_m = 2$ and $\tau_M = 3$ time units.

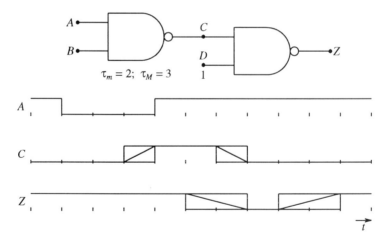

Fig. 6.3 Ambiguity regions in a min-max model

The ambiguity regions are indicated by a rectangle in which a diagonal is drawn. Min-max models are employed in many logic simulators, amongst others, TEGAS [Joh80].

6.2 LOGIC SIMULATION

Inertial Delay. Switching a gate requires a minimum amount of energy of the input signal. When the pulse duration of a gate input signal is too short, no pulse will be produced at the output. The minimum duration of a pulse which is capable of producing a response at the gate output is called the *inertial delay* τ_i. If the pulse duration τ of a gate-input signal is larger than the gate delay τ_p, the reponse at the output is a pulse with a duration τ delayed by the value τ_p. If the pulse duration of the input signal has a value in the range from τ_i to τ_p, the response at the output is a very short pulse, called a *spike*.

Accurate Delay Models

A more accurate specification of a delay model takes into account all kinds of factors, notably the dependence on the power supply, the temperature and particularly the fanout (the number of gates connected at the gate output). Another possibility to model the gate delay is by specifying the delay with a probability density function. A usual assumption is a *normal (Gauss) distribution* of the gate delay. In that case, the gate delay is specified by the *mean delay* τ and the *standard deviation* σ. When N gates are cascaded, the mean delay τ and the variance σ^2 of the total circuit is obtained by summing the mean delays and variances of the individual gates. Logic simulators using these probability models are used in DIGSIM [Mag77] en F/LOGIC [Wil79].

Realistic values of gate delays can be obtained by evaluating the required information for the gate model, including parasitic capacitances, from the mask artwork. The problem of circuit extraction is discussed in Subsection 8.4a.

Three-Valued Logic

Simulation using only two logic states 0 and 1 has serious drawbacks. The ambiguity regions (uncertainty regions) in Fig. 6.3 suggest the introduction of a new logic value or state U (short for Undefined or Unknown). The value U is used here to indicate that a signal is in a state of transition ($0 \to 1$ or $1 \to 0$). Truth tables with the three logic values 0, 1 and U can be constructed. Note that U has no inverse: $\overline{U} = U$.

The three-valued logic is also employed to initialize (i.e., assign initial values to all node signals of) the network to be simulated at the beginning of the simulation procedure. When the initial values of certain node signal are not known, these nodes are assigned the value U until we have found the known values 0 or 1 after a simulation run. The three-valued logic can also be used to detect hazards and races.

Hazards and Races

Hazards and races are unwanted pulses generated in the network. Since these pulses may disturb the logic performance of the circuit, they must be detected and eliminated. When a signal at some node A can follow two different paths to meet again at some node B, we are dealing with a *reconvergent fanout* (Fig. 6.4).

Such a situation is the most important cause of the occurrence of a hazard. When two different delays are associated with the two possible paths from A to B, an event in A may produce an unwanted signal in B, which is called a *hazard*.

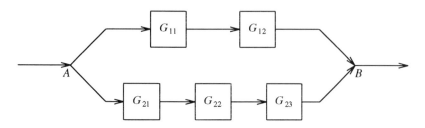

Fig. 6.4 Reconvergent fanout

There are several types of hazard:
Static-0 hazard: an unwanted 1-pulse in a normally 0-signal.
Static-1 hazard: an unwanted 0-pulse in a normally 1-signal.
Dynamic hazard: a sequence of unwanted 0 and 1 values.

When two events at different nodes in the network propagate through different paths and meet together at some node to produce an unwanted signal at this node, we are dealing with a *race*. The terms hazards and races will be used interchangeably.

Example: The occurrence of a static hazard is illustrated by a simple network given in Fig. 6.5a.

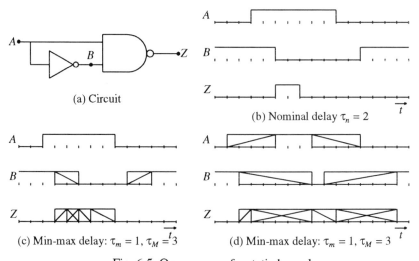

Fig. 6.5 Occurrence of a static hazard

6.2 LOGIC SIMULATION

For the AND gate and the inverter, a nominal delay of $\tau_n = 2$ is used in Fig. 6.5b and a min-max delay with $\tau_m = 1$ and $\tau_M = 3$ in Fig. 6.5c. It is expected from Fig. 6.5a that Z constantly has the value 0. Simulation reveals a static-0 hazard with a duration of two time units. This hazard is produced only when the input signal changes state from logic 0 to logic 1, but not vice versa. When after some simulation steps the ambiguity regions in a min-max model have become sufficiently wide, a hazard will also be detected when a $0 \rightarrow 1$ transition occurs at A (see Fig. 6.5d). This is in contradiction to reality! In this case, the min-max model leads to a worse simulation result than the nominal-delay model.

When a hazard is detected, an attempt should be made to eliminate it by introducing some modifications in the circuit. Sometimes the insertion of an additional gate may remove the hazard.

Suppose that an unwanted pulse occurs as a result of events appearing at about the same time at different nodes in the network. If the unwanted pulse occurs only when the events take place in a given order, we are dealing with a *critical race*.

Example: The simplest example of a critical race is illustrated in Fig. 6.6.

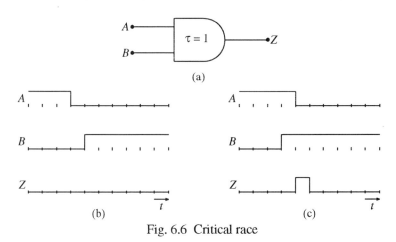

Fig. 6.6 Critical race

Multiple-Valued Logic

Breuer [Bre72] showed that the use of three-valued logic can lead to wrong conclusions. Describing the $0 \rightarrow 1$ transition by $0 \, U \, 1$ and the $1 \rightarrow 0$ transition by $1 \, U \, 0$ [Eic65] may also fail to produce the correct results.

Accurate delay models use more than three logic values. For example, in addition to 0, 1 and U, we may use R ($0 \rightarrow 1$ transition), F ($1 \rightarrow 0$ transition), H (high-impedance state). etc. F/LOGIC, TEGAS3 and SALOGS-IV [Cas78] employ 6-valued, 7-valued and 8-valued logic respectively.

Detection of Static and Dynamic Hazards

Dynamic hazards cannot be detected by using 3-valued logic. Bose and Szygenda [Bos77] proposed a procedure for explaining the possible occurrence of dynamic hazards by using binary truth tables. They use 5-valed logic (0, 1, R, F and U), where R and F are given in a binary code for the current time, a future time and some intermediate time points, while U may be either 0 or 1. By choosing different instants of time for the actual $0 \to 1$ transition of R, e.g., 0111, 0011 and 0001 (and something similar for F), one can derive if a dynamic hazard can occur in the output.

Description of a Logic Network

Prior to the simulation process, a suitable description of the logic network to be simulated must be given. The data structure of this description is usually a linked list structure. Besides the primary inputs and outputs, this structure contains the complete description of each gate of the network and the interconnections, i.e., the name of the gate, the logic state of the gate, the delay model and the (pointers to the) fanin and fanout lists.

Next-Event Simulation or Event-Directed Simulation

An *event* is a change in the logic state of a gate. Only a small percentage of the signals at the various nodes of a logic network will undergo a transition (event) in response to an event at a primary input. To reduce the required computer time as much as possible, we have to examine only those logic gates where an event occurs. In any case of event at one of the inputs of a gate, we have to investigate whether this event produces an event at the gate output, after the appropriate gate delay. This way of simulation, referred to as *next-event* or *event-directed simulation*, implies that an event is traced in its course from primary input to primary output. This is referred to as *selective trace*.

Next-Event Simulation with Unit Delays

Suppose that all gates in a network have the same delay τ. The simulation is carried out at discrete time instants with fixed time intervals τ. When one of the inputs of a gate changes state, then an event may take place at the output of the gate. From the fanout list of this gate it is known to which gate inputs this event has been propagated. Use can be made of *stacks*, which alternately function as current-time stack and future-time stack. The former stack, associated with the current simulation time instant, stores gates which have an event at one of their inputs. These gates are examined one by one to see if the event is propagated to their outputs and hence to the inputs of the gates connected to these outputs. The latter gates are stored in the future-time stack, associated with the next discrete simulation time instant. When both stacks are empty, the simulation can be considered to be terminated.

6.2 LOGIC SIMULATION

Next-Event Simulation with Different Delays

When the various gates in the network have different delays, besides the current-time stack more than one future-time stack are required to store the gates with events. If τ is the fixed time interval between subsequent discrete time instants, then the analysis is facilitated when each delay is a multiple of τ. Let us choose τ equal to the greatest common divisor of the different gate delays. The number of required future-time stacks is equal to the largest delay divided by τ. Suppose that 4, 6 and 8 ns are three different delays which occur in the network. Then, $\tau = 2$ ns and $8/2 = 4$ future-time stacks are needed, associated with four subsequent discrete time instants. The stacks can be structured in a ring form as a time clock. Once the last stack of the ring has been handled, the first stack is again considered, etc. Only when an event occurs, will the gates stored in the current-time stack be handled, while the fanout list is used to store appropriate gates in the proper stacks.

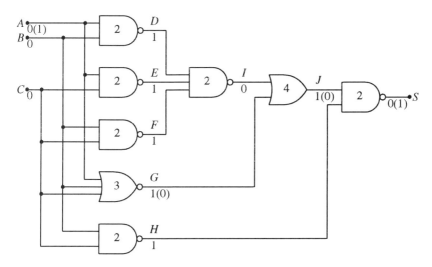

Fig. 6.7 Full-adder as a simulation example

As an illustration, consider the simulation of a full-adder given in Fig. 6.7. The nominal delays in time units are indicated inside the gate symbols. In this example, we need four future-time stacks in addition to the current-time stack. Let us denote the stacks by S_1, S_2, S_3, S_4 and S_5. The stacks form a ring, where S_5 is connected to S_1. There are two types of stack operations: PUSH (i.e., inserting gate data in the stack) and POP (i.e., fetching gate data from the stack). Note that a stack is a Last-In First-Out List, i.e., the data fed in last is fetched first from the stack. The initial states of the network to be simulated is indicated in Fig. 6.7. It is assumed that the initial states for the primary inputs A, B and C have been stored in stack S_1. Suppose we want to know what happens when the input state changes from 0 to 1. PUSH and POP operations are carried out on the five stacks. The simulation process is continued until no more events are inserted in future-time stacks. From Fig. 6.8 it is seen that the event in A propagates through the gates G and J to the primary output S.

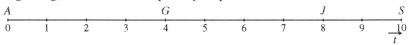

Fig. 6.8 Next-event simulation

Alternative Simulation Methods
Min-max delays require a more complex procedure for the time scheduling and updating of the events. There exists another method of time scheduling different from the one indicated above. Instead of simulation at discrete time instants with constant time intervals τ, where each delay is an integer multiple of τ, arbitrary delay values can be used. The data of the gates are placed in records of a linear list sorted in a chronological order. When a new event must be stored for a time t_i counted from the current time instant, a new record is inserted at the proper place in the time list.

HILO Simulator
HILO, supplied by GenRad Inc., is a logic simulator at logic-gate and higher levels [Blu87]. Integrated circuits to be simulated may consist of simple SSI modules (TTL, CMOS, ECL), MSI modules (TTL, CMOS, NMOS, ECL) and LSI packages (microprocessors, memory ICs). Transfer gates can be modeled as primitive gates.

A special description language is used for describing the signal waveforms which are fed to the circuit. HILO MARK 2 [Fla81] is a hardware-description language which can handle both synchronous and asynchronous models of digital systems, providing structural as well as behavioral information. Three-valued logic (0, 1, U) and a delay value assigned to each module so that HILO can perform event-driven simulation.

Circuit descriptions can be constructed as a collection of subfiles. HILO supports a hierarchical design approach in which circuits are described in terms of simpler circuits. Libraries containing models of a wide range of standard ICs and new subcircuits are supplied with the HILO system.

The HILO simulator can be used in two modes:
a. Fault-free simulation of the circuit under normal conditions. A feature is the location of timing problems.
b. Fault simulation for simulating the operation of the circuit under fault conditions. In some cases, test patterns for detecting faults can be generated.

The HILO simulator can be used in conjunction with programs from the

Silvar Lisco suite [Blu87]. For example, CASS (Computer Aided Schematic System) supports the creation and editing of circuit and logic diagrams by using an interactive graphical menu-driven user interface.

BIMOS

One of the simulation tools within the Silvar Lisco CAD package [Blu87] is BIMOS. This is a gate-level logic simulator for analyzing MOS integrated circuits. The circuit to be simulated is entered through the CASS editor.

A feature of BIMOS is its fast prototyping capability, including a range of facilities, such as fault simulation. The BIMOS simulation output may be viewed by either one of the graphics post-processors PPRG and LOGAN [Blu87].

6.2b Timing Verification

Introduction

Timing verification does not provide a detailed logic simulation. In effect, it focuses on the verification of the correct *timing* of logic systems. In programs for timing verification, all delays of gates and interconnections along a suitable path in a logic system are summed. The total delay of such a path must satisfy certain constraints. Timing verification is also referred to as *delay analysis*. Usuallly, the required delay values are derived from the mask data by the process of circuit extraction and parameter evaluation. Timing verification is primarily meant for clocked sequential circuits.

Definitions of Timing Parameters

Let us first define a number of timing parameters relevant to clocked circuits. The *maximum clock frequency* f_M of the clock pulse depends on the shortest pulse generated in a given device technology and on the potential timing problems. The *clock pulse width* t_w must have a prespecified minimum value. Clock schemes are based on *pulse triggering* (e.g., in the case of master-slave flip-flops) or *edge triggering* (with state changes either at the rising edge or the falling edge of the clock pulse). The edge of the clock pulse at which the state changes is called the triggering edge. The two important timing parameters which relates the data input signal to the clock pulse are the setup time and the hold time. Figure 6.13 explains this by the case in which the rising edge of the clock pulse is the triggering edge.

The *setup time* τ_S is the minimum time interval by which the rising edge of a data signal must precede the triggering edge of the clock pulse. The *hold time* τ_H is the minimum time interval for which the data signal must be held stable after the triggering edge of the clock pulse. The minimum values for τ_S and τ_H must apply to all devices in the circuit. When the above conditions are not satisfied, the final state of a flip-flop will be indeterminate and a wrong state may result.

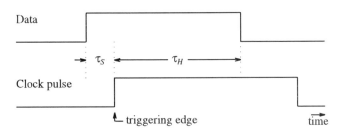

Fig. 6.13 Definition of setup time and hold time

An essential property of clocked logic is that the clock pulses initiate all state changes of the storage elements (flip-flops) in the circuit at the same instant of time. In principle, the triggering edge of the clock signal must arrive at the clock inputs of the flip-flops at the same time instant. A basic requirement of clocked logic is that the whole system must settle at a stable state within one clock period. The time required for the system to settle determines the maximum clock frequency which can be used.

Due to the different time delays caused by different path lengths in the timing circuits, the clock inputs at the different points within the system do not change at the same time. A *clock skew* is defined as the time difference by which two clock signals originating from the same master clock arrive at two different flip-flops in a system. In a complex system, we have to deal with many different clock skews, each of which may cause performance errors.

Example: Figure 6.14a shows a cascade of two D flip-flops, in which the indicated clock skew τ_C causes a wrong result. A state change at the input of the first flip-flop must be transferred in one clock period. When the input signal changes from 0 to 1, the flip-flop states in two successive clock periods must be $Q_1 Q_2 = 1\ 0$ and $Q_1 Q_2 = 1\ 1$. Suppose that we have a clock skew τ_C as indicated in Fig. 6.14b. Assuming that the flip-flop delays are zero, the signal change in Q_1 is immediately followed by a signal change in Q_2, which is wrong, i.e., the state of the flip-flops during the next period is 1 1 instead of 1 0. If the flip-flops have a delay of τ_{FF}, this wrong operation will not occur if $\tau_C < (\tau_{FF} - \tau_H)$, where τ_H is the hold time. A clock skew as indicated in Fig. 6.14c would not be harmful. Assuming that data transfer must take place in both directions, it is desirable that τ_C is as small as possible.

6.2 LOGIC SIMULATION

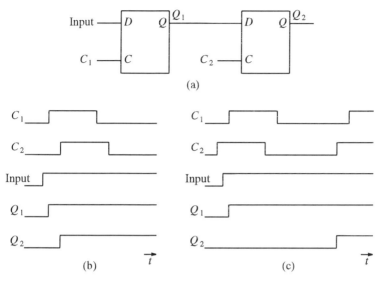

Fig. 6.14 Illustrating the effect of a clock skew

Verification of Timing Constraints

For timing verification of practical circuits, we can use the model depicted in Fig. 6.15.

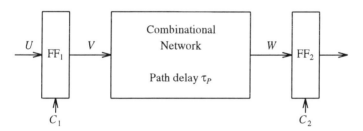

Fig. 6.15 Model for timing verification

It is usually sufficient to consider the path from the input U of a flip-flop passing through the output V of this flip-flop and a combinational network to the input W of the next flip-flop [Ben82]. A signal event in U must arrive at point W within a period τ_{period} of the clock signal. If τ_{FF} denotes the flip-flop delay and τ_{path} the delay in the combinational network from V to W, the time needed for an event at U to arrive at W is equal to $\tau_{FF} + \tau_{path}$. This time must not be too short, but also not too long.

6 DESIGN VERIFICATION AND EXPERT SYSTEMS

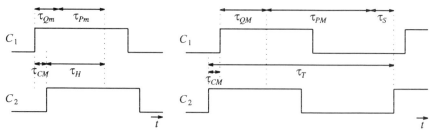

Fig. 6.16 Illustrating the short-path and long-path conditions

Taking into account the setup time τ_S and the hold time τ_H, while the clock pulses C_1 and C_2 have a clock skew of τ_C, the following conditions must be satisfied.

Short-Path Condition:

$$\tau_{FF(min)} + \tau_{path(min)} > \tau_H + \tau_{C(max)}$$

Long-Path Condition:

$$\tau_{FF(max)} + \tau_{path} + \tau_S < \tau_{period} - \tau_{C(max)}$$

The quantities τ_H, τ_S, $\tau_{FF(min)}$ and $\tau_{FF(max)}$ for a flip-flop are usually known beforehand. The clock skew τ_C can be determined by evaluating the total delays of the master clock to the two subsequent flip-flops. The difference between these delays gives the clock skew. The various timing verifiers developed differ in the way in which τ_{path} is evaluated. For the delay of a gate, one often uses the mean value τ and the standard deviation σ. Note that in the case of inverting gates we have to deal with rise delays τ_R and fall delays τ_F.

Delay Calculation

The reliability of timing verifications and logic simulations, which require the necessary delay information, is critically dependent on the accuracy with which the delay values are specified or calculated. For this reason, various methods for evaluating the delays in a logic circuit have been proposed. Particularly in VLSI, the evaluation of parasitic interconnection parameters is essential. Although several circuit parameters can best be measured experimentally with test chips, some theoretical work has been done. In VLSI technlogy with feature sizes on the order of one micron, the effect of interconnection delays dominates that of the gate delays. The simplest structure occurs when two logic gates are interconected by a conducting polysilicon wire. Usually, such a wire can be modeled by an RC transmission line. To evaluate the interconenction delay, capacitance and resistance values must be estimated.

Penfield and Rubinstein [Rub83] presented a computationally simple technique for computing upper and lower bounds of the propagation delays in

MOSFET interconnection nets with a tree structure. The idea is to find rigorous closed-form expressions giving upper and lower bounds for the transient voltage waveforms rather than the actual waveforms. The calculation of these bounds is based on the observation that during a transient the voltage at each node of the circuit is a monotonic function of time. The results have been applied in many timing simulators [Sch87].

6.2c Switch-Level and Timing Simulation

Switch-Level Simulation

In switch-level simulation, which is particularly suitable for MOSFET circuits, each MOSFET is modeled by a controlled switch. A circuit is represented by nodes connected by switches. The source and drain of a MOSFET are bilateral and behave as a short circuit or an open circuit, depending on the voltage at the gate (positive or negative respectively in the case of an N-channel MOSFET). Structurally, the circuit model corresponds to the original network. Important components of a MOSFET circuit are the pulldown-pullup inverter and the pass transistor. A *node* obtains the value 0, 1 or U (Unknown). The high-impedance state is implicitly contained in the possibility of the open circuit between source and drain.

Switch-level models provide more accurate behavioral and structural information than logic gate-level models, while avoiding the high computational cost associated with analog-circuit models [Hay87]. Structurally, a switch-level circuit model corresponds to the original network. The switch-level simulations differ in their methods of representing node voltages, node capacitances, transistor modeling and the sequence of time steps [Bry87].

The Switch-Level Simulator MOSSIM

Bryant [Bry80] developed MOSSIM, which is the first switch-level simulator. MOSSIM distinguishes three types of node: *input nodes* (with a fixed value determined by exernal factors: input voltage, power supply, clock signals, ground), *pullup nodes* (connected through the pullup transistor to the power supply; they are logic 0 if grounded, otherwise they are logic 1) and *normal nodes* (the remaining nodes).

Example: Figure 6.9 shows a part of a circuit, which contains the three types of node. The nodes V_1, V_2, V_{DD}, Φ, $\overline{\Phi}$ and GND (ground) are input nodes, P is a pullup node, while M and N are normal nodes.

MOSSIM has been developed for simulating clocked MOSFET circuits on the assumption that the clock frequency is sufficiently low as to ensure the circuit to be stabilized within a clock period. Conventional circuits operate with two-phase, nonoverlapping clock pulses. A switch-level simulator does not analyze the electrical behavior of the circuit in detail.

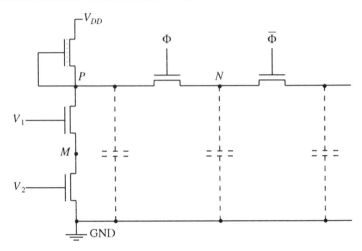

Fig. 6.9 Three types of nodes in MOSSIM

During the simulation process, a sequence of input signals and suitable clock signals are applied to the various input nodes. The network is partitioned into *groups*, which contain external and internal nodes. The external nodes are either input nodes (besides the known input nodes also the gates of MOSFETs) or output nodes (nodes which are connected to MOSFET gates of other groups). In forming the groups, node splitting is performed such that each group contains a separate V_{DD}, Φ and GND. We start with unit-delay simulation, operating on various groups which in fact are unilateral subnetworks.

Let us consider what happens within a group. The states of the nodes must be updated in accordance with the signal flow through the transistor network. An undirected graph is set up, consisting of nodes and branches of all transistors which are in the closed condition (source and drain short-circuited). All disjunct sets of interconnected nodes are called *classes*. The logic value assigned to any class is determined by the "strongest" node. The order of strength of the nodes is: input $\to$ pullup $\to$ normal. When the strongest nodes have the same state, this state will be assigned to the class. Otherwise, we choose the state U. Two or more classes can be combined to *superclasses*, when they are connected by branches with U states. Depending on the particular states being assigned to each of the classes of a superclass, a state U of any of the classes can be propagated to neighboring classes of the same strength. This process is called poisoning. All "poisoned" classes must be set to the state U.

6.2 LOGIC SIMULATION

Switch-Level Timing Simulator SLS

SLS (Switch-Level Simulator) is a logic and timing simulator for MOS digital circuits at the switch level [van86]. It is an extension of Bryant's simulator:
a. The circuit elements are modeled with analog values instead of strength abstractions. The logic states are directly determined from the actual circuit parameters and no classification for the circuit elements is necessary.
b. Each MOSFET is modeled as indicated in Fig. 6.10.
c. Min-max voltage waveforms are used to account for circuit-parameter deviations and model inaccuracies.
d. SLS can simulate the timing behavior of the circuit. The extensions are such that they hardly affect the efficiency of the principle of switch-level simulation.

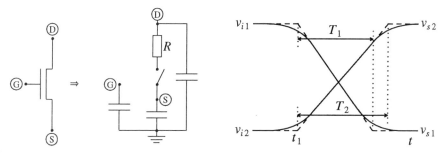

Fig. 6.10 MOSFET model   Fig. 6.11 Elementary waveforms

The real node voltages (solid curve in Fig. 6.11) at a circuit node as functions of time are approximated by *elementary waveforms*, given as piecewise-linear functions (dashed lines in Fig. 6.11). An elementary waveform consists of three parts: an initial stable voltage value v_i, a linearly rising or falling line segment during an interval T and a new stable value v_s. An initial stable voltage is disturbed when an *event* ($0 \rightarrow 1$ or $1 \rightarrow 0$ logic transition) takes place. An event (in Fig. 6.11 at time t_1) usually occurs when a logic transition appears at a transistor gate. An event at a transistor gate affects the source and drain and hence their vicinities.

Nodes are divided into *forced nodes* (with fixed values, e.g., power voltages and input voltages) and *normal nodes* (with network-dependent values). A property of forced nodes is that they block the influence of a given normal node on another node. This, together with the switch in the transistor model, determines the *vicinity* of a normal node n, which is defined as the set of nodes, including n, for which there exists at least one path to n through closed or undefined transistors with no intermediate forced nodes in the path.

Suppose that a specific node n has a stable initial value v_i. When an event occurs at node n, the vicinity of this node is determined. Two elementary quantities of the nodes in this vicinity are evaluated: the new stable state v_s and the time interval T. The stable voltage v_s is evaluated by using two independent methods: voltage division and charge sharing. By alternately assigning the values 0 and 1 to all transistors whose gates have the value U, minimum and maximum values for the stable voltages, $v_{\min}$ and $v_{\max}$, can be obtained. By evaluating the worst-case and best-case resistances in the paths from a node to ground and power source, voltage division gives the values $v_{\min}$ and $v_{\max}$.

For using charge sharing, the vicinity is split into groups of nodes interconnected by transistors in the Closed state. Separate groups are connected by transistors in the Undefined state. For each group consisting of normal nodes, the first minimum and maximum stable voltages are obtained by calculating the total minimum and maximum charge of the group and dividing these values by the total group capacity. The groups are then classified as 0, 1 and U groups according to the stable voltages just calculated. Then, stable voltages are calculated by considering the connection of each group with other possible groups. Here, we use upper and lower bounds of the charges of the 0, 1 and U groups. Finally, network topology is used to classify some groups in the U band. Ultimately, $v_{\min}$ (or $v_{\max}$) is taken as the smaller (larger) one of the two values found when using resistance division and charge sharing.

The time interval T is determined from the layout of the circuit according to Penfield and Rubinstein [Rub83] and Wyatt [Wya84]. T is taken as the average between the minimum and maximum values of the computed values of T.

Example: Figure 6.12 shows an example of the voltages $v_{\min}$ and $v_{\max}$ at a certain node.

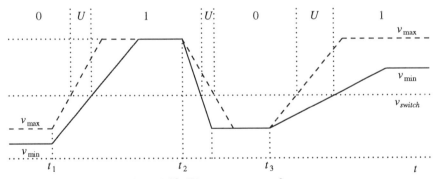

Fig. 6.12 Elementary waveforms

The associated logic values are determined as follows. When the minimum and maximum values $v_{\min}$ and $v_{\max}$ are both greater (smaller) than the threshold voltage v_{switch}, the logic value is 1 (0). When the minimum value is smaller and the maximum value is greater than v_{switch}, the logic value is U. In Fig. 6.12,

6.2 LOGIC SIMULATION

events occur at $t = t_1, t_2$ and t_3. Each event at a node starts up a renewed calculation of voltages in the vicinity of this node.

During timing simulation, an event list is generated and the chronologically ordered events are handled one by one. An event at a transistor gate affects the source and drain and hence their vicinities. A timing simulation step consists of two steps:
a. The transistor whose gate is connected to a node that is currently on the event list is updated.
b. The vicinities of drain and source of this transistor are updated.

SLS is capable of simulating NMOS transistor circuits at three levels: pure switch logic, logic including resistance division and charge sharing, and timing simulation. The simulator is written in C and runs under UNIX. Mixed switch-level/function-level simulation can be performed by incorporating calls to function blocks in the circuit description. See also page 267.

SLS and SNEL
The SLS simulator of Barzilai *et al.* [Bar88] is a fast switch-level simulator. The simulation involves two updating processes: the transistor update in which a new graph is formed by dropping those edges which correspond to nonconducting transistors, and the node update in which nodes belonging to the same group are divided into subgroups of connected nodes. The objective is to evaluate at subsequent time steps the steady state of each node of the circuit by an edge-by-edge traversal of paths connecting the nodes.

Based on SLS, SNEL is an event-driven switch-level simulator, which uses extensive functional abstraction of the circuit behavior prior to the simulation process [Bla90]. This functional abstraction determines the set of switch-level phenomena occurring in individual elements or clusters of circuit elements.

Timing Simulation
When the computational accuracy of the transient responses of transistor circuits is of crucial importance, detailed transistor models must be used and simulation must be performed at the *circuit-element level* (Subsection 6.3a). While this circuit simulation is primarily relevant in analog circuits, it is also required in small subnetworks of a digital system, at least in specific portions of the system where tightly coupled feedback loops exist. The main drawback of circuit simulation is that computer time and memory requirements restrict the size of the circuits that can be analyzed to a maximum of about 200 transistors.

On the other hand, logic circuits containing thousands of logic gates can be handled easily by using logic simulators (Subsection 6.2a). Simulation at the *logic-gate level* has some deficiencies:
a. Only discrete logic levels are considered, devoid of any detail in the analog waveform of the signals.

b. There are circuit structures which cannot be handled by conventional logic simulators, e.g., MOS transmission gates.

There is a need for simulators which combine the advantages of logic simulators (speed by next-event simulation and selective trace) and circuit simulators (accuracy by waveform representation of the signals). The advent of VLSI MOS cirucits has led to the development of efficient *timing simulators* with the objective of filling this need. The following attributes are relevant to *timing simulation*:

a. Timing simulation is aimed at *logic circuits.*
b. In contrast to logic simulation, it provides analog *waveforms* of the logic signals at selected output nodes.
c. To enable cost-effective simulation of large-scale networks, *simplification* procedures of circuit modeling and analysis algorithms are cardinal.
d. The purpose of timing simulation is the verification of the exact *timing* of the logic-signal transitions.

To bring the speed of timing simulation close to that of logic simulation, some measures are taken, notably the following:

a. *Model simplification.* Since device-model evaluation is generally the most expensive part of circuit analysis, timing simulators use macromodels or stored-table models which are looked up at simulation time. Bearing in mind that nonlinear analysis is more costly than linear analysis, it may be effective to avoid nonlinear-function evaluations by using piecewise-linear macromodels. Capacitors, which are usually voltage-dependent, are often taken to be constant.
b. *Simplification of the analysis algorithms.* Latency is exploited in the simulation process by using the selective-trace principle. The Newton-Raphson iterations and matrix-solution procedures inherent in circuit simulation are replaced by explicit, single-iteration vector-product computations.

Many simulators in the range between the logic level and the circuit-element level have been developed [Sch87]. An example of a circuit-level-oriented timing simulator is DIANA [Arn78]. DIANA can be used for the simulation of (linear or switched-linear) analog circuits and of digital logic circuits, but it was primarily developed for mixed analog-digital networks, such as analog-digital converters. The circuit primitives may be the (linear) circuit elements R, L, C and controlled sources for the analog part, and logic gates and JK flip-flops for the digital logic part. An important feature of DIANA is the use of *controlled switches.* Though DIANA is a mixed-mode simulator, the simulation mode is reduced to circuit-element simulation using the MNA method (see Subsection 6.3a).

RSIM, developed by Terman [Ter85], is a logic-level timing simulator which employs a simple MOS-transistor model consisting of a resistor R in series with a switch controlled by the three-valued gate voltage. The effective resistance value R is determined separately for each transistor, depending on the width-length ratio and the type of function of the transistor. Lumped capacitances are incorporated into the model. The logic states of the nodes are determined by an event-driven simulation technique, which includes charge-sharing calculations.

ELogic (Electrical-Logic) is a set of circuit modeling and simulation programs based on the Nodal Analysis Method [KIm89]. The algorithms provide a continuous speed/precision trade-off between the circuit level and the logic/switch level as well as accurate timing information. The ELogic algorithms have been used to implement both a simulator (ELOSIM) and a timing verifier (E-Crystal).

6.2d Expert Systems for Digital-Circuit Verification

Knowledge-Based Simulation
In the preceding subsections, algorithmic methods of digital logic simulation were discussed. From a wide variety of algorithms which are available, a specific algorithm can be selected to suit a particular application. The non-algorithmic activities, which require AI techniques, include reasoning about such things as method selection and the interpretation of the simulation results. Knowledge-based simulation is meant to enhance the system performance by introducing a symbolic decision and control capability.

The knowledge-based approach allows the user to select a suitable model or simulation method and interpret the results of the simulation process. Some models are provided with incomplete, noisy or inconsistent data, e.g., the time of occurrence of an event. In such situations, qualitative reasoning using approximate or constraint-based data may be used to interpret events or infer future behavior. There must be a provision for a mapping between numeric and symbolic data structures and an effective coupling of procedural simulation methods with nondeterministic AI methods.

Two approaches to the coupling of simulation and AI are in common use:
a. Loose coupling of separate algorithmic simulation programs that communicate via data files. The AI program assists in selecting the suitable program and possibly in the analysis and interpretation of the simulation results.
b. Incorporation of simulation programs in a knowledge-based system with communication via shared memory or the passing of data structures. Key attributes of a simulation algorithm may be described by a frame structure or by a set of declarative relations, such as a Prolog rule set.

Heuristic Circuit Simulation

Gullichsen [Gul85] uses Prolog to describe digital logic circuitry. A "backward" simulation method is proposed using theorem proving based on constraint satisfaction. The heuristic circuit simulator can serve as the basis for an implementation of Roth's D-algorithm (see Subsection 7.2a).

DDL/SX Verifier

DDL/SX is a rule-based system for logic synthesis (see Subsection 5.2d). As this system cannot handle timing problems, such as critical-path analysis, the synthesis results have to be changed for delay correction to avoid logic errors. Delay correction should avoid that two input signals which propagate via different paths arrive at the output at the same time.

When a synthesized circuit has to be modified, the logic of only a small subcircuit is changed, leaving the remainder of the circuit unaffected. When a delay correction in a subcircuit is performed, the DDL/SX Verifier checks the logical equivalence of the subcircuit before and after the changes [Man87]. By verifying subcircuits one at a time rather than the whole circuit, the verification task is considerably alleviated.

RDV

RDV (Rule-based Design Verifier) was developed at Stanford University for functional and fault simulation as well as for timing verification [Gho87]. A characteristic feature is the distributed modeling of the circuit components or functional modules. The scheduling and communication primitives as well as the simulation and verification algorithms are distributed among all the models. Signals propagate through the circuit along paths determined by the interconnectivity of the circuit components or modules. RDV uses Ada as the hardware description language and simulaton environment. Models implemented as Ada tasks may, in principle, execute concurrent entities on a multiprocessor Ada machine so that parallelism may be utilized with relative ease.

CRITTER

CRITTER, developed by the VLSI/AI group of Rutgers University, is a verification tool for critiquing digital circuits [Kel84]. The input consists of a hierarchical circuit description, a behavior specification and the input stimuli. Internal signals (e.g., the clock signal) must meet certain conditions. The system builds up a database of the circuit's performance. The circuit is first decomposed into subcircuits in such a way that constraints on the subcircuits can be derived from the global circuit constraints.

The critiquing process involves analysis of the functional correctness, timing, robustness and processing speed of the designed circuit. A specialized simulation facility analyzes if the circuit is correct within the specification. A critical-path

6.2 LOGIC SIMULATION

analysis evaluates the timing behavior and operating speed of the circuit. A design is said to be robust if it satisfies its constraints despite reasonable parameter variations.

The knowledge includes information about circuit diagrams and circuit-analysis techniques, such as subcircuit simulation and path-delay analysis. Circuit diagrams are represented using frames, while other knowledge is in the form of algebraic formulas and predicate calculus. CRITTER is implemented in Interlisp.

The results of various circuit-analysis techniques (simulation, path-delay analysis, etc.) are evaluated and compared with specifications. The system evaluates the circuit performance and measures the degree to which the specifications are met. Circuit sensitivity to changes in device parameters can be evaluated. To this end, design flaws are pinpointed, if any exists. Diagnostic information is given with recommendations for repair.

REDESIGN

REDESIGN, another development of Rutgers University, assists engineers in the redesign of existing circuits, when functional specifications must be altered [Ste85]. CRITTER is used in REDESIGN as a subsystem.

The input is the functional description and the circuit diagram of the existing circuit. Given the desired modifications of the performance, a circuit simulator is used to investigate whether or not the old circuit satisfies the new specification. The system selects circuit parts which have the greatest potential to realize improvements of circuit performance by circuit modifications. Then, the suitable modifications are proposed and their effects on the circuit behavior are analyzed. Possible undesired side effects are determined.

The circuit knowledge is represented as a network of modules and data paths. Causal reasoning involves such questions as what circuit parts must be added or modified, when output Z is required, given an input A. Reasoning about the purpose and specification of modules is also needed.

LEAP (LEarning APprentice] is a tool for the acquisition of design refinement operators [Mit85]. It can be used as a learning tool in REDESIGN.

DEGAS

DEGAS (DEsiGn ASsistant) provides an integrated and efficient environment for the development of expert systems to assist VLSI chip designers [Jac89]. Its knowledge base is structured in terms of rules which use an easily extensible set of primitives to implement the selection of objects and to execute actions upon them.

The DEGAS environment is primarily designed to support expert systems able to diagnose topological, electrical and timing problems in a circuit and to suggest modifications for correction of malfunctions and improvement of the circuit performance. A first prototype of DEGAS was written in Franz LISP.

TLTS

TLTS (Transmission Line Troubleshooting System) is a knowledge-based tool developed to modify a faulty circuit into a fault-free circuit [Sim89]. An evaluation-redesign loop consists of an iterative procedure of circuit simulation using the simulator SPICE, analysis and evaluation of the simulation results, diagnosing the causes of a circuit failure, and modifying the structure and components of the circuit. The knowledge of TLTS is organized in a blackboard architecture, implemented in the ORBS (Oregon Rule Based System) language.

A learning component has been incorporated into TLTS [Sim90]. New knowledge sources in the blackboard system are formed from redesign knowledge by compiling the knowledge contained in redesign plans. The learning component also acquires redesign metaknowledge for selecting the most appropriate knowledge source to redesign a faulty circuit. The system utilizes the problem-solving information acquired by the user in cooperation with TLTS.

DIALOG

DIALOG is an expert debugging system for checking the correctness of logic rules in NMOS or CMOS logic [DeM85]. Knowledge is described in LEXTOC (Language EXpressing TOpology Constraints), allowing for unification (like in Prolog), object creation, property assignment, association of relations, rule formulation, logic or arithmetic evaluation as well as conversational constructs.

Circuit verification is founded on a set of sufficient conditions for a correct electrical behavior [Bol88, Bol89]. To alleviate the problem of large subnetworks, the concept of Unilateral Blocks is introduced. Rules are used to increase flexibility and to manage complexity of the algorithmic verification steps. A symbolic analysis applied to switch-level networks examines the context of local bugs and reports only relevant error messages.

New algorithms for timing verificaton and analysis of synchronous MOS circuits handle a very wide class of timing faults [Van88]. A rule-based method is used to transform the MOS transistor network, extracted from the physical layout, into a network of unidirectional subcircuits. Each subcircuit is characterized by a logic model, which is used to accurately derive the timing constraints. The timing verifier can be used for a wide range of static and dynamic NMOS and CMOS circuit designs.

Static Debugger and Simulation Compiler

Kolodny [Kol85] described two schematic-based tools for validation of VLSI chip designs. The tools represent two complementary approaches to the verification task, i.e., static inspection of the design and efficient functional modeling, derived from circuit schematics. The static debugger uses pattern matching in conjunction with path analysis and delay modeling to detect schematic design-rule violations. Many fatal bugs, which were not discovered by

6.2 LOGIC SIMULATION

simulation, have been traced by this tool. The simulation compiler converts the analyzed network into an efficient source code which verifies the hardware functionality. A switch-level model is used combined with extracted logic expressions and macromodels.

SOCRATES

SOCRATES (Synthesis and Optimization of Combinatorics using a Rule-based And Technology-independent Expert System) automatically synthesizes and optimizes a circuit, given its Boolean logic description [Gar84]. The system consists of four modules, each performing a specific function:

1. Logic-function reduction and synthesis of a technology-independent multiplexer circuit.
2. Circuit-level optimization of the circuit obtained in 1. by successive application of rules describing replacements of one or more circuit elements by other functionally equivalent, but more desirable elements in the target technology.
3. Extraction and comparison. From the circuit generated in 2., a Boolean function is extracted and compared to the original Boolean function for equivalence verification. The objective is to obtain equivalence by successive replacements of subcircuits.
4. Rule generation. When a subcircuit can be replaced by a logically equivalent but more desirable subcircuit, this result creates a new rule.

Automating Timing Design

TDS (Timing Design System), developed by Kara *et al.* [Kar88], is an automatic tool for timing design of interfaces between VLSI chips in microcomputer systems. It is a Prolog-based expert system that interprets the specification sheets of VLSI chips, and can synthesize, diagnose and verify timing charts. A functional model is used, based on timing specifications rather than structural information.

The TDS architecture is shown in Fig. 6.17.

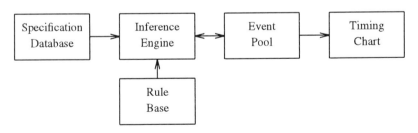

Fig. 6.17 TDS architecture

The specification database specifies all the timing parameters for the chip being considered. The rule base contains design criteria and device-specific rules. The inference engine uses the rules in the rule base to calculate event times. The event pool is a dynamic database containing names and updated event times. Ultimately, TDS generates the timing chart.

6.3 CIRCUIT SIMULATION

6.3a Circuit-Element-Level and Mixed-Mode Simulation

Simulation at the Circuit-Element Level

Analog circuit design takes place at a low abstraction level, notably at the transistor level. Simulation at the circuit-element level is usually referred to as *circuit simulation*. Each transistor is modeled in terms of circuit elements, such as resistors, capacitors and controlled sources. The complexity of the detailed transistor-based structure of the circuit increases rapidly as the circuit size grows. Whenever possible, circuit partitioning is performed so that reasonable circuit sizes can be analyzed.

Over the years, human experts have gained considerable experience in designing various analog circuits. When a new design is wanted, advantage can be taken of this experience. Frequently, a new design is based on an earlier design, which is appropriately modified in order to satisfy the new performance requirements. Verifying the correctness of the design is accomplished by circuit simulation, which boils down to circuit analysis of a suitable model of the circuit. Chip design that combines circuit, device, layout and fabrication-process issues unveils nonideal effects, which can be taken into account in circuit simulators (see Section 8.4).

Circuit analysis is usually based on the *Modified Nodal Analysis* (MNA). Besides the node voltages, appropriate branch currents may be taken as independent variables. Given the circuit elements and their interconnections, the MNA matrix can be set up in a straightforward manner. Nonlinear elements can be handled by the Newton-Raphson method, which is a process of iteratively linearizing, solving linear equations and updating [Sch87].

When an input signal is applied to the nonlinear circuit, the time response can be evaluated by running three programs in a nested structure, as indicated in Fig. 6.18. The main program P_1 reads the input file and constructs the associated data structure, which is stored in a common-block area of the primary memory. When this input processing is finished, P_1 performs the desired circuit analyses

6.3 CIRCUIT SIMULATION

by using the programs P_2 and P_3. Program P_1 has to be executed any time the circuit is modified.

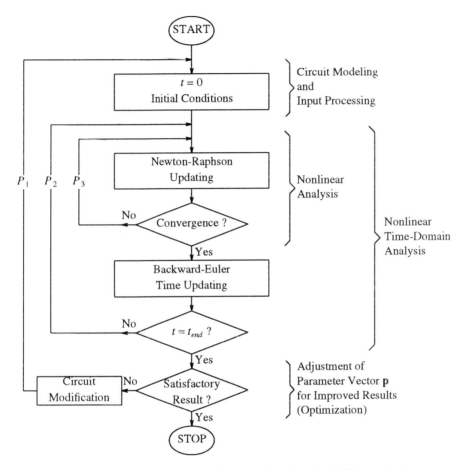

Fig. 6.18 The computational process by using the MNA method

The time-updating program P_2 carries out the time-domain calculations in the specified time interval. Any suitable implicit integration technique, including the Backward-Euler method, can be used. When the circuit is nonlinear, program P_2 includes repeated runs of program P_3. The nonlinear analysis program P_3 performs the necessary Newton-Raphson iterations which must lead to convergence in each time step. When the programs P_3 and P_2 use both Newton-updating and time-updating models, incorporated into the MNA matrix, the analysis problem is reduced to the process of iteratively solving updated systems of linear equations [Sch87]. SPICE (Simulation Program with Integrated

Circuit Emphasis) is a typical simulator program that is based on the MNA method outlined above [Nag75].

SPICE2

SPICE2, developed at the University of California, Berkeley, is probably the most popular circuit simulator. Circuit elements used in SPICE are resistors R, capacitors C, inductors L, coupled inductors K, transmission lines T, independent voltage and current sources V and I, voltage-controlled current sources G, voltage-controlled voltage sources E, current-controlled current sources F, current-controlled voltage sources H, junction diodes D, bipolar junction transistors Q, junction field-effect transistors J and MOS transistors M. Capacitors, inductors and the four controlled sources can be described by nonlinear equations. A circuit description is started by numbering all the nodes of the circuit. The circuit is described by specifying each component by its name, its type (defined by the first letter of the component name: R, C, etc.), the circuit nodes and the values of its electrical parameters.

Examples:
R5 3 4 10K $\Leftarrow$ A resistor R5 of 10 kΩ connected between nodes 3 and 4
C1 7 10 500PF $\Leftarrow$ A capacitor C1 of 500 pF connected between nodes 7 and 10
G3 3 0 4 0 0.2MMHO $\Leftarrow$ A voltage x V between nodes 3 and 0 controls a current source between 4 and 0 whose current equals $2.10^{-4}x$ A.
Q14 5 2 4 QMOD $\Leftarrow$ A bipolar transistor Q14 whose collector, basis and emitter have the node numbers 5, 2 and 4 respectively, and whose electrical parameters are specified in a separate model statement, e.g.:
·MODEL QMOD NPN BF=60 IS=1E-13 VAF=50 $\Leftarrow$ The model under the name QMOD refers to an NPN transistor with a forward beta of 60, a transport saturation current of 10^{-13} A and a forward Early voltage of 50 V.
M11 8 6 7 1 MX L=4U W=2U $\Leftarrow$ A MOS transistor whose drain, gate, source and substrate node are numbered 8, 6, 7 and 1 respectively, and whose electrical parameters are specified in a separate model statement. Channel length and width are also specified.

The transistor specification statements may contain additional information, e.g., the required initial conditions. The power of SPICE lies in the accurate modeling capabilities of the semiconductor devices. In the SPICE version 2G, the parameters which can be specified for the bipolar model and the MOS transistor model amount to 40 and 42 respectively. Parameters which are not specified are given default values. The Gummel-Poon model forms the basis for representing a bipolar transistor [Sch87]. SPICE provides three levels of modeling MOS transistors, the first level being the Shichman-Hodges model [Sch87].

6.3 CIRCUIT SIMULATION

The main types of circuit analysis are small-signal AC analysis for frequency responses (amplitude and phase characteristics) and transient analysis for time-domain responses. DC analysis for calculating the DC operating points of the circuit nodes is usually performed prior to frequency and time-domain analysis. Special features of SPICE are the distortion analysis, noise analysis and Fourier analysis.

The SUBCKT feature of SPICE can be used to compose a circuit from lower-level predefined modules.

Example: A 2-input NAND described as a complete transistor circuit can be considered as a predefined module with 2 input nodes, one output node, GROUND and power-supply nodes. A one-bit-adder with carry-in and carry-out nodes can be composed from eight 2-input NANDs. A two-bit adder is composed of two of such one-bit adders.

Several analog simulators based on SPICE2 have been developed for use with personal computers [Ban89]. The most important ones are outlined below.

SPICE3
SPICE3 is an interactive SPICE version written in C [Qua86]. A post-processor for SPICE3 is NUTMEG. Interactive optimization with DC, AC and transient analysis is also attached to SPICE3.

PSPICE
This personal computer version of SPICE is a SPICE2-based simulator from MicroSim Corporation [Tui88]. It contains the same features as SPICE2, except for distortion analysis. PSPICE has several interesting options. The graphics postprocessor PROBE provides interactive viewing of simulation results with high-resolution graphics and a display of the Fourier transform of waveforms. PARTS allows the creation of model libraries with new devices. Monte-Carlo analysis allows multiple runs of analysis with different component values within given tolerance assignments.

IS_SPICE
IS_SPICE is a version of SPICE2G.6 for PCs sold by Intusoft. In addition to a graphics postprocessor, it has a preprocessor which functions as a general programmer's editor.

Mixed-Mode Simulation
A plethora of timing simulators and mixed-mode simulators have been reported in the literature [Sch87]. Some recently developed methods are outlined below.

VIEWSIM/AD
VIEWSIM/AD (VIEWSIM Analog Digital simulator) is capable of simulating mixed analog and digital systems [Cor88]. It is based on VIEWSIM, a 28-state

logic simulator, and PSPICE. By exploiting multitasking operating system capabilities, the digital and analog simulators run as separate subprocesses under the control of the VIEWSIM/AD simulation control process. This process is responsible for initialization and synchronization of the VIEWSIM and PSPICE simulation subprocesses. A synchronization algorithm schedules the logic events in VIEWSIM and calculates circuit values in PSPICE. Time-step sizes in PSPICE are adjusted only when a signal that affects both digital and analog circuitry changes state.

Explorer Checkmate
This is a fully integrated set of tools for verifying the functionality of integrated circuits and testing layouts [Men90]. It can be used for the verification of digital, analog and mixed circuits in various technologies (CMOS, bipolar, BiCMOS and GaAs). It is developed by Mentor Graphics in cooperation with Texas Instruments. The programs test design rules, measure and store parameters, provide net lists, compare layouts to the circuit schematic with interactive correction of errors.

Since Explorer Checkmate allows distributed processing in a network, a part of the tasks can be performed in the idle time of other workstations. Thus, such a network is optimally exploited and the processing time is reduced. Since English-like expressions are used, the design rules are easy to understand and adapted to new technologies, such as BiCMOS. False and redundant errors are automatically reduced.

PMLS
PMLS (Parallel Multi-Level VLSI Simulator), a multi-level simulator developed by AEG [Loh90], is implemented in the object-oriented language POOL2. It covers four abstraction levels: register-transfer, functional, gate and switch level. Four-valued logic (0, 1, U and Z) is used. At the switch level, the signal has a continuous range of strengths.

For all abstraction levels, PMLS employs just one simulation concept, characterized by using circuit partitioning and distributed discrete event simulation. Subcircuits, which may contain elements at different abstraction levels, are simulated by subsimulator processes. Subsimulators communicate by exchanging event messages. They execute asynchronously, depending on the local simulation time using an effective synchronization technique.

The highly interactive interface allows the user to suspend and restart a simulation and change parameters (stimuli, delays), the circuit structure and the abstraction level. PMLS is flexible in that it can be easily modified for functional enhancements. The use of a general-purpose parallel machine DOOM guarantees a high performance.

6.3 CIRCUIT SIMULATION

Hierarchical Timing Simulation
The hierarchical timing simulator HITSIM, written in Smalltalk, was developed as part of an integrated design system, which incorporates leaf-cell and composed-system design, hierarchical simulation and layout design [Lin86]. The VLSI design can be partitioned into hierarchical levels of semantic cells. A semantic cell in a two-phase synchronous system is recursively defined: it is either a clocked leaf cell (whose inputs are controlled by clocked pass transistors) or a legal composition of semantic cells. The two classes ("LeafCell" and "CompositionCell") contain methods to transform cell specifications provided by the user into suitable classes and methods for performing timing simulation.

Low-Level Mixed-Mode Simulation
Computer-aided design tools at the lower levels of abstraction (process, device, circuit) are collectively referred to as *technology CAD* (TCAD) [Llo90]. Although design tools at each of the three lower levels exist, integration of these levels increases the efficiency of the design process. In a conventional TCAD procedure, a numerical process simulator (e.g., SUPREM-3) drives a robust numerical device simulator (e.g., PISCES-II), which uses a numerical optimization process to generate device model parameters suitable for circuit simulation. Apart from the computational inefficiency in using optimization algorithms, the correlations between device and process parameters get lost. To re-establish these correlations, Green and Fossum [GrE92] proposed an application-specific process that eliminates the need for optimization, as indicated in Fig. 6.19.

Fig. 6.19 Low-level mixed-mode simulation

The process simulator SUPREM-3 is linked to the seminumerical mixed-mode device/circuit simulator via SUMM. The UNIX-based, menu-driven program SUMM evaluates the model parameters from a tabulated one-dimensional doping profile such as that predicted by SUPREM-3. It automatically writes the the parameter values in the circuit file for MMSPICE.

Object-Oriented Simulation
A special approach to simulation employs objects as basic elements. This approach has been applied to switch-level simulation [Roy85] and mixed-mode simulation [Lat85, May87].

6.3b Expert Systems for Analog Circuits

Analog-Circuit Design

A traditional approach to the design of analog circuits starts with an existing design which in some way satisfies the design specification. With an iterative process of appropriate circuit modifications, the design is successively improved, until the design specification is fully satisfied. The use of simulators and optimization procedures is instrumental in achieving this goal [Sch87].

The simulation-optimization iteration cycle can be implemented by a knowledge-based system in which recommended design changes can be provided by the system. There are several knowledge-based approaches to analog-circuit design. One approach uses a knowledge base which contains different alternative circuit configurations for a specific design goal, e.g., for realizing a filter function. After an intelligent choice of one specific configuration, the design parameters are determined so as to meet the desired filter characteristics.

In another approach, the knowledge base contains a variety of standard circuit modules, e.g., operational amplifiers, which can be used to construct the desired analog circuit. In still another approach [Fun88], the expert system selects one of several circuit configurations in the knowledge base after which transistor geometries are determined using analytic equations. If the circuit simulation results do not satisfy the performance requirements, modification in device sizes or circuit topology are made to improve the circuit performance. The analog circuit design can be further automated by using an automatic layout generator. By recognizing critical nodes and circuit components, compact layouts can be achieved. To facilitate automated analog synthesis, CAD tools have been developed [Ant92].

Below, several simple and more advanced expert systems for circuit analysis and synthesis are outlined.

EL

An early expert system for analyzing the DC behavior of analog circuits is EL [Sus75, Sta77]. The basic technique underlying the circuit analysis is *propagation of constraints*. In this technique, the set of possible solutions is gradually reduced by applying a number of rules or operators that impose "local constraints" on the solution. This process proceeds so long until no more rules can be applied. The final solution is found in the remaining manageable solution space (which may be empty).

Constraint rules contained in the knowledge base include Ohm's Law and Kirchhoff's Current Law. Frequently used rules are the following:

if the voltage across a resistor R is known, *then* compute the current through R using Ohm's Law.

if the current through R is known, *then* compute the voltage across R.

6.3 CIRCUIT SIMULATION

if all but one of the currents flowing into a node are known, *then* compute the unknown current by using Kirchhoff's Current Law.

Node voltages and branch currents in a network can be computed by propagating symbolically expressed constraints through the circuit.

Example: The above rules allow the system to evaluate node voltages and branch currents of the simple resistor circuit given in Fig. 6.20. Denote V_{ij} as the voltage between node i and node j, and I_k as the current through resistor R_k. Start at an arbitrary node, e.g., node 3. Let V_{30} be equal to x V. Applying Ohms's Law, $I_4 = x$ A. Kirchhoff's Current Law applied to node 3 yields $I_3 = x$ A and hence we obtain successively: $V_{23} = 3\ x$ V; $V_{20} = V_{23} + V_{30} = 4\ x$ V; $I_2 = V_{20}/R_2 = 4\ x/2 = 2\ x$ A; $I_1 = I_2 + I_3 = 3\ x$ A and $V_{12} = R_1\ I_1 = 3\ x$ V. As a result, $V_{10} = V_{12} + V_{20} = 7\ x$ V = 7 V. Since $x = 1$, all node voltages and branch currents are known.

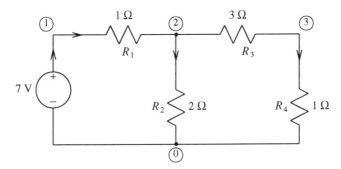

Fig. 6.20 Sample circuit

SYN
SYN (circuit SYNthesis) is an attempt to synthesize a class of electrical circuits by using heuristics [deK80]. It generates symbolic expressions, e.g., an amplifier gain is expressed symbolically in terms of circuit parameters. The symbolic methods used produce solutions which are clear and insightful. Like EL, the synthesis task is based on analysis by propagation of constraints. However, the analysis program EL calculates voltages and currents of a given circuit, whereas SYN's objective is to determine the component values, given the circuit schematic. SYN generates an appropriate model of the circuit for different frequency regions. When the solution process falls short, the system can accept advice from the user in the form of rules of thumb.

ELSYN
ELSYN of IBM [Sus82] has been developed for the analysis and synthesis of analog circuits. Values of circuit parameters are represented and manipulated by applying constraints, such as the Kirchhoff Laws. Starting from one component,

these constraints are propagated throughout the circuit, and parameter values are assigned to the circuit elements.

AUSPICE
AUSPICE gives assistance to the user of the SPICE circuit simulator [Gub88]. The knowledge base contains expert knowledge on circuit and numerical problems which may arise in circuit simulation. AUSPICE is capable of finding a way out of nonconvergence as well as mistakes made during the circuit specification. This expert system, which is built on a goal-driven shell, runs on a personal computer as well as on a DEC VAX. The third version, AUSPICE3, first analyzes the input file and recognizes the troublesome elements or parameters of a possibly nonconverging circuit analysis. Automatic corrections can be made, depending on the type of error. AUSPICE3 is written in C and runs on VMS under ULTRIX.

AUSPICE is transformed into a learning system, called ALEA (Auspice + LEArning) [Zan90]. The new system achieves an autonomous and continuous growth of its knowledge based on the experience gained in solving failed simulations. The learning algorithm establishes relationships between the values of SPICE parameters and the adjustments needed to obtain a correct simulation.

Circuit Analysis Using a Deductive System
Design of electronic circuits can be viewed as a goal-directed composition of basic components. Analysis of circuits involves the process of finding lower-level subcircuits in a functional hierarchy. Tanaka [Tan88] discusses the problem of circuit analysis by using a deductive system. Circuits can be represented in logic. A LISP-based deductive system for writing predicate calculus rules, called Duck [McD83], is used for structural analysis of circuits.

An electronic circuit shows great resemblance to natural language. When a circuit is viewed as a sentence and its elements as words, analysis of the circuit is analogous to parsing a language. Circuit structures are defined by deductive rules analogous to definite clause grammars. Using these rules, a circuit is decomposed into a parse tree of functional blocks.

Filter Synthesis
Electrical filters are designed to realize a prescribed transfer function, i.e., output/input voltage ratios that have prespecified frequency characteristics. There exist a wide variety of electrical filters: passive, active, switched-capacitor and digital. Filter synthesis can be based on standard filter structures. A Prolog-based active-filter synthesis procedure was proposed by Wawryn [Waw89]. A Prolog program searches for an appropriate filter structure that realizes the desired transfer function.

Budzisz [Bud89] described a rule-based methodology for TAC filter synthesis. A TAC filter is composed of exclusively transconductance amplifiers

6.3 CIRCUIT SIMULATION

and capacitors. The synthesis method is based on searching for graphs describing structures of circuits with the desired transfer functions.

Analog-Circuit Diagnosis
Apfelbaum [Apf86] of Teradyne, Boston, developed an expert system for diagnosing faulty analog circuits. The knowledge base contains comprehensive knowledge for each analog component and a set of diagnostic rules for analog circuits. The system reduces and refines test data, automates failure analysis and generates repair-oriented diagnostics. Repair personnel can be guided by the diagnostics messages, while accurate failure data provide feedback on manufacturing quality through real-time warnings and summary reports on defects.

DEDALE
DEDALE is an expert system, developed by ESD/IBM (France), for troubleshooting complex analog circuits [Dev87]. It is based in particular on order-of-magnitude reasoning applied to function models. This kind of reasoning considers both the sign and relative order of magnitude of the voltages in the circuit in relation to the nominal values. On the assumption that significant changes in the circuit voltages are caused by a defective component, function decomposition of a circuit and heuristic rules are exploited in the use of three strategies: top-down (searching the defect in lower levels), horizontal (considering neighboring components at the same level) and bottom-up (considering higher-level blocks if no conflicts have been detected at the same level).

DEDALE is written in (IBM) VM/PROLOG and has four components: an object-oriented language to describe a circuit both functionally and structurally, a library of qualitative models for generic components, a problem solver which performs order-of-magnitude reasoning, and strategic rules.

Circuit Synthesis by Iterative Improvements
A diagnostic expert system in a feedback loop of the design cycle can be used for iterative performance improvements of an analog circuit [War91]. The expert system compares the simulated performance of the circuit with the input specifications. Whenever any discrepancy is detected, the expert system uses the knowledge base to locate those parts of the circuit that are responsible for the discrepancy. Transistor sizes are optimized or topological changes are made with the purpose of removing the discrepancy. A series of iterative improvements will ultimately lead to a circuit that satisfies the design specification.

Facts and rules in the knowledge base are particular cases of Horn clauses. Reasoning for the presented knowledge is based on bidirectional forward and backward chaining. The backward chaining strategy repeatedly uses the modus tollens and pattern-matching method to verify hypotheses. The process begins with a goal as a denial of the hypothesis to be proved, then reasons in a backward manner from conclusion to conditions and finishes successfully when the contradiction is reached. The backward-chaining strategy is descibed by an AND-OR tree (see Fig. 4.3).

Analog-Circuit Design

Sheu *et al.* [She88] described a knowledge-based approach to analog IC design with a fixed circuit architecture. The method is based on iterative improvements of the design which is evaluated by a circuit simulator. The whole cycle repeats until the desired design goal has been achieved. An expert system OP-1 has been implemented in Turbo Prolog to assist designers in designing a CMOS operational amplifier. An improved version of the above system allows design flexibility in that a chosen circuit topology and transistor geometries can be modified [She90]. Portions of the circuit can be replaced during a design process in order to improve the circuit performance.

Klinke *et al.* [Kli92] described a rule-based method for analog-circuit design which is successful when applied to a rather narrow class of circuits. Typical applications are CMOS operational amplifiers and switch-capacitor filters.

The usefulness of cell libraries in digital-system design has led analog-circuit designers to design analog-circuit cells, which can be used to compose complex analog circuits. Knowledge-based systems for designing such cells have been developed [Poh90].

OASE

OASE (Operational Amplifier and analog circuit Synthesis Expert system) is a knowledge-based simulation environment which forms an integrating shell for analog-circuit simulators [Mil90]. The knowledge base contains circuit and technology-specific knowledge. Currently, OASE includes a device-level simulator and opamp-specific circuit knowledge. By exploiting expert knowledge for dealing with circuit-specific aspects, such as test-function generation and simulation clustering, it relieves circuit designers from time-consuming work. Furthermore, it provides intelligent assistance for decision making in the field of evaluation of simulation results.

A blackboard approach was chosen to allow rapid prototyping and modular development. The independent knowledge sources (simulation expert, result-evaluation expert, system manager, environment manager, etc.) read their inputs from the blackboard and store their global results and control information on it. The concept of the system allows an easy extension of both the incorporated knowledge and the functionality, creating a basis for a more complex analog design environment.

The simulation phase in the context of the OASE synthesis process is performed by SILAS (SImulation ASsistant) [Mil91]. SILAS is used to automatically characterize the performance of a designed circuit by a number of

6.3 CIRCUIT SIMULATION

adequate AC, DC and transient simulations. Furthermore, it evaluates the results with respect to the most suited redesign strategy. The prototype system runs on a SUN 4/260 workstation under UNIX.

Expert Essence

Given a set of design specifications, Expert Essence determines the most appropriate circuit block among various alternatives [Fuj87]. Design rules are interpreted in a flexible way. Expert Essence automatically relieves given constraints and assumes a set of additional conditions unless a feasible solution is found. Design knowledge is represented in the form of frames (for specifying design objects) and rules (consisting of "necessary" and "sufficient" conditions on its premise part). Frames and rules may contain quantitative data on speed and power. Heuristic design rules are evaluated under the control of meta rules. A well-balanced trade-off between power and speed (or frequency range) is made.

iJADE

Lai *et al.* [Lai87] described a high-performance CMOS VLSI circuit generator iJADE, implemented in Franz LISP. Important features of iJADE include the following:

a. iJADE combines analytic tools with a rule-based expert system to optimize the circuit performance in a closed-loop form.
b. It can detect latches, trace clock paths, accurately simulate voltage waveforms and tune synchronous circuits to satisfy the clock-timing constraints.
c. The functional pattern-recognition ability is implemented in a rule-based system.
d. Its hierarchical structure for both circuit and programming saves memory space, speeds up computations and can define the potential problem subcircuit clearly.

To optimize the timing performance, iJADE calculates propagation delays in the circuit, searches for critical paths and modifies the circuit iteratively until the specification is met. For this purpose, iJADE uses a switch-level timing simulator JADE [LAi87], a switch-level timing analyzer, algorithms for delay reduction, a rule-based expert system and a frame database.

IIL Circuit Generation

Watanabe *et al.* [Wat86] of Hitachi described a knowledge-based system that automatically translates logic circuit net data into an optimal circuit implemented in IIL (Integrated Injection Logic). The knowledge-based design system is based on logic programming using Horn clauses. The circuit-translation process is realized through an inference process using axioms for semantic interpretation of gate-connectivity extraction, gate translation and redundant gate reduction. An important feature of this knowledge-based system is the possibility of incremental development of the CAD system, ease of system modification through additions and revisions of rules and realization of high-quality circuit translation through the use of domain-specific nonprocedural knowledge of design experts.

RUBICC
RUBICC (RUle-Based Integrated Circuit Critique) is an expert system for critiquing VLSI designs at the cell and transistor level [Lob85]. Its knowledge base is implemented as hierarchical frames in the Hewlett-Packard language HPRL (Heuristic Programming and Representation Language). HPRL supports both forward and backward chaining rules as well as demons and escapes in the underlying LISP language. RUBICC performs at least 45 distinct error checks on a wide variety of MOS circuit configurations.

QCritic
QCritic is a rule-based system for analyzing and reviewing bipolar analog-circuit designs [Ber86]. Good design practices are represented by a set of rules. QCritic serves as a post-processor for the SPICE simulator and can be used to examine a given analog circuit for possible violations of these rules. QCritic is written in OPS83 and C, and runs on an VAX11/780.

Critic
Critic is a knowledge-based system for critiquing circuit designs [Spi88]. The knowledge base consists of process and design-style constants, a set of primitive descriptions, a set of structure descriptions and error-checking rules. The information in the knowledge base is used to find errors and a "bad design style" in circuit designs. The errors range from easy-recognizable errors, such as transistor bulk being connected to the wrong supply or a static gate having non-equal rise and fall times, to hard-to-recognize errors, such as those dealing with charge sharing, timing and testability. Critic is incorporated into the Berkeley Design Environment.

VERA
VERA (VERification Assistant) is a rule-based system for the analysis and verification of electrical, topological and timing properties of circuit designs [KOs88]. It finds errors that are typically not found by simulation. VERA, implemented in Common LISP, is flexible in that it handles both digital and analog designs at different abstraction levels and in different technologies. It is capable of re-sizing improper devices and of adding, deleting or changing design objects. When data compilers or module generators are used, a high-level circuit description is extracted from the network layout, allowing verification at a higher level.

6.3 CIRCUIT SIMULATION

PROSAIC
The decomposing principle in PROSAIC [Bow85] is an approach to combine the systematic synthesis and knowledge-based approach. It is also possible to combine the parameterized and knowledge-based synthesis approaches. This idea requires the availability of an optimization package which can make the proper dimensioning according to the specifications. As an example of decomposition, an operational amplifier is decomposed into an input stage, an intermediate stage and an output stage. A rule-based system is able to build up analog blocks like operational amplifiers by stage.

OASYS
OASYS, developed at Carnegie-Mellon University, is a synthesis framework with the goal to convert performance specification and a process-technology file into an analog-circuit implementation (circuit blocks and their interconnections) [HAr89, Car90]. OASYS is based on three key ideas:
1. *Selection* of circuit topologies from among a set of fixed alternatives.
2. *Hierarchy* of sub-blocks. A circuit topology is an interconnection of sub-blocks, each of which can be represented in terms of lower-level modules with the transistor as the lowest-level module.
3. *Translation* of a performance specification at one level of the hierarchy into a particular topology of sub-blocks. The lowest level produces the transistor circuit that satisfies the original design specification.

The resulting circuit may further be improved by using a circuit-optimization tool, e.g., DELIGHT.SPICE [Nye88]. Like in logic systems, hierarchical decomposition plays an important role in the way human analog-circuit designers attack the design task. OASYS has a knowledge base that contains a library of reusable design knowledge, which is easily updatable.

BLADES
BLADES (Bell Laboratories Analog Design Expert System) works as a separated front end and divides the problem according to the specifications into subproblems [ElT89]. The system works as an experienced circuit builder and consultant. It uses ready simulators like ADVICE and ANSYS for fine tuning the circuit. The BLADES system has some typical features of knowledge-based systems for analog and mixed design area: dedicated working area, a limited number of architectures available and need for bidirectional communication with the simulation programs outside. The system is mainly used as a consultant aid for the designer emulating the way to design analog circuits.

BLADES uses both intuitive and formal knowledge in one program. It is a comprehensive knowledge-based analog-circuit design environment that operates at different levels of abstraction, depending on the complexity of the current design task. BLADES has been successfully implemented and is capable of designing a wide range of functional subcircuits as well as a class of integrated bipolar operational amplifiers.

ESTEPS

ESTEPS (Expert System for TEst Point Selection) is a Prolog-based system for test-point selection to facilitate fault diagnosis in analog circuits [Man90]. ESTEPS utilizes the program SAPTES (Symbolic Analysis Program for TEstability computation), which is able to compute the network testability, when test points are specified. The testability measure constitutes the basic criterion for the selection of a set of test points. However, the choice of a suitable set, starting from the range of all possible candidate test points, is greatly simplified by considering simple rules, derived from experience and heuristic reasoning. ESTEPS is written in Borland Turbo Prolog.

Analog Layout Synthesis

Chowdhury and Massara [Cho90] described an expert system for general-purpose layout synthesis. They indicated that the two main layout tasks (placement of circuit components and wire routing) cannot be isolated since the information compiled during the placement phase is also used during the routing phase. The modular system structure allows additional expert-design rules to be added at any level of the system hierarchy. Each hierarchical module has a set of plans and each plan may have a set of subplans. Each plan or subplan is defined as a set of rules and a decision criterion determines which rule or set of rules is fired. At this level, the additional expert rules can be defined to enhance or tailor the system to a particular need.

Chapter 7
VLSI TESTING AND EXPERT SYSTEMS

7.1 FAULT DIAGNOSIS

7.1a Fault Models

Introduction
This chapter discusses the testing problem of logic circuits, in particular circuits at the gate level. Topics to be discussed include fault models, testability measures, path sensitizing and alternative methods of test generation, fault simulation, design for testability, built-in testing and, of course, expert systems for testing purposes. With the increasing complexity of microelectronic chips and the great demand for ASICs, the testing problem is being aggravated. The vital importance of good chip testability compels the designer to pay due attention and effort to the testing problem [Abr90]. Testing considerations must be an essential part in chip design and not an afterthought.

Fault diagnosis and *test generation* are indispensable activities in designing LSI or VLSI chips [Agr88]. The problem of testing will be more acute as the chip complexity increases by larger gate densities. Testing should be clearly distinguished from simulation. Both testing and simulation are meant to verify whether the logic performance of a circuit is correct. The essential difference is that *simulation* verifies if the circuit *design* is correct, that is, if the circuit performance is in agreement with the design specification, whereas *testing* is needed to detect possible *hardware faults* resulting from *physical defects* (e.g., mask defects, metallization defects and package flaws) and to examine what effect it has on the circuit behavior. To guarantee an acceptable degree of product availability, testing allows failures to be detected and repaired.

Testing has the objective to detect the presence of a fault (*fault detection*) and, if possible, to pinpoint its location (*fault location*). When the circuit is simple, the required testing procedures can be carried out by using hardware testing equipment. The problem with complex chips is that only a restricted number of very many nodes of the circuit are accessible from the outside (inputs and outputs, and possibly testing pads). The question arises: How can a fault at an internal node of a circuit be detected or localized merely by applying signals at the input terminals and by measuring at the output terminals?

Generally, testing amounts to a comparison of the measured responses of the designed circuit to the correct responses (which are assumed to be known in advance). Let us define a *test signal*, *test pattern* or *test vector*, or *test* for short, as a pattern of bits on the primary inputs of the network to be tested such that possible faults can be detected on the primary outputs. When more than one test is required, we have a *test set*. A sequence of tests is called a *test sequence*. The objective of *test generation* is to find a complete test set required to detect or localize a fault. The *fault coverage* is the fraction of all possible faults that can be detected or located by a test set. Test-set verification by using fault simulation (see Subsection 7.2c) is the process of demonstrating that a set of selected tests is sufficient for detecting faults.

In order to devise effective methods of testing, it is important to be familiar with potential sources of circuit malfunctioning. In analog circuits, we usually deal with parametric faults. In digital circuits, the class of stuck-type faults is most frequently used.

Stuck-Type Faults
The generation of tests for detecting or localizing faults in a logic circuit requires the definition of *fault models*. A universally accepted fault model for physical defects are the *stuck-at-0* (s-a-0) and the *stuck-at-1* (s-a-1) *faults*, which mean that a node of the network is permanently at logic 0 and logic 1 respectively. Physically, this may correspond to a permanent connection of the node with ground or the power supply. Another possibility is that a line at the input of a gate is unconnected. For example, suppose that one of the inputs of a NAND gate is unconnected. Then, that input can be put permanently at logic 1, corresponding to a stuck-at-1 fault.

Most methods for test generation are based on the presence of a permanent single stuck-type fault (s-a-0 or s-a-1). This assumption is justified for several reasons:

a. The stuck-type fault is amenable to the use of conventional tools for simulation and analysis of logic circuits, such as Boolean algebra or extensions of it.
b. It is an accurate representation of a large class of physical defects; many physical defects create stuck-at-0 or stuck-at-1 conditions.
c. The single-fault assumption is justified if one assumes that, once a fault-free circuit has operated correctly and is tested at regular times, the probability of occurrence of more than one stuck-type fault at a time is negligibly small.

For these reasons, the *single stuck-type fault* will be assumed first. For simplicity, let us consider combinational networks composed of gates with only one output terminal.

Other Faults
Besides the assumption of a *single fault*, i.e., one and only one fault is assumed to occur at a time, we may consider *multiple faults*, i.e., various faults occur simultaneously. Figure 7.1 shows a few other faults. In Fig. 7.1a, an

7.1 FAULT DIAGNOSIS

interconnection wire may be broken at P, Q or R. In Fig. 7.1b, two *bridging faults* (short-circuits between two different nodes in the network, P-Q and R-S) are indicated.

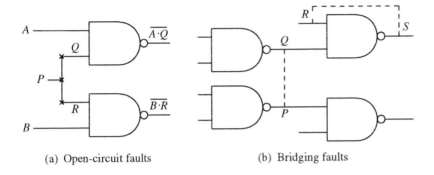

(a) Open-circuit faults (b) Bridging faults

Fig. 7.1 Examples of physical defects

In addition to fault models, which represent physical failures (defects), we may consider *error models*, which represent the errors due to the physical failures in the information at the outputs of circuit modules during normal operation [Abr86]. Concerns, such as the extent of information corruption due to a physical failure, the extent of error propagation, and the time between failure inception and possible error detection (error latency), can be investigated by first developing models of the types of errors generated by physical failures.

The stuck-type faults are actually opens and shorts at gate inputs and can be considered to be independent of the technology used. Special classes of physical defects are characteristic of the specific technology. For example, in CMOS technology, the *stuck-open fault* may be defined [Ban84, Jha90]. The impact of faults in control lines and dynamic MOS technology is in need of further research.

Delay faults cause the logic to switch at speeds lower than normal. Contrary to permanent faults, *intermittent faults*, which are present in some intervals of time and absent in others, are elusive and often difficult to detect.

Fault Analysis

With increased chip complexity, the effectiveness of fault models and the associated testing techniques will be questionable. Some weak points inherent to conventional models are the following:

a. Most models and testing techniques assume the logic-gate level.
b. The models (e.g., stuck-type models) are independent of the semiconductor technology which is used to implement the logic circuit.

c. The faults considered are unranked, i.e., all faults have equal importance.

Chen *et al.* [Che85] pointed out that reliable testing procedures must be based on more realistic models, which can be extracted from the IC layout and the processing characteristics of the technology which is used. They presented a systematic procedure to predict all the faults that are likely to occur in MOS integrated circuits. The procedure, called Inductive Fault Analysis (IFA), consists of three major steps:

1. Generation of physical defects using statistical data from the fabrication process.
2. Extraction of circuit-level faults caused by these defects, and
3. Classification of fault types and ranking of faults based on their likelihood of occurrence.

The fault list can be used to guide the development of testing techniques using realistic models.

Failures in Semiconductor Memories

Effective testing of memories requires knowledge of the organization of the memory under test. A Random-Access Memory (RAM) may consist of address decoders, memory cells, I/O units for data transfer, sense amplifiers and a clock generator. In order to know which faults should be detected, it is necessary to trace the physical causes of malfunctioning of the memory.

During the many manufacturing steps, a wide variety of failures may be distinguished. The most important failure modes are summarized below:

a. *Short circuits* and *open circuits*, resulting from excessive or insufficient metallization and bonds which make no contact with the I/O pads.
b. *Breakdown* in a clamping diode or any other semiconductor junction.
c. *Input* and *output leakage currents* in excess of the tolerable bounds.
d. *Decoder malfunctioning*, resulting from the inability to address some portions of the array due to a defective decoder. A fault in only one bit at the input of the decoder makes half of the cells inoperative.
e. *Multiple writing*, causing data written into one cell to reach other cells. The origin of the failure may be a leaky input or a short circuit.
f. *Pattern sensitivity*, which involves a change in the contents of a cell as a result of READ and WRITE operations in neighboring cells. This fault mode may be a function of the information read or written, the cells which are addressed and the order in which these cells are addressed.
g. *Refresh disfunction* (sometimes called "sleeping sickness"), which manifests itself as a loss of data during the specified minimum refresh time.
h. *Slow access time*, due to the presence of a considerable capacitive charge on the output-driver circuit. It takes an excessive amount of time to reduce the current, thus making the access time slow.
i. *Excessive write recovery*, resulting in a lengthening of the access time when

a WRITE command is immediately followed by a READ command, while both commands use the same data line. The reason is that a higher voltage is required for writing. The higher this voltage is, the more time it takes the data line to fall to its normal level than in the case of a READ command.

j. *Slow sense-amplifier recovery*, i.e., the sense amplifier tends to stay in the same state when an abrupt change of state occurs. For example, a 0 after a long sequence of 1s is read as a 1.

7.1b Testability Measures

Testability
With increasing complexities of logic circuits and systems, the testing and maintenance costs may dominate the other costs in the design cycle. It is therefore imperative to reduce these costs by making the testability of a circuit an important consideration during the entire design cycle. Testability has to do with the effort required for generating tests, the ease with which testing is to be performed, the quality of the tests for fault detection and the possibility to localize the fault. Various quantitative *testability measures* have been proposed. The testability is improved when more input and output nodes i and u are accessible. A simple measure is $M = (i + u)/g$, where g is the number of gates. Preferably, a measure M is defined in the range $0 \leq M \leq 1$, where $M = 1$ and $M = 0$ correspond to a completely testable and completely untestable network respectively.

Many testability measures have been proposed to enable the designer to compare alternative designs in terms of their testability. However, care should be exercized in using quantitative testability measures and interpreting results obtained from testability computations [Agr84, Hui88]. A testability measure usually provides a relatively poor indication of whether or not an individual fault will be detected by a given test.

Controllability and Observability
In searching for methods for the design of testable circuits it is useful to introduce the concepts of controllability and observability. The *controllability C* of a circuit T has to do with the ease with which the inputs of subnetworks S can be controlled from the primary inputs of T. The *observability B* is the ease with which responses of subnetworks S can be determined by observing the primary outputs of T. A quantitative measure for C and B has been given by several authors. Let us discuss the most important ones. Testability measures can be used as controllability and observability of a circuit node from the outer world, as statistical quantities or as cost functions. By evaluating the testability of alternative circuits in the process of circuit design, we can select the most

testable one.

Definitions of controllability and observability are given below by considering the testability program SCOAP.

SCOAP

SCOAP (Sandia Controllability/Observability Analysis Program) calculates for each node of a circuit six testability values which give a measure for the controllability and observability of the node [Gol80]. SCOAP assumes the circuit to be composed of standard cells which are divided into two types: combinational standard cells (such as ANDs, ORs, inverters and buffers) and sequential standard cells (such as flip-flops, latches, etc.). Every standard cell has two cell depths: a combinational cell depth and a sequential cell depth which indicate how easy combinational or sequential information respectively can be propagated through the cell. The combinational and sequential cell depths are defined as 1 and 0 respectively.

The calculation of the testability measures is carried out in two phases:
1. The network is traversed from the primary inputs to the primary outputs, while all controllability measures are calculated.
2. The network is traversed from the primary outputs to the primary inputs, while the observabilities are calculated.

SCOAP assigns to node n of the circuit the four controllability measures: combinational 0-controllability $CC^0(n)$, combinational 1-controllability $CC^1(n)$, sequential 0-controllability $SC^0(n)$, sequential 1-controllability $SC^1(n)$, and the two observability measures: combinational observability $CO(n)$, sequential observabilty $SO(n)$. All these testability measures are non-negative integers.

The controllability measures are defined as the minimum number of nodes that must be set in order to produce either 0 or 1 on the node in question. The observability measures are defined as the number of nodes that must be set in order to propagate the value of that node to the primary output plus the number of combinational or sequential cells on the sensitive path.

The controllability of an output node of a standard cell or gate is calculated from the controllabilities of the cell inputs. Assume that the input values of a cell are known. Then, the combinational (sequential) controllability of the cell output is equal to the minimum of the sum of the combinational (sequential) controllabilities corresponding to the input values. This sum is calculated for all possible sets of inputs which realize the relevant output, augmented with the combinational (sequential) cell depth.

Example: Consider a 2-input NAND gate with inputs X_1 and X_2 and output Y. The output Y equals 0, when $X_1 = 1$ and $X_2 = 1$. Then,
$$CC^0(Y) = CC^1(X_1) + CC^1(X_2) + 1$$
$$SC^0(Y) = SC^1(X_1) + SC^1(X_2)$$
The output Y equals 1, when one of the inputs is logic 0. Then,

7.1 FAULT DIAGNOSIS

$$CC^1(Y) = \min \{CC^0(X_1), CC^0(X_2)\} + 1$$
$$SC^1(Y) = \min \{SC^1(X_1), SC^1(X_2)\}$$

Starting at the primary inputs of a logic circuit with known controllabilities, the controllabilities of the other nodes can be calculated. When there are feedback loops, controllabilities are fed back and an iterative calculation procedure is initiated until a stable state has been established.

The observabilities of the network are calculated from the primary outputs backwards to the primary inputs. The observability of a cell input is equal to the observability of the selected output plus the minimum of the sum of the controllabilities corresponding to the necessary input node assignments plus the cell depth. The sum is calculated for all possible input node assignments which lead to a sensitized path (see Subsection 7.2a).

Example: Consider again the 2-input NAND gate. In order to propagate one input to the output, the other input must be set to 1. Then,

$$CO(X_k) = CO(Y) + CC_1(X_i) + 1$$
$$SO(X_k) = SO(Y) + SC_1(X_i),$$

where $k = 1, 2$, and $i = 1, 2$.

Starting at the primary outputs with known observabilities, we are able to determine the observabilities of all other nodes. When there are feedback loops, it is necessary to iterate so long until a stable state is found.

Testability Programs

Testability and its measures are often related to the controllability and observability of a logic circuit. Analysis tools for evaluating the controllability and observability have been incorporated into systems for fault diagnosis. SCOAP from Sandia Labs is such a system. Other systems include the following: COMET (Controllability, Observability, and Maintenance Engineering Technique) from Bell Laboratories [Cha74], CAMELOT (Computer-Aided MEasure for LOgic Testability) from Cirrus Computers [Ben81], COMET (Controllability and Observability MEasurement for Testability) from United Technologies Microelectronics Center, Colorado Springs [BEr82], VICTOR (VLSI Identifier of Controllability, Testability, Observability and Redundancy) from the University of California, Berkeley [Rat82], IDAS (Integrated Design for testability and Automatic test pattern generation System) from Siemens [Tri84], DTA (Daisy Testability Analyzer) from Daisy Systems Corporation [Wan84], STAFAN (STAtistical Fault ANalysis) from AT&T Bell Laboratories [Jai85], and TIP (Testability Improvement Program) from the University of Southern California [Che85].

Savir [Sav83] showed that a good controllability and a good observability of a circuit do not guarantee that the circuit will be highly testable. Anyway, realizing testability for the purpose of minimizing testing and maintenance costs should always be a primary concern in circuit and system design. An example of a system that helps engineers make logic circuits and systems more easily testable is CATA (Computer-Aided Test Analysis system), developed by Robach et al. [Rob84].

Evaluating testability of large and complex logic circuits is a time-consuming process. Fast algorithms that provide approximate information about the testability of either single faults or the whole design have been proposed [Hui88]. These algorithms involve measures that estimate the random-pattern testability of gate-level faults in designs with combinational logic.

7.1c Expert Systems for Fault Diagnosis

Fault Diagnosis
Fault diagnosis of VLSI systems deals with fault localization and fault identification based on observed system behavior. Expert systems for fault diagnosis of chips may be classified according to the type of knowledge stored: shallow knowledge or deep knowledge (see Subsection 4.5a).

This subsection gives brief descriptions of a number of expert systems which have been developed for diagnosing malfunctioning VLSI chips.

APEX3
APEX3 was developed for diagnosing malfunctions of electronic equipment [Mer83]. Since, in effect, APEX3 is an expert-system shell, it needs a suitable knowledge base for its operation. A diagnostic tree is used with the root node representing the symptom (e.g., detection of a fault during system operation) and the leaf nodes representing specific faults (or underlying causes). Backward chaining involves selecting the most likely goal fault and asking questions to determine whether it is present. If not, the procedure is repeated starting from the next most likely goal. Forward chaining involves starting at the top (root node) and, by asking the user questions and making deductions, working down toward the goal faults. A mixed strategy of backward and forward chaining is used by traversing the tree upwards and downwards.

The APEX3 control mechanism can be split into five elements: mechanisms for backward chaining, forward chaining, deciding whether to backward or forward chaining, handling preconditions, and commands to allow the user to override the default strategies. APEX3 has been written in a highly modular fashion, allowing changes to any particular part of the strategy to be made without affecting other parts. The bulk of APEX3 is written in Prolog. The parts that involve numerical operations are written in the programming language POP-2.

7.1 FAULT DIAGNOSIS

MIND
MIND (Machine for INtelligent Diagnosis) is a rule-based expert system designed to reduce the repair time associated with the diagnosis and repair of VLSI systems [Wil85]. Extensive use is made of expert knowledge in the form of heuristics. The MIND system includes the following components: the test-knowledge database, a diagnostic rule database, causal maps containing information relating symptoms and potential failure sources, an inference engine, the Check program that works at the functional level, test-system hardware with built-in self-test (BIST) features, and a user interface.

The test-system hardware is designed so that individual field-replaceable units can be controlled and observed internally. The diagnostic rules are defined so as to improve the efficiency of the fault-localization process. Based on a set of stimulus-response functions, the Check program stores the symptoms in the test-knowledge database. The diagnostic rules are organized into a tree of separate modules, each of which consists of a cyclic list of rules pertaining to a specific area of diagnosis. This organization enhances the hierarchical search process through the system, subsystem and the component levels. According to the type of symptom observed, a path of the tree is traversed which lead to the faulty component.

ATEX
ATEX (consisting of ATEX.KB and ATEX.RT) is an expert system designed to assist a test technician in detecting and isolating faulty modules in a given electronic circuit [Ben86]. ATEX.KB is used to generate and maintain the knowledge base that includes information about the behavior of electronic systems and general characteristics of electronic tests and their behavior when applied to various types of modules.

The input to ATEX includes an initial set of findings which indicate the existence of faults. In addition, it contains information about the circuit description, the test points, the set of available tests and other relevant data, such as the reliability of various components. ATEX.RT contains the inference engine that operates on the knowledge base and the test results, and offers assistance in probabilistic diagnosis assessment and effective management of the testing process, including goal-setting and test evaluation and testing. An iterative diagnostic process ultimately leads to location of the fault.

Prototyped in Prolog, ATEX was written in C. It can run on a MS-DOS-based personal computer or integrated with a minicomputer dedicated to testing.

MIT System
An expert system, developed by Davis *et al.* of MIT [Dav84], helps technicians in troubleshooting digital electronic hardware. It reasons from "first principles", i.e., from an understanding of the structure and behavior of the circuit components. Structure has to do with the interconnection of modules at several

hierarchical levels. Behavior is defined in terms of input-output relationships of the modules.

A complete specification of a module includes its structural description and a behavioral description in the form of rules. These rules can be distinguished between simulation rules (defining constraints on the electrical behavior of modules) and inference rules (which allow reasoning about the modules).

Example: Consider a NAND gate. A simulation rule is: *if* either input is a 0, *then* the output is 1, *else* the output is 0. An inference rule is: *if* the output is 0, *then* both inputs must be 1.

The fault-diagnosis problem can be formulated as follows. Let a circuit be composed of a number of interconnected modules. Given specific values of the primary inputs of the circuit, the primary outputs of the correctly working circuit has known values. When one or more output values deviate from the correct values, the problem is to find which module is responsible for this discrepancy.

Four issues of central concern in the MIT system are the following:
a. Diagnosis is accomplished via the interaction of simulation and inference. Paths are traced backward in the circuit starting from the discrepant primary output. Inferencing about possible logic values obtained by simulation in forward direction in comparison with the correct values obtained by simulation in forward direction must lead to the location of the faulty module.
b. Paths of causal interaction play a central role in diagnosis. Understanding the interactions by which one component affects another is used rather than using traditional fault models.
c. Constraining and guiding the diagnostic process. In cases where different paths of interaction are possible (such as in bridges), the most restrictive model (the one that considers the fewest paths of interaction) is used first, falling back on less restrictive models, if it fails.
d. The concept of *locality* (or physical adjacency). This concept proves to be useful in understanding why bridge faults are difficult to diagnose.

Westinghouse Diagnostic System

A knowledge-based diagnostic system that combines both shallow and deep knowledge of the circuit under test was described by Havlicsek from Westinghouse Electric [Hav86]. The deep knowledge representing the complete description of the structure and behavior of the circuit is stored in the design database. Rules contained in the knowledge base are used to diagnose the faults. The rules represent relations between test stimuli, symptoms and faults.

The causal reasoning inference engine relates the observed symptoms to the structure and function model in the database. The rule-based diagnostic system is interfaced to an automatic test equipment (ATE). The inference engine receives

7.1 FAULT DIAGNOSIS

symptoms from the ATE, searches the rule base for appropriate rules and reports conclusions and required actions to the test equipment and the operator. Test experts or knowledge engineers can create, check and modify the knowledge base by using the expert-system development tool. The causal reasoning is based on that given in the MIT system.

DART

DART (Diagnostic Assistance Reference Tool) of Stanford University is a maintenance tool for the diagnosis of hardware faults in digital computers [Gen84]. DART uses a device-independent language for describing devices and a device-independent inference procedure for diagnosis. Consequently, it can be applied to a wide class of devices ranging from digital logic to nuclear reactors. Like the MIT system, DART exploits deep knowledge. In diagnosing malfunctioning digital logic systems, the structure (components and their interconnections) and the signal behavior of the circuit (equations, rules or procedures that relate circuit inputs, outputs and states) must be described in a suitable description language. DART accepts a statement of a system malfunction in a formal language, suggests tests and accepts the results, and ultimately pinpoints the components responsible for the failure. The diagnosis is based on the occurrence of a single non-intermittent fault.

The hierarchical levels inherent in complex logic systems are exploited by restricting the diagnosis at one level at a time, thus keeping the diagnosis manageable. Starting at a high level of abstraction to determine the module in which the fault lies, the diagnosis is repeated at lower levels, until a replaceable faulty module is identified. All symptoms are expressed as violations of the expected behavior. At each level of the hierarchy, DART uses a general deductive inference procedure to analyze suspect components and generate discriminatory tests from information about the circuit being diagnosed. The program first computes a set of suspect propositions. If the set contains only one element, the diagnosis is done. Otherwise, DART tries to generate a test to discriminate the suspects. This process is continued until the faulty component has been found.

DART has been implemented in Common LISP and uses MRS as the inference engine. It runs on a VAX-11/780.

GDE

Whereas most diagnostic systems assume the occurrence of a single failure source, GDE (General Diagnostic Engine) diagnoses circuit failures due to the simultaneous occurrence of multiple faults [deK87]. The deep knowledge used by the system represents the behavior and function of the individual components comprising the system under test. The diagnostic task requires two phases. The first phase identifies a difference between actual and fault-free performance. The second phase proposes further tests that should find the real cause of the malfunction.

To obtain the required efficiency in searching for multiple faults, GDE exploits the features of Assumption-based Truth Maintenance System (ATMS) [deK86]. GDE proposes a sequence of measurements which efficiently localize the failing components. To achieve this goal with a minimum number of measurements, GDE need only be provided with the *a priori* probabilities of individual component failures. The combination of probabilistic inference and ATMS enables GDE to apply a minimum-entropy method to determine what measurement to make next. The best measurement is the one that minimizes the expected entropy of candidate probabilities resulting from the measurement.

FDA

FDA (Fault Diagnosis Assistant) is an interactive knowledge-based tool for gate-level troubleshooting of faulty VLSI chips [Pur88]. FDA employs deep knowledge of the system behavior and structure (connectivity file of the circuit under test). The initial symptom of chip misbehavior is revealed by introducing a fault-detecting input pattern and observing the faulty outputs. Given this symptom, FDA recommends test engineers to use an efficient sequence of probes at internal nodes that progressively localizes and identifies single stuck-type faults in combinational logic or combinational partitions within set/scan logic.

FDA solves two complementary problems: fault localization (by detecting discrepancies between normal-behavior signal predictions and actual signal measurements) and fault identification (by observing consistency between fault hypothesis signal predictions and actual signal measurements). FDA is implemented in OPS83, which combines the declarative pattern-driven control of OPS5 and the procedural advantages of C.

PROD

PROD is a diagnostic expert system that identifies both parametric and catastrophic faults [Odr85]. It can be expanded to diagnose faults that cannot be described quantitatively. PROD analyzes the joint probability density function of measured integrated-circuit parameters to determine the source of faults resulting in faulty chips. Fault simulations show the effects of faults with particular probability distributions of circuit parameters (e.g., delay times). The PROD system builds and executes a fault-simulation plan which has the highest probability of matching real faults.

CIRCOR

The rule-based system CIRCOR (CIRcuit CORrection) assists in repairing NMOS and CMOS digital circuits [Kro89]. The system can handle all types of physical failures that occur in integrated circuits (e.g., shorts, opens, missing transistors). The knowledge base consists of recognition rules (R rules)

7.1 FAULT DIAGNOSIS

describing the topology of subcircuits and their interconnections, and diagnosis rules (D rules) for heuristic diagnosis. An extracted connectivity list obtained by using R rules is compared to the error-free connectivity list. D rules can be used to locate the faulty element and recommend the user to repair the fault. Common LISP primtives are used and the system runs under MS-DOS PCs and VMS-VAX.

Sequential Circuit Fault Diagnosis
Fault diagnosis of synchronous sequential circuits can be performed by a deep reasoning approach [Rog89]. Given a set of test input patterns, a set of measurements of the actual behavior is compared with the correct circuit behavior. When a discrepancy is detected, the technique generates a set of plausible candidates which might be responsible for the circuit failure. A deep-reasoning approach is applied to the sequential circuit by considering a combinational circuit. This circuit is created by breaking the feedback loops and introducing probing points to measure state values. Quintus Prolog is used on a VAX11-750.

PESTICIDE
PESTICIDE (Prolog-written Expert System as a Tool for Integrated CIrcuits DEbugging), developed at IMAG, Grenoble, is a Prolog-based system using deep knowledge for debugging combinational as well as sequential circuits [Mar89]. The knowledge base contains static information about the unit under test and dynamic information about the testing procedure, and a control program. The fault detection and localization process requires the following steps: partitioning the circuit structure, comparing observed and simulated logic values, and establishing the diagnosis using PESTICIDE. Data acquisition is performed by a scanning electron microscope used in voltage contrast mode and linked to CAD tools. Single as well as multiple faults may be assumed in the fault diagnosis. Diagnostic results have been obtained using the C-Prolog interpreter (version 1.5) on a SUN 3/160 computer under UNIX (version 4.2).

Diagnosis of Memory Failures
Viacroze and Lequeux [Via89] described a Prolog-based expert system which helps diagnose failures in VLSI memories. The accuracy of the diagnosis depends on the information available from the memory unit under test. To facilitate testing, each type of memory is partitioned into testable blocks. Experience with previous cases of malfunctioning is stored in a database to help solve recurring failure problems. The expert system consists of a database of relevant data, a rule base for diagnosis, a dialog module for user-system communication and an explanation module to justify and explain the inferences.

Abu-Hanna/Gold System
Abu-Hanna and GOLD [Abu88] described an expert system for diagnosing

dynamic systems with feedback loops and synchronous or asynchronous state transitions. The diagnosis process is assisted by a multi-level simulator for the purpose of verification and elimination of hypothesized suspects. The simulator operates at several hierarchical levels using models ranging from high-level coarse qualitative to low-level detailed quantitative models. Accordingly, there is a knowledge base for shallow and deep knowledge, each with its own inference engine. Shallow knowledge contains symptom-to-fault associations in the diagnosed system. Deep knowledge represents both functionality and structure of the system at multiple levels of abstraction.

In a top-down process, shallow knowledge is first used for diagnosis. If it fails, deep knowledge is called upon. Prolog is used to implement the representation language and the two inference engines. Learning develops as a transfer from deep to shallow knowledge. The objective is to improve the efficiency in subsequent cases.

Diagnosis of Bus Failures
A variety of circuit defects and failure modes may lead to a bus failure [Tim83]. Fault diagnosis of bus failures is further complicated since different implementation technologies exhibit different output characteristics. Gonzalez [Gon87] addressed the problem of bus failures during in-circuit testing and discussed a method for testing and diagnosing bus failures using expert systems techniques.

VMES
VMES (Versatile Maintenance Expert System) is an expert system for troubleshooting logic circuits [Sha86]. It uses deep knowledge in the form of structural and functional descriptions of the circuit under test. VMES is embedded in SNePS (Semantic Network Processing System). Structural and functional knowledge is integrated into a single network. The inference engine is a rule-based system implemented in the SNePS language. Its control flow is enforced by a LISP driving function. A hierarchically arranged knowledge base provides abstraction levels and makes the inference engine able to focus on a limited number of objects at any time. A small set of SNePS rules is activated at every stage of the diagnosis process. The main mode of communication between the user and the reasoning mechanism is a graphical interface which displays the given circuit and allows the inference process to be traced graphically.

Fault Diagnosis of Electronic Systems
Wawry and Zinka [WAw89] developed a prototype expert system that assists in diagnosing faulty electronic components, wirings, printed-circuit boards, and recommends appropriate repair. The whole procedure consists of fault detection, localization, identification and correction. The reasoning procedure uses forward and backward chaining. The frame-based system has been written in LISP and

7.1 FAULT DIAGNOSIS

implemented on an IBM PC.

Ng-Chow Troubleshooter

Ng and Chow [Ng89] described an expert system for the diagnosis of electronic equipments. It reasons from first principles using a device-specific hierarchical structural model and is guided by heuristics and a troubleshooting control strategy. The structure of an analog or digital circuit is viewed as a hierarchical interconnection of modules and is represented as a semantic network. A structure analyzer processes the semantic network to infer topological relationships and performs signal propagation.

During execution, the expert system cycles through the three major steps of troubleshooting: candidate selection, check-point recommendation, and inference. The selection and recommendation steps are rule based. The inference process relies on the structure analyzer to provide tolopogical analysis. Basic set operations are performed as the expert system goes through the various diagnostic tasks. The semantic network will continue to expand until the fault is located to the smallest replaceable unit. A personal-computer implementation of the expert system can handle fault diagnosis of a large variety of analog and digital equipments.

ESTC

An automated diagnosis system implemented in Prolog on a compatible IBM PC is ESTC (Expert System for Troubleshooting digital Circuits) [Fan88]. A structurally hierarchical model of the circuit is used. A discrepancy of the expected behavior determined by a simulator is detected. Based on the observed symptoms, a diagnostic procedure is started to locate the component responsible for the malfunctioning of the circuit. A fault simulator verifies the effect of the fault on the behavior.

WEDS

WEDS (Westinghouse Expert Diagnostic System Ada) provides significant improvements over rule-based expert system shells for diagnostic applications, including a diagnostics-oriented development system and inference engine [Buc90]. By fully integrating the WEDS concepts within a standard Ada Automatic Test Equipment (ATE) system, the expert system has gained more sophisticated operator interfaces and better interfaces with external software. In turn, the ATE system achieves an expert-system diagnostic capability that involves little additional training to use and requires a minimal amount of new code generation to implement. The expert system has acquired the ability to communicate (via more sophisticated interfaces) with the operator, developer, external software, and automatic test equipment. This system also provides facilities for automatically acquiring statistical and historical knowledge for both newly designed and previously fielded UUTs.

7.2 TEST GENERATION

7.2a Path Sensitizing and D Algorithm

The Exhaustive Method

The objective of test generation is to find a test set with the largest possible *fault coverage*, i.e., the percentage of the specified or possible stuck-type faults detected by the test set. The simplest method for test generation is the *exhaustive truth-table method*. This will be illustrated through the example given in Fig. 7.2.

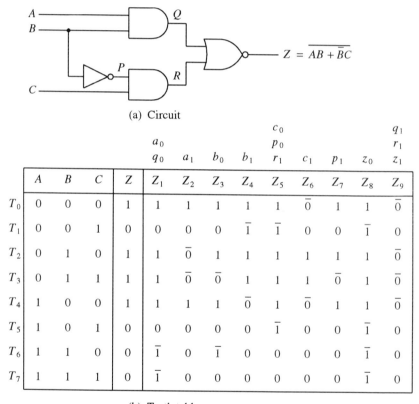

(a) Circuit

	A	B	C	Z	a_0 q_0 Z_1	a_1 Z_2	b_0 Z_3	b_1 Z_4	c_0 p_0 r_1 Z_5	c_1 Z_6	p_1 Z_7	q_1 r_1 z_0 Z_8	z_1 Z_9
T_0	0	0	0	1	1	1	1	1	1	$\bar{0}$	1	1	$\bar{0}$
T_1	0	0	1	0	0	0	0	$\bar{1}$	$\bar{1}$	0	0	$\bar{1}$	0
T_2	0	1	0	1	1	$\bar{0}$	1	1	1	1	1	1	$\bar{0}$
T_3	0	1	1	1	1	$\bar{0}$	$\bar{0}$	1	1	1	$\bar{0}$	1	$\bar{0}$
T_4	1	0	0	1	1	1	1	$\bar{0}$	1	$\bar{0}$	1	1	$\bar{0}$
T_5	1	0	1	0	0	0	0	0	$\bar{1}$	0	0	$\bar{1}$	0
T_6	1	1	0	0	$\bar{1}$	0	$\bar{1}$	0	0	0	0	$\bar{1}$	0
T_7	1	1	1	0	$\bar{1}$	0	0	0	0	0	0	$\bar{1}$	0

(b) Truth table

Fig. 7.2 Detection of stuck-type faults

The truth table of the circuit in Fig. 7.2a with inputs A, B and C is given in Fig. 7.2b. The correct outputs Z are given in the fourth column. Assume that a stuck-at-0 or a stuck-at-1 fault occurs at each of the seven nodes A, B, C, P, Q, R and Z. A stuck-type fault is indicated by a lower-case letter a, b, etc. with a

7.2 TEST GENERATION

subscript 0 or 1 for s-a-0 or s-a-1 respectively. For example, a_1 is node A stuck-at-1. The actual outputs in the presence of a stuck-type fault are given in the columns Z_1 through Z_9 in Fig. 7.2b. From this table we can derive which tests (input bit patterns) produce a deviation from the output of a fault-free network. In the truth table, the faulty logic values are denoted with a bar. For example, P s-a-1, indicated by p_1, is detected by the test $T_3 = 011$.

This exhaustive method of test generation leads to the following remarks. A specific test can detect more than one fault, while in most cases there are more than one test available to detect a given fault. For example, the test $T_3 = 011$ detects the faults a_1, b_0, p_1, q_1, r_1 and z_1. The fault q_1 is detected by four tests: $T_0 = 000$, $T_2 = 010$, $T_3 = 011$ and $T_4 = 100$. *Fault location* requires further examination. The test $T_6 = 110$ can detect any of the faults a_0, b_0, q_0 anf z_0, but it does not specify which fault actually occurs. An additional test $T_7 = 111$ will localize the fault b_0, if $Z = 1$. The test $T_0 = 000$ detects the faults c_1, q_1, r_1 and z_1. The additional test $T_3 = 011$ localizes the fault c_1, if $Z = 0$.

Path Sensitizing

The most important methods for test generation are based on path sensitizing. It is assumed that a prespecified stuck-at-0 or stuck-at-1 fault is present at a certain node of a logic network. We are dealing with *path sensitizing*, when a given test produces opposite logic values for the faulty and fault-free network along a path, the *sensitized path* to a primary output of the network. In order to detect a fault s-a-d, $d = 0$ or 1, the input vector (the test) must be chosen such that the signal at this node in the fault-free network takes on the value $\bar{d}$.

A test-generation method with path sensitizing consists of three steps:

a. The postulation of a specific fault s-a-d and the assignment of the complementary value $\bar{d}$ to the node at which the fault is present.
b. The construction of a sensitized path, so that the logic effect of this fault is propagated to the primary output. To make this *forward tracing* possible, all inputs of the gates, which lie in the sensitized path are assigned a suitable logic value, depending on the gate type (see Fig. 7.3). this is called *implication*.
c. Utilizing the results in a. and b., *backtracing* from the primary output to the primary inputs of the network is performed in order to assign values to nodes, which have no values as yet, without causing inconsistencies.

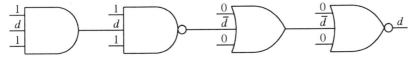

Fig. 7.3 Sensitized with logic values d and $\bar{d}$

Of all gates in the sensitized path, one input (from generally several inputs) of each gate is included. Figure 7.3 shows how a diagnostic signal d traces a sensitized path through four different gates. The other inputs of these gates are assigned values as indicated in the figure, that is, all other inputs of AND or NAND gates are assigned the logic value 1 and all other inputs of OR or NOR gates are assigned the logic value 0.

As an illustration of the sensitizing-path method let us carry out a test generation for the network of Fig. 7.2a by hand. Assume that node Q has the fault stuck-at-0. The test to be generated must produce at Q the logic value 1. The sensitized path to the primary output Z consists exclusively of the OR gate Z so that R obtains the value 0. This condition leads to the following possibilities for the primary inputs: $B = 1$, $C = 0$ or $B = 1$, $C = 1$ or $B = 0$, $C = 0$. The condition $Q = 1$ leads to $A = B = 1$. Combining these conditions while excluding inconsistencies, yields the tests: $ABC = 110$ and 111 for detecting the fault Q s-a-0. In a similar way we find for the fault Q s-a-1 the tests: $ABC = 000, 010, 011$ and 100.

D Algorithm

A basic algorithm for test generation based on path sensitizing is the *D algorithm* of Roth [Rot66, Rot67]. This algorithm implements the necessary tests for the detection of specified stuck-type faults in the way as demonstrated in the simple circuit example of the preceding subsection. For the purpose of tracing, the propagation of the fault along the sensitized path, the *diagnostic signal D* is introduced. By definition, D has the logic value 1 in the fault-free circuit, but the logic value 0 in the faulty circuit. Similarly, $\bar{D}$ represents the diagnostic signal that is normally 0, but becomes 1 when the fault is present. In the sensitized path, a sequence of D and $\bar{D}$ will appear, where D and $\bar{D}$ have opposite values.

In order to make the test-generation process amenable to computer processing, the "calculus of D cubes" is utilized. By *cube* we mean a string of logic values from the set 0, 1, D (or $\bar{D}$) and the don't care value X, assigned to specific lines of a logic network, e.g., the inputs and the output of a logic gate. An example of a cube is the string of logic values of inputs and outputs of a combinational circuit in a truth table. If the cube contains a D (or $\bar{D}$) signal, we have a *D cube*. The ingredients required for performing the D algorithm are: primitive cube, primitive D cube of a fault, propagation D cube and a number of intersection rules.

The *primitive cube* of a logic gate which realizes a specified combinational function F gives the necessary and sufficient conditions on the inputs of the gate to obtain the output value 0 or 1. The PC table is a truth table in compacted form. Figure 7.4 shows the NAND gate, its Karnaugh map and the rows C_i, $i = 1 \ldots 4$, which represent the four primitive cubes. By introducing the don't care variable X, the PC table is more compact than the truth table.

7.2 TEST GENERATION

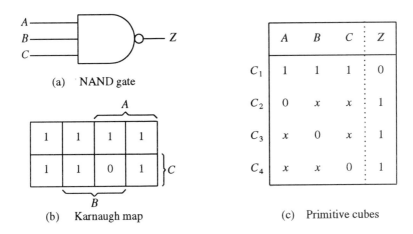

Fig. 7.4 NAND gate and primitive cubes

The PC table can be derived for subnetworks larger than the conventional logic gates. A general procedure for constructing the PC table consists of evaluating the prime implicants corresponding to the output $Z = 1$ or its complement $\bar{Z} = 0$. As an example, consider the gate subnetwork given in Fig. 7.5a. From the Karnaugh map given in Fig. 7.5b we can derive the prime implicants AB, $\bar{B}C$ for $Z = 1$ and $\bar{A}B$, $\bar{B}\bar{C}$ for $Z = 0$. From these prime implicants, the PC table can be derived.

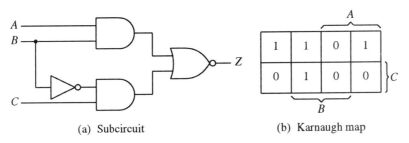

Fig. 7.5 Subcircuit and Karnaugh map

A *Primitive D Cube of a Fault* (PDCF), associated with a gate or subnetwork, provides the minimum input conditions needed to initiate a sensitized path such that a D or $\bar{D}$ appears at the output. For a simple gate, the PDCFs are easy to find. Suppose that one of the inputs of a 3-input NAND is stuck-at-0. Then, we find the PDCF: $D11\bar{D}$, $1D1\bar{D}$ or $11D\bar{D}$, depending on which input is stuck-at-0. The cube $111\bar{D}$ indicates a stuck-at-1 fault at the output of the NAND.

A general procedure for constructing PDCFs consists of bitwise intersections of the primitive cubes of the fault-free and the faulty network. Let the primitive cubes of the fault-free network for $Z=0$ and $Z=1$ be given by G_0 and G_1, and those of the faulty network by F_0 and F_1. In order to produce $D=1 \cap 0$ or a $\overline{D}=0 \cap 1$ at the output, the intersections $\alpha_0 \cap \beta_1$ or $\alpha_1 \cap \beta_0$ respectively must be performed.

Example: The network of Fig. 7.5a and its Karnaugh map are given in Fig. 7.5b. Let us assume a fault consisting of a short-circuiting of the diode. The cubes G_0, G_1, F_0 and F_1 are given in Fig. 7.6.

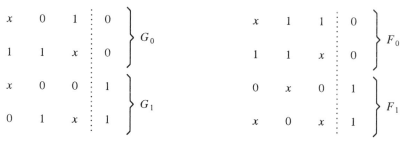

(a) PC table of fault-free network   (b) PC table of faulty network

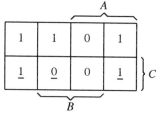

(c) Karnaugh map of faulty network

Fig. 7.6 Primitive cubes of the circuit of Fig. 7.5a

Applying the intersection rules yields:
$G_0 \cap F_1$: $x010 \cap x0x1 = x01D \equiv 001D$ or $101D$,
$G_1 \cap F_0$: $01x1 \cap x110 = 011\overline{D}$

The resulting PDCFs are $001D$, $101D$ and $011\overline{D}$. When the Karnaugh maps of the fault-free and the faulty circuits are known, the PDCFs can be directly read from these maps. Figure 7.6a gives the Karnaugh map of the circuit of Fig. 7.5a, when the diode is short-circuited. Comparing with Fig. 7.5b reveals that the three underlined logic values (which are different in the two Karnaugh maps) correspond to the three PDCFs found above.

The *Propagation D Cubes* (PDCs) of a fault-free gate give the conditions for the propagation of D or $\overline{D}$ along a chain of fault-free gates. PDCs can be derived from the primitive cubes by binary intersection rules introduced by Roth

7.2 TEST GENERATION

[Rot66]. A sensitized path can be constructed by *D intersections*, i.e., suitable intersections of PDCs of successive gates. The goal is to propagate a D or $\overline{D}$ along a sensitized path to the primary output. See Fig. 7.3, where the diagnostic signal D has the value d (= 0 or 1).

In searching for the test at the primary inputs, we have to avoid inconsistencies, i.e., assigning a 1 to a node which was assigned the value 0 in an earlier stage, or vice versa. In applying the D algorithm, there may be a choice of various possibilities. One arbitrary choice is then made. If inconsistencies are met during the execution of the algorithm, we have to return to this point to consider another possible choice. This is called *backtrack*.

The *D algorithm* consists of the following steps:

Step 1: Select a stuck-at-0 or a stuck-at-1 fault at a certain site in the circuit.

Step 2: *PDCF construction*. Construct a PDCF such that it produces a D or $\overline{D}$ representing the given stuck-type fault at the fault site. Generally, there exists a choice. Backtrack may be necessary in a later step.

Step 3: *Implication*. The construction of a sensitized path implies that relevant nodes must be assigned suitable values. When an inconsistency is met, backtrack must be performed.

Step 4: *D drive*. Construct a sensitized path from the site of the fault to the primary output.

Step 3/4: Implication as in Step 3. The D drive for determining the sensitized path must be continued for each subsequent gate in the direction of the primary output by using intersections of the PDCs of these gates. At each subsequent gate the Step 3/4 must be carried out.

Step 5: *Backward implication: Consistency and justification*. When by performing D drives the primary output is reached, the unassigned nodes in the direction of the primary inputs must be assigned values. Inconsistencies must be avoided. Backtrack may be necessary.

Example:

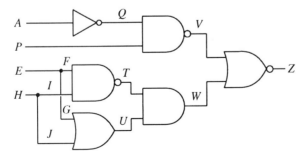

Fig. 7.7 Logic-circuit example

As an illustrative example, consider the gate circuit given in Fig. 7.7 (an Exclusive-OR). Each gate G_X has one output node X. Let us see which cubes are successively generated for the network of Fig. 7.7, when a test is to be found for the fault T stuck-at-1. The node symbols F, G, I and J may be ignored. The D algorithm generates the following steps:
- Introduce the diagnostic signal $\overline{D}$ at output T of gate G_T.
- The initial cube of gate G_T is $EHT = 11\overline{D}$.
- The next gate G_W is an AND; hence, the implication is $U = 1$.
- D drive of gate G_W produces the cube $TUW = \overline{D}1\overline{D}$.
- The next gate G_Z is a NOR; hence, the implication is $V = 0$.
- D drive of gate G_Z produces $VWZ = 0\overline{D}D$.
- The primary output is reached and backward implication follows, that is, working backwards from the primary output to the primary inputs, all nodes in the circuit are assigned a logic value according to the primitive cubes of the gates.
- $V = 0$ yields $QP = 11$ and $AP = 01$.
- $U = 1$ yields $EHU = 1x\,1$ or $x\,11$, but we already found: $EHT = 11\overline{D}$.
- The final result is the test: $APEH = 0111$.

Networks with reconvergent fanouts can cause difficulties when applying the D algorithm. When in a network a node can be reached from another node through two different paths A and B, an attempt is made to construct a sensitized path along path A. When an inconsistency occurs, path B is attempted. If this path also fails, the D-drive is applied to both paths A and B simultaneously.

Example: Figure 7.8 shows a circuit with a reconvergent fanout. Ignore the symbols G and H.

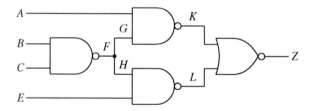

Fig. 7.8 Reconvergent-fanout example

Example: Consider the problem of generating a test for detecting a stuck-at-1 fault at F in the circuit of Fig. 7.8. The following steps are taken:
- Introduce the signal $\overline{D}$ at line $F \Rightarrow$ Initial cube $BCF = 11\overline{D}$. Implications: $A = 1$ and $E = 1$.

7.2 TEST GENERATION

- D drive through gate G_K $\Rightarrow$ Cube $AGK = 1\overline{D}D$. Implication: $L = 0$.
- D drive through gate G_Z $\Rightarrow$ Cube $KLZ = D0\overline{D}$. Primary output reached.
- Backward implication from $L = 0$: Cube $FEL = 110$ $\Rightarrow$ $F = 1$ is inconsistent with $F = \overline{D}$ (i.e., $F = 0$ in the fault-free circuit).
- Backtrack: D drive through gate G_L. Similar steps as above lead to the same inconsistency.
- Backtrack: D drive through gates G_K and G_L simulateously $\Rightarrow$ Cubes $AFK = 1\overline{D}D$ and $FEL = \overline{D}1D$.
- D drive through G_Z $\Rightarrow$ $KLZ = DD\overline{D}$.
- Backward implication: $F = 1$ $\Rightarrow$ Cube $BCF = 110$.
- The final result is the test pattern $ABCE = 1111$.

Note that the introduction of redundant subnetworks for the purpose of improving the reliability has the consequence that certain stuck-type faults can no more be detected. For that reason it is expedient to examine first which parts of the network are redundant before test generation is started.

In the above, the logic circuit to be tested is assumed to be given as an interconnection of logic gates. Jain and Agrawal [Jai83] suggested that the D algorithm can be applied to MOSFET circuits, when they are first converted into equivalent gate-level networks.

Application of the D Algorithm at the Functional Level

The D algorithm has been used as a basis for developing test-generation programs at higher levels than the logic-gate level. Breuer and Friedman [Bre80] considered complex sequential circuits composed of functional primitives, such as counters and shift registers, and a D algorithm was developed to solve the test-generation problem. Abadir and Reghbati [Aba83] also extended the applicability of the D algorithm to logic networks at the functional level. The network is modeled by a set of binary decision diagrams which were introduced by Akers [Ake78]. A *decision diagram* is a concise means for completely defining the functional operation of functional modules.

The D algorithm has been applied to the generation of test patterns for detecting functional faults by constructing a data graph from the register-transfer-level description [Su82]. Levendel and Menon [Lev82] used hardware-description-language constructs (such as *if* ... *then* ... *else*) and functional operators (such as addition) to describe the circuit behavior. Insertion of a fault (stuck-type or functional fault), path sensitization and justification steps, inherent in the D algorithm, are then introduced to derive test patterns. Another behavioral test-generation method was proposed by Son and Fong [Son84]. The tests are extracted from a functional table which has been derived from the hardware-description-language description of the circuit. A reachability matrix is used to guide the path sensitization and eliminate unnecessary searching for unreachable paths.

7.2b Alternative Methods for Test Generation

The PODEM Algorithm

It was pointed out in discussing the D algorithm that backtracks must be performed when contradictions are encountered during the test-generation process. For some classes of networks, notably combinational circuits containing Exclusive-ORs with reconvergent fanouts, the D algorithm turns out to be ineffective due to the many backtracking cycles needed.

Goel [Goe81] developed a more effective test-generation algorithm, called PODEM (Path-Oriented DEcision Making). Any stuck-type fault in the network is eligible for generating a test to detect it, but, different from the D algorithm, values are first assigned to the primary inputs (PI) after which a forward implication is performed. Backtracking can occur only at the primary inputs, implying that backward implication is not necessary. Primary input patterns are examined until a test is found. If no PI pattern can be a test, the fault is undetectable or redundant.

The PODEM algorithm proceeds as follows. First, all nodes are assigned the don't care X. An *initial objective*, symbolically expressed as (D, L) or $(\overline{D}, L)$, is first defined, that is, a stuck-at-0 or a stuck-at-1 fault is selected at a specific node L. An arbitrary input variable must be found such that the diagnostic signal D or $\overline{D}$ will arrive at the site L of the fault. In order to achieve a high likelihood of meeting this initial objective, a *backtrace* is performed, i.e., searching a path starting at L backwards to a primary input variable. Assuming that L is the output of a gate, the initial objective at L is transferred to an objective (κ, K), $\kappa = 0$ or 1, at an unassigned input variable K, which is the output of a gate. In turn, (κ, K) is transferred to an appropriate objective at an unassigned input variable. This procedure stops when an unassigned primary input has been found. This backtrace is followed by the *implication* operation in which the logic values of each node in the circuit is determined as a result of the PI assignments made thus far. If the PI assignment does not satsfy the initial objective, the backtrace is repeated and different unassigned inputs are assigned a value.

The next step is the propagation of the diagnostic signal to the primary output (PO), comparable to the D drive in the D algorithm. The choice of new initial objectives is aimed at propagating the diagnostic signal (D or $\overline{D}$) from the site of the fault to the PO. In determining initial objectives, a look-ahead technique, called *X-path check* is used to determine if a D or $\overline{D}$ can be propagated to the PO, that is, checking that the path being sensitized contains only X nodes and hence is not blocked. Backtrack is performed if all PIs have been assigned a value and the fault effect has not reached the PO. The purpose of backtrack is to choose untried PI bit patterns as long as an inconsistency occurs. If all possible input bit patterns have been tried without success, the fault is said to be undetectable.

7.2 TEST GENERATION

Example: The PODEM algorithm will be illustrated by considering the circuit given in Fig. 7.8. Let us find the test pattern that detects a stuck-at-1 fault at line F. The following steps are taken:
- Introduce the diagnostic signal $\overline{D}$ at line F.
- Initial objective: $(\overline{D}, F) \Rightarrow F = 0 \Rightarrow$ Backtrace leads to objectives $(1, B)$ and $(1, C)$.
- Implication of $B = 1$, $C = 1$ yields $F = G = H = \overline{D}$.
- Initial objective: $(D, K) \Rightarrow$ Backtrace leads to objective $(1, A)$.
- Implication of $A = 1$, $B = 1$, $C = 1$ yields $F = G = H = \overline{D}$ and $K = D$.
- Initial objective $(\overline{D}, Z) \Rightarrow$ Backtrace leads to $L = 0$ and $E = 1$.
- Implication of $A =1$, $B = 1$, $C = 1$, $E = 1$ yields $F = G = H = \overline{D}$ and $K = L = D$ $\Rightarrow Z = \overline{D}$.
- The test pattern is $ABCE = 1111$.

PODEM tries to satisfy the output objective of a certain gate as quickly as possible with the hardest/easiest strategy:

a. If the value κ at the output of a gate, associated with the current initial objective, can be obtained by setting any input of this gate to a controlling state, i.e., 0 for an AND or NAND gate, and 1 for an OR or NOR gate, then choose the input that can be most easily set.
b. If the value κ can be obtained only by setting all inputs of the current gate to a noncontrolling state, i.e., 1 for an AND or NAND gate, and 0 for an OR or NOR gate, then choose the input that is hardest to set since an early determination of the inability to set the chosen input will save the time that would be wasted in attempting to set the remaining inputs of the gate.

Choosing between the hardest/easiest controllable inputs requires a means of determining the relative controllability. Thus, an essential step in the PODEM algorithm is the assignment of relative controllability values using one of the methods given in Subsection 7.1b. The controllability indicates the ease with which a given line can be set to some logic value. Suppose that the objective is to set an input of a NAND gate when the gate output is 1. Though any of the input lines can be set to 0, the input with the best controllability value is selected (easiest strategy). When the output of the NAND gate is 0, the input with the worst controllability value is to set to 1 (hardest strategy).

The PODEM algorithm creates a decision tree with successive assignments of 0 or 1 to primary inputs, while the ordering reflects the sequence in which the current assignments are made. The decision tree can be implemented as a Last-In First-Out (LIFO) stack. A parallel enumeration algorithm based on PODEM has been used to design a multiprocessor implementation of parallel logic verification of combinational logic [Ma89].

FAN Algorithm
While the PODEM algorithm generally needs fewer backtracks than the D

algorithm, FAN (FANout-oriented test generation) was developed by Fujiwara and Shimono [Fuj83] to further reduce the number of backtracks, notably when many gates have fanouts exceeding one. FAN attempts to identify and resolve conflicting situations early in the test-generation process.

When a fault is selected, the D drive is first performed, but every time the diagnostic signal D (or $\overline{D}$) is propagated through a gate, a backward trace is started. This is repeated until the fault signal is propagated to a primary output.

The essence of FAN is that the algorithm adopts strategies that make the D drive and backward trace more efficient. The main goal of these strategies is to reduce the number of choices or time-consuming backtracks as well as the process time between backtracks.

The FAN algorithm uses several strategies to satisfy this objective:
a. In each step of the algorithm, assign unique values (0, 1, D or $\overline{D}$) to as many lines as possible.
b. In a D drive, the set of gates from which a choice can be made is called the D frontier. When the D frontier consists of one gate, a unique sensitization can be performed with unique implications resulting in a reduction in the number of selections and hence the number of backtracks.
c. By using a suitable branch-and-bound algorithm in setting up the decision tree in FAN, inconsistencies are detected earlier than in PODEM.
d. For fanout-free networks, line justification can be performed without backtracks. To avoid fruitless computations, special attention is paid to fanout points. When the occurrence of an inconsistency is expected, the backtrace is stopped.
e. FAN may perform backtrace from two or more lines simultaneously so as to assign values to unjustified lines. This is called multiple backtrace. This is more efficient than tracing along a single path.

Example: Consider the same example as above. Find a test pattern for a stuck-at-1 fault at line F of the circuit given in Fig. 7.8. The FAN algorithm generates the following steps:
- Introduce the signal $\overline{D}$ at F.
- Implication $B = 1, C = 1, A = 1, E = 1$ yields $F = G = H = \overline{D} \Rightarrow K = \overline{D}, L = \overline{D} \Rightarrow Z = \overline{D}$.
- The test pattern is $ABCE = 1111$.

FAN has been applied in the test-generation system FUTURE, developed by NEC [Fun85].

SOCRATES
SOCRATES (Structure-Oriented Cost-Reducing Automatic TEst-pattern generation System), developed by Siemens, generates test patterns for combinational and scan-based circuits [Sch88]. Based on the FAN algorithm, SOCRATES uses an improved implication procedure, unique sensitization and

7.2 TEST GENERATION

multiple backtrace. Compared to FAN, it involves a reduced number of backtracks and an earlier recognition of conflicts and redundancies. A fault-simulation algorithm is part of the system.

An extension to SOCRATES is published by Sarfert et al. [Sar92]. The system is based on predefined high-level primitives, e.g., multiplexers and adders.

Critical-Path Test Generation

The D algorithm generates one or more tests for one stuck-type fault at a time by generating a sensitized path from the fault site to a primary output [Bre76]. The method of *Critical-Path Test Generation* (CPTG) works the other way around by generating critical paths starting from a primary output going back to primary inputs. A *critical path* is a path containing a chain of critical gate inputs. A gate input is called a *critical input*, when a change of its logic value (0 or 1) causes a change in the gate output. Figure 7.9 shows the truth tables of the AND, NAND, OR and NOR gates, with critical nodes (gate inputs and outputs) indicated with a subscript c. For example, consider the second cube in the truth table of the AND gate. The 0 input is a critical input, since a $0 \to 1$ transition directly causes a $1 \to 0$ transition at the output.

AND			NAND			OR			NOR		
A	B	Z	A	B	Z	A	B	Z	A	B	Z
0	0	0	0	0	1	0_c	0_c	0_c	0_c	0_c	1_c
0_c	1	0_c	0_c	1	1_c	0	1_c	1_c	0	1_c	0_c
1	0_c	0_c	1	0_c	1_c	1_c	0	1_c	1_c	0	0_c
1_c	1_c	1_c	1_c	1_c	0_c	1	1	1	1	1	0

Fig. 7.9 Truth tables with critical gate inputs and outputs

Starting with a critical value at a primary output, the CPTG algorithm generates critical paths to the primary inputs. In some cases, a critical gate output corresponds to two critical inputs, leading to a branching into two different critical paths.

Figure 7.10 shows the generation of critical paths in the circuit given in Fig. 7.7, starting from $Z = 0_c$ (lefthand side) and $Z = 1_c$ (righthand side). A sequence of solid arrows represents a critical path. For example, the path *ZVQA* is critical, since all the nodes passed are critical nodes. A characteristic feature of the CPTG algorithm is that each test generated is able to detect a specific stuck-type fault at each of the critical nodes in the path, i.e. a stuck-at-$\overline{d}$ fault at critical nodes with logic value $\overline{d}$, where d = 0 or 1.

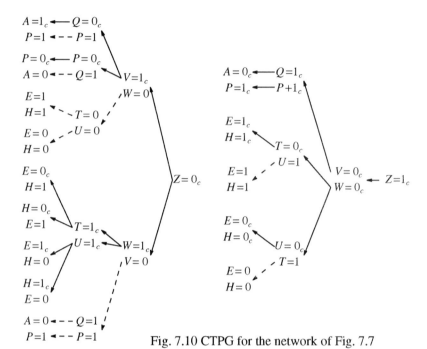

Fig. 7.10 CTPG for the network of Fig. 7.7

	Test				Critical Path	Detected stuck-type faults
	A	P	E	H		
$Z = 0_c$	1_c	1	0	0	AQVZ	a_0, q_1, v_0, z_1
	1_c	1	1	1		
	0	0_c	0	0	PVZ	p_1, v_0, z_1
	0	0_c	1	1		
	0	1	0_c	1	ETWZ	e_1, t_0, w_0, z_1
	0	1	0	1_c	HUWZ	h_0, u_0, w_0, z_1
	0	1	1	0_c	HTWZ	h_1, t_0, w_0, z_1
	0	1	1_c	0	EUWZ	e_0, u_0, w_0, z_1
$Z = 1_c$	0_c	1_c	1_c	1_c	AQVZ, PVZ ETWZ, HTWZ	a_1, q_0, v_1, z_0, p_0 e_0, t_1, w_1, h_0
	0_c	1_c	0_c	0_c	AQVZ, PVZ EUWZ, HUWZ	a_1, q_0, v_1, z_0, p_0 e_1, u_1, w_1, h_1

Fig. 7.11 Tests for stuck-type faults as derived from Fig. 7.10

7.2 TEST GENERATION

The total results of the CPTG algorithm depicted in Fig. 7.10 are collected in Fig. 7.11. The 10 tests for the inputs *APEH* will detect the stuck-type faults as given in the last column. Apart from the single faults detected, multiple faults as an arbitrary combination of single faults in a critical path can be detected.

Statistical Methods of Test Generation
When a circuit contains a specific fault, random patterns can be applied to the circuit input. A pattern that produces an output which deviates from the correct value, is said to detect the fault. It can be shown [Agr82] that the statistical correlation between the testability at the site of the fault and the probability of detecting the fault by applying random patterns may not be high. Statistical methods of test generation are applied in conjunction with fault simulation (see Subsection 7.2c).

Parallel-Processing Techniques
The above methods of test generation are based on serial algorithms which are executed on conventional Von Neumann computers [Kir88]. Multi-processor architectures allow the use of parallel-processing techniques for test generation [Kle92]. These techniques are based on different strategies: fault partitioning (by dividing the fault list among the processors), heuristic parallelization (e.g., using concurrent parallel heuristics), search-space partitioning (e.g., a sort of divide-and-conquer approach), functional or algorithmic partitioning (dividing an algorithm into independent subtasks that can be executed on separate processors in parallel) and topological partitioning (instantiating separate circuit partitions on different processors).

7.2c Fault Simulation

Introduction
Fault simulation is the simulation of a logic circuit (usually at gate level) in the presence of one or more faults. In many advanced logic-circuit simulators, fault simulation is an available option. In addition to the study of the circuit behavior under faulty conditions, fault simulation has two important functions:
a. *Test-set verification*: verifying the usability of tests, which have been generated using conventional test generators.
b. *Test generation*: generating tests for the detection of stuck-type faults.

Test-set verification is indispensable since binary logic is used in test-generation programs. The resulting test sets themselves are simple (sequences of) bit patterns. There is no guarantee that a test set which detects prespecified faults in a two-valued delay-free network will also detect all these faults in a fault-simulation run with realistic circuit models. In other words, the 100% fault coverage of a complete test set obtained by using a test-generation program may be less than 100% when the set is verified by fault simulation. Hence, an important result of a fault-simulation run is the overall fault coverage achieved with a given test set. When the detecting capability of the test sets have been verified by using a simulation program, we may be more confident that the test sets will detect faults in the hardware after the fabrication process or in the maintenance field.

In Subsections 7.2a and 7.2b, algorithmic methods of generating tests have been described. An alternative to test generation is the statistical, or pseudo-random, approach. Test patterns, taken at random or selected by using some heuristic, are applied to a fault simulator. By observing discrepancies at the primary outputs of the circuit, fault-detection tests can be generated.

The classical types of fault simulation are parallel, deductive and concurrent simulation [Sch87]. These three types of simulation will be descussed below.

Parallel Fault Simulation

The earliest fault simulators were based on parallel simulation. This type of simulation has been applied in CC-TEGAS and F-LOGIC. Parallel simulation involves the simulation of the fault-free network and at the same time f modifications of this network, each of which represents a network containing a single stuck-type fault. In addition to one signal at a node, f additional signals of the faulty networks are considered in the data structure of each gate.

When fault simulation is desired, the following extra activities should be incorporated:

a. *Fault specification* to indicate which faults are important. Some faults are more likely to occur than others.
b. *Fault insertion*. A mechanism must be provided in the simulator for the automatic insertion of the specified faults at the appropriate fault sites during the simulation run.
c. *Propagation* and *detection* of the fault. Eventually, the presence of a fault should be identified as a difference in the output responses of the networks with and without a fault.
d. *Postprocessing*. The result has to be presented clearly to the designer, e.g., by displaying the deviations of the expected results.

Deductive Fault Simulation

Parallel fault simulation requires f/w simulation runs for a test when w denotes the total number of faults which can be handled in one simulation run. *Deductive simulation* uses only one simulation run for the fault-free network and all faulty networks simultaneously.

A characteristic feature of deductive fault simulation is the use of a *fault list*, which is associated with each gate input or output in the network. A fault list at a node specifies the faulty circuits whose logic values at this node deviates from the good value. By applying random values, $d = 0$ or 1, to the n primary inputs, n

7.2 TEST GENERATION

faulty networks, each containing a stuck-at-$\bar{d}$ fault, can be introduced first. This defines n fault lists associated with the inputs. The fault list for a gate output is deduced from the fault lists of the gate inputs.

For the sake of comparison, let us consider parallel fault simulation as well as deductive simulation when a logic gate with four inputs K, L, M and N is being handled. See Fig. 7.12. Suppose that five faulty networks, each with a distinctive stuck-type fault, are simulated in addition to the fault-free network. For convenience, 2-value logic is used for the good (fault-free) network G and the five faulty networks F_i, $i = 1 \ldots 5$. A fault list contains the registration of the faulty networks, i.e., networks with signals deviating from the correct value. When the logic gate is an AND, NAND, OR or NOR gate respectively, the output fault list is as given in the rightmost column in Fig. 7.12.

Inputs								Output						
PFS						DFS			PFS					DFS
G	F_1	F_2	F_3	F_4	F_5	Fault list		Gate	G	F_1	F_2	F_3	F_4	F_5	Fault list
0	1	0	0	1	1	{1, 4, 5}	K	AND	0	0	0	0	1	0	{4}
1	1	0	1	1	0	{2, 5}	L	NAND	1	1	1	1	0	1	{4}
1	1	0	0	1	0	{2, 3, 5}	M	OR	1	1	0	1	1	1	{2}
1	0	0	0	1	1	{1, 2, 3}	N	NOR	0	0	1	0	0	0	{2}

PFS = Parallel Fault Simulation; DFS = Deductive Fault Simulation

Fig. 7.12 Deduction of fault lists

The fault list at the output of a gate can be evaluated by using the following intersection rules:

a. For an AND or NAND gate (OR or NOR gate, as the case may be), each fault that appears in all 0-input fault lists but not in any 1-input (0-input) fault list is propagated to the output fault list.

b. In the special case that all inputs of an AND or NAND gate (OR or NOR gate) are 1 (0), each fault that appears in the input fault list is propagated to the output fault list.

Example: Let us consider an OR gate with four input values and the associated fault lists as given in Fig. 7.12. The fault list of the OR gate output Z can be evaluated as follows:

$$L_Z = L_L \cap L_M \cap L_N \cap \bar{L}_K = \{2,5\} \cap \{2,3,5\} \cap \{1,2,3\} \cap \{2,3\} = \{2\}$$

This example reveals that some entries in an input fault list may be absent in the fault list. The objective of deductive simulation is to propagate a non-empty fault list to the primary output. In order to achieve as many detection tests as possible, a new stuck-type fault is introduced each time a new node is reached during the simulation run from primary input to output. In this example, since the good output value of the OR gate is 1, a stuck-at-0 fault is introduced at node Z.

Concurrent Fault Simulation

With deductive simulation, propagation of fault lists is accomplished by operations on fault lists. In *concurrent simulation*, each time a stuck-type fault at a gate is encountered during a simulation run, an additional faulty gate is added to the good gate. It is essential to follow the propagation of fault effects from the fault origin to the primary output. Some fault effects reach the output, while others are blocked along the way.

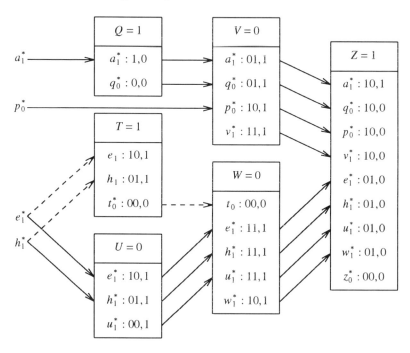

Fig. 7.13 Concurrent simulation applied to the network of Fig. 7.7

Figure 7.13 gives the simulation results when concurrent simulation is applied to the network of Fig. 7.7. An attempt is made to introduce the largest possible number of faults in the network. This is accomplished by introducing a stuck-type fault g_0 or g_1 when the node has a value of 1 or 0 respectively. Thereafter, it is examined if the effect of each individual fault appears at the primary output. In the latter case, we say that the fault being considered can be detected by the test applied to the primary inputs.

In Fig. 7.13, the test vector $APEH = 0100$ is applied to the primary inputs. Each block in this figure represents a gate in the circuit containing the following attributes: the logic value of the fault-free network and a list of fault effects with

7.2 TEST GENERATION

the syntax: $[x_d : j \; k \; \cdots \; m, z]$, where x_d is the stuck-type fault with $d = 0$ or 1, and $j, k, \cdots m$ are the input values of the gate resulting from x_d, while the last symbol z is the resulting output value of the gate. When z deviates from the correct value, the fault effect is transferred to the next gate.

A specific fault x_d, $d = 0$ or 1, can only be detected at the primary output if a fault effect propagates along a sensitized path from the fault origin to the primary output. When this occurs, the path is identified by tagging x_d^* to each node in this path. The asterisk (or any other appropriate mark) attached to the fault symbol x_d indicates the visibility of the fault at a specific node. As long as a fault is visible along a path, x_d^* appears at the subsequent lines. When, on the way to the primary output, a marked fault x_d^* at a gate input becomes invisible at the output, the mark is discarded and the fault (and hence its associated faulty circuit) will be left out of consideration. As seen from Fig. 7.13, the test $APEH = 0100$ applied to the network of Fig. 7.7 detects the stuck-type faults $a_1, p_0, r_0, v_1, b_1, s_1, u_1, w_1$ and z_0.

PODEM-X

A comprehensive testing system developed around IBM's LSSD (see Subsection 7.3a) is PODEM-X, which comprises three test-generation programs, a fault simulator and a test-compaction program [GOe81]. Test generation is performed by three programs:

a. SRTG (Shift Register Test Generator) performs functional tests on the blocks of shift registers which are used in the LSSD partitioning method.

b. RAPS (RAndom Path Sensitization test generator) attempts to generate tests by applying random bit patterns to the primary inputs of the circuits, sensitizing a large number of sensitized paths to the primary outputs and use the fault simulator FFSIM (Fast Fault SIMulator) to determine which faults can be detected by the patterns.

c. Detection tests for the remaining faults are generated by PODEM, which was discussed in Subsection 7.2b.

A compaction program is used to minimize the number of tests to detect a complete set of faults.

Alternatives to Fault Simulation

Fault-simulation costs grow rapidly with increasing circuit complexity. To avoid excessive simulation costs in VLSI chip design, several alternatives have been proposed.

Critical path tracing, as proposed by Abramovici *et al.* [Abr84], determines fault detection without explicitly performing fault simulation. The method consists of simulating the fault-free circuit with a given input test pattern and using the computed signal values for tracing paths from primary outputs toward primary inputs to determine detectable faults. Critical path-tracing has many advantages. The fault-free simulation produces realistic values at each node of the circuit. Backtracing is an efficient test-generation procedure, as shown in connection with the critical-path test-generation method discussed in Subsection 7.2b. Keeping away from using fault simulation means that there is no need for fault insertion, fault enumeration and other operations which are required in the fault-simulation process. In order to make critical-path tracing efficient, the circuit is divided into a fanout-free region (whose backtracing process is simple) and the remainder of the circuit which needs special treatment.

Another attempt to reduce the costs related with fault simulation is *STAFAN* (STAtistical Fault ANalysis), proposed by Jain and Agrawal [Jai85]. Controllabilities and observabilities of circuit nodes are defined as probabilities which are estimated from signal statistics obtained from fault-free simulation for a given test set. The computed probabilities are used to derive unbiased estimates of fault-detection probabilities and overall fault coverage for the given set of input vectors.

Behavioral-Level Fault Simulation

When using conventional fault simulators, simple faults (e.g., stuck-type faults) are introduced into the circuit. The objective is to investigate whether or not the fault can be detected by observing the output. Even when none of the faults introduced into the circuit has been detected, the chip may not function correctly. This leads us to the assumption that the actual faults to be considered must be more complex or at least different from the simple faults inserted during the simulation process. Efforts have therefore been made to use more realistic fault models.

Ghosh [Gho88] described a fault simulator which allows the introduction of behavioral fault models. The models may represent faulty state or timing parameters, a faulty description that is substituted for a part of the good description, or a combination of these.

7.2d Expert Systems for Test Generation

Introduction

The algorithms for test generation discussed in Subsections 7.2a and 7.2b have several drawbacks:

a. They are single-target-fault algorithms, i.e., the test generator only considers one fault at a time.

b. They ignore the semantics of the design and hence may waste search time on some simple task. For example, to set the output of a 4-bit adder to be 10, it suffices to provide 3 values (two for the data input and one for the carry-in) such that the sum of them is 10.

7.2 TEST GENERATION

c. They mainly work on the gate-level description of the circuit and ignore the fact that most complex circuits are designed using hierarchical structures.
d. Although most algorithms use some heuristics to guide the search process (e.g., finding the hardest or easiest objectives in PODEM, multiple backtracing in FAN, the LEARN procedure in SOCRATES), these heuristics are all hard-wired into the program and hence the augmentation of new heuristic rules requires the modification and recompilation of the programs.
e. These algorithms are basically "memoryless", i.e., any knowledge obtained during one iteration of test generation is discarded after that iteration finishes. Thus, the same efforts may be done again and again in different iterations.

A test engineer, however, has little difficulty in dealing with the above problems when he/she has a good understanding of how a circuit operates. Although both a test engineer and the algorithms referenced above may use the same basic sensitization techniques, a person usually knows how to set a circuit into desired states in a more efficient way than that found by a computer. He/she will frequently employ his knowledge dealing with the circuit's functionality and global structures to achieve an efficient test strategy.

Because of this situation, several researchers have proposed using AI techniques to guide the ATPG process and several test generation systems using a knowledge-based technology have been implemented. Each of these systems addresses one or more of the problems listed above. We next describe some of these algorithms or systems.

Partitioning Expert Systems

To facilitate testing of large-scale VLSI circuits, the circuit should be partitioned into testable subcircuits. Delorme *et al.* [Del85] described a functional partitioning expert system for digital circuits as a step to generate test sequences using both algorithmic and AI techniques. When test patterns for the circuit partitions have been generated, an integration process produces the overall test sequence of the entire circuit (or board).

Prolog-Based Test Generation

Several expert systems for test generation are based on Prolog. Svanaes and Aas [Sva84] showed the applicability of Prolog for test generation using the D algorithm and for finding complete tests for acyclic networks.

Varma and Tohma [Var88] described Protean, a Prolog implementation of an automatic test generator for detecting stuck-type faults in scan-designed standard-cell circuits and iterative logic arrays. Protean comprises a cell test generator, which generates test knowledge and propagation characteristics for cells, and a hierarchical test generator, which uses this high-level test knowledge in conjuntion with low-level structural information to generate tests for the circuit.

P Algorithm

The P algorithm, developed by Srinivas [Sri86], is based on the critical-path test-generation technique (Subsection 7.2b) and is applicable to both combinational and synchronous sequential circuits. Whereas the symbol D in the D algorithm represents simultaneously the fault-free and the faulty values on a circuit node, the symbol P is a free variable that is assigned the value of either logic 0 or logic 1. P's complement is N. Whereas the D equations are used to propagate the error signal from the fault site to the primary output, the P equations are used to trace back the symbol P from the output to the inputs, i.e., to sensitize the circuit output to a set of paths in the circuit.

Using the Interactive Theorem Prover developed at Argonne National Laboratories, the P algorithm was implemented for combinational circuits [Sri86]. The language used is based on a clausal form which is an extended subset of first-order predicate calculus. Given the set of clauses that represents the connectivity of an arbitrary combinational circuit, the reasoning system generates expressions from which the test set is derived.

C Algorithm

The C algorithm generates tests for combinational circuits, based on the concept of concurrent test generation [Yau86]. It aims at generating a test vector that covers as many faults as possible. In traditional methods (D algorithm, PODEM, FAN), faults are generated one at a time. The C algorithm considers simultaneously all the faults that have not yet been detected.

For any fault to be detected, the primary output of the circuit is driven to logic 0 or 1 so that a maximum number of faults are covered. The concurrent test generation involves backtracing from the primary output under heuristic guidance complying the theme of backward reasoning. The efficiency of the C algorithm is achieved through adopting a powerful AND/OR graph search formalism.

This greedy algorithm tries to justify a logic value on a line in such a way that many paths through the circuit are simultaneously sensitized, causing a maximum number of faults to be detected.

Brahme-Abraham Test Generator

Brahme *et al.* [Bra87] introduced an algorithm for test generation which uses the knowledge of the data section and the control section of the designed circuit. These sections at each level in the hierarchy store information about how to propagate and backtrace signals through them. The algorithm propagates and backtraces differently depending upon whether the fault is in the control section or in the data section.

SUPERCAT

Test generation can be carried out in two steps:

7.2 TEST GENERATION

a. Automatic test generation of small functional circuits.
b. Integrating the tests generated in step a. to produce the global test program of the whole complex circuit, i.e., deriving the primary inputs which activate the fault at the fault site and propagate the fault effect to a primary output.

The SUPERCAT system [Bel83] was developed to assist the test designer in step b. To that end, the knowledge base comprises three kinds of information:

1. A range of programs for automatic test generation of different types of logic (random logic, RAM, PLA, microprocessor, etc.), as well as the input descriptions (gate level, state table, instruction set) and the types of faults covered (stuck type, pattern sensitive faults, etc.).
2. A library of commonly used test patterns acquired from test engineers.
3. Information on the applicability of design-for-testability techniques.

SUPERCAT can be used for data-driven circuits (represented by a data-flow model) and control-driven circuits (represented by a control-flow model). The data-flow model is based on partitioning the circuit into minimally testable circuit modules. When a data-driven circuit is dealt with, SUPERCAT invokes the CATA system [Rob84], which performs the following tasks: determining the data flows and the test paths in the circuit, and performing a testability evaluation with subsequent proposals for specific testability measures. SUPERCAT is able to determine all the flows for any part of the circuit and, using a given test strategy, identify the test paths to be used from among the set of flows. Controllability, observability and testability of a module in a flow are derived to allow discrimination between flows.

The control model describes a circuit in terms of functional resources, each of which is associated with a timing protocol. A resource can be passive, such as a data path, or a controller. When control circuits are handled, tests are based on the normal operation of the control circuit, which is divided into the controller and the controlled sources.

In connection with its objective, SUPERCAT has to compile a large amount of knowledge related to various sources of information, such as level and type of description, fault models, test methods and strategies, and testability measures. On that account, SUPERCAT relies on the cooperation of dedicated test expert systems. The designer may call such expert systems independently and/or successively, and act upon their results. The CATA system performs three tasks: generation of flows (information paths throughout the system), testability evaluation, and generation of test paths (to be determined among the set of flows according to a test strategy).

The test expert working on the control-flow model is efficient for a (sub)circuit made up of a controller and a set of controlled resources. Two approaches are mainly proposed for the test of the controller [Bel84]: an identification method (testing a set of paths of the symbolic execution tree covering all the edges of the control-flow graph) and a distinction method (making fault hypotheses, defining test patterns at a structural level, and defining a set of paths of the symbolic execution tree allowing to send these patterns to the controller). Testing the controlled resources includes the determination of the input value domain of a controlled resource. An expert first determines which path of the symbolic execution tree activates a given resource, and then which value may be sent to a controlled resource for a given path. The test strategy for testing controlled resources includes a bottom-up as well as a top-down approach.

HITEST
Most test-generation algorithms (see Subsections 7.2a and 7.2b) have been developed for low levels, particularly the gate level. Such algorithms are known to be polynomial complex. Human test programmers have the tendency to view the circuit at a high macroscopic level and thus attempt to incorporate more global aspects into testing systems.

A thorough knowledge of the system under test is essential for effective test generation and fault diagnosis. This is specifically true when human knowledge about some aspects of the circuit operation is not directly suited to automation. In such cases, a knowledge-based approach to testing may be tried. An example of a knowledge-based testing system, called HITEST (Intelligent TEST generator), was developed by researchers at Cirrus Computers, Fareham, with financial and technical assistance from British Telecom and the UK Department of Industry [Rob83, Wha83, Ben84].

HITEST is a test-generation system which permits the use of both human knowledge and algorithmic procedures to aid in test generation. Knowledge is stored using frames and slots, allowing a hierarchical representation. The circuit-definition language CCL (Cirrus Circuit Language, a derivative of the HILO-2 language), is used to describe the circuit under test. CCL allows the circuit to be modeled in a hierarchical manner at the level of functional modules, gates or MOS transistors. Standard device and circuit models and previously compiled subcircuits are stored in a library. CWL (Cirrus Waveform Language) is used for specifying both the stimuli applied to the circuit and the expected responses. Procedures containing frequently used waveforms can be defined. CWL supports constructs as block structures, strict data typing and loops.

HITEST provides both fault-free and fault simulation for 15-valued logic. Fault simulation uses the Parallel Value List method [Moo83], which combines parallel, deductive and concurrent approaches. To enhance the user's understanding of the simulation and testing process, HITEST offers two display systems: a standard visual display unit for monitoring the logic analysis results and a color graphics unit for displaying the signal flow in the circuit schematic.

Since HITEST test generator is a subsystem of the simulator, it shares all facilities (CWL, displays, etc.) of the simulator system. Figure 7.14 shows a

7.2 TEST GENERATION

schematic structure of the HITEST system.

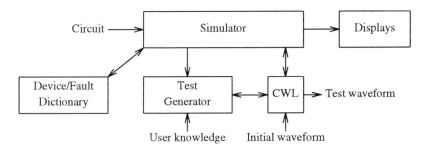

Figure 7.14 The HITEST system

CWL provides the interface between test generation and simulation. The TESTGEN statement, supported by CWL, acts as a subroutine call to the test generator. A single-text argument relates to the actions to be taken by the test generator. These actions may be based both on knowledge supplied by the user and on the current state of the circuit. Depending on the initial waveform, CWL can invoke the test generator which produces a set of assignments for the primary inputs using an extension of the PODEM algorithm and the circuit knowledge described in CCL.

SATURN

The SATURN Test Generation System integrates and improves upon existing test-generation systems [Sin87]. Though inspired by the D algorithm, SATURN is different in that
a. it also generates tests at levels higher than the gate level,
b. it is more flexible permitting the most promising choices in the search space depending on the situation,
c. fault models other than the stuck-type fault model can be specified,
d. the user can specify how to test a class of components,
e. SATURN minimizes computational effort by caching tests and solution to subgoals in achieving results,
f. it reduces the number of tests generated by collapsing tests.

The inputs to the SATURN system consist of specifications of the design, the controllable inputs and observable outputs, the fault models, of how to test a class of components, user-specified test sequences, and the level of abstraction for testing. The testing result is a collection of tests that check all the possible faults in the design, and a list of tests that could not be achieved. The deductive cost estimates for controlling and observing the value of every port are computed to have a measure for the number of inference steps required to control/observe the value of a port. Three sources of information are used to decide how to test any module: its behavior specification, its substructure (submodules and their interconnections) and a user-supplied test sequence. For each module, the user can specify which of these sources of information is to be used in testing it. SATURN uses a dynamic control scheme for propagating test values. It uses this scheme to select the next most promising task. The deductive cost estimates are used to choose the cheapest alternative and in case of failure the system backtracks to the choice that led to the failure and not just the last choice.

The SATURN test-generation system has been implemented in Maclisp, running on a PDP-20 and Symbolics 3600 series machines. An integral part of the reasoning and representation methods of SATURN is built on top of the MRS knowledge-representation system.

KBTA

KBTA (Knowledge Based Test Assistant) is an expert system for test generation, developed at DEC, Hudson [Yor86]. This system is primarily used by test engineers who acquire chips from external vendors. Since the vendor's data book contains functional descriptions of the chips, test vectors are generated at the functional level. The KBTA system contains three basic types of components: knowledge/data acquisition tools, consistency/plausibility checking tools, and program-generation/optimization tools. Other tools include the relational database browser for examining KBTA data/knowledge bases and an XY plotter utility for plotting waveforms stored in the KBTA database. The knowledge base contains schemata and rules to represent properties and languages of automatic test equipments, the operation of (functional networks of) hardware components, and hardware testing techniques. In addition, working knowledge of the unit under test (UUT) is required.

The objective of KBTA is to explore the knowledge-based approach to automatic synthesis of test programs. Components implemented in the prototype system include the wave/signal editor, the block-structure editor, the pin-information editor, the plausibility checker, the timing-consistency checker, the timing-reduction utility, the tabular functional block editor, and the graph-walking portion of the program generator. Other components are being developed to improve the KBTA system. The wave/signal editor derives the timing information from the waveform diagram supplied in the data book and then uses that information for actually coding the program. The function editor provides the user with a means of easily entering the logic function which each component of the UUT performs. Most of the plausibility checking is done at the time of data entry. The timing-consistency checker is invoked as the timing information has been entered. It evaluates timing intervals required for executing a graph-partitioning algorithm. By decomposing the network of timing relationships, smaller graphs can be checked for inconsistencies, while analytic measures of the complexity of the timing network for the unit under test are provided. The timing-consistency checker produces all the information required

7.2 TEST GENERATION

to generate the timing integrity test vectors for the UUT. The test program generator produces test code by matching the specification of a test for the UUT against the "solved problems" in the solved-problem library. A large part of the test program consists of timing-integrity checks. The set of tests may be automatically generated by "walking the timing-consistency graph".

KBTA is written in VAXLisp and runs on VAX 11/785.

Functional Test Generation

The complexity of test-generation problems at the gate level is NP-complete. Not surprising that test-generation methods have been proposed at higher levels of abstraction [Bha89]. Cosgrove and Musgrave [Cos91] suggested that functional approaches to test generation will do well within an expert-system environment. Heuristic functional approaches and knowledge-based systems are gaining interest from test engineers.

Gupta and Welham [Gup88] described a Prolog-based workbench used for generating sufficient conditions for functional testing of a given circuit on a given fault class. The conditions must be chosen so as to yield an output value different from that of the correct value. The behavior of modules is defined in a library identified by the Prolog predicate *unfold*. Behavioral models of the faults are represented in a library with the Prolog predicate *unfold-fault*. Besides the libraries, a CAD tool provides a more natural interface with the user. A lot of information presented to the user or required from the user can be represented in textual or graphical notation.

Lea *et al.* [Lea88] described an expert system capable of producing functional test programs for digital circuit boards. It employs rule-based and heuristic techniques. Various knowledge bases contain goal-driven production rules utilizing certainty factors. A frame-oriented relational database contains device models and register-transfer logic of the circuit. A powerful search algorithm identifies valid test paths through the circuit and thus define functional tests. Once identified, the tests are ordered in terms of increasing test complexity to aid diagnostics. Finally, the test paths are passed on to a low-level test timing generator to produce the actual test vectors required to test the board.

7.3 DESIGN FOR TESTABILITY

7.3a Scan-In/Scan-Out Methods

Testability Aspects

Realizing testability for the purpose of minimizing testing and maintenance costs should always be a primary concern in circuit and system design [Wan91]. Testability measures are defined in Subsection 7.1b. However, care should be exercized in using quantitative measures and interpreting results obtained from testability computations [Agr82]. Savir [Sav83] showed that a good controllability and a good observability of a circuit do not guarantee that the circuit will be highly testable. The testing trouble arising when a circuit with a reconvergent fanout is used is illustrated in Subsection 7.2a.

Design for testability has become customary in chip design. Testing has to be incorporated in the design process from the first step. The desire to design testable circuits has led to many methods which make use of (a) easily testable circuits, (b) universal test sets of logic circuits, (c) test-point insertion, (d) partitioning and (e) scan-in/scan-out methods [Sch87]. In realizing logic functions, it is expedient to choose that one from alternative hardware implementations that is easily testable. It is well known that fanout-free networks (with fanouts of 1) are easier to test than networks with reconvergent fanouts. An implementation, which also requires a small number of tests, is the Reed-Muller expansion form.

The usual problem of test-set generation is to find a complete test set for detecting a specified set of faults in a given hardware implementation of a logic functon. An alternative problem formulation is given as follows. Given the logic function, find a minimal test set that can be used for testing an arbitrary implementation of the logic function. The minimal test set depends on the expression form of the function. A minimal test set can be derived for particular functions, like the *unate function*, which is given by a product of sums or a sum of products in which none of the variables appear in the common or the complementary form. Hence, we must try to change the given logic expression in an equivalent form of a special type, such as the unate function.

A disadvantage of most of the above-mentioned methods is that the total number of gates becomes unpractically large. The number of transistors or gates in a chip may increase, but the number of input and output pins is limited. Consequently, the controllability and observability of internal gates in complex chips will become difficult to evaluate. A practical method to improve the testability is to introduce additional test pins, which require additional chip area. Of importance is how and where these pins must be placed. A suitable network partitioning may also help alleviate the testability problem. Bennetts [BEn84] formulated 25 practical guidelines for designing testable logic circuits.

Scan Paths
Testing sequential circuits has always been a difficult problem. With i inputs and r memory elements (latches or flip-flops), the number of possible input bit patterns and internal states of the network is 2^{i+r}. Apart from the excessively large number of 2^{i+r}, it is very hard, if even possible, to initiate the circuit into a

7.3 DESIGN FOR TESTABILITY

prescribed state from which all future states can be predicted. There is no way of testing the circuit by exercising all possible states. The complexity of the problem would be alleviated when all the memory elements in the circuit could be made controllable and observable. By incorporating appropriate additional hardware into the circuit, the memory elements can be arranged in such a way that they can be set into any desired state. This approach has led to a number of structured design techniques with the objective of separating the sequential and combinational parts of the network.

Williams and Angell [Wil73] were the first to pave the way for the development of a series of scan-in/scan-out methods which prove to be a powerful solution to the problem of testing sequential circuits.

Suppose that the sequential part of a logic circuit consists of a number of latches. All these latches are taken apart and linked together to form a shift register. In order to improve the controllability and observability of the cirucit, a control pin M is used to switch the circuit in one of two possible modes: the *normal mode* and the *test mode*. To this end, additional multiplexer circuitry must be incorporated into the circuit. When $M = 1$, the circuit operates in the normal mode exactly as if the additional logic were absent. When $M = 0$, all the latches are connected in cascade to form a shift register. In this test mode, also called the *shift-register mode*, the latches are disconnected from the remainder, which is a combinational circuit. In this way, test generation is reduced to the relatively easy task of testing a combinational circuit.

In the shift-register mode, the circuit can be set to any desired state by supplying the state variables in serial bit form to the input of the shift-register chain of flip-flops (*scan in*). The sequence of bits that appears on the output of the chain of latches can be checked to see whether the circuit is in a given state (*scan out*). The chain of latches in the test mode is referred to as the *scan path*. After an initial state has been set, the circuit can be switched to the normal mode, then returned to the test mode so that the final state can be checked.

The testing procedure consists of the following steps:
a. Switch the circuit to the test mode and check the shift-register operation.
b. Set the shift register to the desired initial state by shifting the proper test sequence into the latches.
c. Return to the normal mode and apply a test sequence associated with this state to the primary inputs.
d. Switch to the test mode and read out the final state.
e. Compare this state and the primary-input signals with the correct values.

The LSSD Technique

If the sequential part of a circuit consists of latches, each system latch is extended by an additional test latch to form master-slave memory modules. The system latch and the slave latch are clocked by separate clocks A and B. If a shift register consisting of merely system latches will scan out its internal state, half of the information is destroyed. To enable a normal system behavior, the system latch must be extended with one system data input. A multiplexer or an extra system clock can be used. The latter is used in the LSSD technique, introduced by IBM [Eic91]. The technique is called LSSD (Level Sensitive Scan Design) since the two latches in each memory module are designed hazard free, that is, independent of rise and fall times.

Alternative Scan-In/Scan-Out Methods
Several alternative realizations of the memory module have been proposed. Their principal differences are summarized below [Sch87].

Scan-Path System of Nippon Electronics Corporation. The System Data line and the Scan-In line are used as the input line in the normal and test mode respectively. The latches are replaced by raceless D-type flip-flops. Only one shift clock is used in the test mode.

Scan-Set Logic of UNISYS Computers. The shift register is composed of non-system latches. It can sample up to 64 random system points in parallel, which can then be shifted out. The set mode allows one to inject the shift-register contents in the system logic. Then, the system latches can be loaded with the results. Testing can also take place during system operation. A sampling pulse is applied to the shift register to get a real-time sytem "snapshot".

Random Access Scan of Fujitsu. All system latches are controlled and observed by a technique similar to a Random Access Memory (RAM). It has the same result as the LSSD and the Scan Path approach: testing is reduced to testing combinational circuits. Each latch can be uniquely addressed and set by a combination of XY addressing and preset/reset facilities. Also, the output can be observed when a latch is selected.

Boundary Scan
A standard technique that addresses the growing complexity of printed-circuit boards is known under the name *boundary scan* [Mau90]. In this technique, access is required to only four dedicated pins on a board's edge connector: one pin for input, one pin for output, a clock and a control line. In this way, boards with internal nodes, which traditional bed-of-nails testers cannot access, can be tested.

7.3b Built-In Testing

Self-Testing
The circuit may provide its own test patterns, e.g., by means of a built-in random test generator. When these test patterns are applied to the circuit, faults may be detected giving rise to an error message at an observable output. Self-testing is

7.3 DESIGN FOR TESTABILITY

the technique of automatic detection of a fault, eliminating the need of applying externally generated tests to the primary outputs. Specific test patterns could be stored in an on-chip ROM.

Signature Analysis

Signature analysis is a compact testing method introduced by Hewlett-Packard Corporation for maintenance of microprocessor-based systems [Fro77]. In order to provide for testing with a Hewlett-Packard Signature Analyzer, the design of these systems must contain additional circuitry, including a ROM in which a special program for stimulating the necessary test sequences are stored.

When a system is to be tested, the Signal Analyzer is connected to it. Then, a test sequence is applied to the system and the output response is passed through a 16-bit linear feedback shift register which is part of the Signal Analyzer. In effect, the lengthy data streams which represent the output response of the system are compressed into a 16-bit string. This string, referred to as the *signature*, is a particular bit pattern which is associated with a particular data node of the system under test. For easy readability, the signatures are recorded or displayed as four-digit hexadecimal characters.

Example: Figure 7.15 depicts a 16-bit linear feedback shift register which can be used for signature analysis.

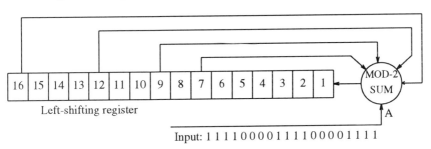

Fig. 7.15 Feedback shift register for signature analysis

The five inputs to the modulo-2 sum circuit consist of the input bit stream A originating from the system under test and the bit streams fed back from the 7th, 9th, 12th and the 16th flip-flop of the shift register. A properly interconnected set of Exclusive-OR gates can be used to implement the modulo-2 sum circuit. The register is clocked by the same clock as the input bit stream. If we apply the input bit stream $X = 1\ 1\ 1\ 1\ 0\ 0\ 0\ 0\ 1\ 1\ 1\ 1\ 0\ 0\ 0\ 0\ 1\ 1\ 1\ 1$ to A and we stop after entering the last bit, the state of the shift register is given by the signature $S(X) = 0\ 0\ 0\ 1\ 0\ 1\ 1\ 0\ 0\ 1\ 0\ 1\ 0\ 1\ 1\ 0$, which in hexadecimal form is 1 6 5 6. If the input bit stream were $Y = 1\ 1\ 1\ 1\ 1\ 0\ 0\ 0\ 1\ 1\ 1\ 1\ 0\ 0\ 0\ 0\ 1\ 1\ 1\ 1$, the state of the shift register would be given by the signature $S(Y) = 1\ 0\ 0\ 1\ 0\ 1\ 1\ 1\ 0\ 0\ 0\ 1\ 1\ 1\ 0\ 0$, or in hexadecimal form: 9 7 1 F, where F represents 1 1 0 0. Suppose that the above bit stream X and Y originate from a fault-free circuit and a faulty circuit respectively. Evidently, a slight difference between X and Y (only a single bit in this case) leads to signatures $S(X)$ and $S(Y)$ which are completely different.

During a production test of the system, the signature at each node is compared to the known good signature for the node and a fault is indicated if there is any difference. By backtracking in the circuit to a node that has a correct signature the location of the faulty component can be found. Signature analysis unites simplicity of testing with an excellent fault coverage. A further advantage is that it allows the system under test to run at its normal clock rate (up to 10 MHz) enabling timing errors to be detected.

A feedback shift register of length n (which is 16 in this case) will detect all errors in data streams of n or fewer bits. It can be shown [Fro77] that any single-bit error, regardless of when it occurs, will always be detected by stopping the register at any time and comparing the signature with the known good signature. This error is detected independently of the length of the input sequence. Hence, when the error has disappeared many clock periods earlier, the effect of the error is still detected.

BILBO

One of the many built-in methods is discussed below. The LSSD/Scan-Path concept can be combined with the technique of signature analysis, yielding the BILBO (Built-In Logic Block Observation) technique. The example which will discussed here concerns an 8-bit BILBO register proposed by Könemann et al. [Kon80].

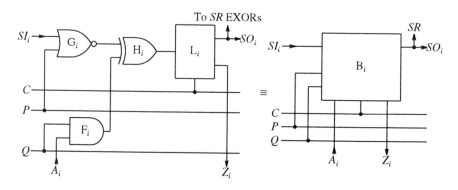

Fig. 7.16 BILBO block

A BILBO block contains a system latch L_i augmented with a NOR gate G_i, an Exclusive-OR gate H_i and an AND gate F_i connected as shown in Fig. 7.16. The input and output pins are A_i (connection points to the combinational part of the network), SI_i and SO_i (Scan-In and Scan-Out), C (Clock signal), P and Q

7.3 DESIGN FOR TESTABILITY

(control pins) and for three specific blocks SR (to ExORs of the feedback shift register). By assigning suitable binary values to the control pins P and Q, the three primary modes of the BILBO register can be initiated.

- $P\,Q = 1\,1$ is the normal basic operation mode, in which each latch L_i is connected to the combinational network via F_i and H_i.
- $P\,Q = 0\,0$ is the scan-in/scan-out mode, in which the latches L_i are connected in cascade via the gates G_i and H_i.
- $P\,Q = 0\,1$ is the linear feedback shift register mode, in which the network takes the form of a linear feedback shift register as shown in Fig. 7.17. Instead of one input, eight unique inputs are used.
- $P\,Q = 1\,0$ can be used for resetting purposes.

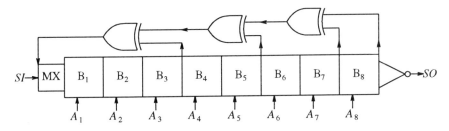

Fig. 7.17 BILBO feedback register

7.3c Expert Systems for VLSI Design Testability

Introduction

As the chip complexity increases, chip testability becomes a more serious problem. In the past, a great number of design-for-testability techniques have been developed. For PLAs alone, more than 20 testing techniques have been proposed. For some important techniques, see [Ost79, Aga80, Hon80, Fuj81, Dae81, Sal83, Bre85, Tre85].

When a specific circuit is being considered, the problem arises as to how to select the appropriate testing techniques which makes a circuit easily testable. The following discusses expert systems which assist the designer in making chips more testable.

Testability Expert

HAT (Heuristic Adviser for Testability) uses a hierarchical structural description of a circuit to refine testability estimations [Abr85]. Realistic testability measures for complex circuits can be evaluated by taking advantage of the hierarchical information in the design. By using the circuit-description language SCALD, HAT recognizes regular structures. In addition, it identifies special structures, such as test points and scan-path registers and treats these like primary inputs and outputs. HAT is designed as a prototype tool to estimate fault coverage versus test length for a variety of testability analysis algorithms and a variety of test-generation styles.

A main concern in chip design is the consideration of a test strategy for debugging and for generating production tests. Crastes *et al.* [Cra89] use testability expertise to determine test points to be inserted for testing purposes. The expertise is based on high-level modeling in terms of basic functions expressed in Prolog. Design modifications aim at increasing controllability and observability features by introducing multiplexers and scan paths.

Another expert system for test-strategy generation was proposed by Lutoff and Robach [Lut87]. The strategy takes into account various parameters, such as testability measures and technological or functional specificities of the circuit unit under test.

TEXAS

TEXAS (Testing EXpert AdviSor) analyzes the testability of VLSI architectures [Cos89]. The purpose is to suggest a suitable testing approach, such as Test Pattern Generation, Design for Testability and Built-In Test. Testability assessment is achieved by the conceptual partitioning of a circuit into a number of testable units. Partitioning is accomplished at the behavioral/functional level or the layout level.

CATA

CATA (Computer-Aided Test Analysis) guides the hardware engineer in designing a digital or analog system that is easily testable both in manufacturing and field maintenance situations [Rob84]. It provides test-program specifications automatically and supplies both the information paths through the system and a top-down organization of test procedures. CATA defines the order in which system functions should be tested for fault isolation. It permits design engineers to evaluate testability and make modifications during the entire design cycle.

Robach *et al.* [Rob89] developed an expert system for the functional specification of test programs. Once the set of module functions has been determined, the next step is to define a test strategy whose aim is to select the subset of functions that is required to ensure the test of all hardware primitives and to provide an order for the application of those functions. The chosen test strategy is based on a Start-Small approach: A first test concerns a minimal set of hardware components and each subsequent test adds more hardware until all components have been covered. When a test detects a fault, the fault part is one of those added for the current test.

DFT System

The DFT (Design For Testability) system, developed at Syracuse University and

7.3 DESIGN FOR TESTABILITY

IBM, consists of a design representation, DFT rules, inference clauses, transformation rules and design clauses, all written in Prolog [Hor83, Hor84]. The design representation describes node functions and interconnections, while design clauses state the behavior of the nodes. Inference clauses include memory-block observabilities. If none of the clauses can be proven, the design is judged faulty. Prolog's ability to modify its own clauses (in other words, its database) gives the system some learning capability. In addition to design for testability, the Syracuse-University/IBM system has facilities to perform functional simulation, fault diagnosis and automatic test generation [HOr84].

Test Generation and Design for Testability
Shirley *et al.* [Shi87] of MIT and GenRad proposed a methodology for ensuring the testability of complex sequential circuits at low cost. This methodology, which contains both test-generation and design-for-testability components, is based on the following considerations:
a. Test generation of the difficult class of sequential circuits is restricted to the easily testable parts of the circuit, avoiding the worst cases by bounding the search done during test generation.
b. The system suggests minimal DFT additions to areas of the circuit where the test-generation system failed so that those areas becomes testable.

This approach improves our ability to trade off testing costs against, for example, performance and area costs. The main goal is to minimize the test-generation time, the test-application time and the degradation of circuit performance due to the required additional hardware for improving the testability. Important features of this approach is the use of symbolic simulation as a means for computing the relationship between circuit and component operations, and the identification of normal behavior as an important concept.

TDES System
TDES (Testable Design Expert System), developed at the University of Southern California [Aba85, BRe89], can be used to modify a chip design such that the chip will be easily testable. The strategy is to imbed testing hardware in the original circuit without affecting the algorithm or function implementation used in the design process. A prototype of the TDES system has been implemented in LISP to be run on a DEC-20. TDES is incorporated into the advanced design automation system ADAM of the University of Southern California [Gra85]. A short description of TDES follows below.

TDES assumes the complete circuit to be composed of the basic structures: combinational logic blocks, registers, PLAs, ROMs, RAMs, tri-state drivers and buses. A graph model is used to represent the circuit. Each node of the graph represents a basic structure, defined by several attributes, such as the type (combinational, register, etc.), the design style (random logic, PLA, etc.), the function (multiplexer, etc.) and the numbers of input and output ports or the number of product terms if it is a PLA. Each directed arc of the graph defines the flow of data between two nodes. Labels are attached to arcs to identify attributes, such as source, destination and type of connections (normal or boundary). Partitioned blocks which may or may not correspond to functional blocks (control, data path, I/O or memory) and which must be made testable are called *kernels*. Each kernel for which at least a *Testable Design Methodology* (TDM) exists is assigned a weight which is related to its testability. TDMs include *ad hoc* techniques (e.g. extra test pins), structured techniques (e.g. LSSD) and built-in test techniques (e.g. BILBO). Kernels are processed with decreasing weights (corresponding to increasing testability). A kernel subgraph is characterized with an I mode (which describes the operation type and the transfer between input and output ports), the activation mode (enable or disable) and gate delays or clock information. An identity transfer path describes the data flow between structures.

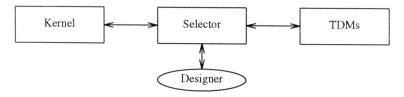

Fig. 7.18 TDES system

Globally, a TDES procedure involves three main units: the kernel which must be made testable, the knowledge base containing the available TDMs, and the selector (Fig. 7.18). Based on a set of requirements specified by the designer, the selector decides which TDM to choose. Since the attributes of a TDM are not equally important, different weights are assigned to the individual attributes (testability characteristics, effects on the original design, requirements on the test environment, design cost). A requirement vector consists of these weighted attributes. The TDM that best approaches the requirement vector is selected. A TDM is evaluated by attaching appropriate values to its attributes.

In order to build up a knowledge base, TDES uses a set of *frames* for describing testable design methodologies. The objective is to assign suitable TDMs to given kernels. To this end, the following information must be made available:

a. The identification of the kernel for which a TDM is needed.
b. The built-in test (BIT) structures associated with a specific TDM to allow the kernel to be tested adequately.
c. The required on-line and off-line hardware and/or software for executing the test.

7.3 DESIGN FOR TESTABILITY

d. A test plan describing the sequence of executions to perform the test.
e. Measures for evaluating the costs and merits of potential TDMs.

Each TDM frame describes its structural, behavioral, quantitative and qualitative aspects and can be divided into three main groups:

1. *TDM template.* This group contains information about the TDM's structural architecture, such as type, style and the size of the kernel to which the TDM is applicable. It also describes which BIT structures are required by the methodology and how they must be connected to a kernel. For example, a kernel to be tested by a BILBO TDM may be of style PLA or random logic.
2. *TDM test plan.* This part of the frame describes which sequence of actions must be performed to execute a test. The circuit is assumed to be synchronous.
3. *TDM measures.* This last group of slots in the TDM frame is used to specify values of the measures. Most of these values are estimates used to make early design decisions between alternative TDMs.

TDES uses several *measures* to evaluate a TDM and to help select a specific global or local embedding process for it. Relevant measures for *local embedding* are the area overhead associated with a particular embedding and the optimal execution time. *Global embedding* is related to global actions to be taken with regard to the embedding process which may lead to an optimal local embedding solution. An example of a global measure is the extent to which BIT structures created for a local embedding in a kernel can be shared with other kernels in the circuit. Another global measure concerns the possibility of testing a large circuit eliminating the need for testing parts of this circuit. Measures which are *TDM specific* are the fault coverage, external test equipment and I/O pin requirements. In most cases, this information can be obtained directly from the appropriate slots in the TDM frame.

With the help of a score function, TDES can evaluate which embedding solution is the best. The score function SF is given by

$$SF = \sum_i M_i * N_i * P_i,$$

where M_i is the value of measure i,
N_i is the normalized weight of measure i,
and P_i is the priority weight assigned to measure i.

The embedding process is a unification process in which the slots in a TDM frame are attempted to match with the actual actors from the circuit. Rules are given to find all possible ways for embedding TDMs in kernel subgraphs. For example:

Rule 1: *if* "a given set C_1 of conditions are met" *then* "carry out the embedding process, generate an optimal test plan and calculate the score of the embedding".
Rule 2: *if* "for a specific type of template mode, a given set C_2 of conditions are met" *then* "apply the TDM for kernel K".

Rule 3: *if* "for a specific type of R, a given set C_i of conditions are met" *then* "match R with the template node".

For example, suppose we want to embed a TDM in a subgraph SGk of kernel K, then Rule 2 is used to check if the BILBO TDM is applicable to kernel K. From the template, we can notice that BILBO TDM is applicable to any kernel of design style PLA or random logic. Suppose that the BILBO has a fanin from B_1 and a fanout to B_2. Rule 3 can now be used to find a match for B_1 and B_2 with proper latches. Then, a multistep test plan can be generated with an optimal test time. When none of the structures in the subgraph of K can be modified to match one of the template nodes of a TDM, another suitable embedding of the scan-path register must be attempted. The solution with the least delay time is then selected.

By examining the score function of various embeddings, the embedding suited to the system is selected. Let us illustrate the use of TDES by considering the following example. First, TDES identifies three major kernels, namely PLA, ROM and C (random combinational). These kernels are processed according to the weight factors assigned to them. Say, TDES explores various embedding solutions for the BILBO TDM. Suppose further that the user selects the best BILBO embedding. This choice may require one or more registers to be modified by a linear feedback shift register (LFSR) to generate pseudorandom test vectors, possibly with a Signature Analysis mode of operation to capture the PLA signature.

Let us now consider the ROM as a kernel. A given register can be modified to an exhaustive LFSR to address all the ROM contents, while another register may operate as a signature analyzer to compact the ROM responses into unique signatures. Finally, a combinational kernel C is considered. The different latches are linked together to form a shift register.

TEST is a knowledge-based system for designing testable VLSI chips being developed at the University of Southern California [BRe89]. It incorporates the two expert systems TDES and PLA-TSS (see below), which were implemented earlier. In addition, new extensions are being incorporated into TEST, including a scan system, a design planner, a functional test-generation system and a graph-modeling production system.

PLA-TSS

PLA-TSS or PLA-TDM Selection System [Zhu88] is a subset of the TDES system discussed above. It assists a designer in selecting the most suitable testable design methodology (TDM) for a given PLA. The input to PLA-TSS is a description of the PLA (inputs, outputs and product terms) along with a specification of requirements, including the effect of using a DFT technique on the original design. The knowledge base contains a frame representation of TDMs, each of which is given by its attributes (e.g. fault model, fault coverage,

7.3 DESIGN FOR TESTABILITY

area overhead, extra delays, extra I/O connections). An attribute has a value which may be integer (e.g. the number of extra I/O connections), real (e.g. the area overhead), binary (e.g. self-testing: yes or no) or a set (e.g. specification of a fault model).

To select test strategies which are optimally adapted to the configuration of the PLA, TDMs are evaluated and compared in a given environment. The system generates an evaluation matrix whose columns correspond to the TDMs and the rows correspond to their attributes. Given the designer's requirements, each attribute is assigned a dimensionless *penalty-credit function* which defines the "distance" between the attribute values and the required values. The quality of a specific TDM is given as a linear combination of weighted penalty-credit functions of the attributes. The system evaluates the influence of specific TDMs on criteria, such as fault detection, circuit overhead, test generation and design costs. The TDM with the highest score provides the closest match to the designer's requirements.

To assist the designer in specifying the initial requirements a ramification analysis is performed in order to determine the impact of changes made in the required values. When the result is still unsatisfactory, a backtracking process tries to make suitable changes in the requirements as to reach a satisfactory solution. When a failure is encountered, PLA-TSS determines the reason for that failure and returns to the point where the failure originates. It resolves the problem to avoid the failure and restarts the search procedure. This is called "reason analysis directed backtracking". There are four types of failures and the most critical failure is handled first. The user may reject solutions, make changes in the requirements and ask for explanations concerning specific choices.

Functional Specification of Test Programs
Robach *et al.* [Rob89] defined the functional specification of test and maintenance programs. This has led to the development of an expert system with the aim to select a test strategy for complex devices, logic printed-circuit boards or systems. The strategy takes into account various parameters, such as testability measures and technological of functional specificities of the equipment under test. The knowledge base contains the functional description of the logic system to be checked, the definition and behavioral representation of the test strategies and the designer's expertise. The expert system is implemented in OPS5 and runs on VAX.

TIGER
TIGER (Testability Insertion Guidance ExpeRt system) is a methodology for evaluating and making intelligent and optimized choices among the various existing testability approaches [Aba89]. Most of the TDES concepts have been refined and incorporated as part of TIGER. The knowledge base of TIGER is built upon the representation system KRAFT (Knowledge representation using

Relations, Attributes, Functions and Types) [DSo89].

Brunel Test Strategy Planner

A test planner, developed at Brunel University, aids a designer in selecting the optimum strategy for testability design [DIs89]. The planning process uses a set of tools, such as the cost model, the area-overhead estimation and the knowledge base of test methods. The inference engine utilizes these tools in forming a test plan for the circuit under test.

Alvey VLSI-BIST System

A knowledge-based system designed for incorporating built-in self-test (BIST) equipment into a VLSI design has been developed under the Alvey VLSI program [Jon87]. The software is written in LOOPS, supporting object-oriented, rule-based and procedural programming.

As an initial step, the system examines the test needs of the subcircuits and identifies which internal test resources already exist and which must be added to the circuit. The principal test resources are registers, which may be converted to other forms with test capabilities, such as linear feedback shift registers and BILBOs. Each test resource is assigned a global weighting which is higher the more blocks can be served by it.

The main step involves proposals for test plans to meet contraints on test time and the maximum area penalty for incorporating BIST equipment into the circuit. The general aim is to find the proper balance between a low BIST overhead and a high degree of concurrent testing.

Silc TESTPERT

A design-for-testability expert system TESTPERT [Fun86] is incorporated in the Silc silicon compiler (Subsection 5.3b). Different testability techniques may be used for different parts of the circuit. The Silc ADFT (Automatic Design For Testability) system has four components:

a. Testability rules are built into Silc's parsing and logic synthesis software.
b. The Testability Evaluator determines the testability of a design, based on design constraints, cost criteria etc., and identifies those parts of the design being synthesized that may be hard to test (testability "bottlenecks").
c. TESTPERT (TEStability exPERT), which is the intelligent component of the Silc ADFT system. The TESTPERT uses the information provided by the Testability Evaluator as well as other pertinent information, such as cost criteria, test environment information, etc. to propose design modifications to improve the testability. to the design.
d. A library of testable structures, which are used to implement the DFT modifications to a circuit.

The designer provides a set of requirements on such attributes as area overhead and test time. Various testable structures are available to TESTPERT.

7.3 DESIGN FOR TESTABILITY

Effects on fault coverage and logic area, when using specific structures, can be evaluated. Testability enhancements are carried through as early as possible during the design process.

Checking of Design Rules for DFT

DFT (Design for Testability) methodologies must satisfy prescribed design rules. Expert systems have been developed to check VLSI circuits for violations of these rules. To explain this point, let us consider LSSD (see Subsection 7.4b). LSSD enhances the testability of a circuit by virtue of two properties: level sensitivity and scan. Level (as opposed to edge) sensitivity of bistable devices ensures freedom from certain timing defects. The scan property (achieved by connecting all the bistable elements in a long shift chain) ensures complete controllability and observability of the internal state of a circuit. These properties are enforced by a set of guidelines or rules. Some typical design rules are the following:

a. Only one clock must control the clock inputs to the memory elements.
b. A latch can feed another latch if and only if their enables are fed by disjoint primary inputs.
c. The test clock (the clocks employed for shifting) terminals of a test point latch or flip-flop cannot be gated by the data output of another latch.

Note that these rules are relatively independent so that a circuit can be checked for compliance with these rules in any order. These rules have such a character that they lend themselves very well as rules in a knowledge-based system.

Horstmann's Prolog-based expert system [Hor83, Hor84] contains knowledge of DFT techniques and the possible trade-offs among them. Using the DFT rules, the system carries out three tasks:

1. It checks the circuit for any LSSD DFT design rule violation.
2. Portions of the circuit that violate rules are modified so as to remove the DFT violation.
3. The system generates controllability and observability information that can be used by an automatic test-pattern generator.

The system is implemented in Prolog and uses the meta-Prolog system as its inference engine. The DFT rules are represented by a set of Prolog clauses. Each rule is given to the DEMO theorem prover available in the meta-Prolog system. DEMO takes any given goal and proves it true or false against a given set of Prolog facts and rules. During the process of checking a rule, the theorem prover generates a search tree, which is used to determine what changes, if any, need to be made to the ciruit description in order to satisfy the rule. These changes can be made either automatically by the system, with nominal assistance from the user, or manually by the user. In case the DFT rule check turns out to be successful, the results of the rule checking are used to derive certain control sequence information for an automatic test-pattern generator.

PROSPECT (PROduction System for Partitioning and Evaluating Chip Testability) is a rule-based system for DFT design with scan-path techniques [Lat86]. It addresses two problems: verifying DFT guidelines for several variations of the scan-path methodology, and providing a partitioning for automatic test-pattern generation of scan-path designs. DFT guideline verification consists of
a. Identifying the DFT constructs, such as scan-path elements.
b. Identifying associated constructs, such as scan-path control logic.
c. Verifying the proper interconnection and use of those constructs.
The partitioning process for test generation must recognize input and output pins of scan-path elements and the functions of the DFT constructs so that only the testable logic is partitioned. PROSPECT is implemented in OPS83 and runs on a DEC VAX 8600.

Son [Son85] of GenRad described a system for testability rule checking and test generation as a part of the HITEST system. Design rules for off-line testability in a scan-design scheme are specified either topologically or semantically in a frame-based system. Thus, rule checking can be performed in either case. The rule-based system evaluates a design with respect to design rules for testability, edits the region violating testability criteria and then generates a set of test patterns for the constrained design. To perform these tasks, the system contains three major components: a testability checker (which evaluates design rules by checking any violation of them for a predefined testability method), a circuit editor (to transform the portion in violation to meet design requirements for the testability criteria), and a test generator (for producing an appropriate set of test patterns under the constraints imposed by the design rules). The combinational test generator of HITEST is used to generate the test patterns.

The rule-based checker, described by Koseko et al. [Kos90], divides the checking process into symbolic simulation and violation detection. Not only the testability rules applied in the violation detection, but also circuit primitive operation rules applied in the symbolic simulator are treated as rule bases. Moreover, circuit primitives and symbolic signal values described in the rules can also be predefined by testability-rule engineers.

Bidjan-Irani et al. [Bid87] developed a Prolog-based system which checks register-transfer descriptions written in CAP/DSDL for compliance with a set of user-specified rules. Rules related to DFT methodologies are expressed in first-order predicate calculus and automatically translated in clausal form.

ESTA (Expert System for Testability Automation) is an intelligent CAD tool for verifying VLSI designs and checking their compliance with well-established DFT techniques, such as LSSD and BILBO [Cam88]. The knowledge base contains LSSD and BILBO methodologies as well as hardware representations. Knowledge is represented in production rules and frames. Hardware devices are described in the Prolog-based language ProTest [Cam87]. The domain-

7.3 DESIGN FOR TESTABILITY

independent reasoning mechanism controls frame activation and rule firing. The user interface has powerful debugging facilities, which assist the user in locating DFT errors and in correcting them.

Design for Testability

Mangir [Man83] suggested to use a Prolog-based expert system to generate testable VLSI designs. The knowledge base must contain information about built-in self-test techniques, the implementation cost, and the applicability at system, board and chip level.

EXTEST is a rule-based tool for designing testable digital circuits at different levels (layout, circuit and logic level) [Vie87]. Design rules are applied to improve the testability of the circuits. Besides the stuck-type fault model, other models, such as stuck-open faults, are considered. Bridging and stuck-open faults are converted into easily testable faults. A new interface enables human experts to modify and improve the rule base [Mat89]. EXTEST is implemented in C and runs on VMS under UNIX.

DFT Expert, developed at the Indian Institute of Technology, Kharagpur, is an expert system which provides testability for both PLAs and random-logic circuits [BHa89]. This system performs the following tasks:
a. It accepts the RTL description of the VLSI system as a network of modules (such as arithmetic-logic units, multiplexers, buses) and identifier classes of modules (data processors and data transporters).
b. It selects a test method (e.g., for testing PLA circuits or random logic).
c. It configures the DFT structure into a circuit so that the selected method can be implemented. The subtasks in this step include identifying the structure, minimizing the logic-area overhead and threading the scan paths.
d. It generates test schedules for the data processors.

Current work on DFT Expert includes the incorporation of knowledge for RAM and ROM design for testability.

Furuya *et al.* [Fur90] discussed a knowledge-based system that realizes the automatic embedding of DFT techniques in VLSI circuits. DFT techniques as well as circuit structures are represented by a structured database, and the rules to be applied are represented by a rule base. The objective is to minimize the cost function. Exhaustive testing, locally exhaustive testing and random testing have been implemented for combinatinal circuits. Universal testing, memory random testing and built-in testing have been implemented for PLAs. The language used for constructing the system is OPS83.

Synthesis of Fault-Tolerant Systems

DEFT (DEsign For Testability system) is an expert system which uses DFT knowledge to modify circuits designed by the IBM MVISA Design Automation System into more easily testable circuits [Sam86]. The modifications generally consist of the addition of hardware to facilitate testing. The circuit function of the modified circuit remains unchanged. The DEFT system uses a Prolog-based knowledge representation scheme. The Prolog interpreter is written in Franz LISP and incorporates a special demon mechanism allowing it to interface with LISP. After DEFT has presented the user with a list of the DFT-related modifications that were made to the circuit, the user may request the system for an explanation of the reasoning steps underlying the suggested changes [SAm86]. A mixture of text and graphics is used to explain its behavior.

The automatic generation of fault-tolerant chips is facilitated by using intelligent tools. An important consideration is the optimization of design parameters, such as area, delay and reliability. Optimization with multiple and conflicting objectives can be achieved by using simulated annealing. This approach has been implemented in the synthesis system SWIFT (Silicon compiler With Inclusion of Fault Tolerance), developed at the University of Southwestern Louisiana [Sab89]. Algorithmic and rule-based approaches are used in the process of inserting BILBO blocks.

A block schematic of the SWIFT system is shown in Fig. 7.19.

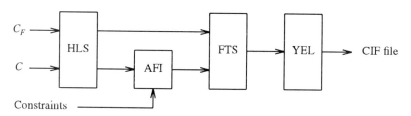

Fig. 7.19 The SWIFT system

The high-level synthesis tool HLS accepts C_F, a C program with additional constructs for specifying fault tolerance and generates an optimized data path and control. When HLS accepts an ordinary C program, the fault tolerance is included by using AFI, an autmatic fault-inclusion tool. AFI consists of a domain expert interacting with a fault-tolerant technique selection and inclusion algorithm that uses simulated annealing. FTS (Fault-Tolerant System) synthesizes the design by using a set of prefedined fault-tolerant building blocks. The YEL (Yield Enhancement Layout) tool performs the placement and routing phase and generates a CIF file.

Chapter 8
LAYOUT DESIGN AND EXPERT SYSTEMS

8.1 PLACEMENT AND ROUTING

8.1a Partitioning and Placement

Partitioning

System partitioning implies the assignment of sets of gates to particular functional modules, while *placement* is the assigment of locations to these modules on the chip surface in a way that satisfies certain optimization criteria connected with the total wire length and the chip area occupied. *Layout partitioning* is the process of dividing the layout into a number of smaller, manageable layouts so that one or more objectives are satisfied. Layout partitioning may be based on the same functional principle as adopted in the design, simulation and testing stages of the design cycle. However, there are schemes which have the special purpose of being applied to layout partitioning only. Partitioning, placement and routing are strongly interrelated.

A main objective of partitioning is to minimize system design and maintenance costs and to maximize system performance. It is difficult to realize a partitioning which satisfies all of these objectives. A proper partitioning is subject to many constraints, including power limitations, signal-propagation delay and the testability of chips. To help choose a proper partitioning, the following general guidelines may be observed:

a. Minimize the number of chips and the number of chip types.
b. Minimize the number of interconnections between chips.
c. Minimize the occupied chip area needed.
d. Optimize the system testability and maintainability.
e. Minimize the package costs.

Some guidelines are mutually exclusive. For example, a smaller number of partitions would reduce the capability of fault location and detection and hamper standardization. Judicious trade-offs or compromises between these guidelines must be made to meet the partitioning objectives. The maximum permissible chip size defines roughly the total number of chips required.

A limiting factor in IC layouts is the number of bonding pads that can be placed on a single chip. These pads occupy a relatively large area (typically 130 times the area occupied by a MOS transistor) and are subjected to damage during the bonding process. Therefore, the number of connections between chips and the number of bonding pads must be minimized. The number of bonding pads depends on the package chosen to house the chip. Common packages provide 16, 24, 48 or more I/O pins.

Due to the complexity of integrated circuits, partitioning algorithms are necessarily based on heuristic rationale. Algorithms developed for partitioning are based on the constructive principle or the exchange principle.

The Placement Problem

Placement in integrated-circuit layout design is the assignment of each circuit primitive (gate, cell, circuit block, etc.) and each bonding pad to a location in such a manner that the routability of the circuit layout is optimized. *Routability* is the ease with which the routing of the interconnections between the circuit primitives can be effectuated. Routability is an intangible goal since it is highly dependent on a number of factors. Figures-of-merit which are being used as measures for routability include:
a. The total length of all interconnection wires.
b. The number of signal cuts.
c. The chip area occupied by the layout.

Signal cuts will be discussed later. Usually, figure-of-merit a. is employed, giving rise to the following *placement problem*: find the optimal placement of circuit modules and bonding pads by minimizing the total interconnection length without violating any design rules. When some nets are more critical than others, a weight may be assigned to each net so that the weighted total interconnection length is considered.

In most cases, interconnections are restricted to horizontal and vertical segments, which usually are called *orthogonal* or *Manhattan connections*. The *Manhattan distance* between two points (x_1, y_1) and (x_2, y_2) in the plane is given by $d = | x_1 - x_2 | + | y_1 - y_2 |$. The length of a *net* (equipotential set of interconnections) is frequently defined as half the perimeter of the smallest rectangle which circumscibes the net's terminals. This approximation is the lower bound equal to the Steiner minimal tree in orthogonal routing when each net can be routed without being blocked by obstacles.

The choice of minimizing the total wire length as a placement criterion is motivated by the assumption that some detrimental effects, such as signal delay, capacitance loading and crosstalk, are simultaneously reduced. Other criteria which have been used include the minimization of the number of corner-turns and the number of wire crossovers.

Heuristic Placement Procedures

Several heuristic techniques have been proposed for the placement of components and cells. Frequently, a placement procedure consists of a constructive procedure to produce an initial placement, followed by an iterative procedure to improve the initial placement. A *constructive* method operates

8.1 PLACEMENT AND ROUTING

iteratively on the subset of unplaced modules and selects one of them to adjoin the subset of already placed modules. In an *iterative-improvement* method, modules are repositioned in some systematic way so as to improve the placement.

A constructive and initial-placement procedure as implemented in the LTX standard-cell layout design system [Per77] is outlined below. The cell layout consists of a number of rows of standard cells. The constructive placement algorithm places unplaced cells in two steps: we first select the next cell to be placed and then the block in which it will be placed. The decision as to which cell is to be connected next is based on the *inside-outside connectivity (IOC)*, which is computed for all unplaced cells. Suppose that an unplaced cell has p nets which are to be connected to already placed cells and q nets which are to be connected to unplaced cells. Then, $IOC = p - q$. The cell with the highest IOC is selected as the next cell to be placed. This cell is assigned a trial placement in each block with a sufficient capacity remaining. An initial placement consists of placing the cell in the block that results in the minimal total wire length. In case of a tie, the block with the highest remaining capacity is chosen.

The result obtained in a procedure for constructive and initial placement of cells in a number of blocks depends strongly on which cell is placed first. When an initial placement has been completed, iterative trial exchanges of cells may improve the initial placement. The placement algorithm may consist of a repeated sequence of initial placements and iterative improvements to achieve an optimal solution.

Another widely used heuristic procedure is *min-cut placement*. When a cut line c_1 divides a chip into two blocks B_1 and B_2, as shown in Fig. 8.1a, signal lines connecting modules in one block to modules in the other block will cross the cut line at points, called *signal cuts*. The total number of these signal cuts is a measure of the density of the interconnecting signal lines. In min-cut placement, the number of signal cuts is used as the minimizing criterion instead of the total interconnection length [Bre77]. This number can be changed by moving a module from B_1 to B_2, or vice versa.

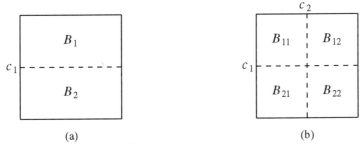

Fig. 8.1 Principle of min-cut placement

Min-cut placement is a global approach with the objective of minimizing the number of signal cuts in the following way. After the cut line c_1 has been introduced, the number of signal cuts is minimized by moving modules one at a time from one block to the other. This procedure is repeated by introducing a second cut line c_2 (see Fig. 8.1b), which divides block B_1 into B_{11} and B_{12}, and block B_2 into B_{21} and B_{22}, after which the number of signal cuts in c_2 is minimized. The procedure is continued until each block contains just one module, or a satisfactory solution has been obtained. Min-cut placement, which is an extension of the Kernighan-Lin partitioning algorithm, can be used to advantage in macroblock structures [Lau79].

In a cell-based design, a mirror image orientation of adjacent cells may yield a reduced routing length.

Example: Figure 8.2a shows a part of a standard-cell layout: two adjacent cells P and Q with its neighboring wire layout.

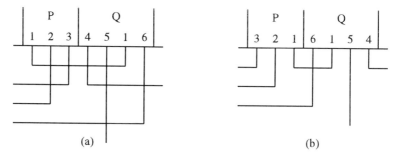

Fig. 8.2 Effect of mirroring standard cells

Figure 8.2b shows the situation when both cells *P* and *Q* are mirrored. It is seen that a smaller area is occupied by the wire layout.

A review of current placement techniques was given by Preas and Karger [Pre86].

8.1b Global Routing

The Routing Problem

Once the placement of all circuit and system primitives (transistors, gates, cells or blocks) has been completed, the next step in the layout-design cycle is *interconnection routing*. When two or more pins on the chip layout plane must be interconnected, conducting wires must be laid in one or more interconnection layers so that the pins have the same electrical potential. An equipotential set of

8.1 PLACEMENT AND ROUTING

conducting wires which connect two or more pins on a chip is called a *net*.

When all nets in a circuit consist of straight-line segments, which are restricted to be horizontal or vertical, the net structure is referred to as a *Manhattan structure*. Two Manhattan solutions for a net which connects four fixed pins are shown in Fig. 8.3. It is assumed that one interconnection layer is available.

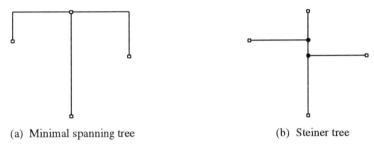

(a) Minimal spanning tree   (b) Steiner tree

Fig. 8.3 Manhattan structures

A solution with a minimal total length of line segments is provided by the *minimal spanning tree* of Fig. 1a. When additional nodes are allowed, the optimal solution is a *Steiner tree*. Steiner trees can also be constructed for Manhattan structures as in Fig. 8.3b.

The *routing problem* can be formulated as follows: Given the components of a logic circuit or system and the interconnection list, i.e., a list of which sets of pins have to be interconnected. The task is to lay out the nets which connect the pins according to the interconnection list such that some optimizing criterion (e.g., a minimal total interconnection length) is satisfied.

Many algorithms have been proposed for interconnection routing. The layout structure is an important factor affecting the kind of wire routing in ICs. An irregular layout structure of a bipolar circuit usually allows for only one interconnection layer with crossunders, where necessary. *General routers* are suitable for such structures. This subsection discusses the general routing problem in which parts of the area between pins to be connected may be forbidden regions, e.g., because of the presence of predefined circuit blocks.

When the pins to be interconnected are restricted to a small area, for example, a routing channel in a standard-cell structure, the routing is called *local routing*. This subject will be treated in the next subsection. In ASIC design, the routing is directly related to the standard layout structures, such as the standard-cell structure, gate array or PLA.

Two general routers which find the shortest Manhattan path between two points in a plane in the presence of obstacles are described below.

Lee's Routing Algorithm

Lee's *wave-front* or *maze-running router* [Lee61] is a general router based on a grid structure. The algorithm will be illustrated by an example.

Example: The goal is to find the shortest Manhattan path between the starting square S and the target square T of the grid structure in Fig. 8.4. The white region represents obstacles and should be avoided.

10	9	8	7	6	5	4	3	4	5	6	7	8	9	10	11	12	13	14
9	8	7	6	5	4	3	2	3	4	5	6	7	8	9	10	11	12	13
8	7	6	5	4	3	2	1	2					9	10	11	12	13	14
7	6	5	4	3	2	1	S	1					10	11	12	13	14	15
8	7	6	5	4	3	2	1	2					11	12	13	14	15	16
9	8															15	16	17
10	9															16	17	18
11	10															17	18	19
12	11	12	13	14	15	16	17	18				21	20	19	18	19	20
13	12	13	14	15	16	17	18	19				T	20	19	20	21
14	13	14	15	16	17	18	19	20					21	20	21	

Fig. 8.4 Example of Lee's routing

Lee's algorithm is based on the generation of successive wave fronts expanding from S to T. First, all unoccupied squares immediately adjacent to S are labeled 1. The squares next to these are labeled 2, and so on, until the target square T is reached. Then, a path is traced back from T to S by always selecting squares labeled one less than the current square as indicated by arrows in Fig. 8.4. This algorithm guarantees the establishment of a connection path if one exists.

To save computer storage and running time, many speed-up techniques have been proposed. Some of these techniques employ methods that avoid labeling of irrelevant squares, e.g., by framing the working area within a rectangle surrounding the two points. A saving in computer storage is further achieved by using an efficient coding, i.e., by using alternately 1 and 2 instead of the ordinal numbering considered above, restricting the number of bits to be stored for each square to two regardless of the path length measured in the number of squares from S to T [Ake67]. Other versions of Lee's algorithm have been published [Rub74, Hoe76]. In the case of unsuccessful connections, the necessary completion by hand often requires the removal of already found connections.

8.1 PLACEMENT AND ROUTING

Hightower's Routing Algorithm

In contrast to Lee's algorithm for determining a Manhattan path between two points, Hightower's algorithm is a *line-search technique* [Hig69]. There is no guarantee that the shortest path will be found or even that a path will be found at all. However, Hightower's algorithm has important advantages:

a. It is extremely fast; it tries to find a path with the minimum number of bends rather than the shortest paths.
b. Its accuracy is only determined by the ability of a computer to store small numbers.
c. It is applicable to a surface whose dimensions correspond to the largest number that the computer can handle.

Figure 8.5 illustrates the principle of Hightower's method for finding a path between two points S and T.

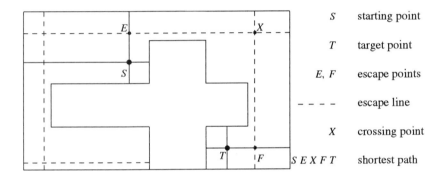

Fig. 8.5 Principle of Hightower's algorithm

First, two perpendicular line segments are drawn through S extending in both directions from boundary to boundary. Then, we attempt to find an escape point from which we draw an escape line extending beyond the previous boundaries of S. When such an escape point has been found, it becomes the new point S'. The same process is started from T. The algorithm restarts alternately from S and T until ultimately a line segment which is connected to S crosses a line segment connected to T. In this way, a connection path is constructed which consists of all line segments which run from S through escape points to the crossing point and from this point through escape points to T.

The speed of Hightower's algorithm is obvious since the basic operations consist of constructing lines through given points until a boundary is reached on both sides. A suitable data structure must be used to minimize the number of searches. A modification to Hightower's algorithm provides a guaranteed solution for general routing problems [Hey80].

Routability

In IC-layout design, it is important to be able to predict the routability of the interconnections and the space needed for efficient wiring. Agrawal [Agr76] evaluated the probability of success of a Lee's routing procedure. Patel *et al.* [Pat84] described TEWAS (Theoretical and Experimental Wirability Analysis System) for predicting wiring-space requirements of VLSI chips. Logic generation, placement and approximate routing procedures can be used to further refine the prediction estimate.

Global Routing

Placement and routing are interrelated. Placement involves the assignment of topographical positions to the modules in such a way that routability is optimized. The regions in the plane which are not occupied by any module constitute the routing area, which is fully available for the interconnecting wires. With modules restricted to the rectangular form, the routing area can be partitioned into a set of rectangular regions, called routing channels. To alleviate the intricacy of the placement-routing interdependence, an intermediate phase, called *global routing*, is inserted between placement and detailed routing.

Global routing is concerned with topological aspects and global allocations of interconnection nets to the different routing channels. Global routing attempts to avoid the need for excessive channel widths or congestions in certain channels. Only when the global-routing phase has been completed, can the individual channels be routed one by one. Global routing is essential in both block structures and gate arrays. In standard-cell structures, various procedures, such as clustering and linear placement, may be viewed as prewiring activities similar to global routing with the objective of improving the routability.

Global routing has the primary objective to facilitate the subsequent phase of detailed channel routing. In the majority of channel routers, wire segments must be aligned on a regular grid of horizontal and vertical lines. This restriction implies a uniform width of all wires, which is neither realistic nor efficient in terms of area occupation. An improved solution is provided by a gridless router which uses a set of contours instead of the grid as the framework [Gro91].

Another routing problem is how to determine the order in which the different channels are to be wired [Sch87]. An ideal situation is achieved, when full-custom modules are interconnected without any routing at all, that is, neighboring modules are aligned in such a way that pins to be interconnected abut. This case is discussed in Subsection 8.2b.

Optimization by Simulated Annealing

Layout design, like virtually all designs, involves optimizations in some form or other. The following describes a recently developed optimization procedure which is applicable to many problems in layout design. By making a connection between statistical mechanics and multivariate optimization, Kirkpatrick *et al.*

8.1 PLACEMENT AND ROUTING

[Kir83] introduced a general-purpose method of optimization, called *simulated annealing*, which can successfully be applied to problems with an objective function involving conflicting goals and a great many degrees of freedom. Finding the global minimum of such an optimization problem is assumed to be NP-complete.

The heuristic method of simulated annealing has a paradigm in the annealing process of a hypothetical fluid which is brought into a low-energy state, such as in growing a large single crystal. Suppose that at some high temperature a lump of material is in a liquid form. The most effective way to bring it into the lowest energy state is careful annealing by lowering the temperature slowly, spending a relatively long time at temperatures near the freezing point to allow defects to anneal out of the growing crystal, and then cooling the crystal more rapidly to bring the atoms to rest. Although the analogy between optimization and cooling of a fluid is not perfect in all respects (e.g., ideal fluids contain equal atoms, whereas many interesting optimization problems involve many distinct, non-interchangeable elements), optimization by simulated annealing follows a sequence similar to that outlined above. A pseudo-temperature is used to function as a control parameter in the same units as the objective function.

Optimization by simulated annealing requires four ingredients [Kir83]:
1. A concise description of a system configuration.
2. A random generator, which provides random changes in the elements of the configuration.
3. An objective function, which contains the trade-offs that have to be made.
4. An annealing schedule of the successive temperatures and the length of time for which each temperature is to be held constant.

Simulated annealing is a modification of the method of iterative improvements in that, in addition to the downhill moves inherent in conventional optimization procedures, there is a possibility of permitting uphill moves. The latter is important, since it enhances the probability of finding the global minimum of the objective function.

Simulated annealing has been applied to, among other things, the partitioning problem [Kir83], module placement or floorplan design [Kir83, Cas87, Sia87, Vai87, Sec88, Dur89], logic optimization [Fle85, Lam86], PLA folding [Moo85, Won87, Won88], PLA design for testability [Lig86], and the wire-routing problem [Kir83]. For example, a design goal is to minimize the silicon area required for implementing given logic functions and, at the same time, to achieve a layout that meets the requirement of wirability.

Since simulated annealing is slow in solving combinatorial problems, such as VLSI placement, many researchers have attempted to speed up the process by using parallel algorithms. Durand [Dur89] argued that temporary errors in cost-function evaluations have a weak effect on the convergence of simulated annealing, whereas the cumulative errors must be minimized.

8.2 LAYOUT DESIGN OF BLOCK STRUCTURES

8.2a Block Structures and Symbolic Layout

Block Structures

The hierarchical layout structure, schematically shown in Fig. 8.6, is the most general structure for very complex networks. The complete layout may consist of blocks of standard gates, blocks with regular layout structures (gate array, ROM, PLA), blocks with irregular structures and even separate gates.

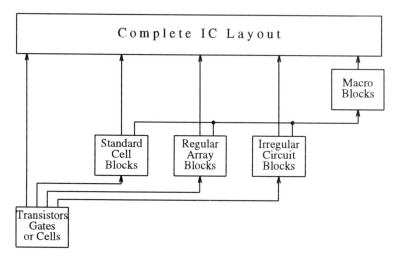

Fig. 8.6 Hierarchical block structure of a complex system

Suppose that a complex system consists of a number of functional blocks, whose layouts have been carefully designed in an earlier stage according to one of the methods mentioned before. For the case of convenience, let us assume that the blocks have a rectangular form with known dimensions. The layout problem is then to place the blocks on a plane in such a way that the chip area is minimal. We should take into account that the wire interconnections still have to be laid.

For the optimal placement of rectangular blocks use can be made of a *block graph*. Figure 8.7a shows an example of the optimal placement of four blocks A, B, C and D. The layout of the blocks can be described by using a pair of mutually dual block graphs: the *horizontal block graph* $G_x = (V_x, E_x)$ and the *vertical block graph* $G_y = (V_y, E_y)$.

If the block structure is planar, i.e., if there are no overlaps, each of the block graphs will be planar, acyclic and directed from left to right or from top to bottom. The graphs G_x and G_y define the relative positioning of the blocks with respect to one another. A rectangle R is described by a pair of edges $\{x_E, y_E\}$

8.2 LAYOUT DESIGN OF BLOCK STRUCTURES

with edge weights which correspond to the horizontal (or vertical) dimension of the rectangle. Each vertex V_x (or V_y) corresponds to a vertical (or horizontal) boundary between blocks. The above graph representation of a block structure stems originally from the technique of dissecting a rectangle, first described by Tutte *et al.*

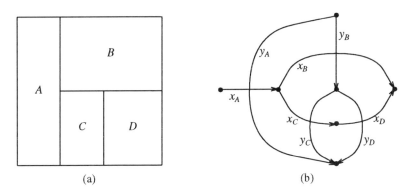

Fig. 8.7 Dissected rectangle and block graph

The block layouts considered above and exemplified by Fig. 8.7 represent an ideal case in which the number of rectangular blocks collectively occupy a minimal rectangle without leaving any area unused. When an arbitrary number of rectangular blocks is given and a planar arrangement of blocks with minimal area occupation is to be found, unused area between the blocks will in general arise. So far, the rectangles are assumed to represent functional blocks of a chip. In circuit-layout problems, routing channels between the blocks are needed to allow the wires to be routed. In the above, the routing channels are indicated by the vertices of the block graphs. Since a channel area is as important as a block area, we may define a *block-channel graph* which contains an edge for each block as well as for each routing channel in the block-graph assembly.

The *general block-placement problem* can be formulated as follows: Given a number of rectangular functional blocks which must be interconnected, find a placement of these blocks of prescribed fixed dimensions along with the necessary wire-routing blocks with flexible dimensions such that a minimum (preferably square) layout area is occupied. Clearly, block placement and wire routing are strongly interrelated.

Symbolic Layouts
An approach to layout design may involve partitioning into two sequential steps:
a. The production of a rough layout of a transistor circuit ensuring the correct relative placement of the layout primitives according to the given circuit topology.

b. The manipulation of the layout primitives such that a minimum chip area is utilized.

Since the objective of the second step is the reduction of the total chip area, this step is referred to as *layout compaction*. When the compaction procedure is performed in such a way that the layout-design rules are satisfied, exhaustive design-rule checking in later phases is unnecessary. This is a great advantage since design-rule checking is a time-consuming activity.

When the layout primitive or coherent sets of primitives are represented by *symbols* rather than by their actual geometrical shapes, the layout is said to be a *symbolic layout*. A symbolic layout is independent of the target technology. Transistors, via contacts and wires can be represented by specific symbols. The primary aim of a symbolic layout is to have a conceptually clear presentation of the topology of the circuit by using simple, recognizable symbols for the layout primitives. There is an obvious reason for using a symbolic layout. It provides an intermediate step in the process of converting the circuit diagram to the ultimate mask artwork. The symbolic layout markedly resembles the given circuit diagram. As a result, in contrast to the geometric mask artwork, the circuit connectivity can easily be verified so that the occurrence of errors can be eliminated in an early stage. When modifications of the layout are necessary, it is far easier to manipulate with a symbolic layout than with the geometrical layout at the various mask levels. An optimal initial layout is extremely important, since it directly affects the ultimate compaction capability. The translation of the symbolic layout to geometrical patterns is an automatic process which can be entrusted to the computer. MOS technology lends itself very well to symbolic layouts.

Layout Compaction

A non-optimal placement of the functional blocks entails wasted areas which are not occupied by blocks of interconnection wires. This problem can be handled by manipulating the edge weights in the block-channel graph. Optimization of the block placement amounts to minimizing the wasted areas. The maximum length and width of a block assembly is determined by the longest paths of the graphs G_x and G_y. The total block area is defined as the area of the smallest rectangle which encloses the blocks and routing channels. This is equal to the product of the longest paths of G_x and G_y. Hence, in terms of the graph models, optimization of the block placement is reduced to the minimization of the longest paths in G_x and G_y.

Example: Figure 8.8a depicts an initial floorplan of five modules A, B, C, D and E with connections from A to B and D, from B to A and C, from C to B and D, and from E to all the other modules. A floor plan is desired with a rectangular bounding box whose length-width ratio approaches unity. After shifting and rotating suitable blocks, the ultimate result is the floor plan given in Fig. 8.8b.

8.2 LAYOUT DESIGN OF BLOCK STRUCTURES

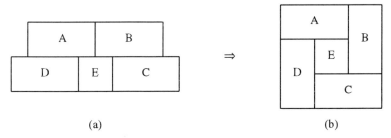

Fig. 8.8 Optimizing the block placement

A classical layout-compaction method involves the search for horizontal and vertical strips of unused area which extend from side to side. Such strips are called *compression ridges*. An example is shown in Fig. 8.9a.

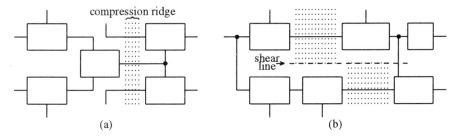

Fig. 8.9 Compaction by removing compression ridges

When compression ridges are removed, a compacted layout is obtained. A compression ridge may consist of two or more partial strips which are shifted with respect to each other along a shear line, as shown in Fig. 8.9b. In this case, parts of areas are succesively shifted to remove the compression ridge. This compaction method is therefore often referred to as the *shearing method*. Care should be taken to satisfy design rules in all stages of the compaction process.

MULGA [Wes81] allows layout compaction to be performed. The layout primitives (leaf cells) are placed on a *virtual grid*, which is an array of dimensionless grid lines. The virtual grid establishes a relative placement of the primitives. The actual physical placements and the geometrical dimensions of the circuit elements are determined during the compaction process. Compaction in MULGA is performed by contracting (or expanding) the grid in a non-uniform manner. Dead spaces (unused areas) can be removed as in the shear-line compaction method in both the X direction and the Y direction. A review of symbolic layout-compaction methods was given by Boyer [Boy88].

VIVID
The VIVID (Vertically Integrated VLSI Design) system is a symbolic layout

design system based on a virtual-grid methodology [Rog85]. It reduces design time needed to obtain a working custom VLSI circuit. Unique features of VIVID are:

a. Technology independence over a wide range of MOS types, including CMOS and NMOS.
b. An open architecture that simplifies both the use of existing tools and the incorporation of new tools. Fast layout debugging by using a symbolic-level circuit extractor and an interactive circuit simulator.
c. The translation of the symbols in the symbolic layout into the geometric layout is made flexible by associating with the symbols a number of parameters which represent a broad range of structures.
d. Fully automated mask generation and automated chip assembly by a hierarchical and a leaf-cell compactor ensuring correctness by construction.

8.2b Heuristic Approaches to Floorplanning

Rectangular Duals of Planar Graphs

Suppose that a system is to be designed by interconnecting a number of predesigned functional modules. Let the layouts of the modules be given as rectangular layout blocks. An important design criterion is to arrange the modules in such a way that the total area occupied by the modules and the wire interconnection regions is minimized.

Since the optimized modules have prefixed areas, the problem can be reduced by minimizing the total area occupied by the wire interconnections. The optimal solution is achieved, when the final layout of the complete system is a dissected rectangle whose dissected pieces are formed by the rectangular functional modules and virtually no area is occupied by interconnections. This solution is only possible when the modules are placed in such a way that interconnections are needed exclusively between neighboring modules. Such an optimal solution is more easily obtained when the length-width ratios and the pin positions of the functional blocks are allowed to be variable. Layout design using such parameterized blocks is called *floorplanning*.

Example: A simple example is the six-module system shown in Fig. 8.10a. The interconnective pattern is given by the *connectivity graph* (CG) shown in Fig. 8.10b. The nodes in the CG correspond to the individual rectangular modules in Fig. 8.10a. The edges between the nodes in Fig. 8.10b indicate the presence of interconnecting wires between the corresponding neighboring modules. The numbers attached to the edges represent the numbers of interconnecting wires between adjacent blocks as indicated in Fig. 8.10a. Since the CG specifies the adjacency of blocks, the CG is also referred to as the *adjacency graph*.

8.2 LAYOUT DESIGN OF BLOCK STRUCTURES

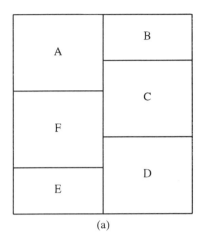
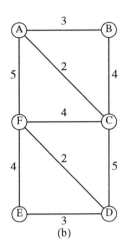

Fig. 8.10 Dissected rectangle and connectivity graph

Let us examine the relationship between the dissected rectangle and the connectivity graph more closely. The dissected rectangle is the block layout which is wanted and whose connectivity graph is given. The dissected rectangle can be considered as the rectangular dual of the connectivity graph if the latter is extended to its 4-completion. The meaning of 4-completion will be explained below.

In a dissected rectangle, four or less than four component rectangles can be considered as corner rectangles. In the example of Fig. 8.10a, the corner rectangles are A, B, D and E. These corner rectangles define four sets of neighboring modules on four sides of the dissected rectangle. In the example of Fig. 8.10a, we have AB on the top side, ED on the bottom side, AFE on the left side and BCD on the right side. Four additional nodes of the CG are introduced, one on each side, connected to the nodes of the corresponding block array, as shown in Fig. 8.11a.

Let us now explain the duality of the CG and its rectangular dual RD. Consider a cube with the dissected rectangular drawn on the front face as shown in Fig. 8.11b. When the top, bottom, left and right faces of the cube are turned over, the six dissected rectangles on the front face are extended by four contiguous rectangles. Each rectangle has a node as its dual. Two nodes of adjacent rectangles are connected by an edge crossing the bordering line shared by the two rectangles, as shown in Fig. 8.11c. The extended connectivity graph thus obtained is called the 4-completion of the connectivity graph. In order to avoid routing problems in + crossings, the nodes in the dissected rectangle are T-crossings. As a consequence, the CG is a triangulated graph, that is, each internal mesh in the CG consists of three edges. Non-overlapping rectangles in the rectangular dual requires the CG to be planar.

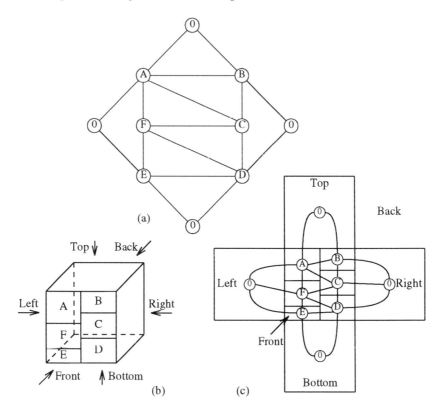

Fig. 8.11 4-completion of a connectivity graph

Theorem: A plane graph G with only triangular meshes has a rectangular dual if and only if there exists a 4-completion of G. The outer cycle contains at least four edges. Internal edges have degree ≥ 4, i.e., there are at least four edges incident to an internal vertex. See, for example, Fig. 8.11b.

The Maling-Mueller-Heller method

Given a number of functional modules (whose layouts are known as rectangular blocks) and their interconnections, let us consider the search of (sub)optimal placements of the modules with minimal area occupied by the blocks. The outer boundary of the assembled module arrangement should take the form of a rectangle. An example, given by Maling *et al.* [Mal82], will be worked out.

The starting point is the Planar Original Graph (POG), which is the functional block diagram shown in Fig. 8.12a. The nodes of the POG represent the functional blocks and the edges represent the interconnections for data and

8.2 LAYOUT DESIGN OF BLOCK STRUCTURES

control signal transfer between the modules. These edges indicate that two neighboring modules are adjacent to one another. The numbers attached to each edge indicate the numbers of interconnecting lines between the modules and hence the length of the bordering line. The numbers within the circle nodes give a measure of the module sizes in area units.

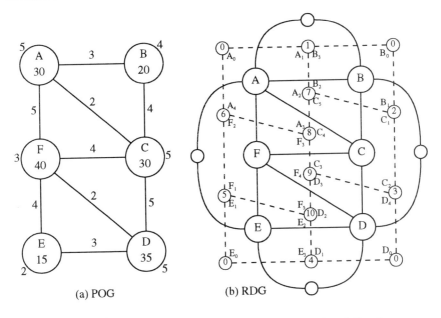

Fig. 8.12 Planar Original Graph and Rectangular Dual Graph

A *Rectangular Dual Graph* (RDG) which can be derived from the POG is shown in Fig. 8.12b by dotted lines. Duality implies that nodes (meshes) in the POG are mapped as meshes (nodes) in the RDG. The objective is to generate a planar module layout based on the RDG with the following properties:

a. The complete layout fits into a rectangle which is dissected into rectangular blocks each representing a functional module.
b. A POG node (module) is mapped into a layout rectangle and a POG edge is mapped into (a part of) a side shared by adjacent rectangles.
c. Corner nodes in the POG are mapped in corner rectangular blocks. This implies the introduction of nodes in the RDG indicated by the number 0.
d. Nodes on the perimeter of each rectangle are labeled by X_i, where X corresponds to the dual node in the POG and i has the subscript 0 if it is a corner node, and further 1, 2, ..., n, where n is the number of nodes on the perimeter.

e. A node in the RPG is a connection point of two edges (when it is a corner node), three edges (at a T crossing) or four edges (at a + crosssing).

Several RDGs may correspond to one POG, differing in alternative orientations of the T crossings. These RDGs are found in three phases:

Phase 1: Setting up layout equations, which are node equations (for summing up the number of right angles at each of the 10 nodes, i.e., 2 or 4) and mesh equations (for enumerating the number of right angles at the nodes within the meshes, i.e., 4). The result for the example being considered is shown in Fig. 8.13.

Angles	$A_0B_0D_0E_0$	$A_1A_2A_3A_4$	$B_1B_2B_3$	$C_1C_2C_3C_4C_5$	$D_1D_2D_3D_4$	$E_1E_2E_3$	$F_1F_2F_3F_4F_5$	
(1)		1 0 0 0	0 0 1					2
(2)		1 0 0		1 0 0 0 0				2
(3)				0 1 0 0 0	0 0 0 1			2
RDG (4)					1 0 0 0	0 0 1		2
node (5)						1 0 0	1 0 0 0 0	2
eqs (6)		0 0 0 1					0 1 0 0 0	2
(7)		0 1 0 0	0 1 0	0 0 0 0 1				2
(8)		0 0 1 0		0 0 0 1 0			0 0 1 0 0	2
(9)				0 0 1 0 0	0 0 1 0		0 0 0 1 0	2
(10)					0 1 0 0	0 1 0	0 0 0 0 1	2
corners	1 1 1 1							4
RDG |A|	1 0 0 0	1 1 1 1						4
mesh |B|	0 1 0 0		1 1 1					4
eqs |C|				1 1 1 1 1				4
|D|	0 0 1 0				1 1 1 1			4
|E|	0 0 0 1					1 1 1		4
|F|							1 1 1 1 1	4

Fig. 8.13 Node and mesh equations

The five corresponding layouts are shown in Fig. 8.15. In addition to the dissected rectangle as depicted in Fig. 8.15a (corresponding to the dashed rectangular dual graph in Fig. 8.12b), four alternative dissected rectangles can be found by properly shifting the T crossing and hence the angle distribution at the nodes 7, 8, 9 and 10 in Fig. 8.12b. This leads to a score of 0, when the angle of a node in tracing a mesh is 180^0. See Fig. 8.13. The mesh equations of the five solutions are collected by the table in Fig. 8.14.

8.2 LAYOUT DESIGN OF BLOCK STRUCTURES

↓	$A_0B_0D_0E_0$	$A_1A_2A_3A_4$	$B_1B_2B_3$	$C_1C_2C_3C_4C_5$	$D_1D_2D_3D_4$	$E_1E_2E_3$	$F_1F_2F_3F_4F_5$
a	1 1 1 1	1 0 1 1	1 1 1	1 1 1 0 1	1 0 1 1	1 1 1	1 1 1 0 1
b	1 1 1 1	1 0 1 1	1 1 1	1 1 0 1 1	1 0 1 1	1 1 1	1 1 0 1 1
c	1 1 1 1	1 1 0 1	1 1 1	1 1 1 1 0	1 0 1 1	1 1 1	1 1 1 0 1
d	1 1 1 1	1 1 0 1	1 1 1	1 1 1 1 0	1 1 0 1	1 1 1	1 1 1 1 0
e	1 1 1 1	1 0 1 1	1 1 1	1 1 1 0 1	1 1 0 1	1 1 1	1 1 1 1 0

Fig. 8.14 Five layout solutions

The first row (solution a) of Fig. 8.14 corresponds to the dissected rectangle given in Fig. 8.15a. The second through fifth rows of Fig. 8.14 correspond to the dissected rectangles given in Fig. 8.15b, c, d and e respectively.

Phase 2: Optimization of perimeter constraints. Phase 1 gives all possible solutions with different relative positions of the individual rectangular modules. The objective of Phase 2 is to minimize the sum of the areas of the individual modules subject to two sets of constraints (see Fig. 8.15f which corresponds to the dissected rectangle shown in Fig. 8.15e). Since the complete layout must have a rectangular form, the sums of the lengths of subblocks in horizontal or vertical directions must be equal. This leads to the first set of constraints in the form of equality equations (see Fig. 8.15f):

$$x_A + x_B = x_A + x_C = x_F + x_C = x_E + x_D$$
$$y_A + y_F + y_E = y_A + y_F + y_D = y_B + y_C + y_D$$

The second set of constraints comprises lower bounds of individual areas as specified by the POG in Fig. 8.12a.

The optimization problem can be formulated as the minimization of

$$x_A y_A + x_B y_B + x_C y_C + x_D y_D + x_E y_E + x_F y_F$$

under the two sets of constraints given above. To alleviate the solving process, this nonlinear problem is reduced to two sets of linear equations which can be solved by the Simplex method [Sch87]:

Minimize $x_A + x_B$ subject to the constraints.

$$
\begin{aligned}
x_A - x_B & = 0 \\
x_A + x_B - x_C \quad\quad - x_F & = 0 \\
x_A + x_B \quad - x_D - x_E & = 0 \\
x_A & \geq 5 \\
x_B & \geq 4 \\
x_C & \geq 4 \\
x_D & \geq 5 \\
x_E & \geq 2 \\
x_F & \geq 5 \\
-x_C \quad\quad + x_F & \geq 1
\end{aligned}
$$

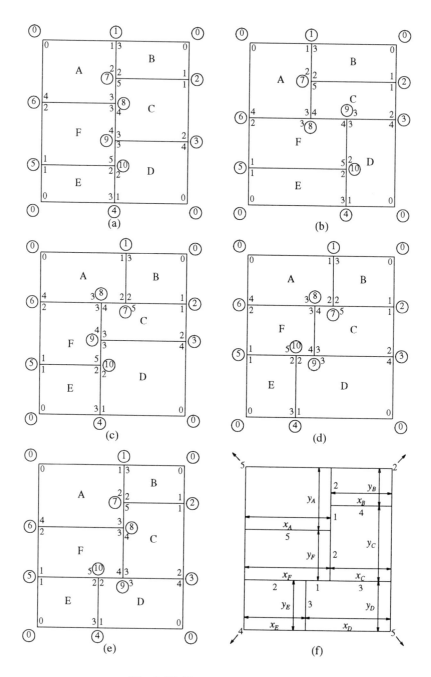

Fig. 8.15 Five different layouts

8.2 LAYOUT DESIGN OF BLOCK STRUCTURES

Minimize $y_A + y_E + y_F$ subject to the constraints

$$
\begin{aligned}
y_D - y_E &= 0 \\
y_A - y_B - y_C + y_F &= 0 \\
y_A &\geq 5 \\
y_B &\geq 4 \\
y_C &\geq 5 \\
y_D &\geq 5 \\
y_E &\geq 5 \\
y_F &\geq 2 \\
y_A - y_B &\geq 1
\end{aligned}
$$

The lower bounds in the above inequalities are in accordance with the border lengths of neighboring modules and indicated in Fig. 8.12a by numbers beside the circular nodes.

Phase 3. The Simplex program may produce several solutions. If area constraints are satisfied, the solution is accepted. A solution may be as given in Fig. 8.16.

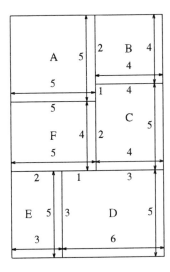

Fig. 8.16 Final solution

Functional modules in floorplanning can be parameterized allowing appropriate values to be assigned to performance parameters, in addition to variable geometric dimensions.

8.2c Expert Systems for Layout Design

Introduction

Designers of CADCAS systems are faced with the problem of how to control the increasing complexity of VLSI circuits. The solution lies in the hierarchical top-down approach which has several advantages:

a. The computational complexity is reduced by avoiding the combinatoric explosion of increasing number of alternative solutions.
b. Design on a level-by-level basis has become computationally manageable.
c. Functional submodules, designed earlier and stored in the database, can be recognized and reused in new designs.

AMBER

Any chip design is aimed at attaining a reasonable production yield. An important parameter in layout design is the area required to implement a system on a chip. The chip area is related to the yield and manufacturing costs. It is therefore expedient to let the layout design phase precede by a reliable estimation of the expected chip area.

AMBER [How86] is a knowledge-based system developed to produce area estimates of chip layouts. The knowledge base contains the information required to perform effective area estimation. Several experts are available to perform specific subtasks. A module-generator expert produces area estimates of function blocks. A library expert manipulates knowledge of existing layouts and their restrictions. A module generator expert produces area estimates of function blocks.

AMBER permits the designer to decompose the circuit into blocks whose areas can be obtained from the knowledge base or estimated by the designer. With the aid of the recomposition expert, a placement of the blocks is attempted with a minimum routing area. Aspect ratio and area estimations are performed for different floorplans. An approximation expert performs a set of transformations based on heuristics acquired from interviews with human experts. A controller expert functions as a guide in the problem-solving process.

PARAID, a Design Partitioning Tool

PARAID (PARtitioning AID) is an interactive design tool to assist designers in partitioning their designs [GUp88]. It is also useful when existing designs must be implemented in a new technology. PARAID produces an acceptable register-transfer-level block structural description of each implementable partition, given an RT-level description of a digital system. The knowledge base contains partitioning knowledge which can be used to guide the process of exploring the partitioning solution space.

AI-Based Placement

An AI-based cell placement technique, developed by Ho and Cha of the

8.2 LAYOUT DESIGN OF BLOCK STRUCTURES

University of Southern California [Ho89], integrates the principle of consistent labeling, abstraction level planning and meta-level control. The experimental results indicate a significant improvement in placement quality over the pair-linking and min-cut placement methods.

Knowledge Acquisition Tool
KAT (Knowledge Acquisition Tool) acquires knowledge of placement strategies for the layout of printed wiring boards [Fuj89]. Special consideration is given to bus routing. KAT's main task is the acquisition of clustering and placement rules. The acquisition is performed in two consecutive steps:
a. Rule extraction, in which rule candidates acquired from expert designer's drawings are inserted into a Temporary Database.
b. Rule verification, in which candidates are compared with verified rules in the Working Database.

CLASS
CLASS (Chip Layout ASSistant) is a knowledge-based system that assists designers in floorplanning and produces block-size estimates, block placement and global routing [Bir85]. The input specifications to CLASS includes: technology and design constraints, desired chip size and aspect ratio, the functional blocks to be used and the interconnectivity between the blocks. In designing the floorplan, CLASS interacts with the layout system TALIB (see below) and the router WEAVER (see Subsection 8.3b). Human intervention has the purpose of improving the quality of the floorplan.

Global Routing Optimizer
Traditional VLSI layout design consists of three phases: placement, global routing and detailed (or local) routing. Since more than half of the available chip area is allocated to routing, the quality of global routing has a great impact on the final detailed routing result. A rule-based system has been developed for global routing optimization [Cai87]. The system is written in XLISP, an experimental object-oriented language [Fla87]. A frame mechanism has been used to represent the working data. Rules are also implemented in frames. Meta rules are used to influence the control flow to improve the efficiency. During the optimization process, nets in congested channels on critical paths are selected for rerouting. A reroute is attempted to reduce the number of times a net crosses the critical paths.

TALIB
The rule-based system TALIB [Kim83, Kim84, Kim86], implemented in OPS5, originates from Carnegie-Mellon University, Pittsburgh, and is suitable for designing *orthogonal layouts* of small transistor circuits in single-metal, single-polysilicon NMOS technology. TALIB generates a mask description output in CIF, where technology-dependent rules are satisfied.

The input to TALIB consists of two parts:
a. A schematic description of the circuit components (transistors or logic gates) and their interconnections. The transistor sizes can be specified, although default values are available.
b. Topological and geometric constraints around the boundary of the circuit layout, including the order in which the external connections must appear on this boundary.

TALIB generates default values for the transistor sizes, unless specified otherwise. Specific cells can be transformed in a form more suitable for implementation in NMOS technology. For example, an AND gate connected to a NOR gate is changed into an AND-OR-Invert gate, which needs fewer transistors. After such transformations, all logic gates are expanded in their components, which are transistors. In the transistor circuit, appropriate clusters of transistors can be identified as subcircuits which are amenable to planar layout. TALIB builds up the layout based on such subcircuits.

TALIB combines heuristic AI techniques with algorithms. The latter can be called upon as subtasks in the rule-based expert system. TALIB employs a great amount of built-in knowledge. There are 100 subtasks, including five really important ones: the pattern recognizer (which recognizes and manipulates frequently occurring subcircuits), the area splitter (which divides the layout area into smaller blocks representing subcircuits), the topological placer (which assigns the subcircuits to relative places according to the boundary constraints and the subcircuit interconnections), the layout generator (which generates the geometric layout of each subcircuit), and the compactor (to minimize the area occupied by the complete circuit). While these tasks are performed, TALIB's controller sees to it that all constraints are satisfied. TALIB uses a "least commitment" approach, that is, only strictly necessary decisions are made.

TALIB generates floorplans in a top-down manner. Global placement of components and preliminary allocation of wires are followed by establishing exact locations of individual components and wires. When an unworkable situation is encountered at any level, bottom-up backtracking is attempted to remove the obstacles. Since reasoning is used at the lowest level with its large solution space, TALIB is not suitable for large layouts.

Four types of task-specific knowledge are present:
− analytical (for recognizing general situations)
− synthetical (for laying relations between different abstraction levels)
− axiomatic (for controlling the consequences of design decisions)
− control (for controlling the execution order of the tasks).

HOPE

HOPE (Hierarchy Optimizing Placement Expert) is an expert system for partitioning and placement in gate arrays [Lai86]. It contains a tool for analyzing

8.2 LAYOUT DESIGN OF BLOCK STRUCTURES

and manipulating hierarchy. A routability measure is evaluated to judge the quality of the partitioning result. Given a specific placement of cells, modifications can be made to alleviate wiring congestion and routability.

HOPE's expert system shell is written in Prolog. Procedures written in another language can be called. The knowledge base consists of different independent experts (called "advisors") with knowledge about partitioning and placement. Experts can be consulted by the highest-level expert on partitioning and placement issues, e.g., whether or not a module can be placed on specific rows of the gate array.

Other experts which are invoked include the partitioning advisor (which suggests splitting a module or merging with another module), the min-cut advisor (which suggests how the external net count can be reduced), the 2-D advisor (which consider the effect of folding a linear array into two or three rows), the row-matching advisor (for matching module sizes to row sizes) and the congestion advisor (for relating connection to available channel densities). Rules include the way in which layout algorithms work in relation with the array structure, heuristic rules about physical structures and metarules which combine results with other rules.

Chip Planner

The expert chip planning system of the University of Genova [Ant84] consists of three subsystems: the symbolic layout assembler HEICLAS, a topological evaluator which delivers data to the layout assembler, and a reasoning system. Only the important decisions are left to the user. The topological properties of the functional blocks for the detailed layout are evaluated by the topological evaluator. This subsystem consists of various evaluating functions, belonging to a certain functional block in a VLSI network. It is based on a deterministic algorithm.

HEICLAS is an expert system that handles placement and routing. The layout process takes place incrementally, that is, a module is added one by one to the design. The system manipulates rectangular blocks of arbitrary dimensions and pins. The user can choose a strategy with which the system must generate solutions. The process is highly iterative. The user can store, compare and investigate intermediate solutions.

CADRE

CADRE, developed by AT&T Bell Laboratories, is an automatic layout design system that converts a hierarchical *structural circuit* description of interconnected cells into a hierarchical *symbolic layout* description which is independent of process and design rules [Ack85]. The input may be the result of some logic synthesis tool, e.g., DAA. The symbolic layout description of the CADRE output is passed on to the MULGA system for translation into the mask artwork. The CADRE approach simulates the human physical design process, in which several skills, such as floorplanning, leaf cell layout, critical path analysis and global and local routing are merged together. An important feature is that optimzation decisions are made globally.

CADRE works with a number of cooperating experts, each of which captures specialized knowledge in order to handle different aspects of the design process independently. See Fig. 8.17. The experts may represent algorithms, rule-based systems or even humans. A "manager" coordinates the expert operations by allocating appropriate subtasks to the experts at appropriate times, interpreting their results and integrating these results to yield a global optimal solution. Communication between experts primarily takes place through global constraints, e.g., "cell A above cell B" in a floorplanning case. While the experts solve locally constrained problems, the manager provides the overall design strategy. The manager decides which constraints should be given to an expert and when. A big advantage of the CADRE approach is that new experts can be added in the future. For example, parts of human intervention may be replaced by new rule-based experts.

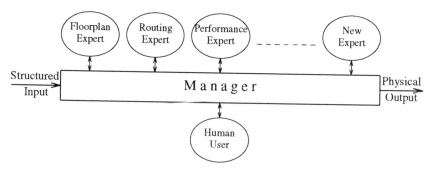

Fig. 8.17 CADRE architecture

The design process progresses top-down by floorplanning of block layouts, followed by detailed layout of blocks and leaf cells, and bottom-up by a backtracking facility. If some intermediate result proves to be unsatisfactory, the manager reverts to redesign at a higher level. This backtracking process may be repeated an arbitrary number of times. A more detailed description of this process follows below.

A top-down floorplanning process is initiated with constraints supplied by the user. The structural input specification can be viewed as the initial set of constraints. At the beginning, an unconstrained layout is made of each cell. This gives a lower bound area estimate for each cell (bottom-up). Then, the manager cycles in a loop between two processors: the *search strategist* and the *stabilizer*. The search strategist decides which cell should be designed and placed. This decision is governed by criteria, such as external interconnections, cell area and

8.2 LAYOUT DESIGN OF BLOCK STRUCTURES

the cells placed so far. When certain constraints may no longer be valid, the stabilizer corrects the result by either changing some constraints or by modifying some parts of the design. Then, the search strategist is invoked again to decide which leaf cell is designed next. The layout is complete, when no cells are left.

The manager consults the *floorplan evaluator* if not all the constraints can be resolved. For example, the floorplanner introduces a routing area to make certain connections possible, but this leads to a violation of area constraints. The evaluator tries to localize the constraints that appear to be the cause of the trouble. It then gives recommendations as to how this problem can be resolved, possibly by suggesting a new set of constraints.

Experts, which can be used in the CADRE system, are described in more detail below. These include the floorplanners Flute, Floyd and Fork, and the leaf-cell layout tools Topologizer and Coordinator.

FLUTE

Flute [Wat87] is a floorplanner used in CADRE for the optimal layout of rectangular modules by using heuristic knowledge. Its input is a structural block-level description with internal and external constraints (area, aspect ratio, pin positions, connectivity graph). Its output is a rectangular floorplan which hopefully meets the input constraints. Input constraints are specified by the designer or generated during the CADRE design cycle. Flute uses a mixture of algorithms, state control and rule-based programming.

Flute works in two phases:
a. A *topological* phase in which the connectivity graph is transformed into a planar rectangular grid graph. Routing modules are introduced where intersections occur in order to produce a planar graph.
b. A *geometric* phase in which area constraints are taken into account. Linear and quadratic inequalities, which express these constraints, are solved in order to produce a minimal area solution.

FORK

Fork [Yu87] is another rule-based floorplan expert designed for the CADRE system. Structural constraints include the I/O pin positions and the aspect ratios of the rectangular circuit modules. The main difference between Fork and the two preceding experts is the strict separation between *rules* and *control strategies*. Rules are implemented inside a *cost function*, while the control strategies are independent general state machines. The cost function is a weighted sum of the costs attached to the selected architecture, module connectivity and shapes, as defined in the respective rules.

Fork uses three distinct phases:
a. A *constructive* phase, in which the modules are added one at a time. An initial floorplan is generated by a greedy algorithm.
b. An *improvement* phase, in which a more general search strategy is used to

examine many different topologies.

c. A *fine-tuning* phase, in which the floorplan is evaluated by solving structural and geometrical constraints and some minor topological changes are made.

Each phase has its own control strategy, implemented as a set of states controlled by a state machine controller. The separation between the rules and control has several advantages. Traditional rule-based systems have problems when rules are added to an existing set of rules. There is a possibility of interaction between the rules. Because the rules in Fork are built into a cost function, this problem does not occur. Also, it is quite easy to add new phases to Fork, simply by defining a new control strategy and implementing a new state machine. In this way, the *Knowledge of Implementations* and *Floorplan Structures* used in the Floyd system could be added. This approach is flexible and the easiest to maintain when compared to the floorplanners Flute and Floyd.

FLOYD

Floyd [Dic86] is an Australian production system, written in OPS5, containing about 120 production rules. It runs under UNIX on VAX 11/780. Furthermore, Floyd is written in LISP and C. Floyd constructs custom floorplans in a hierarchical manner. As input, it uses relatively small, rectangular modules ("leaf cells"), which are fixed *a priori*, that is, sizes and pin positions are not variable. Requirements to the final result are: a module placement with the smallest possible space, the shortest possible wiring between the different modules (= leaf cells).

Floyd works hierarchically, that is, the design task is divided into increasingly smaller partitions. The floorplan domain knowledge is deduced from other projects and is represented in three forms: (a) rectangular graphs, (b) classes (of module layout implementations) and (c) production system. The production system (c) has (a) and (b) as subsystems. The knowledge used by Floyd can be subdivided into four categories:

a. *Reasoning with rectangles*. The ability to place and insert rectangles that sufficiently abut connecting rectangles without overlap in such a way as to minimize the required area. A rectangular graph represents the relative positions of the rectangular modules. The graph is built up by adding new nodes to the incomplete graph one by one.

b. *Knowledge of implementations*. Some modules, such as PLAs, have a distinct layout structure and therefore have highly constrained connections, whereas other modules, such as random logic, may encompass a wide variation of interface constraints. This knowledge is represented with *classes*: a general category of module, such as adder, register or random logic. It can be used, e.g., to preplan.

c. *Floorplan Structures*. For example, the regular structure of arrays can be recognized from the structural input description and exploited to facilitate the

8.2 LAYOUT DESIGN OF BLOCK STRUCTURES

design procedure. This category of knowledge and the next one are used in a *production system*, which manipulates the above knowledge categories a and b.

d. *Control strategies*. Techniques, such as least commitment (i.e., if there is inadequate information to solve a problem at a certain level, it is best not to overcommit oneself as this will affect the next decisions), planning and abstraction are used mainly with backtracking.

Note that the Flute system only uses the first category of knowledge.

Floyd begins with a list of modules and their interconnections. Modules are placed one at a time. The order in which modules are selected for placement is determined by using all kinds of criteria (e.g., area, number of connections). A position is chosen for the module. Here, the search space is restricted by the planning task. This task contains rules which recognize certain structures. If this is not successful, connections are broken or other placements are tried with backtracking. When successful and several alternative solutions are found, a quality factor F is calculated in the "class" knowledge representation. The placement with the best F value is accepted. The others are removed from the working memory.

FLAIR

FLAIR (FLoorplanning using an Artificial Intelligence appRoach), developed at the University of Dortmund, is a floorplanner based on flexible block realizations [Bru88]. It provides different design styles and performs explicit planning of wiring areas between the blocks. FLAIR solves the floorplanning problem in two steps: topology planning and geometry planning. Both steps can be realized by expert systems based on the blackboard architecture.

TOPOLOGIZER

Topologizer [Kol85] is a rule-based expert system that translates a *transistor connection description* into *symbolic cell layout*. The input consists of a CMOS-circuit description (graphically, schematically or in text form) and a description of the environment (aspect ratio of cells, external connections, abutment, load). The output is a symbolic layout, which is independent of process rules. The symbolic layout can be converted to a mask level description by MULGA, which allows layout compaction to be performed (see Subsection 8.2a). A language ICDL (Intermediate Circuit Description Language) captures both geometric placement and circuit connectivity. MULGA provides the designer with the capability of interactive editing, layout compaction, circuit-connectivity extraction and timing simulation of MOSFET circuits.

Typically, when a hierarchical top-down procedure for designing a complex system has reached the level of leaf cells (transistors, wires, vias), the system can be assembled using the MULGA system. In a number of bottom-up steps, subcircuits, representing functional modules, are combined to create higher-level modules until the target system is reached.

The architecture of the Topologizer system is depicted in Fig. 8.18.

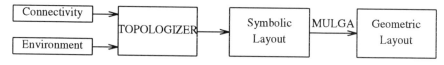

Fig. 8.18 The Topologizer system

Characteristics of Topologizer are:
- Rather than using the algorithmic approach, rules are used to be able to solve ill-defined problems.
- Rather than making a direct translation of high-level input to mask level, Topologizer produces an intermediate symbolic cell layout output which is process independent.
- Rather than using a standard library of predesigned layouts and trying to connect these together to achieve the required functionality, Topologizer tries to custom fit a cell to its external environment.

Topologizer allows user intervention. It uses two separate phases, each with its own set of rules:
a. *Transistor placement*, where interconnected transistors are placed closely together. This usually results in two rows of transistors, segregated by transistor type. Topologizer groups P-type transistors in rows parallel to the VDD line and N-type transistors in rows parallel to the VSS line. Techniques such as exchange and rotation are used to improve the placement result.
b. *Routing*, where the necessary connections between transistors are made. Initially, a wasteful but correct design is generated, but this design is iteratively improved according to a set of rules.

COORDINATOR

Coordinator [Kol86] is a layout expert, where *symbolic* layout data exist together with *geometric* mask data. Coordinator's goal is to optimize the design. It achieves this goal by making the design more compact. The user is allowed to edit both categories of data before and after compaction. Both categories of data are represented by a single layout representation, which results in an efficient use of memory. This layout representation is implemented using a database, which combines the circuit connectivity of symbolic design with the graphic details of geometric design as described below.

Each leaf cell contains a number of *elements* (transistors, wires and contacts). Each element points to a number of *terminals* through which elements may be symbolically connected. In addition to this, elements also contain geometric information. Terminals point to *virtual segments* (ordered lists of terminals distributed along a fixed axis), which prefer to move together to maintain

8.2 LAYOUT DESIGN OF BLOCK STRUCTURES

connectivity. Each terminal lies on a horizontal segment and a vertical segment. The X and Y locations of a terminal are determined by the X and Y locations of its horizontal and vertical segment. Segments do not necessarily extend across the entire cell and are therefore different from a virtual grid. Because segments cannot cross a cell boundary, segments from two adjacent cells are grouped together into *virtual segment groups* (ordered lists of segments distributed along a fixed axis), which prefer to move together to maintain connectivity. The X and Y locations of a segment are determined by the X or Y location of its horizontal or vertical segment group.

The compaction itself operates in one direction at a time. This means that an X compaction (where all vertical segment groups are coordinated) and a Y compaction (where all horizontal segment groups are coordinated) are necessary to compact a cell. Consequently, the first direction of compaction is usually optimized at the expense of the other. This leaves room for improvement in the future. the compaction is performed on a *segment group by segment group basis*.

Coordinator is a layout expert that performs a compaction scheme using a number of cells, and thus is quite different from Topologizer that operates on a single cell. In the CADRE system, both experts could be used together, where Topologizer's output is further optimized by Coordinator.

PIAF

PIAF (Package for Intelligent and Algorithmic Floorplanning) is a floorplanner, which consists of an algorithmic part and a knowledge-based part [Jab88]. It partitions the floorplanning task in a way that allows efficient use of heuristics and specialized design knowledge in the generation and pruning of the solution space. A program determines the floorplan of a full-custom integrated circuit by successive refinements [Jab89].

PIAF uses several knowledge representation schemes, including predicates, procedures, rules and frames. The inference engine is a simple procedure written in Prolog [Jab90]. The floorplanning process resembles the algorithm described in [Mal82]. See Subsection 8.2b.

Fuzzy Logic Application

Fuzzy logic is extremely useful in solving problems with multiple objectives [Zim87]. One such problem is cell placement, which is an important step in VLSI chip design. Lin and Shragowitz [Lin92] applied fuzzy set theory to the multiple-objective placement problem. Constructive placement consists of two major steps: select a cell, and select a position for placing the selected cell. An unplaced cell is a good candidate for next placement if it has strong connectivity with already placed cells. Selecting a cell position is governed by three criteria: timing delay, wasted area, and the effect on future utilization of chip area. A simple example of a fuzzy rule has the following form: If a cell position produces small timing delays *and* small wasted area *and* good border evenness, *then* it is a good position.

SHEDIO

SHEDIO is a knowledge-based environment for VLSI system design, covering the entire hierarchy from high-level system specifications to low-level physical description and layout [Bou88]. A multilevel frame evaluates and optimizes the topology of the components that compose the digital system. This is done by examining a predefined set of algorithms, which the system must execute efficiently.

Placement of the components according to priorities and physical constraints is performed by an expert system, called DESIGNER. This system accepts as input a weighted-graph, retrieves the rules and constraints of the particular semiconductor technology from the internal database and accesses the knowledge base in order to select any available knowledge of any related previous design. Then, the VLSI system layout design, using the DESIGNER package, follows two major steps:

a. A multilevel independent placement of the system components.
b. A hierarchical geometrical transformation for optimum VLSI layout design.

In particular, the interconnection lengths between system components as well as the chip area are minimized by optimizing the placement of the system components.

LES

LES (Layout Expert System) generates layouts for one-metal silicon-gate CMOS VLSI modules [Lin88]. It is programmed in OPS83 and runs on a VAX-11/750 with UNIX. The input description for LES consists of two parts:

1. A list of components and their interconnections. Components may vary from a transistor to a leaf cell composed of several transistors.
2. Environmental constraints which are related to the interface with neighboring modules. These include module size and pin specifications.

LES produces a symbolic layout which implies that exact component sizes and positions are not fixed. The symbolic output is handled by ICOMP, an optimizing compactor, which converts the layout into a mask artwork compatible with the CMOS technology and changing design rules.

Instead of routing through channels, LES tries to abut interconnected leaf cells. Connections between distant leaf cells are made across other cells in the way as usual in the sea-of-gates approach. The transistor sizes are given with the input specification. The loading and parasitics may depend on the external environment. An optimal layout can be achieved in an iterative process with a timing analyzer and a circuit optimizer.

The layout area is divided into a number of parallel strips delimited alternately by VDD and VSS lines. In this way, each set of related leaf cells which are placed in a particular strip can be connected to both the VSS and VDD

8.2 LAYOUT DESIGN OF BLOCK STRUCTURES

lines. As a consequence of LES's two-dimensional layout style, the average wire length grows proportionally to the square root of the number of components.

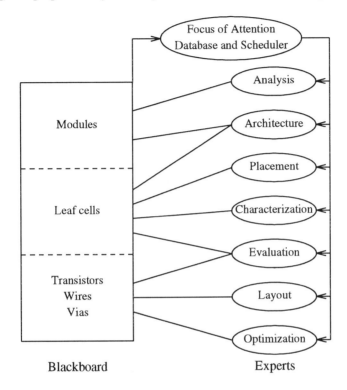

Fig. 8.19 Blackboard architecture of LES

LES employs a flexible floorplanning technique which allows the user to specify the shape and pin positions freely. The objective is to achieve better component densities. The leaf cells in the strips are laid out starting from the most complex leaf cells followed by less complex neighbors. Thus, complex cells have more freedom in pin positions than simpler ones.

The subsequent steps taken in the process of generating a layout are controlled in a rule-based system with a blackboard architecture supported by seven experts each of which performs a specific subtask. Intermediate results are updated on the blackboard and, whenever necessary, an appropriate expert is invoked through a Focus of Control Database and Scheduler (see Fig. 8.19).

A short account on the seven experts is given below.

1. From the input description, the *Analysis Expert* derives useful information, e.g., the module shapes and pin distributions. This expert constructs a graph representation of the circuit. A path is defined as a chain of different modules

from an output port backwards to an input port or a feedback point. From all possible paths, a set of "seed paths" is selected. The seed paths determine the layout architecture to be used.

2. The *Architecture Expert* decides on the layout architecture. It determines the number of parallel strips, their orientation (horizontal or vertical), and the distribution of VSS and VDD lines. The number of strips is related to the number and length of the seed paths and the complexity of the complete circuit.

3. Given the net list and module pin distribution, the *Placement Expert* places the modules in a certain order, determined by the components in the seed paths. Then, the remaining components (not covered by any seed path) are placed in appropriate positions, depending on their connectivity and sizes. The placement task consists of two phases: initial placement and refined placement. Initial module sizes are looked up from a table. An optimal result is achieved by pairwise exchanging components and recalculating wire lengths of connections. A cost function, which is a measure of placement quality, is calculated.

4. The *Characterization Expert* is a sort of global router, which adds routing information to the leaf cells. Given leaf cell placements and the net list, this expert examines and selects possible connections among leaf cells in such a way that a uniform routing pattern with short wires is obtained, while avoiding congestion. The leaf cell specification comprises wiring information, including pin positions and the number of pass-through wires.

5. The *Layout Expert*, called FLEX, can perform three tasks: (a) estimation of the leaf cell size, (b) providing LES with information about optimal pin positions, and (c) generating symbolic layout [Lin87]. FLEX's major task is to determine transistor positions, orientations and routing through the cells. If the layout is not satisfactory, LES backtracks by modifying the cell involved or its neighboring cells. FLEX determines the positions and orientations and places the wiring within the cell.

6. The *Evaluation Expert* evaluates characteristic parameters, such as size, delay and pin positions of all successfully generated leaf cells. If the result is not satisfactory, LES backtracks the design process. Irregularities in the area usage are removed. With the extra information rendered to the Analysis and Architecture Experts, an "optimal design" may be achieved.

7. Depending on the number of virtual rows and columns at the symbolic level, the *Optimization Expert* optimizes the geometric layout. This expert, which contains heuristic rules derived from human experience in layout optimization, does not work in one run on a complete design, but always on a leaf cell and its neighbors.

The following step-by-step procedure may illustrate the design approach adopted by LES:

8.2 LAYOUT DESIGN OF BLOCK STRUCTURES

a. Given the input description, a graph representation is constructed.
b. The Analysis Expert defines a list with possible paths and selects "seed paths" and remaining components.
c. The Architecture Expert analyzes the environmental constraints (aspect ratio, pin distribution) and selects an architecture with a number of strips, their orientation and the distribution of VDD and VSS.
d. The Placement Expert places seed paths and remaining components in the strips and then determines the relative positions within every strip.
e. The Characterization Expert adds wiring information to the cells.
f. FLEX generates the layout of leaf cells.
g. The Evaluation Expert determines if backtracking is necessary. If so, then go back to b.
h. The Optimization Expert optimizes the symbolic layout.
i. ICOMP, an intelligent geometry layout compactor, is applied to the symbolic output of LES to obtain the final layout.

Layout-Compaction Expert

In Subsection 8.2a, it was indicated that layout design of block structures consists of three steps: symbolic design, conversion to geometric layout, and layout compaction. Compaction is effectively performed by scanning the layout from one side to the other. For example, a vertical scan line, crossing a set of (usually rectilinear) layout primitives grouped along this line, is shifted from left to right. Layout primitives at the right of the scan line are shifted as far as possible to the left, while maintaining conformity to the layout design rules. In this way, the dead areas between the primitives are minimized.

KOCOS (KnOwledge-based COmpaction System) is a rule-based system for layout compaction [Bru86]. The input is a CIF representation of the layout from which transistors, interconnection nets and vias are extracted. The compactor uses heuristic rules for evaluating and recognizing improvable structures, such as compression ridges.

Another knowledge-based program for layout compaction has been developed by Hsiao *et al.* [Hsi91]. The layout constraints are expressed by about 200 production rules written in OPS5. Rectilinear polygons are cut into a set of adjacent rectangles so that the mask layout is made up of a set of individual rectangles. Active compaction operations are performed as soon as the scanning line encounters an edge of a rectangle. The compaction system runs on VAX8200 under UNIX. The newest version runs on a SUN3 workstation [HSi91]. The basic architecture comprises a blackboard and a set of experts for system control scheduling. The blackboard has partitions for mask layout representation, (sub)goal triggering representation, working memory, and a knowledge base containing layout constraints. The experts include the scanner, planner, edge checker, layout compactor and I/O interface.

442   8 LAYOUT DESIGN AND EXPERT SYSTEMS

Radenković et al. [Rad92] proposed a different method of layout compaction. It is assumed that the layout modules are rectangular blocks. The blocks are moved downwards one at a time, starting with the bottommost blocks, successively followed by their upstairs neighbors until all the blocks have been moved, leaving minimal area between the blocks. Each block movement contains an algorithmic zone-refining and a knowledge-based part. Zone refining involves both horizontal and vertical movements of blocks. While zone refining has the objective of local compaction, the knowledge-based tool uses expert knowledge to find the best candidate places for the blocks.

PEARL

PEARL (Power-supply Expert Assisted Rule-based Layout) of Digital Equipment Corporation is an expert system that generates the layout of power-supply circuits for printed wiring boards (PWBs) [DeJ86]. Power-supply layout is a subset of the larger analog PWB design problem. It differs in many respects from the digital-circuit layout which has rectangular shapes placed in rows and columns. Power-supply routing requires varying conductor widths and spacings for the differing voltages and currrents. Inductance and signal noise due to inadequate grounding separation should be avoided. The primary side of the power-supply circuit must be kept away from the secondary side. Both horizontal and vertical segments of conductors may run on the same layer.

An OPS5 rule-based system is built on top of a VAX-layout system, which is largely written in BLISS. PEARL provides a power-supply placement tool within the existing layout system, while routing possibilities are being explored.

Knowledge-Based GaAs Overlay Shell

Harris Semiconductor, Milpitas, developed a CAD system tailored to the company's GaAs depletion-mode MESFET technology and the techniques used to develop GaAs chips [Jan87]. The Harris system is based on an integrated design environment from SDA Systems, which has an open-system architecture and offers schematic capture, layout, simulation and place-and-route software on a common database. A Knowledge-Based Overlay Shell automatically handles all GaAs-specific design issues. In the overlay shell, properties are used to transfer information from one representation, such as a schematic, to another, such as a place-and-route representation. The shell initializes layout tools, including placement (of ground-shield pads, I/O pads and standard cells), power bus routing, channel generation, feedthrough insertion, global and detailed routing. A layout-verification system extracts parasitics for simulation, design-rule checking and layout-versus-schematic comparisons.

GM_Plan

The *gate matrix* of rows and columns in polysilicon CMOS technology can be considered as an attempt to layout the transistors in a compact form. The

8.2 LAYOUT DESIGN OF BLOCK STRUCTURES

columns of the gate matrix are implemented as polysilicon line and provide the wiring as well as the polySi gates of the transistors. The rows are implemented as diffusion lines which form transistors at the crossings with the polySi lines. The CMOS gate-matrix layout problem can be formulated as an AI planning problem.

GM_Plan is an algorithm that combines gate placement and net routing into a single, incremental problem-solving loop [Hu90]. The "goal" in the solution algorithm (the gate-matrix layout) consists of many subgoals, each of which corresponds to the placement of a gate to a slot and to the routing of nets leading to that gate. Artificial Intelligence comes into play when several alternative subgoals must be handled, particularly when different nets compete for track or resource usage.

A hierarchical subgoal organization facilitates an objective classification of the subgoals into priority classes according to a proposed distance measure of connectivity. The search process is guided by two control policies:

a. The *most-constraint* policy gives priority to the subgoal whose solution has most constraints, that is, one that has the smallest solution space.
b. The *least-impact* policy selects, among alternative solutions, the one that consumes the least amount of resources.

Prolog-Based CMOS-Circuit Design

Hill and Roy [Hil85] described a silicon compiler that transforms Boolean expressions into CMOS layouts. The compiler consists of three Prolog programs that run sequentially under UNIX:

a. Building a network of transistors,
b. Placement and routing,
c. Generating a virtual grid in *i*.

The Prolog format is converted into the layout language *i*, which is a hierarchical hardware-description language capable of expressing both geometric and electrical information symbolically. The original version is limited to producing CMOS gate-matrix-style layouts.

ELF

A basic problem of technology-dependent software tools for VLSI design is that these tools become outdated as soon as new technologies emerge. Setcliff and Rutenbar [Set90] suggest that for some special applications, such as maze-running routers (see Subsection 8.1b), software tools can be synthesized automatically from high-level specifications accomodating new technologies.

ELF is a prototype implementation which transforms flexible, high-level specifications into working maze routing software for a wide variety of technologies, ranging from multiple-layer printed-circuit boards to semicustom integrated circuits. ELF uses synthesis strategies aimed at transforming an abstract, non-executable description of a software tool into real, working space.

To this end, it is essential that ELF combines maze-router knowledge with program synthesis knowledge in order to guide the search among algorithms and data structures. The ELF prototype system is a rule-based program, mainly written in OPS5 and running under UNIX4.3. The routers produced by ELF are written in C.

ELF builds up a router in three stages:
1. Input stage. Router knowledge and technology constraints are read from the user interface to facilitate the selection in the next stage. A set of usable algorithms is generated.
2. Selection stage. The appropriate data structure and algorithms are selected to implement the required routing task from candidates proposed by the input stage.
3. Code generation stage. The executable output code for the maze router is generated using router domain knowledge, a set of selected high-level algorithm descriptions and a set of data structure implementations.

8.3 LOCAL ROUTING

8.3a Algorithms for Local Routing

Channel Routing and Switchbox Routing
The layout design of a chip consists of four stages: partitioning (Subsection 8.1a), placement (Subsection 8.1a), global routing (Subsection 8.1b) and finally local routing (Section 8.3). The problem of *local routing* can be formulated as follows: Given a rectangular region which is available for interconnecting wires. Pins located at equidistant positions on the periphery of the region are numbered 0, 1, 2, etc. The pins numbered 0 remain unconnected. Other pins with the same numbers must be interconnected such that some optimizing criterion is satisfied. Usually, the objective in local routing is to minimize the total interconnection length and the number of vias.

There are two important types of interconnection region associated with local routing: the routing channel and the switchbox. The most important example of a routing channel is found in standard-cell structures. Rows of standard cells are provided with pins which must be connected to other pins. Two rows of pins are located on both sides of a rectangular region, called the *routing channel*. The programs used to design the interconnection pattern in the channel region are called *channel routers*.

8.3 LOCAL ROUTING 445

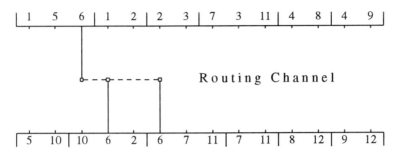

Fig. 8.20 Channel routing

To become familiar with the terminology used in channel routers, let us consider Fig. 8.20. A rectangular routing channel region is bounded on both the top side and the bottom side by a row of equidistant pins. Each pin is assigned a number. Pins with equal numbers must be interconnected, except for the number 0 which remain unconnected. The set of wiring segments connecting equal-numbered pins is called a *net*. In the simplest but frequently used case, each net consists of horizontal and vertical line segments which are restricted to the horizontal and vertical lines of a regular grid placed on the channel. Usually, horizontal segments are laid in the metal layer and the vertical segments in the diffusion layer. Transitions from a horizontal (metal) to a vertical (diffusion) line segment of the same net are accomplished by a *via*. Vias implement the required connections of net segments in the metal and diffusion layers. The pins lie on vertical grid lines which are successively numbered: 1, 2, 3, ... The horizontal grid lines which may contain the horizontal line segments of the nets are called *tracks*. An example of an interconnected net is shown in Fig. 8.20 for net number 6.

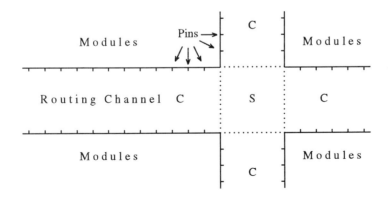

Fig. 8.21 Routing channels C and switch box S

A *switchbox* is a square or rectangular region for interconnecting pins which are located on all four sides of the region. For example, such a region in found in gate arrays, where two routing channels cross each other (see Fig. 8.21). The pins should be viewed as terminals of nets which have pins in the channels meeting the switchbox on four of its sides. Programs which solve the problem of interconnecting pins with nets in the switchbox are called *switchbox routers*.

The terminology used in switchbox routers is explained by a simple example. Pins located on the grid lines on the four sides of the switchbox are numbered such that sets of pins with equal numbers form individual nets. Number 0 is reserved for unconnected pins. The horizontal and vertical grid lines are called the *rows* and *columns* of the switchbox interconnections.

Channel Routers

Most channel routers are derived from the *left-edge channel router* developed by Hashimoto and Stevens [Has71]. In principle, the routing algorithm proceeds as follows. We start with implementing the interconnection net for the leftmost pin. In case of a tie, the upper pin is selected. The horizontal segment of this net is assigned to the uppermost (first) track. The next pin to be connected is the first pin to the right of the point where this segment ended. This process is continued until the first track is fully utilized. Then, the second track is used for interconnections by selecting unconnected pins from left to right. As many tracks as are needed for interconnecting all pins are used.

Example: The pins, given by identical numbers in Fig. 8.22, must be interconnected to form nets.

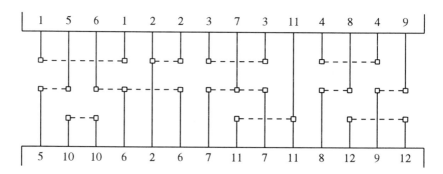

Fig. 8.22 Application of the Left-Edge algorithm

Application of the Left-Edge algorithm, as outlined above, corresponds to placing the interconnection nets in the order 1, 2, 3, ... Three tracks are needed in this case. Instead of selecting pins from left to right, the reverse direction may be chosen as well. This would generally lead to a different interconnection pattern.

8.3 LOCAL ROUTING

Horizontal and Vertical Constraints

When two pins which must be connected have different abscissas, the net consists of one horizontal and two vertical segments. The routing of the interconnections is subjected to horizontal and vertical constraints. The *horizontal constraint* imposes the condition that overlap of horizontal segments of different nets is never allowed. The *vertical constraint* implies that overlap of vertical segments of different nets on the same vertical grid line is forbidden.

The horizontal constraint can be accounted for by dividing the channel length into a number of zones. A *zone* is an interval in the horizontal channel range containing a set of horizontal segments of different nets which are not allowed to overlap, i.e., two horizontal segments in the same zone are not allowed to be placed on the same track. The vertical constraints can be represented by a *vertical-constraint graph* which determines the relative positions of the segments in a zone with respect to one another. This constraint graph for the example of Fig. 8.22 is shown in Fig. 8.23.

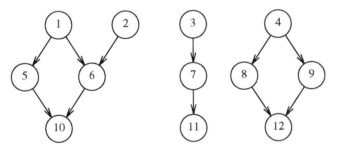

Fig. 8.23 Vertical-constraint graph

For example, the horizontal segment of net 1 must be placed in a track above the horizontal segments of net 5 and net 6. This is indicated in the graph by an arrow from 1 to 5 and 6.

When all edges in the vertical-constraint graph are derived from top to bottom, the relative positions of all horizontal segments in a zone are uniquely defined. When two nets are equivalent with respect to the vertical constraint, we have a *cyclic constraint* [Has71]. An example of a cyclic constraint is illustrated in Fig. 8.24a and b, where the pins of the nets 1 and 2 have the same abscissas. When either net 1 or net 2 is placed first, the other net cannot be routed as a net with a single vertical segment. This problem may be resolved by mirroring one of the cells which contain a pin of 1 or 2. In Fig. 8.24c, the cell containing the lower pin of 1 is assumed to have been mirrored, after which placement of the nets 1 and 2 has become feasible. Such a result may also be achieved by interchanging logically equivalent pins or by interchanging cells. As an alternative routing, one of the nets is allowed to have two horizontal segments and hence a third vertical segment requiring two additional vias has to be made (see Fig. 8.24d).

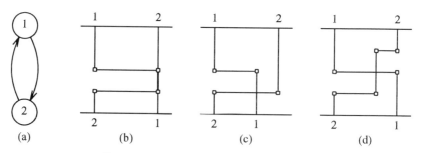

Fig. 8.24 Example of a cyclic constraint

Dogleg Router

The vertical-constraint graph pertaining to Fig. 8.25a does not contain a loop. The solution obtained by using the left edge algorithm requires three tracks for implementing the interconnection of the nets 1, 2 and 3. This is not an optimal solution with respect to the chip area occupied.

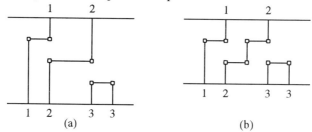

Fig. 8.25 Left-edge and dogleg router

A better solution for such cases is provided by the *dogleg router* [Deu76]. Such a solution for the example being considered requires two tracks by routing net 2 in a "doglegging" fashion, as shown in Fig. 8.25b.

Efficient Algorithm

Yoshimura and Kuh [Yos82] introduced the basic operation of merging of nets with the purpose of reducing the ultimate number of tracks required. The objective is to minimize the critical path length in the vertical-constraint graph. An optimal set of mergings is selected by using a bipartite graph. The search for this optimal set corresponds to the problem of maximum cardinality matching [Law76]. The best column at which to start the algorithm appears to be a density column, i.e., a column with the maximum local density.

Chan's Channel Router

Chan's router makes use of restricted doglegging, i.e., two horizontal segments

8.3 LOCAL ROUTING

for a net are permitted [Cha83]. All multi-pin nets are decomposed into two-pin subnets. The selection of a subnet to be wired is based on both the horizontal and vertical constraints. This helps reduce the chances of requiring extra tracks. Chan's algorithm consists of the following steps:

a. The central one of the density columns is selected and each associated net is assigned a track.
b. The subnets nearest to the already wired nets, called the front-line subnets (FLSs), are determined for both sides such that they satisfy the condition that each subnet of these sets is bound by either a horizontal or a vertical constraint with the subnets already wired.
c. Left and right bipartite graphs are set up and FLSs are assigned the proper tracks by using maximum cardinality matching [Law76]. To enable a good choice, when several matching solutions are found, heuristics are introduced into the algorithm by assigning appropriate weights to the edges in the bipartite graphs. Since the best matching is the one for which the minimal weights of the edges are maximal, we are dealing with max-min matching [Law76].
d. Edges in the vertical-constraint graph which cause a cyclic constraint are removed.
e. To avoid blocking of further net placements due to earlier placements, the actual assignments of tracks to the nets are delayed as long as possible. A definitive assignment is made when the rightmost X coordinate of FLSs is smaller than the leftmost X coordinate of non-FLS subnets.

Greedy Channel Router

The *Greedy Router* of Rivest and Fiduccia [Riv83] places the nets in the channel column by column, from left to right. This router tries to complete the wiring in each column using "greedy" heuristics, before proceeding to the next column. At each column being handled, a top and/or bottom connection is made by running a vertical segment to the nearest track, which either already contains the corresponding net or is empty. Segments of the same net in two (or more) tracks must be connected by means of a vertical segment as soon as possible in order to make one or more tracks available for placing other nets. The greedy router always succeeds, although more tracks than would correspond to the lower bound may be needed.

Horizontal and vertical wire segments are laid in different layers. The wire segments are laid in a heuristic manner, i.e., not all combinatorial possibilities are considered in order to find the optimal solution. Instead, connections are made by using simple rules. Nets are allowed to occupy more than one track. The greedy router scans the channel column-by-column from left to right, completing the wiring within a column before proceeding to the next column.

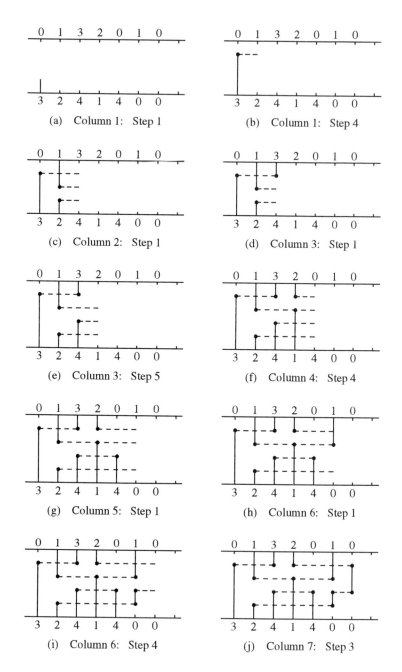

Fig. 8.26 Greedy routing

8.3 LOCAL ROUTING

Initially, a suitable starting number of tracks is assumed. Rules are established in the following iterative sequence of five steps of the routing algorithm:
1. Make connections to pins at the top and bottom of a column in an empty track or in an already existing track of the same net which is nearest.
2. Try to connect split nets in different tracks by making vertical jogs reducing the number of tracks occupied by the net in subsequent columns.
3. If a split net occupies more than one track and connecting the subnets is not possible, try to bring the subnets closer to each other by vertical jogs.
4. If the next connection to a net is on the top side (bottom side), try to move this net to a higher (lower) track to avoid future conflicts.
5. If the connection in step 1. cannot be made in the channel, one more track must be added between existing tracks, preferably somewhere in the center of the channel. The tracks must be renumbered.

The order of these steps determines the priority of the rules. Let us consider a simple example.

Example: Figure 8.26 shows the pins on both sides of the channel to be connected by four nets, numbered 1 through 4. The results of relevant steps to be taken in subsequent columns are illustrated in Fig. 8.26. The final result shown in Fig. 8.26j appears to be non-optimal. Some reshuffling of nets leads to the optimized result given in Fig. 8.27. Note that application of the left-edge algorithm to this simple example is not possible.

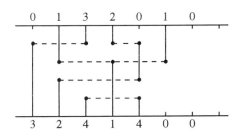

Fig. 8.27 Optimized result

Deviations from Conventional Channel-Routing Techniques
In the conventional channel-routing technique, two interconnection layers are available, each wiring net generally consists of horizontal and vertical line segments with horizontal segments in one layer and vertical segments in the other, and a via is used to connect a horizontal segment with a vertical segment. The wiring segments are restricted to the horizontal and vertical lines of a predefined regular grid placed on the channel. Deviations from this technique have been proposed.

Marek-Sadowska and Kuh [Mar82] allow horizontal and vertical wire segments of a net to be implemented in each of the two interconnection layers. A constraint is that different nets are not allowed to cross or overlap. It is assumed that only two-terminal nets are considered: multi-terminal nets can be split into several two-terminal nets. The nets are classified into five types: A, B, C, D and E, as shown in Fig. 8.28. Solid wire segments are wired in one interconnection layer and dashed segments in the other.

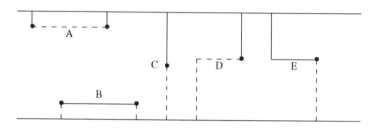

Fig. 8.28 Five net types

YACR2 (Yet Another Channel Router 2) is a channel router that minimizes the number of vias in a two-layer channel [Ree85]. Wire segments are assigned to horizontal tracks by using a modified Left-Edge Algorithm. Unused parts of horizontal tracks are exploited to eliminate vertical contraint violations. To this end, vertical segments are allowed to continue as hozontal segments in the same layer. A virtual grid is used.

When during a routing process further wiring is blocked, two main techniques are available to cancel the blockade: (1) *rip-up and reroute*, i.e., removing net segments and then rewire the removed blocked segments, and (2) *shove aside*, i.e., shift specific net segments and then wire the blocked connections. *Mighty* is a two-layer router which is capable to avoid blockades by using both methods mentioned above [Shi87]. *SILK* is another successful router based on the rip-up and reroute technique [Lin89].

Routing in three or more interconnection layers has also been proposed [Bra88]. A gridless variable-width channel router has been developed by several authors [Che86, Gro91].

A Greedy Switchbox Router

Luk's "greedy" switchbox router [Luk85] has been derived from the greedy channel router described above. The routing procedure proceeds in a similar manner, that is, the switchbox is filled columnwise from left to right by using a small number of simple rules. These rules have to be adapted to allow the connections left and right to be made. Like in the channel router, horizontal and vertical wires are implemented in separate layers.

The basic idea of the extension to switchbox routing is that steps 3 and 4 of the Greedy algorithm (putting split nets closer to each other and placing

8.3 LOCAL ROUTING

rising/falling nets higher/lower) are of no vital importance, although they may improve the result. These rules can be adapted in such a way that connections on the right side of the rectangular routing area can be made. The connections on the left side can always be made inside the area without adapting the Greedy algorithm.

In summary, when the lefthand connections have been entered, the algorithm is carried out one column at a time as follows:

1. Bring the connections on the top and bottom sides of the column, like in the channel router, in an empty row or in an already existing row of the same net depending on which is nearer.
2. Try to make as many rows as possible empty by taking together horizontal segments of the same net.
3a. Try to bring nearer to each other nets that have no righthand connection and occupy more than one row.
3b. For nets that do have a righthand connection, do the following:
    1. Try to place rising/falling nets higher/lower, or
    2. Try to place the nets in the same rows as the righthand connections.
4. When the nets are close to the right side, split the nets that have several connections to the right side (fanout).
5. When in step 1 the connection cannot be made in the channel, add a new row.

The steps 1, 2, 3a and 5 need no explanation, since they occur in the Greedy channel router. Steps 3b and 4 are new.

In step 3b, the wiring is placed in two rows. Either the net is laid above or below when it is rising or falling respectively (step 3b1) or the net is laid in such a way that it comes at the same row (or nearer to) the righthand connection (step 3b2). These two possibilities cannot be applied simultaneously since they can conflict each other. Figure 8.29 illustrates this problem.

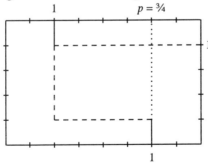

Fig. 8.29 Wiring problem

Step 3b1 will bring net 1 to the lowest net. Step 3b2, however, will hold net 1 in the same row as the righthand connection. In order that both steps can be

performed at the same time, the net must be split, but this leads to unnecessary occupation of the rows. Therefore, if p is the horizontal coordinate of the lower pin 1 in terms of percentage of the total width, the following is proposed:
- If the rightmost connection of the top or bottom side is to the right of p, carry out step 3b2.
- If the rightmost connection of the top or bottom side is left of p, carry out step 3b1 until the rightmost connection is passed and then carry out step 3b2. Here, p represents the portion of the wiring area, e.g., in Fig. 8.29 we have assumed that $p = \frac{3}{4}$.

BEAVER

BEAVER is a fast two-layer switchbox router which minimizes via usage as well as wire length [Coh88]. It considers several alternatives before assigning any wire to a track or column. Though one of the two layers has priority, assignment of wires to layers is delayed as long as possible. BEAVER consists of three subrouters which are run in a given order: a corner router (for making single-bend corner connections), the line sweep router (for straight-line, single-bend connections), and a maze-type thread router (for connections of arbitrary form). When not every wire has been assigned a layer, a layerer completes the switchbox routing by a layer assignment in a way that minimizes the introduction of additional vias. BEAVER is based on the use of heuristics.

River Routing

River routing refers to the special case of wire routing in which a group of nets (e.g., a bus structure) must be routed on one plane, i.e., a single-layer routing is required [Pin83]. Usually, a group of n pins of a circuit module is to be connected to a corresponding group of n pins of another module. A major requirement for routability is the proper order of adjacent pins in a group on the boundaries of the modules. In the special case of Fig. 8.30, planar routing requires opposite orders of the pins on the boundaries of the two modules to be connected.

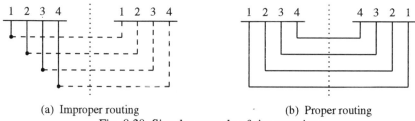

(a) Improper routing   (b) Proper routing

Fig. 8.30 Simple example of river routing

8.3 LOCAL ROUTING

8.3b Expert Systems for Local Routing

Introduction
In Subsection 8.3a, a channel router is described based on the greedy algorithm. This router and other comparable routers generally have the following drawbacks:
- The routers are only applicable to specific types of area (channel, switchbox).
- The routers make inefficient use of the area by placing horizontal and vertical connections in different layers.
- No interaction with a human designer is possible and important nets cannot be laid in advance.

For these and other reasons, there is a need for expert systems which use heuristic rules, where desired, so that the design style of a human expert can be imitated.

WEAVER
WEAVER [Joo86], developed at Carnegie-Mellon University, is a rule-based expert system designed for channel and switchbox routing. Given the pins on the sides of the rectangular routing area, the nets which must connect equal-numbered pins are to be laid on the horizontal or vertical lines of a grid pattern. The routing area is described by two overlapping grids, each representing one of the interconnecting layers (metal and polysilicon). The routing process is executed in a "greedy" manner in that connections are made simultaneously either by line routing or maze routing.

The WEAVER system works with a blackboard architecture in cooperation with a number of experts, each capable of handling a specific subtask. The language used is OPS5. The system employs more than 435 rules, which are acquired by interviewing several human experts.

Several consulting and planning subtasks in the routing process are assigned to a number of experts. One of these experts, the Focus of Attention Expert, controls the routing process and decides which expert must be activated. This decision is made on the basis of a priority list of the experts and the last decision of the preceding expert. Other experts can give their comments or may propose a suggestion. The blackboard functions as a communication medium for the individual experts. Figure 8.31 shows WEAVER's blackboard architecture with its built-in experts.

The blackboard consists of three major sections:
a. The Problem Representation section, which can be accessed by all experts, stores the subsequent wiring results during the routing process.
b. The Decision Representation section, which contains the decisions as to which net should be routed next. This section can be accessed by all experts, which may suggest alternative decisions.

c. The Scratch-Pad section, which contains a number of private subsections, each of which can be used by a specific expert.

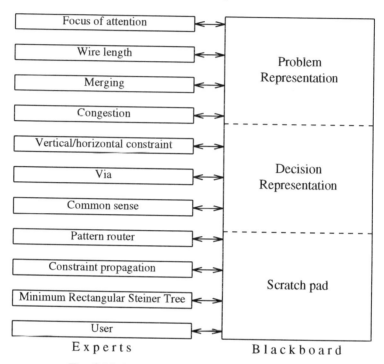

Fig. 8.31 WEAVER's blackboard architecture

The individual experts, which may decide what should be done next on the basis of their own knowledge and criteria will be mentioned below.
- The Focus of Attention Expert decides which of the other experts must be activated.
- The Constraint of Propagation Expert places the wiring depending on the wiring already laid, when there are no other alternatives.
- The Wire Length Expert decides which net must be routed next with the smallest possible wire length as a criterion, e.g., close to the border of the routing area.
- The Vertical/Horizontal Constraint Expert makes its decisions on the basis of a vertical constraint graph, which indicates the order in which the nets have to be placed in rows from top to bottom.
- The Merging Expert indicates which nets can be placed in the same column or row without affecting each other.
- The Congestion Expert sees to it that nets do not occupy more than one track in the portion of the channel which must pass the most wires.

8.3 LOCAL ROUTING

- The Via Expert gives advice as to possible reductions in the number of vias.
- The Common Sense Expert employs rules of thumb whenever other experts fail to give unambiguous advices.
- The Pattern Router Expert chooses nets which can be easily routed.
- The Minimal Rectangular Steiner Tree Expert tries to seek the shortest possible connection between the pins of a net.
- A Human User Expert can be activated to interfere during the process, whenever desired, that is, the system allows user interaction throughout the entire routing process.

Example: A Minimal Rectangular Steiner Tree (MRST) is a net which gives the shortest possible connection between given points. Only horizontal and vertical directions are allowed. There are several algorithms to generate an MRST. Figure 8.32a shows one solution for a net that connects three given points a, b and c. The MRST expert generates the solution of Fig. 8.32b, in which the number of vias is reduced by one, while the total wire length is still minimal.

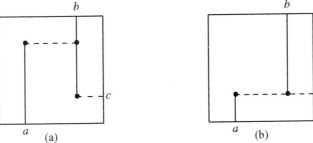

Fig. 8.32 Generation of an MRST

WEAVER has the following advantages:
a. It can be used for channel and switchbox routing, but it can be extended for other routing regions.
b. During the routing procedure, all requirements (such as minimal wire length, routing area, 100% routing) are considered at the same time.
c. Wiring in each of the two layers is allowed in horizontal as well as in vertical directions.
d. WEAVER permits pre-routed nets and restrictions.
e. It allows interactions of the designer during the routing process.

Prolog-Based Multi-Layer Channel Routing

The knowledge-based multi-layer channel router, developed by Vakil *et al.* [Vak88], uses the blackboard model, is implemented in Prolog, and runs on the shared-memory multiprocessor Sequent Balance 8000. The blackboard architecture operates with constraint formulation experts (vertical-constraint expert, wire-length expert, clustering expert, minimal-cutting expert, crossing-graph expert, congestion expert), a constraint-satisfaction expert, and a focus-of-attention expert. The routing objective is to achieve simultaneously: 100% routability, minimum routing area, minimum wire length and a minimum number of vias. The inherent parallelism in this approach is utilized together with the OR-parallelism supported by the Argonne Prolog package.

Global and Switch-Box Router

Keefe and Kendall [Kee86] developed the B&D router, an expert system for global and switch-box routing in VLSI chips. The system starts with a chip description with an associated channel graph. Global routing creates Steiner trees for each net using the channel lattice. In the final step, the B&D router performs the switch-box routing using a blackboard architecture, as employed in WEAVER. The blackboard deals with such concepts as channel, terminal, wire layout, Steiner tree, vertical and horizontal constraint graphs, congestion, convergence and suggested net information.

NOP

The architecture of a CAD-system, developed at the University of Manchester [Kah85], is shown in Fig. 8.33.

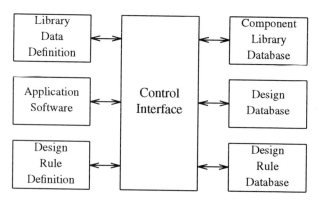

Fig. 8.33 The Manchester University CAD system

The system contains three databases (the Component Library, the Design Database and the Technology Design Rule Database), which can be used by application-directed software via the control interface. This system has a similar function as the blackboard architecture as applied in WEAVER. The Component Library contains information about all aspects of the design components or cells used in the design, e.g., physical dimensions, electrical behavior, etc. The Design Database holds semantic information about the current design. The Design Rule Database stores the technological process data, such as the minimal width of metal wires.

An expert system shell, written in Prolog, has been developed to support the

8.3 LOCAL ROUTING

knowledge-based processing tasks. By allowing data from the databases to be accessed by the application software, the software can be made generally applicable. Applications may be procedural or based on rules. An example of application software is the router NOP for routing the wires in various technologies, including CMOS and BIPOLAR.

NOP (Net Order Processor) is an expert system which assists the designer in executing various tasks in wire routing. A focus-of-attention controller activates the following tasks:

a. *Technological query.* Whenever specific data are not available in the database, the user is asked to offer these.
b. *Topological class selection.* Depending on which conditions in rules are satisfied, NOP selects a link, chain-link, bus, star or tree as the wiring structure.
c. *Point identification.* This step identifies the type of connections in a net, e.g., start or end points. For each net (set of directly connected wire segments), the points to be connected are derived.
d. *Point ordering.* The order in which the nets are routed is determined.
e. *Router selection.* The most appropriate router is selected.
f. *Knowledge updating.* The routing result, obtained thus far, is evaluated. Is it satisfactory? How many track segments and vias were used? If necessary, this information can be used by changing the order of routing nets in task d.
g. *Routing failure diagnosis.* In the case of a failure, NOP analyzes why the routing fails and how to avoid this in the future.

WIREX PROLOG

Fujita and Goto [Fuj83] described an experimental routing system to aid in the manual completion of the wiring in a circuit layout. The routing knowledge is described in a predicate logic form. The system gives a satisfactory reduction in the design time.

The routing system WIREX PROLOG of Fujita *et al.* [Fuj86] is an interactive system, that is, a designer resolves deadlock problems, whenever the automatic built-in router fails. Specific solving strategies can be incorporated in WIREX. The structure is shown in Fig. 8.34.

The system contains two databases: the knowledge base (which contains rules and solution procedures) and the CAD database (which contains general information about the physical properties of the wiring area, such as design rules). Advantage is taken from the inference mechanism inherent in Prolog. A Prolog interpreter functions as the interface system. It accepts predicates which correspond to procedures in the CAD database. These procedures manipulate the physical data in the CAD database and execute design primitives. When the designer gives the name of a rule or solution procedure, the Prolog interpreter translates the relevant program.

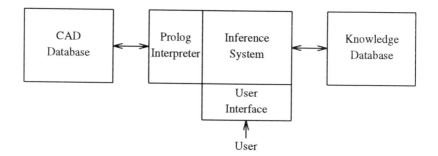

Fig. 8.34 The WIREX system

The part of the knowledge stored in the knowledge base is written in Prolog, but may contain calls of procedures in FORTRAN, stored in the CAD database. The Prolog interpreter then links the FORTRAN procedures, which are stored in the CAD system, to the sequence of operations to be executed. The use of FORTRAN routines speeds up the execution time.

Example: A simple WIREX PROLOG operation is illustrated by an example. The nets A-A and B-B are to be connected in one layer. Suppose that A-A in Fig. 8.35a has been laid so that B-B is blocked. A human design would resolve the problem by first connecting B-B and then A-A, as shown in Fig. 8.35b.

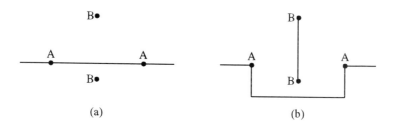

Fig. 8.35 Routing example in WIREX PROLOG

A suitable Prolog rule can be defined and incorporated in the knowledge base to handle this type of problem.

OCR

OCR (Optimal Channel Routing) is an algorithm that searches for an optimal solution of the channel routing problem in VLSI chip design [Wan88]. The algorithm uses good heuristics and dominance rules to terminate unnecessary nodes in the searching tree. All possible topological sequences based upon the vertical constraint graph are enumerated. For each sequence, a linear scan is conducted, and using the horizontal constraint graph, nets are assigned on the

8.3 LOCAL ROUTING

same track. An optimal assignment can be found after this enumeration process by a tree-searching method. The algorithm is implemented in Pascal and runs on a VAX/750.

Switchbox Router

Ho [Ho86] presented a switchbox router based on a constructive process using heuristic rules, which determine the next wire to route. Non-conflict nets are routed first without expending much effort. A route plan is then constructed in Franz LISP on a VAX 11/780 under UNIX. If the next net cannot be routed without conflicts, the conflicts are resolved by patching the almost-correct route plan. This repair process can be regarded as a planning problem in which the overall goal is to resolve all the conflicts and the individual interacting subgoals are to correct individual conflicts. Meta knowledge is introduced to control how, and how far, to undo conflict nets.

CARIOCA

CARIOCA is a switchbox router implemented with a blackboard architecture [Dub90]. Each net is first partitioned into a set of subnets. The routing is performed on a step-by-step basis, while new information becomes available at each step. This information is examined by a set of expert systems in the blackboard architecture in order to decide which subset should be routed next. CARIOCA has been written in LISP.

BLOREC

When the metal interconnections on printed circuit boards are dense and complex, automatic routers may fail to complete all interconnections. Some unroutable interconnections encounter a blockage caused by wires which were placed earlier. Blockage recovery, i.e., rearranging placed paths in order to remove the blockage, is usually a task of a human designer who relies on experience, intuitive insight and an understanding of geometrical relationships.

BLOREC (BLOckage RECovery system) is an expert system which is used to complete partially routed printed circuit blocks [Jos85]. It is a rule-based system implemented in LISP on a VAX11/780.

IPDA

IPDA (Interconnect Performance Design Assistant), developed at Hewlett-Packard, integrates a finite-difference numerical simulation method, a linear interpolation algorithm, an interactive performance synthesis methodology, and the lossless/lossy transmission line SPICE modeling capability into a spreadsheet-style graphical user interface [Cha92]. It assists users in selecting interconnect technologies and optimizing interconnect perfomance for the full hierarchy of packaging, including IC / MCM / PCB / lead bonding / via interconnect designs. The user interface of IPDA has been implemented using a spreadsheet running on UNIX, with the bulk work written in C.

Mask Pattern Generator for CMOS Cells
Shirakawa *et al.* [SHi89] described a rule-based expert system using OPS85 for generating mask patterns of CMOS logic cells. The cells are of the complementary type, in which the lower NMOS network has a connection graph dual to that of the upper PMOS network. The entire process consists of three phases:
1. Positioning of the center lines of source, drain and gate of each MOSFET. The design rules concerning clearances between source or drain and gate, and between source and drain, must be satisfied.
2. Positioning of the I/O terminal of a cell subject to certain design rules.
3. Routing on the first metal layer by a routing expert, which consists of three subexperts: (a) one to select a wire section of an incomplete net, (b) one to determine whether a selected wire section is on the side of the power-supply line or on the side of the I/O terminals, and (c) one to lay out an interconnection route in the assigned region.

8.4 LAYOUT VERIFICATION

8.4a Design-Rule Checking and Circuit Extraction

Introduction
The layout of an integrated circuit must satisfy a set of *design rules*. Two classes can be distinguished:
a. *Mask-design rules*, which define the geometrical and topological requirements, and hence are essential for the chip-layout designer.
b. *Process-design rules*, which define the process parameters, fixed by the chip manufacturer.
  Layout verification consists of two different activities:
a. *Design-rule checking*, that is, verifying if the mask artwork satisfies the mask-design rules.
b. *Circuit extraction*, that is, the reconstruction of the implemented network from the mask artwork.

For design-rule checking, only the mask-design rules are needed, e.g., minimum distance between two wires, minimum overlap of a polygon over another polygon, etc. Process-design rules (e.g., thickness of oxide layer, diffusion thickness, conductivity of diffusion and polysilicon layers, threshold voltage, etc.). are required for circuit extraction.

8.4 LAYOUT VERIFICATION

Design-Rule Checking

Design-rule checking (DRC) has to do with the geometry of the layout primitives (e.g., polygons) in each mask and the topology of these primitives in different masks. Parameters to be checked in DRC include minimum width of each polygon, minimum distance between polygons, minimum or full overlap (enclosure) of polygons.

An important class of efficient DRC methods is based on one-dimensional scanning, where a scan line is shifted from left to right and design rules are checked locally. In advance, the layout primitives are first sorted by increasing X and Y parameters.

Circuit Extraction

From the complete mask artwork and a number of physical data, the original logic circuit can be reconstructed. The reconstructed electrical circuit is a more realistic model of the actual implemented circuit than the designed circuit whose layout was constructed. This *circuit extraction* is therefore essential since it allows to verify whether the actual implemented chip satisfies the design specification. By analyzing the extracted circuit by using a circuit or logic simulator, the correctness of the design can be verified.

Circuit extraction consists of three parts:
a. Identification of transistors and other (desired or parasitic) circuit components.
b. Connectivity checks by comparing the interconnections of the reconstructed network to that of the original network.
c. Calculations of electrical parameters, such as MOS channel length and width, resistance and capacitance values.

Algorithms for the extraction of transistors depend on the implementation technology. For example, a MOS transistor requires 10 or more Boolean combinations. To allow simulation at circuit level, it is necessary to calculate electrical parameters. This generally requires the determination of geometrical parameters, such as distances and areas of polygons.

Layout Analysis

With the emergence of VLSI technology, the signal delays caused by the diffusion and polysilicon interconnection network between the transistors must be calculated in addition to the transistor delays. Formulas have been derived for the minimum and maximum values of the delays in an interconnection net with a tree structure [Rub83]. The analysis of the effects of parasitic interconnection resistances and capacitances allows one to minimize the distributed RC delays of VLSI wires [Bak90]. Particularly in submicron technology, both resistance and capacitance effects in the wiring nets interconnecting the transistors have to be considered [BRE89]. The properties of RC networks and the resulting propagation delays have been studied extensively in the literature [Bak90,

DEw90]. Closed-form expressions for the upper and lower bounds of the propagation delays in MOSFET interconnection nets with a tree structure have been derived [Rub83]. For practical use in timing simulators, it is usual to take the average between the lower and upper bounds for the delay.

Dewilde and Ning [DEw90] use a variant of the Galerkin boundary finite-element method and simplifications of it to analyze models of submicron integrated circuits. In order to make computations very efficient only the surface of conductors is subdivided into a mesh making the coefficient matrix of the network considerably smaller than in the conventional finite-element method, while the capacitance matrix is approximated by a band matrix. Parameters calculated are sheet resistance, transistor threshold and three-dimensional interwire capacitance. SPIDER is an efficient method for capacitance modeling of VLSI interconnections [Nin88].

Circuit Extractors

Circuit extractors can recognize devices, derive their connectivity, and discover violations of tolerance rules. RC network calculations in IC layouts are required to provide information for the purpose of reliable circuit simulations. The complexity of present-day transistor circuits is so large that practical use of circuit models in simulators requires the model to be reduced adequately. In their program SPACE (Submicron Parasitic Artwork to Circuit Extractor), Van Genderen and Van der Meijs [vAN88] remove circuit nodes in the model by local transformations. They use capacitance distribution and Gaussian elimination, which removes the internal nodes one by one. This method has the advantage of maintaining a Π type section between two output nodes. It preserves the total capacitance in the line and, more importantly, the delays.

8.4b Expert Systems for Layout Verification

CCOMP/EX

CCOMP/EX, developed at Toshiba Corporation, compares the interconnecting wires extracted from the circuit layout with the interconnections of the circuit produced in the circuit-design process [Tak88]. Many circuit-synthesis procedures include transformations of functionally equivalent networks with the objective of simplifying the circuit and hence reducing the chip area required. For that reason, the comparison process may generate a large number of false errors, which arise when the circuits are functionally isomorphic, but topologically non-isomorphic.

CCOMP/EX performs a reduction process in which series-parallel groups of transistors are transformed into AND/OR gates. A comparison is then performed on graph isomorphism. A rule-based system is used to remove any of the

8.4 LAYOUT VERIFICATION

inconsistencies that would produce false errors. The comparison process comprises performing a pattern match between the subcircuit being considered and a set of rules on functional isomorphism.

DARSI
DARSI (DAta Reduction System for Interconnects) is a reduction method, which removes circuit elements and nodes in the model, depending on the accuracy requested by the user [Van91]. Starting at the node before the end of the line, it moves backwards to remove nodes one by one. The reduction strategy is implemented as a set of production rules residing in the general-purpose rule-based verification environment DIALOG [DeM85].

CV
CV (Connectivity Verification) checks the correspondence of the interconnections of the elements in an NMOS circuit layout to those in the designed circuit [Spi85]. For some elements, the terminals are logically or electrically equivalent and hence may be interchanged. An example is provided by the source and drain of a MOSFET. CV takes a global approach by comparing groups of elements. The hierarchy in the design is used to form rules that match equivalent groups of elements.

Design hierarchy is used to set up patterns which correspond to equivalent groups of elements in the layout. These patterns form rules in a rule-based system with the circuit description as data. The CV procedure consists of two phases: rule generation and a comparison process. In the generation phase, a hierarchical description of the circuit is translated into a set of definitions and rules for a LISP-based rule interpreter, written in OPS5. In the next phase, the patterns in the rules are compared to the original circuit.

VERCON
VERCON (VERify CONnectivity) is a connectivity verification program implemented in Prolog [Pap88]. Like CV, it uses a global comparisons approach. VERCON uses the designer's hierarchical decomposition of the circuit to form Prolog rules describing the connectivity of all different subcircuits. Each of the rules is then matched against the transistor-level description stored in the database. Matching means that a particular combination of objects corresponds to one of the user-defined subcircuits so that this subcircuit can be extracted from the database. Verification of the layout is completed when the highest-level object in the hierarchy can be extracted.

Integrated Layout Verification
Simoudis and Fickas [Sim85] developed an integrated layout-verification system, which consists of a set of layout-design tools. The prototype system, called *Hephaestus*, consists of three tools:

1. *Rule-based cell-template layout tool.* To speed up the layout development, a set of expert systems has been implemented. Each expert system is dedicated to performing a specific design task. All expert systems share a common database which contains cell templates (e.g., for one-bit adders and random logic gates) to be used for generating layouts for the user.
2. *Goal-directed, rule-based layout editor.* This is based on some elements of Glitter, a knowledge-based system that translates a formal software specification and a set of design goals into a lower-level circuit design [Fic83]. Each method to reach a goal consists of three parts:
   a. The ultimate goal.
   b. A set of preconditions to determine if the method is suited to the specific goal.
   c. A set of actions: (2.c1) indicating a more specific subgoal as an intermediate step to the final goal, and (2.c2) calling the transformation which will change the cell and then the final goal.
3. *Rule-based, design-rule checker.* As soon as the circuit editor proposes a change in the layout, the checker is activated. The design-rule check verifies if the modified layout still satisfies the design rules of the implementation technology which is used.

The language ORBS (Oregon Rule Based System) has been chosen for several reasons:
a. Each circuit can be described in a frame-based representation.
b. This representation can be shared by all tools.
c. Each tool may have its own ORBS system.
d. ORBS have good facilities for passing messages.
e. ORBS has rule-based as well as procedural computational abilities.

Wenin et al. [Wen89] described a rule-based circuit-extraction program which, in addition to conventional electrical data, allows modeling of functional and timing behavior. The circuit extracted from the layout is provided with parasitic effects so that the desired circuit performance can be verified. The layout verification proceeds in several stages: design-rule check of the mask layout, circuit topology comparison with the designer's schematics, extraction of a circuit in a SPICE-compatible circuit description and a functional gate-level description. The functional analyzer recognizes transistor gate topologies. In the final stage, verification of the extracted electrical and functional views may report on errors, e.g., in critical paths and charge sharing.

Chapter 9

MODERN DESIGN METHODOLOGIES

9.1 HUMAN-COMPUTER INTERACTION

9.1a Interactive Graphics Tools

Graphics Tools
The graphics-display facility is one of the most attractive features of modern design systems. It renders immediate visibility of interim results of a design process in a way that allows the designer to interact with the system in a user-friendly and effective way [Min85, Shn87, Suc87].

At the heart of an interactive graphics system is the *visual display unit* (VDU) which mostly takes the form of the familiar cathode-ray tube with a screen to display all the graphical information needed for monitoring the design. This visual information, for which we use the general term *picture*, may contain a clarifying text, circuit schematics, timing characteristics, a mask artwork or any other graphic information which may be useful as the design evolves. A careful choice of phosphors coated on the inside of the screen makes it possible to transform the energy of the electron beam focused on it into visible light. Usually, a display screen is partitioned into a *menu area*, containing command names and design parameters, and a *data area* for displaying pictures which contain graphical design information. A cursor can be moved on the screen under the control of the user.

In order to enable human-computer interaction, we need interactive I/O devices. Several *graphics input devices* are now available to allow the user to specify input information to the design system. One classic device is the *alphanumeric keyboard* which permits the user to enter the specification of a design description (in terms of text and numerical data) as well as a set of commands. Keyboards are usually extended by *function keys* to select different types of functions or subroutines, ranging from a simple operation, such as controlling the cursor position or a set of many different operations. Keyboards and function keys are the common sources of text input and choice input to a graphics-display system. Input operations include the indication of locations on the screen, the selection of entities displayed on the screen and the specification of input values.

Special input devices can be used to map a cursor on the screen. A cursor pointed inside the menu to select a command from the menu list. The cursor on the screen can be defined in several ways. A *mouse* is a small mechanical box containing two orthogonal wheels whose position is converted to the X and Y coordinates of the cursor on the screen. A *track ball* contains a sphere which can be rotated freely with the palm of the hand. The box with *thumb wheels* provides separate control of the X and Y coordinates of the cursor. A *joy stick* is an upright handle which may be turned freely around two axes for controlling either the position of the cursor or the speed at which it moves.

A *light pen* is a pointing device which allows the user to work directly on the screen of a refresh VDU by sensing light on the face of the screen and then generating an interrupt. The light pen provides an immediate identification of graphical information by a direct optical contact with the screen. For example, a designer may use the light pen to determine the coordinates of the point on the screen at which the light pen is pointed or it may assign a value to a particular parameter by pointing it to a displayed table.

A *digitizer* has the form of a draftman's drawing board. It allows the designer to draw a rough schematic which is digitized, that is, translated into digital format for further processing by the design system. The picture to be drawn is entered by a probe which is moved over the surface. A switch associated with the probe enables the user to register X and Y coordinates at any desired position on the surface and hence on the display screen.

A *graphics tablet* is a flat panel used as a sketch pad across which a pen-like stylus can be moved to define a desired path and/or location of the cursor on the display screen. It is a small low-resolution digitizer, often used as an alternative to the light pen or in association with storage-tube displays. The tablet provides a one-to-one mapping to the display surface of the screen. The position of the stylus over the surface of the tablet cause the cursor to take the corresponding X-Y position on the screen. Hence, a picture drawn on the tablet is directly echoed on the screen.

Intermediate results of a design process can be directly made visible on the screen of a visual-display unit. Besides the interactive display devices, there is a need for *output devices* that produce hard copies of interim design results for different purposes. Printing and storage peripherals are surveyed by Hobbs [Hob84]. The most important plotting devices, many of which have a color capability, will be briefly outlined below.

A *drum plotter* contains a roller on which the drawing paper is attached with a sprocket-wheel mechanism. While the roller can be rotated back and forth, a pen can perform a one-dimensional motion, producing a two-dimensional picture. In fact, the picture consists of sequences of small line increments in a number of predefined directions. Hewlett-Packard's *drafting plotter* is a roll-feed plotter with eight pens in various colors and line width for drawing pictures, including non-orthogonal lines, circles, arcs and other geometric primitives. This plotter allows the user to position labels, change character size, slant and direction, while plots can be rescaled, rotated or windowed (that is, only a portion of the original picture is plotted).

In a *flat-bed plotter*, pens of different colors can move in two orthogonal directions to produce two-dimensional layout patterns. A microprocessor may be integrated to control the scaling, speed and the use of special character fonts. An *electrostatic plotter* produces pictures at a high speed. The time needed to complete the picture, which consists of small dots in a predefined raster printed row by row, is independent of the complexity of the picture. Whereas an electrostatic plotter uses an electron beam for writing, the *laser plotter* or *printer* writes the picture line by line by a laser beam (see also page 42). Other hard-copy plotters include the *inkjet plotters* and devices operating on a photographic or electrical basis, directly derived from the pictures on the screen.

Interactive Graphics Displays

A *Visual Display Unit* (VDU) may contain a Cathode-Ray Tube (CRT) or a non-CRT device, such as the plasma and the matrix displays. CRTs may be of the *storage type* or the *refresh type*. Important considerations of a display screen are the *resolution* (the smallest discernable picture detail) and the absence of flicker in the picture.

The *Direct-View Storage-Tube VDU* is based on the electrostatic charging of the phosphor coating on the inner surface of the display screen. Green pictures can be generated by emitting high-energy electrons from a writing-beam gun. Low-energy flood-gun electrons keep the picture visible by maintaining a low-energy electron stream across the entire screen. In addition to the low luminosity and contrast, the storage-tube VDU has the disadvantage that the entire picture must be erased and rewritten when one wishes to delete even a small part of the picture. The picture is erased by hitting the screen with a positive charge pulse.

Refresh displays have the characteristic that the pictures on the screen must be written repeatedly at a rate of at least 25 cycles per second in order to avoid flicker. This implies that extra hardware is needed to perform the refreshing task. In a *Random-Positioning Vector-Refresh VDU* (or *Random-Scan VDU*), a picture is generated on the screen of a CRT by directing an electron beam only on those parts of the screen where the picture is to be drawn. A display generator provides the appropriate voltages for deflecting the beam in the X and Y directions and for controlling the beam intensity. A display file stores the data needed for displaying vector, character and other graphical information. A high resolution can be attained with this random-scan method. The quality and costs of random-scan displays vary in a wide range. Special hardware can be used to speed up the generation of special figures, such as line and characters.

In *Raster-Scan VDUs*, a picture is generated by an electron beam which scans a raster of picture cells on the phosphor-coated screen, much like a television picture is obtained by a scanning procedure. The picture cells, called *pixels*, are arranged in a regular pattern on the screen and are either illuminated or not as the electron beam is scanning the raster line by line. A set of three beams can excite red, green and blue emissions from different phosphors to enable displays in color. The display generator usually consists of a display processor (which performs several functions, such as cursor display, windowing, clipping, etc., as commanded by the user), a display controller (which performs the interface with the VDU), a refresh memory and a video driver associated with the VDU. The beam intensity can be controlled at a very high rate. Raster-scan displays permit fast local changes of the picture. A limitation is the resolution, which is not so good as that of random-scan displays.

A more recent development is the *plasma display*. The screen is composed of a sandwich of two glass plates with neon gas in between. The inner surfaces of the glass plates are covered with grids of horizontal and vertical lines, which produce a two-dimensional array of pixels in the interconnection points. DC or AC voltages applied to the two grids allow one to selectively write or erase a desired pattern of pixels. Since the neon pixels remain on until turned off, they can store the picture data, eliminating the need for extra refresh hardware. Besides the plasma display, other flat-panel displays are gaining acceptance: the *flat CRT*, the *electroluminescent display* and the *liquid crystal displays*.

Graphics Software

It is evident that an adequate amount of software is required for the proper operation of a graphics-display system [ROg85]. Picture-generating software has turned out to be more intricate than was first anticipated. For example, even such an ostensibly simple operation as drawing a straight line segment through distinct pixels is not trivial, particularly when the number of pixels is relatively small. Various graphics languages have been proposed, e.g., EULER [New79]. Special-purpose routines for graphics programming have been developed. A typical example is GINO-F (Graphical Input and Output in FORTRAN), originated from the CAD center at Cambridge, England.

A central issue in computer graphics is the use of appropriate data structures [Fol82]. Data structures for picture processing include linear lists, tree structures, graph structures and recursive structures. Particularly, dynamic arrays suitable for representing tree structures are used. Sofware modules can be stored and called when they are needed. Command languages with a great flexibility in format are available to the user for data entry and action control. Commands are available for data definition and transportation, data-flow control and program execution.

SIGGRAPH is an ACM (Association for Computing Machinery) Special-Interest Group which has been very active in the field of Computer Graphics since 1974. One problem in computer graphics is the release of different, nonstandardized graphics equipment by different vendors. The need for *standards* in computer graphics is evident and many committees have labored on defining requirements for sofware that should be device independent. To interface this software to different input devices or different display devices, one could use a suitable device handler or device driver respectively. A better solution, however, is to agree upon suitable graphics standards.

National and international standards organizations have proposed at least six standards: IGES (Initial Graphics Exchange Specification), NAPLPS (North American Presentation Level Protocol Syntax), GKS (Graphical Kernel System), PHIGS (Programmer's Hierarchical Interface to Graphics Standard), VDM (Virtual Device Metafile) and VDI (Virtual Device Interface). See Computer, Oct. 1985, pp.63-75.

9.1b Intelligent User Interfaces

Developments in Human Interaction
There are five major types of communication between the expert system and the human user [Lug87]:
1. *The batch approach.* A typical example is provided by the computer configurer XCON (see Subsection 4.5c). Given a set of specifications derived from the purchaser's order, along with the hardware constraints (bus, disk size and controllers, main memory, etc.), XCON builds up the desired computer configuration.
2. *The question-and-answer approach.* An example is the medical diagnosis program MYCIN (see Subsection 4.5b). Preliminary data from the patient (name, age, sex) and symptoms (headache, dizziness, etc.) are asked and answered. The questions are focused on trying to establish whether the patient has meningitis. This approach requires large amounts of typing and is therefore rather slow and hardly used.
3. *Direct access of data.* An example is ONCOCIN, designed for recommendation of cancer therapies [Sho84]. This approach overcomes the cumbersome interface of MYCIN by directly accessing the records of the patient for relevant information (from age, weight, to the results of various tests).
4. *Menu-driven approach.* This approach is used by many PC-based shells. The use of a mouse and menus in the question-and-answer sequence proceeds considerably faster than in the MYCIN approach. This approach is also less prone to errors in that the rule designer can specify exactly what answers are expected from each query.
5. *Graphic or Direct Manipulation.* This is the most user-friendly interactive approach to human interaction. Its roots are the object-oriented programming approach, as implemented in Smalltalk. Moving the cursor and clicking the

mouse to change parameters or values gives a comfortable way of user interaction.

An advanced form of computer use is the bidirectional interactive human-computer communication. A vital part of interactive systems is the interface mechanism, which translates user inputs into the internal format and, conversely, converts the system's conclusions into graphics or natural language.

In the past, user interfaces forced the user to descend to the level of computer language, that is, the user must learn computer-understandable code to communicate with the computer. The future will bring more and more intelligent interfaces, which allow human users to communicate intelligently with the computer, preferably by graphical means. The clumsy keyboard is making way, at least partly, for the mouse and menu clicking on the display screen. In a more distant future, speech may be used as the interactive medium [All85].

The increasing size and complexity of expert systems, including the use of deep knowledge models, entails an increasing amount of resources to implement user interfaces. Alternatives to the cumbersome and error-inducing keyboard have been sought. Mice, icons, menus and display windows have become virtually standard components for a graphical interface. The mouse is the pointing device. Icons symbolize application programs and data. Menus allow choices of action to be made. Windows divide up the screen to provide an overview of related subproblems. The display screen filled with icons, menus and multiple windows controlled by a mouse is a recognized improvement, which is generally introduced in personal computers and workstations. A current development is the use of speech as an input/output vehicle [Cul86, Voi90]. It implies the synthesis and recognition of continuous speech. In the voice-input concept, machines recognize and interpret spoken commands, after which preprogrammed assignments are performed in response.

For the computer professional, an intelligent interface machine may help solve the software crisis by providing hardware support for the burgeoning new trend of visual computing [Gra85]. Interface systems have been developed enabling users to access databases in natural language without having knowledge about query languages and databases [Ish87]. An interesting development in this respect is *Hypertext* or *Hypermedia* (see Subsection 4.4a).

X-Window System

A window system has become a standard software component on general-purpose workstations [HOp86]. To bring more unity in the Window concept, MIT designed the X-Window system, X for short [Sch86, Jon89]. X presents to the user a hierarchy of resizable overlapping windows providing device-independent graphics. This system distinguishes itself fundamentally from other graphical user interfaces, because it is based on a network-oriented client-server model. Because of the application possibilities and the wide portability, the

9.1 HUMAN-COMPUTER INTERACTION	473

interest in the X-Window system is increasing. Application programs which employ X to control a display screen can be applied without many problems to a multiplicity of workstations and X terminals of an increasingly number of manufacturers.

The X server is the part of the software that supports the reproduction of the Windows. The clients are the applications which make use of the Windows. The way in which the X server must be started up depends on the workspot system of the user. Four workspot systems can be distinguished: the PC + network card + X server, the X terminal, the X station and the X workstation. In an X terminal, the server is usually built in, otherwise it is read in by starting up from a computer via the network.

The X terminal and the PC (with network card and X server) cannot themselves manipulate client programs, but they do display the graphical output. On the other hand, the X station (often referred to as diskless workstation) has the possibility to manipulate client programs and the X workstation unifies the client and server functions. X-Window System is an open standard available on numerous architectures, allowing packages to be freely exchanged between different systems with considerable ease.

Computer Communication Networks

Complex design projects require the cooperative efforts of a number of human experts. When these experts work at different locations, there is a need for an efficient communication network to facilitate real-time exchange of ideas. Data communication capabilities are expanding gradually. High-speed LANs can be interconnected into larger entities as metropolitan-area networks or even the larger wide-area networks.

An improvement in data rate is manifest in the evolution of high-speed networks to supercomputer and ultragigabit networks in the nineties (see Subsection 2.3c). Fiber-optic networks have a profound impact on information transmission in the future. The transmission of messages as pulses of light along an almost invisible thread of glass is effectively taking over the role of guided transmission by copper and, to a modest extent, the role of unguided free-space radio and infrared transmission [Gre91, GRe92].

The markets for text, data and image communications will increase in the near future. This multimedia transmission is established by accessing gigabit-per-second public networks on a switched basis. The main elements of the new networks will be based on a series of emerging international recommendations known as Broadband Integrated-Services Digital Network (B-ISDN). The series is being developed by the CCITT (International Telegraph and Telephone Consultive Committee) to describe techniques for packet-switching data, voice and video at very high speeds [Kap91, Sta92].

New services are being introduced in the telecommunication industry (see

IEEE Spectrum, 29, p. 37, Jan. 1992). The Switch Computer Application Interface (SCAI) will allow telephone subscribers to design SCAI applications, while the network is protected against applications that might damage the integrity of the public network. The Intelligent Network will be a complete infrastructure in the telecommunication network for creating, testing, deploying and operating new services using personal communications equipment.

Human Factors in Using Graphics Displays

Usually, the blessings of visual-display systems are brought abundantly to the fore. However, the less beneficial aspects of such systems should also be considered. Human factors engineering aims at designing environments, equipment and tools to accomodate human needs and limitations [Min85, Shn87, Suc87].

Prerequisites for good working conditions are the right lighting and the most appropriate physical layout of the workstation in the workspace. Eye and muscle fatigue of VDT operators and possible harmful effects of extremely low-frequency magnetic fields should be avoided. Taking human factors into consideration means that ergonomic, health and safety aspects of working with computers, notably with the graphics displays, are of vital importance.

The combination of functional and ergonomic requirements should constitute the starting point for designing and selecting interactive computer systems. A great deal of research concerning human-computer interaction has been conducted by cognitive psychologists and human-factors specialists. The ergonomic requirements, including psychological aspects and the working environment, have been studied by several authors. Efforts are being made to improve the quality and the ease of human-computer interaction.

Ergonomic Aspects of User Interfaces

In line with the WYSIWYG requirements, the emergence of window systems with WIMP features have made computer sessions more attractive. However, the combination of functional and ergonomic requirements should constitute the starting point for designing and selecting interactive computer systems.

The ergonomic aspects of using a workstation for solving CADCAS problems are mainly related to the human-machine interface. Apart from relieving the skilled designer from routine work, a good interface should be user-friendly [Car83, Shn87, Nor86]. In the past, human designers had to make changes in their problem-solving styles in order to accomodate the computer. In particular, the Von Neumann architecture forces the programmer to implement algorithms in which a problem-solving process is broken down into a sequence of instructions, one after the other. Presently and in the future, specific characteristics and requirements associated with the cognitive and behavioral styles of human brains are exploited to adapt computer systems to providing easy access to computers.

9.2 WORKSTATIONS FOR CHIP DESIGN

9.2a Engineering Workstations

Turnkey Systems

An interactive graphics system may be a local stand-alone system or a graphics terminal of a communication network. Starting in the mid-1960s, completely self-contained hardware and software systems have been developed which can perform a specific design task. Since the user simply has to "turn the key" to get the system running, they are referred to as turnkey systems. Gradually, more design tasks are relegated to the system and the designer is allowed to interact during the design process.

Intelligent graphics terminals of a telecommunication network are distinguished by the degree of "intelligence", indicating how many local functions can be performed, independently of the host computer. To this end, terminals have a local minicomputer or a personal computer and various types of memory devices and a set of input/output devices with more or less sophisticated editing facilities. Advances in microelectronics have led to the development of intelligent microprocessor-based workstations suitable for chip design. With hardware costs steadily dropping and personnel costs rising, workstations are cost-effective tools for chip designers who want to keep control of the whole design cycle.

CADCAS Workstations

The workstation has become an indispensable tool in chip design. It increases the efficiency of the design process and reduces the turnaround time. Improved workstations are being developed by using advances in semiconductor technology and computer architectures, design process and microelectronics price/performance ratios. Workstations are high-performance systems which address desktop computing problems.

The main characteristics of workstations with a view to CADCAS are outlined below.

General characteristics. The minimal operating speed is one million instructions per second. In general, the machine has a 32-bit architecture. The CPU is assisted by a coprocessor to allow heavy computations. The internal RAM contains minimally 16 Mbytes. The hard disk has a storage capacity of at least 100 Mbytes. The operating system, preferably UNIX, has a networking capability. A high-resolution graphics display screen is obtained by a bit-mapping system with at least 1024 by 1024 pixels.

User Interface. A sophisticated user interface is obtained by providing overlapping *windows*, controlled by a window manager program. The user interacts by using the keyboard and a pointing device, notably a mouse. File handling is facilitated by the use of icons and menus. Workstations support graphical applications, such as hardware accelerators, for graphics manipulation and standard graphics software packages, such as GKS.

Multitasking. Though each workstation can be used by a single user, it supports a multitasking operating system. Separate processes can be assigned to different windows so that the user can perform several tasks concurrently. Other tasks may be run in the background, as in conventional time-shared systems.

Networking. Modern workstations provide networking facilities. Workstations may be grouped in clusters by a Local Area Network (LAN), such as the Ethernet bus system and the Apollo DOMAIN token-passing ring. Such LANs provide the ability to access a remote machine (e.g., a supercomputer) and various user facilities. Files can be transferred from one machine to another. Wide-Area Networks can be accessed via a gateway.

Distributed Filing. The power of workstations is increased through their communications capability, including accessing and transferring files, processing data entries and making use of query facilities. Accessing remote databases ensures us of using the newest updated information. A convenient feature is to access files on remote machines without any special command syntax or explicit copy operations. Diskless workstations share a single large disk, often attached to a remote dedicated machine (file server). Examples of distributed filing systems are Apollo's Domain File System and Sun Microsystem's Network File System.

Open-System CAD Framework. Chip design benefits from advanced tools, including schematic capture and hardware accelerators for special tasks, such as simulation, testing and layout. The trend is to use an integrated design system operating in an open system environment.

Multiwindowing

The concept of *windows* may relate to three different levels of an information system:

a. *Database and data-communication level.* Windows may be a measure of the capacity of a specific workstation to exploit external sources for extending its capabilities. In particular, IBM defines windows as the maximum number of mainframes or other computers which can be connected to a particular personal computer or workstation.

b. *Data-processing level.* Windows can be defined in connection with different programs running under a single command structure. We distinguish closed and open systems. Closed systems are really integrated software systems, as exemplified by Lotus 1-2-3 and Symphony. The interface of an open system

9.2 WORKSTATIONS FOR CHIP DESIGN

allows different applications to coexist with a common command structure and file reference.

c. *End-user interface level.* Several windows can be displayed simultaneously on the screen. Pop-up or pop-down menus can be used to select specific applications (see Subsection 2.1a).

Graphics Supercomputers

The developments of graphics tools have been incited by the demands posed by computer-aided design. Conversely, their availability has been a tremendous help in the development and proliferation of CAD systems.

By combining the arithmetic power and wide-bus structures of supercomputers and built-in graphics of workstations, a new class of computers has emerged. They are referred to under various names: graphics supercomputers, personal supercomputers, advanced graphics workstations, etc. The main feature is their ability to display time-dependent information about complex systems in real time.

9.2b Advanced CADCAS Workstations

Engineering Workstations for CADCAS

The use of an engineering workstation has become essential in the Computer-Aided Design of VLSI Circuits and Systems (CADCAS). For our purpose, an *engineering workstation* (EW) will be defined as an integrated design system which provides the chip designer all hardware and software tools necessary for performing the complete design task interactively at a single location. The main goal of an EW is to improve the product quality and at the same time to increase the productivity of the chip designer. An example of a supercomputer workstation for VLSI CAD was described by Fiebrich [Fie86]. The connection machine offers its users general-purpose acceleration capabilities and high interactivity.

Characteristic of engineering workstations is some form of human-computer dialog. An EW should contain the following essentials:

a. A graphics-display system, which provides the chip designer a pleasing facility to interact with the design system by graphical means.
b. A complete set of CADCAS tools for the synthesis, simulation, test generation and layout of the intended circuit design.
c. A database system, which interrelates and integrates the data of application programs at different levels of abstraction.

The graphics display facility is one of the most attractive features of modern design systems. It renders immediate visibility of interim results of a design process in a way that allows the designer to interact with the system in a user-friendly and effective way. The graphics-display system provides the designer with a means for schematic capture and design monitoring on a display screen.

Graphics schematic entry is the specification of the circuit to be designed in the form of a schematic, which is an interconnection of circuit primitives represented by appropriate graphic symbols. Sophisticated pictures may be composed of various geometric primitives, such as lines, circles, ellipses, parabolas, conics, splines, etc. A computer program translates the graphic symbols into corresponding machine-oriented descriptions. Circuit schematics can be constructed by inputting three classes of commands:

a. Identifying picture primitives, such as squares, circles and even complete graphical symbols of circuit primitives,
b. Scaling, rotating and positioning in space, and
c. Combining the primitives by operations, such as union and intersection.

A CADCAS system usually employs an established library of predefined transistors or circuit modules. A circuit or system can be composed by calling out the components which are needed, positioning the components and indicating the interconnections to obtain the desired circuit or system. During the design process, transistor and other symbols can be moved to any desired location and scaled to any size. An alternative way to specify the design is to use a hardware-description language (e.g., VHDL or Verilog). The design description is similar to a computer program.

Design monitoring is concerned with the use of the graphics-display system with its input/output devices to monitor and manipulate the design process as the net lists for circuit analysis and design. By using *schematic editors*, the designer is able to modify the circuit and add necessary information, e.g., rise and fall delays for the purpose of logic-circuit simulation.

Layout editors allow the designer to completely design the circuit layout on the screen. Layout modifications can be entered and violations of design rules may be flagged. The graphics system may also be used for documentation purposes. Circuit and logic schematics may be produced, while layout mask plots and pattern-generation data files can be derived from the approved mask artwork. Model information can also be entered by programmed extraction of information from a database.

Commercial Workstations

Workstations include computer systems ranging from modern powerful personal computers based on 80386 or 68030 chips or better to Precision-Architecture Reduced Instruction Set Computer (PA-RISC), such as the SUN, SPARC, and IBM RS/6000. Advanced workstations, such as Hewlett-Packard/Apollo computers and Sun Microsystems, contain floating-point coprocessors. PA-RISC is a load-store architecture with fixed 32-bit instruction size. The Intel 80486 and Motorola 68040 microprocessors have built-in coprocessors enabling number-

9.2 WORKSTATIONS FOR CHIP DESIGN

crunching activities. Some workstations, including HP/Apollo 9000, support the use of object-oriented languages, e.g., C++2.0. SUN Microsystems offers computers based on SPARC and the UNIX AT&T Release 5.4 operating system.

There is a wide variety of options for backing up design information, ranging from reels of half-inch tape to Write-Once/Read-Many (WORM) cartridges. In performing computation-intensive tasks, care should be taken that time-consuming transfers of data do not cancel out the advantage of using high-speed coprocessors. Using faster and larger RAMs reduces the time spent moving between disk storage and the RAM. Adding a cache memory (small high-speed virtual memory placed between the CPU and the RAM) may also speed up the program execution. Most workstations are provided with hard-disk drives.

A typical workstation is divided into four systems: PA-RISC processor, memory, built-in IO and graphics. The PA-RISC processor consists of a CPU, a floating-point coprocessor and separate data and instructions caches. Rachowitz *et al.* [Kap91] reported on an integrated design system, which includes a Cray supercomputer under Unicos, Amdahl mainframes under MVS/XA and VM/XA, DEC minicomputers under VMS and Ultrix, HP/Apollo workstations under Domain, IBM RS6000 workstations under AIX, Sun workstations under SunOS, IBM PCs and compatibles under MS-DOS and Macintosh personal computers under MacOS.

To enhance teamwork in the design process, workstations must be tied together into a local-area network that optimizes the use of shared resources. A trend is the migration of certain functions from large centralized mainframes and minicomputers to a distributed LAN-based computing environment. Such an environment can be used to exchange design information among cooperating designers in real time.

If necessary, workstations can be customized to individual user's needs. Schematic entry, editing and generation are useful aids in model building, provided by several current software systems, such as Silvar-Lisco's CASS program. Continuing efforts are being made to improve the workstation capabilities is various aspects. The aim is at improving the ergonomic aspects, the resolution of the displayed images and the storage capabilities.

Most CADCAS workstations offer schematic entry and circuit or logic simulation capability. Xerox Corporation added a subsystem SIZING that automatically calculates transistor sizes for NMOS and CMOS circuits, based on user-specified delay requirements. The SIZING program understands electrical requirements. For example, if a gate drives or is driven by a pass transistor in NMOS circuits, the pullup/pulldown ratio should be 8. The system also includes a program that calculates path delays for standard-cell circuits.

A knowledge-based management system (KBMS) may consist of special components: a set of dedicated expert systems, a database management system, processors for handling special types of data and suitable interface equipment.

Advanced KBMSs which combine features of database management systems and expert systems (see Subsection 4.4a), can handle special types, such as images and speech.

9.3 PROBLEM-SOLVING TOOLS

9.3a Machine Learning

Learning Methods

A major characteristic of intelligence is the ability to learn, that is, to acquire new knowledge and understanding by study, instruction or experience. The objective of learning is to extend and improve the knowledge acquired thus far, as new elements of knowledge are presented. The ability to learn depends on the capacity for generalizing, for drawing analogies and for selectively discarding irrelevant information. Considering the benefits gained from learning processes, there has been a continuing effort of AI scientists for developing machine learning methods [Mic83, For89, CLa90]. The difficulty of acquiring knowledge from human experts is an additional incentive to build learning machines, which automatically enhance the global body of knowledge in knowledge bases and communicate this knowledge to people. Knowledge representation, knowledge acquisition and learning methods are interrelated subjects. Machine learning may help knowledge engineers update the knowledge base of an expert system [For86].

Being an application of Artificial Intelligence, a learning machine is an adaptive system, which makes choices about the most appropriate course of action to take and modifies its choice-making behavior if a choice turns out to be inappropriate. By using adequate knowledge models of the problem at hand, the outcomes of specific choices can be evaluated. Depending on the responses obtained, the knowledge model is updated accordingly. In fact, a learning system has to perform a kind of self-diagnosis with the purpose of expanding the knowledge base. This is achieved by searching a space of representations to find what best fits known observations.

Michie [Mic88] distinguishes three criteria for machine learning:
1. Weak criterion: The system uses a training set to generate an updated basis for improved performance on subsequent data. The problem-solving performance is of primary importance rather than such properties as explanation of its reasoning. Examples: Statistical, genetic and connectionist methods of learning.

9.3 PROBLEM-SOLVING TOOLS

2. Strong criterion: The system must additionally be able to "communicate its internal updates in explicit symbolic form" and thus be able to explain in an understandable way what it has learned.
3. Ultra-strong criterion: Internal updates must be communicated in "operationally effective" as well as explicit form, i.e., in a form that additionally allows the expert to improve his or her own performance as well as to understand the machine's behavior.

A learning machine must have some essential abilities: memorizing (storing) training examples of specific problems and already known solutions to them, assessing the examples for the purpose of classification, and generating new knowledge to improve the knowledge which has been acquired thus far. Though there exist a wide variety of techniques used for machine learning, two main approaches may be identified: *inductive learning* and *explanation-based learning*.

Inductive Learning

The most widely used learning methods are based on the rule-induction paradigm. Inductive learning is often called concept learning or learning from examples or concept learning from examples. Though various definitions of concept have been proposed, with origins in philosophy and cognitive psychology, a workable definition should be adopted for its use in machine learning [Ber91].

For our purpose, we conform to the heuristic view and consider a *concept* as a set of objects subject to conditions sufficient to decide whether or not a given object is an element of the set. The set of objects (or the concept) belongs to a known class described in terms of its attributes and values. Then, the task of a learning machine is to induce generalized descriptions of concepts by analyzing the training examples for the purpose of improving the current knowledge.

Inductive learning implies inducing classification rules from a set of training examples. An *example* is a true instance of a given concept, while a *counterexample* is not an instance of it. The existing classes can be used to recognize an example as an instance of a particular concept. A new example is classified by comparing its attributes to class attributes. Induced descriptions of concepts can be used for constructing new classes.

Inductive learning involves the generation of rules and patterns, given a set of specific training examples. The progress of the generalization is tested by applying the rules to a new set of examples, called the test set. Any inductive method has two aspects: the ability to *represent* the concept of interest and to *learn* an optimal representation from the examples.

The learning system searches a space of rules to find those which "best" classify the training examples, where "best" is defined in terms of accuracy and comprehensibility. Inductive rule learning is based on a simple pattern-recognition model of learning, in which correlations between observable features and some final classifications are sought for.

The initial features used to describe examples are chosen manually, and classification is made into one of a fixed number of user-defined classes. The inductive system does not make use of any other domain-specific information beyond that of the training examples themselves. Induced descriptions of different classes of concepts are divided into two different formalisms, with classical representatives being AQ [MIc83] and ID3 [Qui83].

The rules which AQ and ID3 produce constitute a simple "model" of the world, automatically generated from the observations with which they have been presented. Although the structure of this model is simple, the operations they perform in generalizing, compressing and organizing data are fundamental to learning.

AQ Algorithm

Essential in inductive learning is how to characterize a concept and how to classify it. In AQ15, Michalski [MIc86] uses classification rules, which are reminiscent of the production rules in a rule-based system. The rules have the form of logical expressions represented in the propositional language VL_1. Training examples in AQ are represented as a conjunction of selectors, i.e., relational statements that typically contain a predicate descriptor, variables or constants as arguments, and a list of values. A concept-description language VL_2 allows conjunction, disjunction, internal disjunction and numerical quantification over single selector formulas.

A decision rule induced by AQ has the form:

if <condition> *then* predict <class>

where <condition> is a conjunction of selectors or a disjunction of conjunctions. A new example is classified by checking which of the induced rules have their conditions satisfied by the example.

ID3 Algorithm

The "rules" that ID3 (Interactive Dichotomizer 3) learns are represented as a decision tree, which is produced in stages [Qui83]. A node of the tree involves a test on an attribute and each outgoing branch corresponds to a possible result of this test. The leaves of the tree are labeled by class names. To classify a new example, a path from the root of the decision tree to a leaf node is traced. At each internal node reached, the branch corresponding to the value of the attribute tested at that node is followed. The class at the leaf node represents the class prediction for that example.

The ID3 learning algorithm proceeds as follows. A decision tree is generated in stages, starting as a single node containing all the training examples. An attribute test can be used to sort the examples in classes. If all these examples are not of the same class, the test leads to different values of the attributes of the

9.3 PROBLEM-SOLVING TOOLS

examples and the node generates a number of outgoing branches corresponding to these values. To choose which attribute test to use, consider what the new tree would look like for each possible attribute and choose the best.

If all the examples at a leaf node are of the same class, this node does not need to be expanded further. If it contains more than one class, it must be expanded further. The procedure continues expanding the unfinished nodes until no more need expanding.

ID3 uses a function called *entropy* to measure how well an attribute test sorts examples into classes: The expansion of the node with the lowest entropy is the "best" sorting of the examples at the node. The entropy measure used to select the best attribute test is defined as follows:

$$\text{Entropy} = \sum_i w_i E_i$$

where w_i is the weight of the ith branch defined as the number of examples in branch i divided by the total number of examples at the parent node, and E_i is the entropy of the ith branch given by

$$E_i = -\sum_j p_j \log_2 p_j$$

where p_j is the probability of the jth class in this branch, estimated from the training data.

An improved inductive knowledge acquisition tool is C4 [Qui87], which is a descendant of ID3. The algorithm C4 constructs decision trees from objects described in terms of a fixed set of attributes. It is capable of dealing with continuous or discrete attributes, noisy data and missing attribute values. C4 is implemented in C and runs under UNIX.

Limitations of Inductive Learning

Induction involves a series of generalizations from an instance to a class. Several generalizations are possible. From the applications point of view, rules are more flexible, while trees are more efficient for classification. The reliability of inductive learning results depends on the quality of the training examples. Recent developments seek to extend the applicability of rule-induction learning and overcome the difficulties presented by real-world applications.

When attempting to find a satisfactory solution, the designer must make a trade-off between flexibility and tractability of the learning method. Allowing the system a greater flexibility to learn by introducing a more powerful representation language for embodying learned knowledge immediately creates a major search problem, which can easily become computationally infeasible to perform. Current research handles this trade-off.

Clark [CLa90] indicates two classes of limitation with the inductive algorithms:

a. There are a number of deficiencies with the mechanisms they use. This can be overcome by refining the algorithms.
b. There are restrictions imposed not by the particular algorithms themselves but by the whole paradigm of rule induction. These restrictions are caused by the inherent limitation of learning rules with such a simple structure (simply looking for input-output correlations).

The weak points in the algorithms described above are connected with several limiting assumptions of which the three most important are given below [CLa90]:

a. There is no noise in the training data, that is, any genuine regularity in the data is perfect (i.e., without counterexamples). Consequently, the systems will return only rules which are completely consistent with the training examples. They reflect genuine correlations between attributes and classes in the domain and consequently perform well.
b. The features used to describe examples are adequate, i.e., correct classification rules can be formed by Boolean combinations of tests on an example's features.
c. The number of training examples is small enough for the algorithms to run to completion in acceptable time.

Explanation-Based Learning

Several authors introduced a new learning method that has been named *explanation-based learning* [DEJ86, Mit86]. Unlike the induction method, which assigns each trained example to a proper class, the explanation-based approach focuses on explanations to confirm new knowledge. An explanation of why a new example is an instance of a goal concept validates this example to be an improvement of the knowledge base.

A deductive approach is characteristic of explanation-based learning with extensive use of the knowledge in the domain of the concepts to be learned. This requires a complete and correct domain theory so that for every example it can provide an explanation of why it is actually an instance of a goal concept. The domain theory is capable of explaining a training instance and learning occurs by generalizing an explanation of the training instance. Representative methods of explanation-based learning are EBG [Mit86] and EBL [DEJ86].

An application of explanation-based learning is LEAP (LEarning APprentice) [Mit85], which is employed in the VLSI design assistant VEXED [MIt85]. LEAP is a learning-apprentice system that directly assimilates new knowledge by observing and analyzing the problem-solving steps performed by a user. Suppose a rule in the knowledge base is missing, thus preventing VEXED from solving a problem. When the user interactively presents a solution to the problem, LEAP verifies the solution (this being an explanation), and a new rule is produced by LEAP for solving new problems of the same kind.

9.3 PROBLEM-SOLVING TOOLS

Another learning method of automatic knowledge acquisition in a VLSI design base uses a theorem-proving technique to find a new module's functionalities, while its behavior is added to the module library of the system [Wu90].

A variant to explanation-based learning is known as *apprenticeship learning*, which is a powerful method to refine the knowledge base of expert systems for heuristic classification problems. This approach is motivated by the experience that an apprenticeship is a powerful method that human experts use to refine their expertise in knowledge-intensive domains. For that reason, explanation-based learning is also referred to as apprenticeship learning [Wil87]. This approach is a form of learning by watching and observation in which learning occurs by completing failed explanations of human problem-solving actions.

Odysseus [Wil87] is an apprenticeship-learning system that observes a human expert solve a problem. The first step of the learning process is the detection of a knowledge-base deficiency. Explanations are constructed for each of the observed problem-solving actions. If no explanation can be found, the knowledge base is suspected to be incomplete. The second step is to conjecture a knowledge-base repair. The task is to search for the missing knowledge. If new knowledge is justified, the knowledge base is modified.

9.3b Neural Networks

The Nervous System and the Brain

The human nervous system consists of the central nervous system, located within the skull and the spinal column, and the peripheral nervous system with its many branches to all parts of the body. It maintains the communication between various parts of the body by receiving and interpreting stimuli from some parts of the body and transmitting stimuli to other parts of the body. The *central nervous system*, which consists of the brain and the spinal cord, supervises and coordinates the activity of the entire nervous system. The brain, enclosed within the skull, is the portion of this system that receives stimuli from the sense organs, interprets and correlates these with stored impressions in order to control all vital activities.

Structurally, the nervous system is an interconnected network with *neurons* as basic cells. The average number of neurons in a human brain is $0.5 * 10^{11}$. There is a large variety of neurons with chemical, physical and morphological features. A great number of branches (ranging from 10^3 to 10^5) converge into one fiber, called *dendrite*. One or more dendrites are connected to a neuron, while each neuron has only one output connection, which is a long fiber, called *axon*. See Fig. 9.1.

An axon is in functional connection with other neurons in the central nervous system and with muscles, glands and other organs in the body. An axon leads to a junction area, called *synapse*. On one dendrite with its many branches, axon terminals from, say 10^4, neurons are connected.

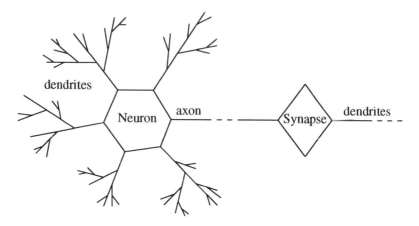

Fig. 9.1 Neuron, axon and dendrites

An essential function of a neuron is to transmit a unidirectional impulse from the dendrites (originating from all parts of the body, e.g., sense organs, muscles, the skin, joints) towards an axon (e.g., a gland or muscle). A neuron can transmit electrochemical impulses to another neuron via this axon. On its way to another neuron, an impulse reaches a synapse, where a decision is made if the target neuron is excited or inhibited. If the combined contributions from several synaptic inputs exceed a certain threshold value, the target neuron generates an impulse of its own.

Why Neural Networks?
Expert systems are hardware/software systems that solve problems by mimicking the way in which human experts solve problems. The knowledge is stored in the knowledge base and hence is an imperfect model of the knowledge about a specific domain. The problem-solving methods used in an expert system are models of solving procedures which human experts normally use.

The human brain solves complex problems easily with creativity. Therefore, it has sense to have an artificial system that realistically mimics the brain in order to have a means of problem solving beyond that possible with expert systems, that is, a system that has a learning ability. The neurons with their connecting fibers form a network for which various kinds of models have been proposed. These models will be referred to as *neural networks* [Hop88, Mul91].

Thanks to research of neurobiologists and psychologists, we are gradually getting a better idea of how the brain works [Joh88]. There has recently been a

9.3 PROBLEM-SOLVING TOOLS

resurgence of interest in investigating neural networks as brain models with the purpose of implementing brain-like computers or other problem-solving machines. Appropriate neural networks can hopefully serve to further the understanding of brain functions.

Neural Networks as Brain Models
Neural networks are artificial systems that are required to model human brains with the objective to handle complex problems in a way a human would do [Vem88]. Like any other artificial model of real-world objects, a neural network has to be a representation which is satisfactorily accurate and at the same time simple enough for avoiding unnecessary computing effort. In addition to the requirement that the neural model is reliable and capable of high-speed processing, it has to achieve the desired knowledge processing. The degree to which the performance of such a model conforms to that of the brain determines the success of neural networks.

A neural network must adequately exhibit the performance characteristics of the brain. One obvious solution is to base the neural network architecture to that of the brain. The main task of a neuron is signal amplification and processing. In accordance to the structure of the brain, neural networks imply massive parallelism. The study of neural networks may lead to a new kind of architecture for computing systems which are able to achieve human-like performance in a specific field.

To be most generally useful, a neural network must prove itself capable of learning and/or retrieving patterns in a noisy, changing environment. Another important property is the ability to avoid chaotic activity and converge to a stable pattern. Convergence behavior is a function not only of environmental noise, but also of the neural network architecture. A great number of models for neural networks have been proposed in the past. Some models consider the threshold character of signal transmission via neurons. Based on the concept of energy functions, others consider states of minimal energy as representing solutions to complex multi-dimensional optimization problems. A problem is then solved through a relaxation process, in which the energy state falls to an energy minimum.

Information coding and memory can be considered in three levels: (1) the genetic code, (2) the microtubule code, and (3) the connectionist code. The first two codes rely on biomolecular structures and provide the basis for a chemical theory of memory and intelligence. The third layer is based on the neural network structure with its synaptic interconnections. Synapses may be of the excitatory or the inhibitory type, when they support an amplification or an attenuation respectively of the potential neuron activation. This view has led to the connectionist theory [Gro82, Heb49]. The term "connectionist" refers to attempts to embody knowledge by assigning weights (numerical strengths) to the connections inside a network of interconnected nodes.

A general function of a neural network is to map some well-defined set of inputs into a well-defined set of outputs by assigning appropriate values to the connection weights. This learning process is generally called *training*. Usually, inputs as well as outputs have binary values (0 or 1).

Grossberg [Gro82] suggests that the inhibitory connections are hardwired, while the excitatory connections have variable coupling strengths according to the associative learning law originally introduced by Hebb [Heb49]. This law was formulated as follows: If neuron A repeatedly contributes to the firing of neuron B, then A's efficiency in firing B increases. By changing the strengths of connections, the neural network learns and memorizes.

Neural Network Architectures

The brain constitutes the center of thought and includes the higher nervous centers from which the activities of the nervous system are coordinated [Joh86]. Neural network implementations mimic, at least partially, the structure and functions of the brain. A first thought could be to consider a realistic computer implementation of neural networks. However, digital computers process data encoded as bit patterns, whereas brains deal with analog signals with a continuous range of values. Furthermore, weighted decisions are made on the basis of fuzzy, incomplete and contradictory information.

A nervous system with its billions of neurons has a highly parallel and distributed architecture. The way in which neurons are interconnected allows us to memorize and compute. Therefore, by implementing neural networks we hope to avail ourselves with a powerful tool that could replace, or at least assist, our brain in solving problems.

It has been proven that in case of a brain lesion due to an accident, the level of brain performance sustained depends on the amount of the damage and to a much lesser extent on the exact location of the lesion. Hence, neural networks are highly fault-tolerant in that a malfunction of a few processing elements does not significantly affect the overall performance. Implementations consist of a large number of processor elements (representing the neurons) with appropriate interconnections. Each element is capable of only very simple actions, such as threshold logic. Implementations of neural networks possess an active, self-programming, associative memory and are robust, since a few degraded or nonfunctional elements will not greatly affect the overall operation of the neural network. Processors can be dead and wires missing or not correctly connected and the network will still work fine. The information related to a given behavioral task seems to be distributed. Neural networks exhibit massive parallelism with a distributed architecture [RUm86, McC88].

Neurons are usually arranged in layers which are of three kinds: input, output and hidden layers. Input elements have only the function to feed signals to

9.3 PROBLEM-SOLVING TOOLS

processing elements connected to them. Output elements emit signals to the external world. The hidden layer contains a large number of processing elements interconnected in a complex network. Input, output and processing elements in the neural network will collectively be referred to as *nodes*. Generally, neurons and processing elements are identical. In a feedforward network, each element can send its output only to elements closer to the output layer.

There are two standard network architectures:

a. *Layered feedforward* (or *associative*) *architecture*. Computation proceeds from the input layer of neurons through the hidden layers to the output layer. There is a unilateral signal transfer from neurons in one layer to all neurons in the next layer.

b. *Feedback architecture*. The network contains at least one feedback loop. A special case is the fully interconnected (*auto-associative*) network in which all neurons are mutually interconnected.

A special case of a feedforward architecture is the single-layer feedforward network which contains all the neurons in the single layer available. There is only one node between any input and any output. In general, every input is connected to each node in the single layer. Such a network presents few problems in training. In the case of multiple layers, the one with a single hidden layer is most frequently used. A multiple-layer network is harder to train than a single-layer network.

To be able to analyze and synthesize artificial neural networks that mimic real neural networks, we have to work with suitable models. Below, some important models of neurons and neural networks are summarized.

McCulloch-Pitts Neuron Model

This is the first model proposed for a neuron [McC43]. The artificial neuron is considered as a multi-input single-output switching device. When the sum of the input signals exceeds some threshold value U_T, the output state has the value TRUE; otherwise it is FALSE. When the output is TRUE, the neuron is said to be fired or activated. The firing of a neuron is the emergence of a pulse train (a burst of pulses) at a certain frequency. This agrees with the property of a neuron which will or will not be activated, depending on the input values.

The McCulloch-Pitts model represents the neuron as a logic device. Hence, a neural network is represented by a logic network which could easily be implemented by using standard elements. However, in reality, a neuron has analog features, such as integration and delay properties.

The signals propagated through the network consist of pulse trains which follow each other with certain intervals. Because the synaptic junction takes a much larger time to process these pulse trains, new pulses pile up on the remains of the previous pulses. The summed result depends on the pulse frequency, that is, the number of pulses of a pulse train per second. At some time, the integrated summation of all the electrical potentials at the inputs will reach a threshold value leading to an activated state of the next neuron. In addition to this temporal summation of pulse trains sent by one specific neuron, spatial summation is used for the summation of pulse trains originating from different neurons.

Perceptrons

Many models have been developed within the frame of connectionist theory. These models have limited features of learning and classifying patterns. A representative example is Rosenblatt's *perceptron* [Ros62]. A simple perceptron model is shown in Fig. 9.2.

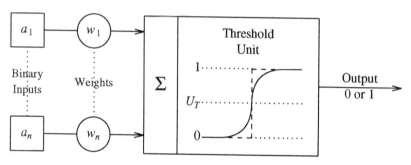

Fig. 9.2 Perceptron model

The basic element is the Threshold Logic Unit, which represents the neuron. Binary inputs a_i, $i = 1 \ldots n$ (representing dendrites), each multiplied by an analog weighting factor w_i, $-1 \leq +1$, $i = 1 \ldots n$, are summed, yielding the sum $S = \sum_{i=1}^{n} w_i a_i$. The output F (representing an axon) is 0 or 1, when $S < U_T$ or $S \geq U_T$ respectively, where U_T is the threshold potential. The threshold unit is given by a sigmoid function

$$1/(1 + e^{-(a_i + U_T)}/U_0)$$

where U_0 is a constant that determines the shape of the sigmoid.

In a learning process, a desired input-output mapping is given as a truth table $F = f(a_1 \ldots a_n)$. When this table is given, learning boils down to evaluating the proper values of the weights w_i. Learning algorithms haven been developed, e.g. the *Widrow-Hoff rule* or the *delta rule* as described by Sutton and Barto [Sut81].

Minsky and Papert [Min69] pointed out some of the limitations of the perceptron model. Particularly, the perceptron cannot be used for complex logic functions. It is shown that some simple recognition tasks, such as "connectedness" and "parity" cannot be performed by perceptrons. Patterns with connectedness have fully interconnected elements. Parity corresponds to the EXCLUSIVE-OR or the modulo-2 sum function: an odd (even) number of 1s in

9.3 PROBLEM-SOLVING TOOLS

the inputs results in a 1 (0) in the output.

Backward Error Propagation

An extension to the perceptron model in Fig. 9.2 arises when each input a_i provided with an appropriate weight is applied to each input of the processing threshold unit. Training rules for such networks are described by Rumelhart *et al.* [Rum86]. Given a set of weights w_{ij}; $i, j = 1 \ldots n$, errors measured by comparing the calculated output values to the desired values are used to determine the weight changes. Suppose a neural network consists of three successive layers: an input layer, a hidden layer and an output layer. A vector sent to the input layer passes through the hidden layer to emerge from the output layer. The deviation of the output vector from the desired vector is an error vector which is propagated back through the network in reverse direction in order to adjust the weights to the error. After a few iterations, the network will have learned the proper output response for a given input vector.

This back-propagation algorithm has been used in a neural network speech synthesizer, called NETtalk [Sej87]. The three layers of the network contain neurons, as indicated in Fig. 9.3. All neurons of the first layer are connected to all neurons of the second layer, which in turn are connected to all neurons of the third layer.

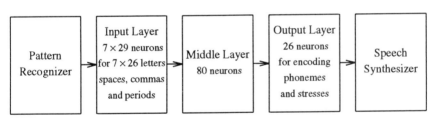

Fig. 9.3 NETtalk system

NETtalk is trained in stages by using Digital Equipment Corporations's DECtalk, which consists of a speech synthesis expert system and a sound generator. The rule-based expert system contains hundreds of rules and represents a large body of accumulated linguistic knowledge. It converts strings of characters to a command which is sent to the sound generator. According to this command, the sound generator produces the appropriate sound through a loudspeaker. Each attempt of NETtalk at reading an English text is compared with a phonetic transcription of a person reading the text. After a training process using a back-propagation algorithm, NETtalk is able to read and speak intelligent English in a few hours.

Hopfield Network

Hopfield [Hop82] described a neural network in which collective interaction of neurons can produce an entire image from a substantial fraction of that image. This system of content-addressable memory consists of numerous basic components, the neurons, each of which has the value 0 or 1. A Hopfield network consists of a single layer of completely interconnected neurons with a symmetric weight matrix.

Let an interconnected network of N neurons be given. Each neuron is connected to each of the other $N-1$ neurons. The synaptic weight of the connection between neuron i and neuron j is denoted by w_{ij}. A weight matrix **W** is constructed with $w_{ii} = 0$, $i = 1 \ldots N$. Denoting the threshold by U_T, the output v_i of the ith neuron depends on the outputs u_j of the other neurons according to the formula:

$$v_i(t+1) = sgn \sum_{j=1}^{N} w_{ij} v_j(t) - U_{Ti}, \quad i=1 \ldots N, \quad w_{ii} = 0$$

where sgn denotes the sign function. The output $v_i(t+1)$ can be interatively computed from the (initial) inputs $v_i(t)$. It can be proven that when the connection matrix is symmetric, every random state will stabilize after a certain number of iterations.

The Hopfield network can be viewed as a dynamical system. The state of the network is given by the vector made up of the N neuron outputs. Given a specific state, the network can proceed synchronously (the next state calculated from the previous state) or asynchronously (the state is calculated as soon as the input changes). In this way, the initial state changes gradually via intermediate states to the final state. Unfortunately, spurious states, which are not in the learned set, are produced in the final state. To minimize the occurrences of spurious states, Hopfield introduced two algorithms [Hop84]:

a. The algorithm of *unlearning* can be applied to states which do not have a close match with any template vectors. With the basic structure unchanged, neurons hold values of 1 and -1 instead of 1 and 0. The purpose is to eliminate non-optimal local minima by slightly decrementing the connection matrix T_{ij} by the amount ΔT_{ij}, when a system stabilizes in a spurious state. Whenever this situation arises, we have $\Delta T_{ij} = \varepsilon\, u_i\, u_j$, $0 < \varepsilon \leq 1$, where u_i and u_j are neurons in the spurious state.

b. A *learning* algorithm is added to the network. This is essentially the same as the unlearning algorithm except that it increases T_{ij} by ΔT_{ij}. Whenever the system makes an erroneous decision, it is necessary to invoke the learning algorithm on the correct vectors to increase the probability of the system stabilizing at the correct final state.

Boltzmann Machine

The Hopfield model usually converges to a local minimum rather than to the global minimum of an energy function. The Boltzmann model bears much structural resemblance with the Hopfield network [ACk85, Hin86]. It takes a

9.3 PROBLEM-SOLVING TOOLS

different approach on updating a random starting state with the objective to maximize the probability of a state of escaping from a local minimum to reach a global minimum. To this end, the principle of simulated annealing [Aar89] is applied (see Subsection 8.1b). The probability that a neuron assumes the value 1 is given by $P = 1/(1 + e^{-\Delta E/T})$, where ΔE is an energy related to the changes in input potentials and T is the annealing temperature. For temperatures approaching zero, the system behaves as a pure Hopfield network, moving down into the nearest local minimum. For any given temperature, the system will reach a thermal equilibrium.

Gaussian Machine

This model of a neural network is described by four equations relating three parameters: the reference activation level a_0, the temperature T and the discrete time step Δt [Aki88]. The Gaussian machine, represented by $GM(a_0, T, \Delta t)$ can be regarded as a generalization of the McCulloch-Pitts model $GM(0, 0, 1)$, the Hopfield model $GM(a_0, 0, \Delta t)$ and the Boltzmann machine $GM(0, T, 1)$.

Gaussian machines have been implemented on specially fabricated neural chips, realizing variable conductance on Hopfield models by deterministic and stochastic switched resistor circuits [Aki88].

Classification of Neural Networks

Two categories of neural networks can be distinguished:
a. Feed-forward classifiers, e.g., the threshold-logic network and back-error propagation. They bear most resemblance to classical perception models.
b. Feedback-driven content-addressable memories, e.g., the Hopfield network, the Boltzmann machine and the Gaussian machine.

The Learning Paradigm with Neural Networks

Neural networks can be trained by successive examples in a real-world environment. As the neural networks adapt to the changes in their environment, they develop their own internal rules. One advantage of neural networks is their ability to handle fuzzy or incomplete data. The speed and robustness of neural networks make them very attractive for a variety of applications, such as pattern recognition, robotic control and combinatorial optimization.

Generally speaking, a neural network comprises a large number of interconnected processor elements (representing neurons). Each connection between neurons has a prescribed weight. The learning process is realized by modifying the weights of connections between neurons. The objective is to generate the desired output as the response to a given output. For the interconnection pattern, there exists a variety of topologies: single-layer feedback networks (e.g., Hopfield networks), multi-layer feed-forward networks (e.g., perceptron-like networks), locally or fully interconnected networks. An initial network model is based on the iterative algorithm for the *search phase*.

The weighting coefficients for the final model are usually derived or trained during the *learning phase*.

Search phase: All the neurons update their new activation values a_i iteratively based on the following equations (1) and (2), until they converge.

$$u_i(k+1) = \sum_j w_{ij}(k)\, a_j(k) + \theta_i(k) \tag{1}$$

$$a_i(k+1) = f(u_i(k+1), u_i(k), a_i(k)) \tag{2}$$

where the activation function f may be deterministic (e.g., step, sigmoid, squashing function) or stochastic.

The converged activation values represent the desired state which is to be "retrieved" from the neural network. In the case of a Hopfield network, $a_i(k+1)$ is defined as

$$a_i(k+1) = \tfrac{1}{2}\left(1 + \tanh\left[(u_i(k) + \eta\, u_i(k+1))/u_0\right]\right) \tag{3}$$

where the nonlinear *sigmoid activation* function $f(x) = \tfrac{1}{2}(1 + \tanh[x/u_0])$, approaches a unit step function, when the parameter u_0 tends to zero. Equations (1) and (3) together define the Hopfield network.

Learning phase: The self-adaption and self-organization capability involves adaptive updating of the synaptic weights by certain "learning rules". Learning algorithms may be: (1) supervised, e.g., delta rule, and (2) unsupervised, e.g., competitive rule. These two classes are meant for very different kinds of applications.

Following the learning phase, one has a *retrieval phase* (or *recall phase*) in which the activation states of the network evolve according to some dynamical rule for fixed weights. The activation dynamics is essentially feed-forward by nature, that is, once the network has learned, its operation in the retrieval phase simply consists of the unidirectional flow of a static input pattern through one or more layers leading to a stable pattern at the output layer.

Learning by Examples

Learning by examples is a form of inductive learning (see Subsection 9.3a). From a set of examples of a specific problem domain, the learning system induces a general description of that domain. This type of learning can be a one-step process (with all examples presented together) or incremental (one-by-one or in small groups). The system proposes a tentative hypothesis consistent with the data available at a given step and subsequently refines the hypothesis after considering new examples. The inductive learning system then generalizes these examples to form a general description of a concept.

Knowledge is distributed over the network rather than represented locally.

9.3 PROBLEM-SOLVING TOOLS

Learning procedures are concerned with the weights w_{ij} in the network which are determined adaptively by some learning scheme. Suppose that at a given time a difference E between the actual output and the desired output is observed. The learning procedure is then to update iteratively each weight w_{ij} by an amount proportional to $\partial E/\partial w_{ij}$, where ∂E is the change in the global error and ∂w_{ij} is the change in the weight w_{ij}. If the learning succeeds, the network produces the desired output for any given input. The learning paradigm is divided into three types: supervised, unsupervised and reinforcement [Bre91]. These types are defined by the type of error signal used to train the weights.

In *supervised learning*, the network is trained by an external "teacher" who presents inputs to the neural network (notably, a multilayer perceptron network with backward error propagation), compares the resulting outputs with those desired and then adjusts the weights in such a way as to reduce the differences. The network is trained using a set of input-output pairs which are examples of the mapping that the network is required to learn to compute. The learning process may, therefore, be viewed as fitting a function. Its performance can be judged on whether the network can learn the desired function over the interval represented by the training set and to what extent the network can successfully generalize away from the points that it has been trained on. Perceptron training forms the basis of many more sophisticated supervised learning algorithms. The original perceptron training with binary outputs was extended by Widrow and Hoff [Wid60, And88] to continuous outputs using the sigmoid function (see Fig. 9.2). Supervised learning algorithms can be implemented by a three-layer feedforward network using a back-propagation algorithm. Applications can be found in control, signal processing and pattern recognition.

In *unsupervised learning*, no external error signal is provided. Through a process of self-organization, internal errors are generated between the neurons which are then used to modify the weights. The network learns to form internal representations of the inputs which produces certain outputs. Hebbian learning is a form of unsupervised learning. Kohonen [Koh84] reported on self-organizing algorithms used for pattern recognition tasks.

In *reinforcement learning*, the network is given only a punish/reward signal. The only information to train the network may be a global evaluation of the overall system's performance. Learning schemes which use this sort of information to train a network are examples of reinforcement learning. The central problem in these schemes is one of distributing the credit or blame from a global evaluation around the network. Applications include stochastic automata and game theory.

A major problem in neural networks is to record new information and yet retain the stability needed to ensure that existing information is not erased or corrupted during a learning process. In a back-propagation network, training elements of a set are applied sequentially until the network has learned the entire set. If, however, a new training set must be learned, it may disrupt the weights so badly that complete retraining is required.

When constantly changing training sets must be learned, the back-propagation network becomes inadequate as a learning network. Grossberg *et al.* [Gro87, Gro88, Die90] explored special configurations, called ART (Adaptive Resonance Theory) networks, which learn new information, while preventing the modification of information learned previously. ART is an unsupervised learning model based on a statistical technique, known as clustering analysis. To be effective, prior knowledge of some aspects of the form and structure of data is necessary. The centroid of a cluster or the long-term memory representative of a classification may be updated whenever new data make it necessary.

9.3c VLSI Applications of Neural Networks

Introduction

The software approach using simulation software running on traditional sequential or parallel computers provides a major class of methods for implementing neural networks [Kun88]. This subsection gives a survey of hardware implementations, followed by a brief account of applications of neural networks, including their use as CADCAS tools [Ram91, Sam91, San92].

The objective of hardware implementations is to cast a realistic model of a neural network into an appropriate electronic circuit. The complexity of the implementation depends on the model on which it is based. It should be borne in mind that a hardware implementation can never be an exact replica of the human brain. Somehow, the brain harnesses chaos to solve problems, whereas electronic implementations use straightforward procedures. In general, more complex and realistic models give more reliable and accurate solutions to the problem being handled. Neural network implementations may be of the analog type, the digital type or a mixture of these types [MOr90]. Any model selection involves a trade-off between simplicity and accuracy.

Hardware Implementations

The advances in VLSI technology have enabled one to develop methods to implement neural networks in hardware [Lip87, SHa89, Sil89, Mor90, Ant91]. The simplest implementations of neural networks involve fixed functionality whose connection weights and architecture are set during fabrication. There is a need for flexible implementations with modifiable weights and learning capabilities.

In the Hopfield-Tank model [Hop86], n neurons are represented by n operational amplifiers, as shown in Fig. 9.4. Resistors, whose values represent weights, connect each neurons's output to the inputs of all others. The weights

9.3 PROBLEM-SOLVING TOOLS

are symmetrical, i.e., $w_{ij} = w_{ji}$ and $w_{ii} = 0$. Generally, the amplifiers have both normal and inverting outputs, permitting the generation of negative weights by using positive resistors. The value R_j of the resistor connected between two neurons i and j determines the synaptic weight T_{ji}.

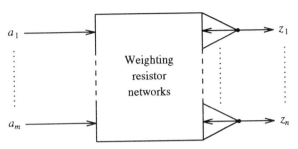

Fig. 9.4 Hopfield-Tank model

The state of neuron j is determined by the effective input potential u_j and the output firing rate $f_j(u_j)$. The synaptic current into the postsynaptic neuron j due to a presynaptic neuron i is proportional to the product of u_i and T_{ji}. Assuming that a grounded capacitor C_j is associated with the neuron j, the total dynamical current of neuron j with contributions of N neurons is described by the following nonlinear differential equation:

$$C_j \frac{du_j}{dt} = \sum_{i=1}^{N} T_{ji} f_i(u_i) - \frac{u_j}{R_j} + I_j; \quad T_{ii} = 0, \; i = 1, \ldots, n,$$

where u_j/R_j is the leakage current and I_j is the external input current.

The output voltage V_j models $f_j(u_j)$. In the high-gain case, where V_j is either near minimum output (logic 0) or near maximum output (logic 1) for all j, the computational energy function is defined as

$$E = -\tfrac{1}{2} \sum_j \sum_i T_{ji} V_j V_i - \sum_i - I_i V_i$$

The basic use of the Hopfield-Tank model is to transfer a problem into an optimization problem. Hence, the problem specification must be put into a form as given above for E. The optimal solution is found when E has been minimized.

The nature of neural networks gives indications as to which requirements must be imposed on the hardware implementation:

a. A densely connected network of a huge number of simple processors.
b. A large number of scalar products $w_{ij} v_j$ must be computed simultaneously.
c. An output function (e.g., sigmoid) must be applied to the result of each scalar product.
d. An output vector must be taken from one device and communicated to other devices.

Both binary and analog algorithms exist for the learning and the recall case.

With analog circuitry, a huge number of units with a high connectivity and extremely high speed can be implemented [And88]. Programmable analog parameterized cells and MOS circuit modules (such as integrators, summers and multipliers) can be configured to develop hardware implementations of neural networks [Bib89]. A problem is the accuracy achieved at fabrication. Graf et al. [Gra89] proposed a CMOS circuit implementing a connectionist neural network model. This model consists of an array of 54 simple processors, fully interconnected with a programmable connection matrix. The circuit was tested as an associative memory and as a pattern classifier.

Implementing fixed-value resistors in integrated circuits for realizing the synaptic weights requires a relatively large chip area. Foo et al. [Foo90] proposed analog components for building neural networks in which variable synaptic weights are implemented by uisng junction field-effect transistors or MOSFETs, operating as linear voltage-controlled resistors. An alternative implementation of variable weights is achieved by using switched-capacitor circuits, which behave as variable resistors. Suitable CMOS transistor circuits can also be used for implementing the sigmoid function.

Mead and Koch [Mea88] implemented a number of resistive networks in analog VLSI subthreshold CMOS technology to reduce the power requirement. These networks implement an almost fixed functionality, where bias voltages can be used to change weights by a few percent. Greater flexibility of devices requires that the weight at each connection is implemented with some form of storage device. Various technologies have been suggested, including static memory cells (for binary networks), digital storage with D/A converters, capacitors with external refresh, CCDs, EEPROMs and photosensitive amorphous silicon. The recall phase of a number of small networks has been implemented with devices such as a 16-node implementation of Kohonen's associative network algorithm.

A neural network architecture can be implemented by a regular MOS structure in which each neuron receives a weighted summation of pulse streams from n other neurons and operate upon this activity to decide a (firing) state [Mur89]. The synaptic weights are stored in a memory. The weights are introduced by appropriately chopping each input stream of excitatory or inhibitory pulses. An alternative is the use of pulses with variable widths.

A vast number of digital implementations of neural networks have been proposed by several authors. Burr [Bur91] gave an overview of existing digital VLSI implementations, including TI's NETSIM, Hirai et al.'s and Duranton-Sirat's digital neurochips, Quali-Saucier's Neuro-ASIC, Neural Semiconductor's DNNA, North Carolina State's STONN and TInMANN, and Adaptive Solution's CNAPS.

Quali and Saucier [Qua89] proposed a flexible architecture for neural

9.3 PROBLEM-SOLVING TOOLS

networks. The basic processor in a distributed, synchronous architecture is associated to one neuron and is able to perform autonomously all the steps of the learning and relaxation phases. Data circulation is implemented by shifting techniques. Customization of the network is implemented by fixing identification data in dedicated memory elements.

Masaki et al. [Mas90] designed a complex neuron circuit consisting of synaptic circuits, dendrite circuits and a cell-body circuit with excitatory and inhibitory inputs. A synaptic circuit produces an impulse whose density is proportional to the synaptic weight. The dendrite circuits are needed to separate the excitatory and inhibitory impulses. A digital neurochip implementing this was fabricated at Hitachi using 1.3-μm, 24k-gate, CMOS gate-array technology. These neuro-chips are used to construct a general-purpose system that can simulate a wide range of neural networks, including Hopfield-type and back-propagation networks. The system consists of several neuro-boards, each containing 72 neuro-chips, and a host computer that can read and write various registers in the neuro-boards. Learning algorithms can be executed and synaptic strengths can be easily updated. By using approriate software, a CMOS wafer-scale integrated 576-neuron chip solves the 16-city traveling-salesman problem in less than 0.1 seconds.

Digital implementations are based on the type of approach, ranging from general-purpose parallel computers (e.g., a Distributed Array Processor), through purpose built boards to custom digital VLSI [Atl89]. The reported speeds of machines carrying out back-propagation learning have reached 10^7 connections per second in both learning and recall nodes, and simulation analyses suggest that rates approaching 10^{11} may be attainable.

A digital systolic architecture was proposed by Kung and Hwang [Kun88]. This design is claimed to be more general purpose than most other architectures. However, the arrays have feedback connections which grow linearly with the size of the neural network. To overcome this bottleneck, Zubair and Madan [Zub89] proposed an alternative systolic architecture.

To match the technological trends toward single-chip mixed analog-digital VLSI neural systems, the development of effective analog CAD tools in conjunction with the existing digital tools is urgently needed for shortening the total chip design cycle. In particular, the development of automated analog layout tools to allow a quick generation of high-quality design-rule-correct neural chip layouts is a crucial step [Che90].

As is conventional VLSI chip design, the testability issue in neural chip design should not be ignored. It can be shown that the conventional techniques to test electronic circuits fall short when neural networks are considered [War89]. It is proposed to test directly on the macro properties of the neural network. The effect of device faults and built-in fault tolerance of the network are captured in so-called *fault-tolerance curves*. These curves, obtained by simulation, link the various macro and micro properties together and allow an information-driven test of the neural network.

The high connectivity of a neural network suggests that an optical approach may have advantages. So far, purely optical approaches are suitable for small-scale circuits using binary units. More success is expected with opto-electronic devices, where optical connectivity is combined with electronic computation [Far89, Szu90]. This approach has led to systems consisting of arrays of photo-sensitive cells selectively masked by a spatial light modulator (SLM). The electronics perform the scalar product and logistic function computation and drive incoherent light sources whose connectivity to the cells is determined by the SLM. Whilst a number of small networks have been implemented using this approach, precise orientation of the various components is required, and programmable spatial light modulators are currently only binary, thus limiting the type that can be implemented.

Applications of Neural Networks
Considering that neural networks mimic human brains, it is obvious to state that neural networks are to be applied to carry out tasks which usually require the use of brains. Two main classes of tasks are (1) the logic reasoning process, including computation, and (2) the process of memorizing and learning. Let us first consider (1). Reasoning and decision-making are processes which can be performed by (knowledge-based) expert systems. Since the performance of a rule-based system is limited by how many rules can be processed in parallel, neural networks are being considered as potential tools for solving these problems.

Let us consider the differences between the knowledge-based approach and the neural-network approach in problem solving. The knowledge in a rule-based system lies in rules, whereas the knowledge of a neural network lies in its connections. A knowledge-based system reasons through symbol generation and pattern matching, but a neural network processes information by propagating and combining activations through the network. The knowledge-based approach emphasizes knowledge representation, reasoning strategies and the ability to explain, whereas the neural-network approach does not. The knowledge-based approach can reason at various levels of abstraction, but the neural-network approach cannot. Applications of neural networks include functions such as associative memory, adaptive learning from examples and combinatorial optimization. Neural networks can be used to fine-tune fuzzy systems and thus produce adaptive systems [KOs92].

Widrow and Hoff's Adaline [Wid60] is a two-layer network similar to Rosenblatt's perceptron. It has been used for adaptive signal processing and control systems, for example, as an adaptive equalizer in high-speed digital modems. Multilayer networks have produced a wide variety of applications. A

9.3 PROBLEM-SOLVING TOOLS

multilayer perceptron has been applied to solve the match-phase problem in the recognize-act cycle of rule-based systems [SAr92].

Many practical problems can be formulated as an optimization problem in which some energy function is minimized. Outstanding applications of neural networks are learning to speak [Sej87] and medical reasoning [Gal88]. A neural network has been used for implementing an image-compression system [Cot87]. Other applications include vision, optimization, signal processing [KOs92], robotics and power systems. Verleysen *et al.* [Ver89] proposed an algorithm for programming a Hopfield neural network as a high-storage content-addressable memory. A potential application is pattern recognition.

The *neocognitron* is a neural-network model, proposed by Fukushima [Fuk82] for visual pattern recognition. A positive feature of the neocognitron is its capability of shift-invariant and deformation-resistant pattern recognition, which is particularly useful in character recognition tasks. The neocognitron performs unsupervised learning. The basic architecture is a multilayer neural network, with a two-dimensional input layer connected to a cascade of layer consisting of planes of different neurons. Interconnections to a specific plane have modifiable weights, determined during learning. Others are fixed and adjustable for specific purposes. The *digi-neocognitron* [Whi92] is a digital version of the neocognitron. For assigning the weight values, multiplications and divisions are replaced by simple shifts by conversion of multiplying or dividing factors to powers of 2. Shifters occupy a reduced silicon area and have smaller propagation delays compared to multipliers. Complex functions are replaced by lookup tables, implemented by simple combinational logic or memory arrays.

Researchers have begun to explore a new computer architecture, called the *neural computer* [Hop86], which simulates the human brain in structure and behavior. Such computers hold the promise of solving hard problems faster than current computers by many orders of magnitude.

Neural Networks as CADCAS Tools

Analog-Digital Converters. The simple Hopfield-Tank model of a neural network (Fig. 9.5) with appropriate values assigned to the weights has been used for the design of analog-digital converters [TAn86]. The analog-digital conversion (ADC) is formulated as an optimization problem. Finding a solution at a local minimum rather than the global minimum makes this method not so useful. Several authors have proposed improved neural methods for analog-digital conversion [Mar91]. Chiang *et al.* [Chi91] realized an adaptive neural network circuit that uses only a single MOS transistor for implementing each of the active elements in the network. This configuration has been applied to a 4-bit analog-digital converter, based on the Hopfield model.

CADCAS Applications. The use of the computational capability of neural networks to solve CADCAS problems has proven to be effective in various CADCAS applications. However, exploiting the Hopfield network for solving CADCAS problems should be taken with care [Yu89].

Testing. In the neural network model of a logic gate, the gate behavior is expressed as an energy function. Using this model, a mathematical technique for test generation can be developed [Cha91]. It can be shown that two equivalent models of a digital circuit are useful in parallel test-generation methods: the Boolean satisfiability model (in which a logic gate is represented by a truth table) and the neural network model.

Layout Applications. Jabri and Li [Jab92] use neural-network techniques to predict the mask layout dimensions (width and height) of rectangular, full-custom circuit blocks, given their net list and port number/type information. In most routing algorithms, the nets are routed sequentially one at a time [Oht86]. There are also global routers which try to route all the nets simultaneously. A typical class of this kind of routers make use of simulated annealing (see Subsection 8.1b). Another design tool that takes all constraint requirements at once is the neural network.

The use of the computational capability of neural networks to solve CADCAS problems has proven to be effective in various CADCAS applications. For the module orientation/rotation problem [Lib88] and the circuit-partitioning problem [Yih88], satisfactory results have been reported.

Routing. Shih *et al.* [Shi91] used a neural network based on the Hopfield-Tank model. The global-routing design is reduced to two optimization problems with the objective to minimize the total sum of the path lengths of all the nets and the objective to make the distribution of the paths in the routing channels as uniform as possible. Only 2-point connections and L-shaped paths along vertical and horizontal grid lines are allowed. Since the Hopfield network usually finds a local minimum, the solution is not guaranteed to be the best solution. The back-propagation training algorithm has been used for channel routing [Gre89]. Once trained, the networks can be assembled using special software which generates the software simulator for the overall neural system. Dedicated hardware can be used to exploit the parallelism in the process.

Expert Networks. Expert networks are event-driven, acyclic networks of neural objects derived from expert systems. Lacher *et al.* [Lac92] developed back-propagation learning for such networks and derived a specific algorithm for learning in EMYCIN-derived expert networks. This method offers automation of the knowledge acquisition task for certainty factors, often the most difficult part of knowledge extraction.

9.3d Toward the Sixth Generation

Semiconductor Technologies

The increasing complexity of VLSI chips with rising numbers of functions implemented on a chip requires larger and larger chip areas, ultimately leading to *wafer-scale integration* [Swa89, Tew89]. This means that the complete system, composed of different subsystems, covers the entire surface of a semiconductor wafer, unlike the usual procedure in which a wafer contains a large number of identical chips.

The improvements in performance of Si and GaAs transistors and the growth in the applications are expected to continue in the next decade. In addition to conventional VLSI applications, there will be a growth in the development and use of high-power transistors as well as GaAs or InP microwave devices and systems.

The continued growth of digital integrated circuits, particularly CMOS circuits, will be made possible by further improvements in fabrication technology that produce reductions in feature size. The highest transistor densities are obtained in static and dynamic RAMs and in electrically erasable ROMs. Technological progress has influenced research on VLSI applications. Besides the numerous ASICs for various dedicated applications, there is an upsurge in the increasing development and use of VLSI signal-processing systems (DSP systems).

ASICs have been surveyed in Section 5.3. Programmable styles can not only be used in ASICs, but increasingly in VLSI signal- end image-processing systems (PDSP systems). Catalog PDSP design employs programmable DSP subsystems stored in a catalog. A DSP core allows a design to be modified with interface hardware to accomodate different applications. Many real-time systems that require high speed, such as speech, mobile radio, radar and video signal-processing applications are usually implemented using dedicated circuits.

Analog integrated circuits and signal-processing experience a growth in applications, such as advanced systems for telecommunications, robotics, automotive electronics, image processing and intelligent sensors. The analog character of neural networks is an additional reason for the growing interest in analog ciruits. Fabrication technologies include BiCMOS, GaAs, GaAS/Si, heterojunction, floating gate superconducting technology.

The most widely used computer is implemented in semiconductor technology. Though silicon is the major semiconductor used, GaAs is also employed, particularly in high-speed applications. The information carriers in the semiconductor material are electrons or holes. A challenge is to build computer systems with 100-year mean time to failure and one-minute repair times [Gra91].

Besides semiconductor technologies, other technologies are being studied for constructing a computer. A typical example is the optical computer.

Advanced Computer-Aided Chip Design

Major trends in design automation are the development of more advanced CAD tools, the integration of individual tools into a coherent CAD framework, and the use of appropriate workstations which share a central database. Automatic data flow between tools alleviates the designer's task. The personal workstations are linked to each other and to a mainframe computer for CPU-intensive tasks.

In the future, a complex chip design project will be performed by a group of expert designers, each sitting in front of an advanced workstation. When workstations are located in different buildings, a Local Area Network must connect the members of the design group to enable real-time consultations at any stage of the design process. Each chip designer can use a framework of a large variety of design tools. An advanced Local-Area Network allows the designer to acquire useful design information by accessing a heterogeneous set of hardware (database systems, supercomputers, mainframes, minicomputers, workstations, personal computers) with different operating systems.

Besides the well-developed algorithmic programs, which solve local problems of logic synthesis, simulation, testing and circuit layout, the designer will avail oneself of a wide variety of expert systems, which handle those problems not amenable to algorithmic techniques. The expert systems assist the human designer in a large number of intelligent tasks spread over the entire design cycle.

Non Von Neumann Computers

VLSI technology and multi-processor capabilities influence the advancement of improved parallel processing. The use of relatively simple, identical processors, each provided with a small memory, reduces the price/performance ratio, while a variety of different computational tasks can be performed by using proper interconnection structures. The efficient distribution and management of the computational tasks taking place on a large number of processors requires the proper interconnection pattern of the processors. There is a need for suitable methods that deal with program composition.

Progress in parallel computing is hampered by a number of factors:
a. There has been a considerable investment in imperative programming with many sequential-programming tools available to aid in the software-development cycle, including testing and debugging. Much effort is needed to convert programs developed for Von Neumann computers to programs which execute efficiently on parallel computers.
b. Developing efficient programs for parallel computers is more difficult than for sequential computers. Completely new algorithms for parallel processing must be conceived.
c. To allow the possibility of writing programs at a high (user-friendly) level, the proper compilation technique must bridge the gap between high-level

9.3 PROBLEM-SOLVING TOOLS

language and machine language. A drawback is the lack of a single, predominant parallel architecture. The problem of program transportability cannot be secured as long as no standard architectures are defined.

Parallel programs fall into two broad categories: transformational and reactive programs [Sei91]:

1. Transformational programs read input data which are processed to deliver output results, as in conventional third-generation programs. Many parallel programs are based on imperative programs which are adapted to operate on parallel computers.
2. A reactive program interacts with its environment, while the program is executing. Its behavior depends on the program's state when the input arrives. In fact, reactive programs are operating, command, control, and real-time systems. Reactive systems are usually nondeterministic. Examples of reactive systems are object-oriented programs (Section 3.6) and knowledge-based systems (Chapter 4). Reactive systems will play a dominant role in fifth-generation and sixth-generation computers.

Operating systems relieves the programmer of the responsibility for resource management, including process allocation and load balancing. Virtual memory management across distributed memories will become commonplace. Operating systems for parallel computers are complex because architectures vary widely and managing resources for multiple heterogeneous processors is difficult. The Mach operating system [Acc86] for shared-memory computers supports simple load balancing and memory management. Local-Area Networks facilitate cooperative processing, in which teams of designers collaborate on a common project.

More widespread acceptance of formal methods for reasoning about a program's correctness will be important.

Computers, Expert Systems and Neural Networks

With regard to the potential applications of modern tools, conventional computers and neural networks complement each other very well. Whereas computers are suitable for computational purposes, neural networks are best in such applications as pattern recognition. Expert systems occupy an intermediate position. When a specific problem is to be considered, either one of the three tools may be most appropriate to use. In complex application domains, such as in VLSI chip design, the use of a suitable combination of all three tools may expedient.

Historically, four approaches to problem solving in VLSI chip design can be identified:

1. *Programmed computing.* Well-established algorithms for solving problems in chip design, analysis and optimization can be applied by using conventional computers. This category encompasses the majority of chip-

design tools. In the future, computers different from the conventional ones based on semiconductor technology (e.g., optical computers) may be useful.
2. *Supercomputing.* Much effort is spent in speeding up the processing of computer operations. Parallel techniques are instrumental in satisfying this goal. Supercomputers (see Subsection 3.2a) are particularly useful for performing scientific calculations.
3. *Using Expert systems.* The use of Artificial Intelligence in chip design will manifest itself in an abundant use of expert system in many parts of the chip design cycle. In this comprehensive guide, an extensive overview is given of expert systems, intended to assist the chip designer in many design tasks. A continuing effort is being given to the development of more advanced expert systems. Applications of expert systems are extending to areas related to chip design. These areas include mainframe management [Sea91], software development [Asl91], telecommunications [Roo87, SAC88, Rus89, Shi92], and computer network design, management and diagnostics [Eri89].
4. *Neurocomputing.* A new trend is the revival of AI research with the ultimate goal to design intelligent computer systems with reasoning and learning capabilities [CiZ90]. In this respect, the fifth-generation computer will evolve to the sixth-generation computer.

VLSI chip design will predominantly be based on the first approach to problem solving, with gradually more contributions of the other approaches. Initially, expert systems were meant to replace human experts so that inexperienced designers could handle sophisticated design problems. However, successful design of very complex systems requires a wide variety of expert systems and qualified designers are needed to take full advantage of their own experience and insight, assisted by a framework of CAD design tools, expert systems and even neural networks. A knowledge source in future blackboard systems may be an algorithm, a rule base, a neural network or any other kind of problem-solving system.

Future expert systems manipulate huge knowledge bases involving complex interrelationships in a particular application domain. Such systems rely heavily on the use of efficient knowledge representations and inferencing mechanisms.

Optical Computer

With decreasing feature sizes in VLSI chips, the limiting factor of the operating speed will be the interconnection delays rather than the gate delays. This encourages researchers to investigate the possibility of using optical interconnections rather than the wiring in conventional VLSI chips. As processor clock speeds approach the maximum sustainable by conventional metal interconnections, designers must consider optical interconnections, greater parallelism and use of locality to boost performance [Sto91]. Lasers, particularly semiconductor lasers, are being applied in a wide range of fields.

9.3 PROBLEM-SOLVING TOOLS

Laser amplifiers of high performance are being developed. A developing computer architecture is the optical computer [Abu87]. Beams of laser light are used for transmitting signals and switching light-activated transistors. The fundamental carriers in the solid material are photons which are particles of electromagnetic radiation. An important feature is that photons can switch with the speed of light. Light may relieve bottlenecks in electrical hardware interconnects. While initially optical interconnects will supplement existing electronic systems, later on the computation as well as the interconnections may be implemented with light.

Optoelectronic circuitry offer many advantages, including increase in data rates, low power consumption, no troubles with crosstalk and electromagnetic interference and no need for impedance matching. High-capacity local-area networks can exploit the high bandwidth of fibers through wavelength division and two-division multiplexing.

Fifth-Generation Computer
The *fifth-generation computer* deal effectively with masses of information, make logical inferences, use speech and images to interact with humans and perform many tasks previously thought to require human expertise. An essential characteristic of a fifth-generation computer is that, if a user presents a problem to the computer (in some cases, by speech), the computer determines its own solving procedure, after which the solution is presented to the user. Even the reasoning used during the solving process can be explained to the user.

Succesful development of fifth-generation computers will be made possible by exploiting advances in VLSI technology, innovative computer architectures, expert systems and the use of sensory data (e.g., speech). New-generation computer system projects in the world include Japan's Fifth Generation Computer System (FGCS) project, MCC's Advanced Computer Architecture Program and DARPA's Strategic Computing Program.

The FGCS project distinguishes three different categories, (1) problem solving and inference functions, (2) knowledge-based management functions, and (3) intelligent interface functions [Mot82, Fei83]. It uses a Prolog-type language equipped with high-power inference mechanisms built into hardware. Parallel processing hardware is realized with dataflow-oriented techniques for parallel processing. The knowledge-based management functions based on the relational model is enhanced with functions for processing semantic data. The intelligent interface function will consist of a collection of systems that process, recognize and synthesize many kinds of information representations, such as figures, characters, voices and images.

The MCC's Program includes fifth-generation systems of which the Database Program is an integral part. A goal is to design a knowledge-based supercomputer with powerful built-in inference mechanisms and the ability to manage large knowledge bases and databases at very high speeds. The Database Program consists of the Advanced Database Systems (ADBS), based on logic programming, and the Object-Oriented Database Systems (ODBS), with object-oriented programming tools.

DARPA's Program, which is specialized in military applications, covers frontier research in fifth-generation functions. To accelerate the process of exploiting advances in AI, computer architectures and microelectronics, DARPA (Defense Advanced Research Projects Agency) of the US Department of Defense started its Strategic Computing Initiative in November 1983 [DAR91]. A major objective is to realize large-scale, evolvable, heterogeneous intelligent systems.

A multi-level architecture and environment, called ABE (A Better Environment), is used to support the evolutionary development of intelligent systems, with emphasis on assembling and reusing both knowledge-processing and conventional software components [Hay91]. Explicit representation of the knowledge underlying a system's design facilitates the explanations of the system's conclusion. A description language is used instead of frames [Mac91]. Inferences between descriptions are derived by using a description classifier. This circumvents troubles in reasoning when a mixture of frames and rules are employed in an expert system.

Intelligent Machines

We may safely state that problem solving in the future will involve the cooperation of humans and the computer with a proper allocation of work. See Subsection 9.1b. A user-friendly interface to the intelligent machine, e.g., an expert system, is of vital importance. Common devices facilitating the user-machine communication include the keyboard, mouse and touch screens. The user should be allowed to input questions or specific information to the expert system. Responses from the expert system may include screen-displayed recommendations, explanations or requests for more information. The requirement for mutual understanding implies that the machine will need to be capable of adjusting its understanding in line with that of a human being.

Far beyond human capacities, supercomputers can manipulate vast volumes of information and perform mathematical calculations accurately with a speed of millions of floating-point operations per second. On the other hand, the computer performs some simple tasks (such as recognizing someone's face) so poorly that a child can outstrip computers with ease. Human's capacity to see and recognize objects, to communicate with one another, and many other things, are unsurpassed by any one machine which may ever be developed. For all that, many efforts are devoted to intelligent machines which on the one hand perform tasks which go beyond human capabilities and on the other hand attempt to mimic human brains in solving problems.

9.3 PROBLEM-SOLVING TOOLS

Two parallel lines of computer research will continue to focus on supercomputers as well as intelligent computers. Supercomputers remain to be a useful tool for performing scientific calculations. The goal to be achieved in current research is to relegate as many tasks, including intelligent tasks, to a machine. Intelligent machines may be based on the integration of expert systems, hypermedia and database technologies [Bie91]. In the course of time, an intelligent machine will be able to take over more and more tasks (drudgery as well as intelligent actions) which previously could only be performed by humans. In developing intelligent machines, we hope to improve our understanding of cognition, learning and reasoning in humans.

Neural Computer

A conventional computer simply maps input data to output results by executing a sequence of instructions according to a given computational algorithm. A different type of computer that holds promise for the future is the *neural computer* [Was89, Ale90, Hec90]. It requires the use of a neural network in which activation is propagated through the network from neuron to neuron. Studies of the human brain have learned that myriads of neurons work together in parallel to create intelligence. This observation propounds a complete rethinking of the way information is processed and hence of the way computers should be built. There is a trend to design computers that mimic human brains which actually process a large number of neurons in parallel. By mimicking some of the brain's processing strategies, researchers hope to build computers capable of reproducing human intellect and intelligence. Neural networks are being applied to a wide variety of applications, including medical diagnosis, fault diagnosis, sonar signal processing, robotics, data communications and man-machine systems [Mar90].

The brain's neuron circuitry processes information at the rate of 10 billion operations per second. Yet, the nerve impulses travel not faster than 250 miles per hour. Considering these facts, novel neural computers rely on a high degree of parallelism. Whereas conventional computers are poor in dealing with recognition problems, human brains can perform such tasks remarkably well by using a large number of relatively slow neurons in parallel. This observation suggests that neural computers must be based on parallel architectures.

A neural computer requires little or no software programming, is inherently fault-tolerant and has a learning capability. This type of computer is a model of the human brain. The electrical and chemical processes taking place in the brain must be modeled with a neuron architecture. Mind that the human brain contains approximately $50*10^9$ neurons each of which has about a hundred connections with other neurons via neural fibers, that is, there are about $5*10^{12}$ connections. The compexity exceeds that of the world-wide telephone network. A computer model cannot be but a simplified replica of the human brain.

Like the conventional computer, the neural computer is a general-purpose machine. The "computing" process is a sequence of states, where each state is updated in a network with variable weights. Users of a neural computer do not have to write complex programs neither to know how the computer solves the problem posed. The neural computer appears to be the preferred tool for a large variety of tasks that conventional computer do poorly, e.g., in pattern-recognition tasks. Though many procedures involve the simulation of neural networks on conventional serial computers, more efficient solution techniques can be realized by using hardware implementations of neural networks. Computational tasks can best be relegated to the conventional (super)computer performing the mathematical and computing algorithms that have shown their power.

Neural networks are usually based on classical dynamic system theory. As a consequence, they exhibit a rigid behavior compared with even the simplest biological systems. To overcome this limitation, a research is conducted on a dynamic neural-network architecture that takes advantage of the notion of treminal chaos to process information in a way that is phenomenologically similar to brain activity [Zak91].

Fuzzy logic can be used in neural networks for improved control in industrial applications. In particular, Japan has developed a variety of neuro-fuzzy products, including camcorders, air conditioners and washing machines.

Learning Machines

Most expert systems, developed in the past years, did not incorporate a learning mechanism. An early expert system with a limited learning capability is AM (Automatic Mathematician), which was designed to discover mathematical laws [Len82]. Initially given the concepts and axioms of set theory, AM was able to induce such important mathematical concepts as cardinality and integer arithmetic and many of the results of number theory. Amongst other things, AM discovered that
- all numbers with exactly three divisors were perfect squares (for example, the number 49 has at its divisors: 1, 7 and 49);
- the square root of a number with three divisors is a number with only two divisors, known as prime numbers.

AM conjectured new theorems by modifying its current knowledge base and used heuristics to pursue the most "interesting" of a number of possible alternatives. AM did not come up to the initial expectations. As the learning process advances, the heuristics prove to be no longer adequate.

A *learning machine* may be viewed as a machine that exploits its learning capability to improve its performance. The learning ability is one of the essential attributes connected with intelligence. That is the reason why learning machines have always been one of the main research topics in Artificial Intelligence.

9.3 PROBLEM-SOLVING TOOLS

People learn by memorization, by example, by analogy and by reasoning from the specific to the general and vice versa. In Subsection 9.3a, the most important artificial learning methods were discussed. Learning is a typical activity of the human brain. That is the reason why attempts are being made to design learning machines by mimicking the brain.

Neural networks (Subsection 9.3b) have the capability of learning by example and therefore serve useful purposes. Teaching algorithms can be used to iteratively modify the network's weights in order that the network responds properly to a set of input patterns. One important learning algorithm for multilayered feedforward neural networks is the back-propagation rule [Rum86].

Learning networks have interesting properties:
a. Even when the training set contains noisy or inconsistent elements, the central tendency of the set is extracted during the learning phase. For example, when the input-output mapping can be obtained by applying some type of rule, the network tends to discover the rule instead of memorizing the input-output pattern pairs, though it memorizes exceptions, if any.
b. As a consequence, correct responses are obtained, even when the input patterns are not included in the training set. Thus, the performance is widely insensitive to noise corrupting the input patterns. Only in the case of very noisy or contradictory inputs will the network performance decay gradually.
c. Information is distributed throughout the network. Parallel and distributed processing makes neural networks widely insensitive to unit and weight deficiencies or disconnections.
d. Neural networks can treat Boolean and continuous entities simultaneously. Hence, neural networks seem highly suitable for handling problems dealing with both symbolic and numerical operations.

Neural networks have some drawbacks:
a. Designing efficient neural networks for learning applications is a very complex problem [Jud 90].
b. Neural networks retain a degree of unpredictability, particularly when the number of possible inputs is large and exhaustive testing is impractical.
c. Neural networks are not able to explain how they solve the problem in the way expert systems use to do.

Sixth-Generation Machine

As the intelligent machines develop and the prices fall, the user community will grow. Similar to the rapid growth in personal computing, there will be a boom in small-scale personal expert systems for use in the home and office. Personal expert systems will be used particularly in various professions: in medical institutions, judicature, universities, and policy making in national and local government. In the long run, intelligent machines may meet or surpass humans in cognitive skills.

The goal of the sixth generation is the availability of machines for intelligent computing, reasoning, measuring and control. The human user has to state the problems and receives the solution to these problems and provides the flexibility and higher thought processes. It is not likely that a specific type of computer, e.g., a neural or optical computer, is to be a panacea for all known problems in computing and information processing. Besides its specific potentialities, each computer type has its weaknesses and limitations. By now, it is not clear what should be defined by sixth-generation computing. It may be an intelligent machine that mimics the human brain, both in architecture and functions. Probably, it will be a mix of different types of computer so that each problem can be attacked by the most appropriate computer type.

The main features of a sixth-generation computer may be:

a. A sixth-generation machine stores and manipulates vast amounts of information.
b. It comprises a combination of the latest developments in computer science: among other things, a supercomputer for performing the high-speed computing tasks, a number of expert systems for solving specific problems needing expert knowledge, and brain-like computers for performing intelligent tasks.
c. Neural-network software and hardware will take their place alongside other tools on the workstation and be used as needed for the specialized class of problems for which they are best suited [Sou89, Sou91].
d. A sixth-generation machine is usually an intelligent real-time system. To allow real-time operation, it incorporates parallel machine architectures.
e. It has the capacity to learn with principles of adaptive programming. The system emulates or matches human behavior, with the elementary features of perception, learning and cognition [Wei91].
f. Programming a sixth-generation computer is much easier than conventional computers. Systems are based on automatic programming and knowledge engineering. For obtaining a solution, it may suffice simply by formulating the problem.
g. Sophisticated interface tools, including the use of voice or speech, are essential. A highly user-friendly interface exploiting advanced image processing will be common practice.
h. The incorporation of intelligent terminals into a computer communications network enhances the capabilities of utilizing distributed databases and software programs.
i. Sixth-generation applications include intelligent robots, computerized cars and translating telephones.

Problem solving in the future requires a synergism between humans and the sixth-generation machine. To assist in solving problems, humans have at their disposal computers, expert systems and neural networks, each of which executes

9.3 PROBLEM-SOLVING TOOLS 513

a dedicated task according to their individual capabilities. A supercomputer is the "number cruncher" that executes the computation-intensive tasks. Expert systems with appropriate knowledge bases in specific domains may automate the process of reasoning, characteristic to human decision making. Neural networks, whose structures resemble the human brain, have a learning capability. Efforts are being made to further improve the performance of expert systems and neural networks.

Real-life problems range from simple ones having a unique solution to very complicated, unsolvable problems. Computationally simple are the problems for which an algorithmic procedure leads to a unique solution. Characteristic for this type of problems are numerical problems, which require numerical data as input. Another example of a simple problem is the query from a database, which contains a collection of *facts*. Data and facts are the simplest forms of information. A higher degree of information which is prerequisite for making intelligent decisions is *knowledge*, e.g., in the form of rules in addition to facts. Expert systems provided with the appropriate knowledge base are meant to assist humans in the reasoning process to arrive at decisions which go beyond the merely deterministic input-output mapping procedure of problem solving.

At a higher level of intelligence, a *learning* process is needed to enhance a human's knowledge and experience. Neural networks are being developed to mimic the brain's behavior, including the learning capability. Many social problems face with logically contradictory propositions and personal prejudices. In such cases, any machine would fall short. For handling social and theological dilemmas, we need superscientific qualities, which is known as *wisdom*. This highest level of human state seems to be reserved to a restricted group of humans.

Anyway, the moral of the above is that any hardware or software tools should be designed to assist humans, i.e., humans should always have control over these tools and never can it be so that intelligent machines overrule humans.

Whatever far the advances in intelligent machines may progress, there will always remain a set of human functions which cannot be accomplished by a machine. Such functions include respect, understanding, empathy or love between humans, as well as psychiatric help. Intelligent machines cannot deal with ethical issues.

REFERENCES

[Aar89] E.Aarts, J.Korst, *Simulated Annealing and Boltzmann Machines*. Wiley, New York, 1989.
[Aba83] M.S.Abadir, H.K.Reghbati, LSI testing techniques. *IEEE Micro*, 3, pp. 34-51, Feb. 1983.
[Aba89] M.S.Abadir, TIGER: Testability insertion guidance expert system. *Int. Conf. CAD*, Santa Clara, pp. 562-565, Nov. 1989.
[Aba85] M.S.Abadir, M.A.Breuer, A knowledge-based system for designing testable VLSI circuits. *IEEE Design & Test*, 2, pp. 56-68, Aug. 1985.
[Abr86] J.A.Abraham, W.K.Fuchs, Fault and error models for VLSI. *Proc. IEEE*, 74, pp. 639-654, May 1986.
[Abr85] J.A.Abraham, W.A.Rogers, HAT: A heuristic adviser for testability. *Int. Conf. Comp. Des.*, Port Chester, pp. 566-569, Oct. 1985.
[Abr90] M.Abramovici, M.Breuer, A.Friedman, *Digital Systems Testing and Testable Design*. Computer Sci. Press, New York, 1990.
[Abr84] M.Abramovici, P.R.Menon, D.T.Miller, Critical path tracing: An alternative to fault simulation. *IEEE Design & Test*, 1, 83-93, Feb. 1984.
[Abu88] A.Abu-Hanna, Y.Gold, An integrated, deep-shallow expert system for multi-level diagnosis of dynamic systems. In [GEr88], pp. 75-94, Aug. 1988.
[Abu87] Y.S.Abu-Mostafa, D.Psaltis, Optical neural computers. *Sci. Amer.*, 256, pp. 88-95, March 1987.
[Acc86] M.Accetta *et al.*, Mach: A new kernel foundation for UNIX development. *Unix Review*, pp. 37-39, Aug. 1986.
[Ack85] B.Ackland *et al.*, CADRE. A system of cooperating VLSI design experts. *Int. Conf. Comp. Des.*, Port Chester, pp. 99-104, Oct. 1985.
[ACk85] D.H.Ackley, G.E.Hinton, T.J.Sejnowski, A learning algorithm for Boltzmann Machines. *Cogn. Sci.*, 9, pp. 147-169, 1985.
[Aco86] R.D.Acosta, M.N.Huhns, S.L.Liuh, Analogical reasoning for digital system synthesis. *Int. Conf. CAD*, Santa Clara, pp. 173-176, Nov. 1986.
[Ado85] Adobe Systems Incorporated, *PostScript Language Reference Manual*. Addison-Wesley, Reading, 1985.
[Ado86] W.S.Adolph, H.K.Reghbati, A.Sanmugasunderam, A frame based system for representing knowledge about VLSI design: A proposal. *23rd Des. Aut. Conf.*, Las Vegas, pp. 671-676, June 1986.
[Afs86] H.Afsarmanesh, D.Knapp, D.McLead, A.Parker, Informaton management for VLSI/CAD. *Int. Conf. Comp. Des.*, Port Chester, pp. 476-481, Oct.1986.
[Aga80] V.K.Agarwal, Multiple fault detecton in programmable logic arrays. *IEEE Trans. Comp.*, 29, pp. 518-522, June 1980.
[Agr76] P.Agrawal, On the probability of success in a routing process. *Proc. IEEE*, 64, pp. 1624-1625, Nov. 1976.
[Agr87] P.Agrawal *et al.*, MARS: A multiprocessor-based programmable accelerator. *IEEE Design & Test*, 4, pp.28-36, Oct. 1987. See also: *IEEE Trans. CAD*, 9,

pp. 19-29, Jan. 1990.
[Agr82] V.D.Agrawal, M.R.Mercer, Testability measures. What do they tell us? *Int. Test Conf.*, Cherry Hill, pp. 391-396, Nov. 1982.
[Agr88] V.D.Agrawal, S.C.Seth (eds), *Test Generation for VLSI Chips.* IEEE Com. Soc. Press, Los Alamitos, 1988.
[Ake67] S.B.Akers, A modification of Lee's path connection algorithm. *IEEE Trans. El. Comp.*, 16, pp. 97-98, Feb. 1967.
[Ake78] S.B.Akers, Binary decision diagrams. *IEEE Trans. Comp.*, 27, pp. 509-516, June 1978.
[Aki88] Y.Akiyama, A.Yamashita, Gaussian Machines: A general neuron model. *Proc. SICE'88*, pp. , Aug. 1988.
[Alb92] J.S.Albus *et al.*, Project reports: Research in the architecture of intelligent machines. *Computer*, 25, pp. 56-79, May 1992.
[Ale90] I.Aleksander, H.Morton, *An Introduction to Neural Computing.* Chapman and Hall, London, 1990.
[Ale84] J.H.Alexander, M.J.Freiling, Troubleshooting with the help of an expert system. Tech. Rep. CR-85-05, Aug. 1984; Building an expert system in SMALLTALK-80. Tech. Rep. CR-85-06, Art. Intell. Dep., Computer Research Lab., Tektronix, Nov. 1984.
[All85] J.Allen (ed.), Man-Machine Speech Communication. Special Issue of *Proc. IEEE*, 73, pp. 1539-1676, Nov. 1985.
[All78] J.Allen, *Anatomy of LISP.* McGraw-Hill, New York, 1978.
[All87] P.E.Allen, D.R.Holberg, *CMOS Analog Circuit Design.* Holt, Rinehart and Winston, New York, 1987.
[Alm89] G.S.Almasi, A.Gottlieb, *Highly Parallel Computing.* Benjamin/Cummings, Redwood City, 1989.
[Alt84] J.L.Alty, M.J.Coombs, Reducing large search spaces through factoring-heuristic DENDRAL and Meta-DENDRAL. In: *Expert Systems. Concepts and Examples.* NCC, Manchester, 1984.
[Amb87] T.Ambler, P.Agrawal, W.Moore (eds), *Hardware Accelerators for Electrical CAD.* Adam Hilger, Bristol, 1987.
[Anc83] F.Anceau, E.J.Aas (eds), *VLSI83. VLSI Design of Digital Systems.* North-Holland, Amsterdam, 1983.
[ANc83] F.Anceau, CAPRI: A design methodology and a silicon compiler for VLSI circuits specified by algorithms. In [Bry83], pp. 15-31, 1983.
[And88] J.A.Anderson, E.Rosenfeld (eds), *Neurocomputing: Foundations of Research.* MIT Press, Cambridge, 1988.
[And87] J.M.Anderson *et al.*, The architecture of FAIM-1. *Computer*, 20, pp. 55-65, Jan. 1987.
[And91] G.R.Andrews, *Concurrent Programming: Principles and Practice.* Benjamin/Cummings, Redwood City, 1991.
[And83] G.R.Andrews, F.B.Schneider, Concepts and notations for concurrent programming. *ACM Comp. Surv.*, 15, pp. 3-43, March 1983.
[ANd87] T.Andrews, C.Harris, Combining language and database advances in an object-oriented development environment. *SIGPlan Notices*, ACM, 22, Dec. 1987.

REFERENCES

[Ann88] J.Annevelink, *HIFI. A Design Method for Implementing Signal Processing Algorithms on VLSI Processing Arrays*. Doct. Diss., Delft Univ. Technology, 5 Jan. 1988.
[Ann90] J.K.Annot, P.A.M. den Haan, POOL and DOOM: The object-oriented approach. In [Tre90], pp. 47-79, 1990.
[Ant92] B.A.A.Antao, A.J.Brodersen, Techniques for synthesis of analog integrated circuits. *IEEE Design & Test*, 9, pp. 8-18, March 1992.
[Ant84] P.Antognetti, A. De Gloria, F.Ellena, L.Repetto, An expert chip planning tool. *IEE Conf. Electronic Des. Aut.*, Warwick, pp. 174-178, March 1984
[Ant91] P.Antognetti, V.Milutinović (eds), *Neural Networks. Concepts, Applications, and Implementations*. Vol. II. Prentice-Hall, Englewood Cliffs, 1991.
[Apf86] L.Apfelbaum, Improving in-circuit diagnosis of analog networks with expert systems techniques. *Int. Test Conf.*, Washington, pp. 947-953, Sept. 1986.
[App88] K.Appleby et al., Garbage collection for Prolog based on WAM. *Comm. ACM.*, 31, pp. 719-741, June 1988.
[Arn78] G.Arnout, H. De Man, The use of threshold functions and Boolean-controlled network elements for macromodeling of LSI circuits. *IEEE Jour. Solid-State Circ.*, 13, pp. 326-332, June 1978.
[Ars91] K.Arshak et Als, Statistical expert systems for process control and error detection for VLSI. *Microelectronic Engng*, 13, pp. 541-546, March 1991.
[Asl91] M.J.Aslett (ed.), *A Knowledge Based Approach to Software Development: ESPRIT Project ASPIS*. North-Holland, Amsterdam, 1991.
[Atl89] L.E.Atlas, Y.Suzuki, Digital systems for artificial neural networks. *IEEE Circ. Dev. Mag.*, 3, pp. 20-24, Nov. 1989.
[Aug89] M.C.August et al., Cray X-MP: The birth of a supercomputer. *Computer*, 22, pp. 45-52, Jan. 1989.
[Aye91] M.Ayel, J.P.Laurent, *Validation, Verification and Test of Knowledge-Based Systems*. Wiley, Chichester, 1991.
[Bac84] J.Bachant, J.McDermott, R1 revisited: Four years in the trenches. *AI Mag.*, 5, Fall 1984; J.McDermott, R1: A rule-based configurer of computer systems. *Art. Intell.*, 19, pp. 39-88, Sept. 1982.
[Bac78] J.Backus, Can programming be liberated from the von Neumann style? A functional style and its algebra of programs. *Comm. ACM.*, 21, pp. 613-641, Aug. 1978.
[Bae83] J.L.Baer, Whither a taxonomy of computer systems. *IEEE Int. Workshop Comp. Sys. Org.*, New Orleans, pp. 3-9, March 1983.
[Bak90] H.B.Bakoglu, *Circuits, Interconnections, and Packaging for VLSI*. Addison-Wesley, Reading, (pp. 194-225) 1990.
[Bal80] R.Balzer, L.Erman, P.London, C.Williams, HEARSAY-III: A domain-independent framework for expert systems. *AAAI-80*, pp. 108-110, Aug. 1980.
[Ban90] F.Bancilhon, P.Buneman (eds), *Advances in Database Programming Languages*, Addison-Wesley, Reading, 1990.
[Ban84] P.Banerjee, J.A.Abraham, Characterization and testing of physical failures in MOS logic circuits. *IEEE Design & Test*, 1, pp. 76-86, Aug. 1984.
[Ban89] W.Banzhaf, *Computer-Aided Circuit Analysis Using Spice*. Prentice-Hall, New York, 1989.

[Bar89] V.E.Barker, D.E.O'Connor et al., Expert suystems for configuration at Digital: XCON and beyond. Commun. ACM, 32, pp. 298-318, March 1989.
[Bar82] J.G.Barnes, *Programming in ADA*. Addison-Wesley, 1982.
[Bar81] A.Barr, E.A.Feigenbaum et al., *The Handbook of Artificial Intelligence*. Four volumes. Pitman, London, 1981, 1982, 1984.
[Bar84] H.G.Barrow, VERIFY: A program for proving correctness of digital hardware designs. *Art. Intell.*, 24, pp. 437-491, 1984; H.G.Barrow, Proving the correctness of digital hardware designs. *VLSI Design*, 5, pp. 64-77, July 1984.
[Bar88] Z.Barzilai et al., SLS. A fast switch-level simulator. *IEEE Trans. CAD*, 7, pp. 838-849, Aug. 1988.
[Bat87] J.S.J.Bate, K.Wilson-Davies, *Desktop Publishing*. BSP Professional Books, Oxford, 1987.
[Bat89] A.Bateman, W.Yates, *Digital Signal Processing Design*. Computer Sci. Press, Rockville, 1989.
[Bat88] D.Batory, GENESIS: An extensible database management system. *IEEE Trans. Software Engng*, 14, pp. 1711-1730, Nov. 1988.
[Bat85] D.S.Batory, W.Kim, Modeling concepts for VLSI CAD objects. *ACM Trans. Database Sys.*, 10, pp. 322-346, Sept. 1985.
[Baw79] A.Bawden et al., The LISP machine. In [Win79], pp.343-373, 1979.
[Beg84] V.Begg, *Developing Expert CAD Systems*. Kogan Page, London, 1984.
[Bel87] F.Belli, W.Görke (eds), *Fault-Tolerant Computing Systems*. Springer, Berlin, 1987.
[Bel83] C.Bellon, C.Robach, G.Saucier, VLSI test program generation: a system for intelligent assistant. *Int. Conf. Comp. Des.*, Port Chester, pp. 49-52, Oct./Nov. 1983; An intelligent assistant for test program generation: The SUPERCAT system. *Int. Conf. CAD*, Santa Clara, pp.32-33, Sept. 1983.
[Bel84] C.Bellon, G.Saucier, Intelligent assistance for test program generation. *IEE Conf. Electronic Des. Aut.*, Warwick, pp. 166-170, March 1984.
[Ben86] M.Ben-Bassat, D.Ben-Arie, J.Cheifetz, ATEX, a diagnostic expert system embedded in an ATE system. *Autotestcon*, San Antonio, pp. 153-157, Sept. 1986.
[Ben83] J.B.Bendas, Design through transformation. *20th Des. Aut. Conf.*, Miami Beach, pp. 253-256, June 1983.
[Ben84] M.J.Bending, Hitest: A knowledge-based test generation system. *IEEE Design & Test*, 1, pp. 83-92, May 1984.
[Ben82] L.C.Bening et al., Developments in logic network path delay analysis. *19th Des. Aut. Conf.*, Las Vegas, pp. 605-615, June 1982.
[BEn84] R.G.Bennetts, *Design of Testable Logic Circuits*. Addison-Wesley, London, 1984.
[Ben81] R.G.Bennetts, C.M.Maunder, G.D.Robinson, CAMELOT: A computer-aided measure for logic testability. *IEE Proc.-E*, 128, pp. 177-189, 1981.
[BEn83] N.F.Benschop, Layout compiler for variable array-multipliers. *Cust. Integr. Circ. Conf.*, Rochester, pp. 336-339, May 1983.
[BEr82] W.C.Berg, R.D.Hess, COMET: A testability analysis and design modification package. *Int. Test Conf.*, Philadelphia, pp. 364-378, Nov. 1982.

REFERENCES

[Ber91] F.Bergadano, A.Giordana, L.Saitta, *Machine Learning. An Integrated Framework and its Applications*. Ellis Horwood, New York, 1991.

[Ber86] S.Bergquist, R.Sparkes, QCritic: A rule-based analyser for bipolar analog circuit designs. *Custom IC Conf.*, Rochester, pp. 617-620, May 1986.

[Ber77] K.J.Berkling, Reduction languages for reduction machines. In: *Future Systems*, pp. 79-116, Infotech, Maidenhead, 1977.

[Ber82] P.B.Berra, Some architectures for database management. In: F.Sumner, *Supercomputer Systems Technology. State of the Art Report*. Pergamon Infotech, Maidenhead, pp. 171-186, 1982.

[BEr91] E.Bertino, L.Martino, Object-oriented database management systems: Concepts and issues. *Computer*, 24, pp. 33-47, April 1991.

[Bet85] D.Betz, Xlisp: An experimental object oriented language (version 1.4). Manual. Manchester, 1985.

[Bha89] D.Bhattacharya, B.T.Murray, J.P.Hayes, High-level test generation for VLSI. *Computer*, 22, pp. 16-24, April 1989.

[BHa89] S.Bhawmik, P.Palchaudhuri, DFT Expert: Designing testable VLSI circuits. *IEEE Design & Test*, 6, pp. 8-19, Oct. 1989.

[Bib89] S.Bibyk, M.Ismail, Issues in analog VLSI and MOS techniques for neural computing. In [Mea89], pp. 103-133, 1989.

[Bid87] M.Bidjan-Irani, U.Glässer, F.J.Rammig, Knowledge based tools for testability checking. In [Bel87], pp. 119-128, Sept. 1987.

[Bie91] L.Bielawski, R.Lewand, *Intelligent Systems Design. Integrating Expert Systems, Hypermedia, and Database Technologies*. Wiley, New York, 1991.

[Bin90] P.Bingley, P. van der Wolf, A design platform for the NELSIS CAD framework. *27th Des. Aut. Conf.*, Orlando, pp. 146-149, June 1990.

[Bir85] W.P.Birmingham et al., CLASS: A chip layout assistant. *Int. Conf. CAD*, Santa Clara, pp. 216-218, Nov. 1985.

[Bir88] W.P.Birmingham, A.Brennan, A.P.Gupta, D.P.Siewiorek, MICON: A single-board computer synthesis tool. *IEEE Circ. Sys. Mag.*, 4, pp. 37-46, Jan. 1988.

[Bir89] W.P.Birmingham, A.P.Gupta, D.P.Siewiorek, The Micon system for computer design. *IEEE Micro*, 9, pp. 61-67, Oct. 1989. See also: *26th Des. Aut. Conf.*, Las Vegas, pp. 135-140, June 1989.

[Bir84] W.P.Birmingham, D.P.Siewiorek, MICON: A knowledge based single board computer design. *21st Des. Aut. Conf.*, Albuquerque, pp. 565-571, June 1984.

[BIr89] W.P.Birmingham, D.P.Siewiorek, Capturing designer expertise. The CGEN system. *26th Des. Aut. Conf.*, Las Vegas, pp. 610-613, June 1989.

[BIr88] G.Birtwistle, P.A.Subrahmanyam (eds), *VLSI Specification, Verification and Synthesis*. Kluwer Acad., Norwell, 1988.

[BIR89] G.Birtwistle, P.A.Subrahmanyam (eds), *Trends in Hardware Verification and Automated Theorem Proving*. Springer, New York, 1989.

[Bjo89] A.Björnerstedt, C.Hulten, Version control in an object-oriented architecture. In [Kim89], pp. 451-485, 1989.

[Bla90] D.T.Blaauw et al., SNEL: A switch-level simulator using multiple levels of functional abstraction. *Int. Conf. CAD*, Santa Clara, pp. 66-69, Nov. 1990.

[Bla89] U.Black, *Data Networks: Concepts, Theory, and Practice*. Prentice-Hall, Englewood Cliffs, 1989.

[Bla85] T.Blackman, J.Fox, C.Rosebrugh, The Silc silicon compiler: Language and features. *22nd Des. Aut. Conf.*, Las Vegas, pp. 232-237, June 1985.

[Bla84] T.Blank, A survey of hardware accelerators used in computer-aided design. *IEEE Design and Test*, 1, pp. 21-39, Aug. 1984.

[Blu87] B.G.Blundell, C.N.Daskalakis, N.A.E.Heyes, T.P.Hopkins, *An Introductory Guide to Silvar Lisco and HILO Simulators.* Macmillan, London, 1987.

[Bob83] D.G.Bobrow, M.Stefik, *The LOOPS Manual.* Xerox PARC, Palo Alto, 1983.

[Bob77] D.G.Bobrow, T.Winograd, An overview of KRL, a knowledge representation language. *Cognitive Science*, 1, pp. 3-46, 1977. See also: *Cognitive Science*, 3, pp. 1-28, pp. 29-42, 1979.

[Bob88] D.G.Bobrow et al., Common Lisp Object System Specification. X3J13 Tech. Rep. 88-002R, June 1988.

[Bol88] I.Bolsens, W. De Rammelaere, L.Claesen, H. De Man, Expert analysis of synchronous digital MOS circuits using rule based programming and symbolic analysis. In [ESP88], pp. 269-282, 1988; Electrical verification using rule based programming and symbolic analysis. In [Sau89], pp. 211-228, 1989.

[Bol89] I.Bolsens et al., Electrical verification using rule based programming and symbolic analysis. In [Sau89], pp. 211-228, 1989.

[Bol90] M.Bolton, *Digital Systems Design with Programmable Logic.* Addison-Wesley, Wokingham, 1990.

[Boo91] G.Booch, *Object Oriented Design. With Applications.* Benjamin/Cummings, Redwood City, 1991.

[Bor80] S.A.Borkin, *Data Models: A Semantic Approach for Database Systems.* MIT Press, Cambridge, 1980.

[Bos77] A.K.Bose, S.A.Szygenda, Detection of static and dynamic hazards in logic nets. *14th Des. Aut. Conf.*, New Orleans, pp. 220-224, June 1977.

[Bou88] N.G.Bourbakis, I.N.Savvides, Specifications of a knowledge-based environment for VLSI system architectural design. *IEEE Workshop Languages for Automation*, Maryland, pp. 212-218, Aug. 1988.

[Bou90] J.M.Bournazel, J.O.Piednoir, Design assistant: An expert tool for ASIC design. *IEEE Custom Integr. Circ. Conf.*, Boston, pp. 29.3.1-29.3.4, May 1990.

[Bow91] W.Bower, C.Seaquist, W.Wolf, A framework for industrial layout generators. *IEEE Trans. CAD*, 10, pp. 596-603, May 1991.

[Bow85] R.J.Bowman, D.J.Lane, A knowledge-based system for analog integrated circuit design. *Int. Conf. CAD*, Santa Clara. pp. 210-212, Nov. 1985.

[Boy91] G.A.Boy, *Intelligent Assistant Systems.* Acad. Press, London, 1991.

[Boy88] D.G.Boyer, Symbolic layout compaction review. *25th Des. Aut. Conf.*, Anaheim, pp. 383-389, June 1988.

[Bra79] R.J.Brachman, On the epistemological status of semantic networks. In: N.V.Findler (ed.), *Associative Networks: Representation and Use of Knowledge by Computers.* Acad. Press, London, pp. 3-50, 1979.

[Bra85] R.J.Brachman, H.J.Levesque (eds), *Readings in Knowledge Representation.* Morgan Kaufmann, San Mateo, 1985.

[Bra86] W.B.Bradbury, ART^{TM}-FUL diagnosis: The rule-based go-chain. *Autotestcon*, San Antonio, pp. 37-43, Sept. 1986.

REFERENCES

[Bra87] D.S.Brahme, J.A.Abraham, Knowledge based test generation for VLSI circuits. *Int. Conf. CAD*, Santa Clara, pp. 292-295, Nov. 1987.

[BRa86] I.Bratko, *Prolog Programming for Artificial Intelligence*. Addison-Wesley, Reading, 1986.

[Bra88] D.Braun et al., Techniques for multilayer channel routing. *IEEE Trans. CAD*, 7, pp. 698-712, June 1988.

[Bre87] G.L.Bredenkamp, Use of spreadsheets in electrical engineering. *IEEE Circ. Dev. Mag.*, 3, pp. 27-35, Sept. 1987.

[Bre89] R.Breitl et al., The Gemstone Data Management System. In [Kim89], pp. 283-308, 1989.

[Bre91] P.C.Bressloff, D.J.Weir, Neural networks. *GEC Jour. Res.*, 8, pp. 151-169, 1991.

[Bre90] F.Bretschneider et al., Knowledge based design flow management. *Int. Conf. CAD*, Santa Clara, pp. 350-353, Nov. 1990.

[Bre72] M.A.Breuer, A note on three valued logic simulation. *IEEE Trans. Comp.*, 21, pp. 399-402, April 1972.

[Bre77] M.A.Breuer, A class of min-cut placement algorithms. *Jour. DA and FTC*, 1, pp. 343-362, Oct. 1977.

[Bre76] M.A.Breuer, A.D.Friedman, *Diagnosis and Reliable Design of Digital Systems*. Computer Science Press, Woodland Hills, 1976.

[Bre80] M.A.Breuer, A.D.Friedman, Functional level primitives in test generation. *IEEE Trans. Comp.*, 29, pp. 223-235, March 1980.

[Bre85] M.A.Breuer, X.A.Zhu, A knowledge based system for selecting a test methodology for a PLA. *22nd Des. Aut. Conf.*, Las Vegas, pp.259-265, June 1985.

[BRe89] M.A.Breuer et al., AI aspects of TEST: A system for designing testable VLSI chips. In [Sau89], pp. 31-76, 1989.

[Bre86] F.D.Brewer, D.D.Gajski, An expert-system paradigm for design. *23rd Des. Aut. Conf.*, Las Vegas, pp. 62-68, June/July 1986.

[BRe90] F.Brewer, D.Gajski, Chippe: A system for constraint driven behavioral synthesis. *IEEE Trans. CAD*, 9, pp. 681-695, July 1990.

[BRE89] J.R.Brews, Electrical modeling of interconnections. In [Wat89], pp. 269-331, 1989.

[Bri75] P.Brinch Hansen, The programming language Concurrent Pascal. *IEEE Trans. Software Eng.*, 1, pp. 199-207, June 1975.

[Bro88] H.Broman, Global and local control of processing using knowledge-based signal processing techniques. An application. *Int. Symp. Circ. Sys.*, Espoo, pp.2375-2379, June 1988.

[Bro83] H.Brown, C.Tong, G.Foyster, Palladio: An exploratory environment for circuit design. *Computer*, 16, pp. 41-56, Dec. 1983.

[Bro82] J.S.Brown, R.R.Burton, J. de Kleer, Pedagogical, natural language and knowledge engineering techniques in SOPHIE I, II and III. In [Sle82], pp. 227-282, 1982.

[Bro85] L.Brownston, R.Farrell, E.Kant, N.Martin, *Programming Expert Systems in OPS5. An Introduction to Rule-Based Programming*. Addison-Wesley, Reading, 1985.

[Bru86] R.Brück, J.Herrmann, KOCOS. An expert system for VLSI-layout compaction. *Int. Conf. Comp. Des.*, Port Chester, pp. 298-301, Oct. 1986.

[Bru88] R.Brück, K.H.Temme, H.Wronn, FLAIR. A knowledge-based approach to integrated circuit floorplanning. *Int. Workshop AI Industr. Appl.*, Hitachi City, pp. 194-199, May 1988.

[Bry80] R.E.Bryant, An algorithm for MOS logic simulation. *Lambda*, 1, pp. 46-53, Fourth Quarter, 1980.

[Bry83] R.Bryant (ed.), *Very Large Scale Integration*. Springer, Berlin, 1983.

[Bry87] R.Bryant, A survey of switch-level algorithms. *IEEE Design & Test*, 4, pp. 26-40, Aug. 1987.

[Bu90] J.Bu, *Systematic Design of Regular VLSI Processor Arrays*. Doct. Diss., Delft Univ. Technology, 22 May 1990.

[Buc78] B.Buchanan, E.Feigenbaum, DENDRAL and Meta-DENDRAL: Their Applications. *Art. Intell.*, 11, pp. 5-24, 1978.

[Buc84] B.G.Buchanan, E.H.Shortliffe (eds), *Rule-Based Expert Systems. The MYCIN Experiments of the Stanford Heuristic Programming Project*. Addison-Wesley, Reading, 1984.

[Buc90] F.L.Buck, Integrating expert system diagnostics within ATE system software. *IEEE Autotestcon*, San Antonio, pp. 85-92, Sept. 1990.

[Bud89] H.Budzisz, Searching for TAC filter structures using rule-based system. *Eur. Conf. Circ. Th. Des.*, Brighton, pp. 372-375, Sept. 1989.

[Bun79] A.Bundy et al., Solving mechanics problems using meta-level inference. In [Mic79], pp. 51-64, 1979.

[Bur83] M.R.Buric, C.Christensen, T.G.Matheson, Plex: Automatically generated microcomputer layouts. *Int. Conf. Comp. Des.*, Port Chester, pp. 181-184, Oct./Nov. 1983.

[Bur91] J.B.Burr, Digital neural network implementations. In [Ant91], pp. 237-285, 1991.

[Bus87] W.R.Bush et al., An advanced silicon compiler in Prolog. *Int. Conf. Comp. Des.*, Port Chester, pp. 27-31, Oct. 1987.

[Bus89] W.R.Bush et al., Layering expertise in s full-range hardware synthesis system. In [Oda89], pp. 147-152, 1989.

[Bus88] M.L.Bushnell, *Automated Full-Custom VLSI Layout Using the ULYSSES Design Environment*. Acad. Press, 1988. See also: *IEEE Trans. CAD*, 8, pp. 279-287, March 1989.

[Bus85] M.L.Bushnell, S.W.Director, ULYSSES: An expert-system based VLSI design environment. *Int. Symp. Circ. Sys.*, Kyoto, pp. 893-896, June 1985. See also: *23rd Des. Aut. Conf.*, Las Vegas, June/July 1986.

[Buz85] G.D.Buzzard, T.N.Mudge, Object-based computing and the Ada language. *Computer*, 18, pp. 11-19, March 1985.

[Cai87] H.Cai, A rule-based global routing optimizer. *Int. Symp. Circ. Sys.*, Philadelphia, pp. 43-46, May 1987.

[Cam87] P.Camurati, P.Prinetto, ProTest: A Design for Testability oriented Prolog Hardware Description Language. *Int. Conf. Comp. Des.*, Port Chester, pp. 297-300, Oct. 1987.

REFERENCES

[Cam88] P.Camurati et al., ESTA: An expert system for DFT rule verification. *IEEE Trans. CAD*, 7, pp. 1172-1180, Nov. 1988.

[Can83] R.R.Cantone, F.J.Pipitone, W.B.Lander, M.P.Marrone, Model-based probabilistic reasoning for electronics troubleshooting. *8th IJCAI-83*, Karlsruhe, pp. 207-211, Aug. 1983.

[Car83] S.K.Card, T.P.Moran, A.Newell, *The Psychology of Human Computer Interaction*. Lawrence Erlbaum, Hillsdale, 1983.

[Car90] L.R.Carley, Automated design of operational amplifiers: A case study. In [Ism90], pp. 45-78, 1990.

[Cas78] G.R.Case, J.D.Stauffer, SALOGS-IV, a program to perform logic simulation and fault diagnosis. *15th Des. Aut. Conf.*, Las Vegas, pp. 392-397, June 1978.

[Cas87] A.Casotto et al., A parallel simulated annealing algorithm for the placement of macro-cells. *IEEE Trans. CAD*, 6, pp. 838-847, Sept. 1987.

[Cha86] M.Chadwick, J.A.Hannah, *Expert Systems for Personal Computers*. Sigma Press, Wilmslow, 1986.

[Cha91] S.T.Chakradar, V.D.Agrawal, M.L.Bushnell, *Neural Models and Algorithms for Digital Testing*. Kluwer Acad., Dordrecht, 1991.

[Cha89] H.T.Chang, K.W.Wu, A knowledge-based approach to design task configuration. In [Oda89], pp. 21-27, 1989.

[Cha74] H.Y.Chang, G.W.Helmbigner, Controllability, observability and maintenance engineering technique (COMET). *Bell Sys. Tech. Jour.*, 53, pp. 1505-1534, Oct. 1974.

[Cha92] N.H.Chang, K.J.Chang, J.Leo, K.Lee, S.Y.Oh, IPDA: Interconnect Performance Design Assistant. *29th Des. Aut. Conf.*, Anaheim, pp. 472-477, June 1992.

[Cha85] E.Charniak, D.McDermott, *Introduction to Artificial Intelligence*. Addison-Wesley, Reading, 1985.

[Cha83] W.S.Chan, A new channel routing algorithm. In R.Bryant (ed.), *Very Large Scale Integration*. Springer, Berlin, pp. 117-139, 1983.

[Che90] D.J.Chen, B.J.Sheu, Automatic layout generation for mixed analog-digital VLSI neural chips. *Int. Conf. Comp. Des.*, Cambridge, pp. 29-32, Sept. 1990.

[Che86] H.H.Chen, E.S.Kuh, Glitter: A gridless variable-width channel router. *IEEE Trans. CAD*, 5, pp. 459-465, Oct. 1986.

[Che76] P.P.Chen, The entity-relationship model. Toward a unified view of data. *ACM Trans. Database Sys.*, 1, pp. 9-36, March 1976.

[Che83] S.Chen, On intelligent CAD systems for VLSI design. *Int. Conf. Comp. Des.*, Port Chester, pp. 405-408, Oct./Nov. 1983.

[Che85] T.H.Chen, M.A.Breuer, Automatic design for testability via testability measures. *IEEE Trans. CAD*, 4, pp. 3-11, Jan. 1985.

[Che84] E.K.Cheng, Verifying compiled silicon. *VLSI Design*, 5, pp. 70-74, Oct. 1984.

[Che89] S.C.Cheng, J.Comella, P.L.Law, Expert's Toolkit: An expert system shell for building diagnostic systems. *SPIE, Vol. 1095, Appl. Art. Intell. VII*, pp. 164-171, March 1989.

[Chi91] M.L.Chiang, T.C.Lu, J.B.Kuo, Analogue adaptive neural network circuit. *IEE Proc.-G*, 138, pp. 717-723, Dec. 1991.

[Chi89] L.Chisvin, R.J.Duckworth, Content-addressable and associative memory: Alternatives to the ubiquitous RAM. *Computer*, 22, pp. 51-64, July 1989.

[Chl90] I.Chlamtac, W.P.Franta, Rationale, directions, and issues surrounding high speed networks. *Proc. IEEE*, 78, pp. 94-120, Jan. 1990.
[Cho90] M.F.Chowdhury, R.E.Massara, An expert system for general purpose analogue layout synthesis. *33rd Midwest Symp. Circ. Sys.*, Calgary, pp. 1171-1174, Aug. 1990.
[Chu84] K.C.Chu, R.Sharma, A technology independent MOS multiplier generator. *21th Des. Aut. Conf.*, Albuquerque, pp. 90-97, June 1984.
[Chr87] K.Christian, *The UNIX Text Processing System*. Wiley, New York, 1987.
[CiZ90] Y.Ci, C.Zhang, C.Sun, *New Generation Computing. Recent Research*. North-Holland, Amsterdam, 1990.
[Cla89] L.Claesen (ed.), *Applied Formal Methods for Correct VLSI Design*. 2 volumes, Elsevier, Amsterdam, 1989.
[Cla90] L.J.M.Claesen (ed.), *Formal VLSI Specification and Synthesis. VLSI Design Methods*. North-Holland, Amsterdam, 1990.
[Cla73] C.R.Clare, *Designing Logic Systems Using State Machines*. McGraw-Hill, New York, 1973.
[Cla85] G.Clark, R.Zippel, SCHEMA. An architecture for knowledge based CAD. *Int. Conf. CAD*, Santa Clara, pp. 50-52, Nov. 1985.
[Cla80] J.H.Clark, A VLSI geometry processor for graphics. *Computer*, 13, pp. 59-68, July 1980.
[Cla84] K.L.Clark, F.G.McCabe *et al.*, *Micro-PROLOG. Programming in Logic*. Prentice-Hall, Hemel Hempstead, 1984.
[CLa90] P.Clark, Machine learning: Techniques and recent developments. In [Mir90], pp. 65-93, 1990.
[CLa84] B.D.Clayton, *ART Programming Primer*. Report, Inference Corp., 1984. See also: *ART Reference Manual, Version 3.0*. Inference Corp., Los Angeles, 1987.
[Cle91] A.Clements, *The Principles of Computer Hardware*. Oxford Univ. Press, Oxford, (2nd ed.) 1991.
[Clo87] W.F.Clocksin, C.S.Mellish, *Programming in Prolog*. Springer, Berlin, (3rd ed.) 1987.
[CLo87] W.Clocksin, A Prolog primer. *Byte*, 12, pp. 147-158, Aug. 1987.
[CLO87] W.F.Clocksin, Logic programming and digital circuit analysis. *Jour. Logic Progr.*, 4, pp. 59-82, 1987.
[Cob84] D.Cobb, *Mastering Symphony*, Sybex, Berkeley, 1984.
[Cod70] E.F.Codd, A relational model of data for large shared data banks. *Commun. ACM*, 13, pp. 377-387, June 1970
[Coh81] J.Cohen, Garbage collection of linked data structures. *Computing Surveys*, 13, pp. 341-367, Sept. 1981.
[Coh88] J.P.Cohoon, P.L.Heck, BEAVER: A computational-geometry-based tool for switchbox routing. *IEEE Trans. CAD*, 7, pp. 684-697, June 1988.
[Col88] A.Collins, E.E.Smith (eds), *Readings in Cognitive Science. A Perspective from Psychology and Artificial Intelligence*. Morgan Kaufmann, San Mateo, 1988.
[Con87] J.Conklin, Hypertext: An introduction and survey. *Computer*, 20, pp. 17-41, Sept. 1987.
[Cor88] T.Corman, M.U.Wimbrow, Coupling a digital logic simulator and an analog circuit simulator. *VLSI Sys. Des.*, 9, pp. 38-47, Feb. 1988.

REFERENCES

[Cor87] D.Cormier, Mastering the BiCMOS mix. *Electronic Sys. Des. Mag.*, 17, pp. 47-55, Nov. 1987.

[Cor80] W.E.Cory, W.M. van Cleemput, Developments in verification of design correctness. *17th Des. Aut. Conf.*, Minneapolis, pp. 156-164, June 1980.

[Cos89] S.Cosgrove, G.Musgrave, The types of knowledge required for an expert test advisor. In [Sau89], pp. 131-144, 1989.

[Cos91] S.J.Cosgrove, G.Musgrave, Test generation within an expert system environment. *IEE Proc.-E*, 138, pp. 36-40, Jan. 1991.

[Cot87] G.W.Cottrell, P.Munro, D.Zipser, Image compression by backpropagation: An example of extensional programming. *Advances in Cognitive Science*, Vol.3, Ablex, Norwood, 1987.

[Cox86] B.J.Cox, *Object-Oriented Programming: An Evolutionary Approach.* Addison-Wesley, Reading, 1986.

[Cra89] M.Crastes de Paulet, M.Karam, G.Saucier, Test expertise from high-level specifications. In [Oda89], pp. 77-84, 1989; M.Crastes de Paulet et al., Test planning from high-level specifications. In [Sau89], pp. 111-129, 1989.

[Cra84] J.D.Crawford, An Electronic Design Interchange Format. *21st Des. Aut. Conf.*, Albuquerque, pp. 683-685, June 1984; J.D.Crawford, EDIF: A mechanism for the exchange of design information. *IEEE Design & Test*, 2, pp. 63-69, Feb. 1985.

[Cul86] R.E.Cullingford, *Natural Language Processing: A Knowledge-Engineering Approach.* Rowman & Littlefield, Totawa, 1986.

[Cul82] R.E.Cullingford, M.W.Krueger, M.Selfridge, M.A.Bienkowski, Automated explanations as a component of a computer-aided design system. *IEEE Trans. Sys., Man, Cybernetics*, 12, pp. 168-181, March/Apr. 1982.

[Dae81] W.Daehn, J.Mucha, A hardware approach to self-testing of large programmable logic arrays. *IEEE Trans. Comp.*, 30, pp. 829-833, Nov. 1981.

[Dan91] J.Daniell, S.W.Director, An object oriented approach to CAD tool control. *IEEE Trans. CAD*, 10, pp. 698-713, June 1991.

[DAR91] DARPA's Strategic Computing Initiative. *IEEE Expert*, 6, pp. 7-64, June 1991.

[Dar81] J.A.Darringer, W.H.Joyner, C.L.Berman, L.Trevillyan, Logic synthesis through local transformations. *IBM Jour. Res. Dev.*, 25, pp. 272-280, July 1981.

[Dar84] J.A.Darringer, D.Brand, J.V.Gerbi, W.H.Joyner, L.Trevillyan, LSS: A system for production logic synthesis. *IBM Jour. Res. Dev.*, 28, pp. 537-545, Sept. 1984.

[Das82] H.W.Daseking, R.I.Gardner, P.B.Weil, VISTA: A VLSI CAD system. *IEEE Trans. CAD*, 1, pp. 36-52, Jan. 1972.

[Dat86] C.J.Date, *An Introduction to Database Systems.* Addison-Wesley, Reading, 1986.

[Dav83] M.Davio, J.P.Deschamps, A.Thayse, *Digital Systems with Algorithm Implementation.* Wiley, Chichester, 1983.

[DAv82] R.Davis, *Teiresias: Applications of Meta-Level Knowledge.* In [Dav82], pp. 227-484, 1982.

[Dav82] R.Davis, D.B.Lenat, *Knowledge-Based Systems in Artificial Intelligence.* McGraw-Hill, New York, 1982.

[Dav84] R.Davis, Diagnostic reasoning based on structure and behavior. *Art. Intell.*, 24, pp. 347-410, 1984; R.Davis, H.Shrobe, Representing the structure and behavior of hardware. *Computer*, 16, pp. 75-82, Oct. 1983.
[Daw86] J.L.Dawson, EXCIRSIZE. An expert system for VLSI transistor sizing. In [SRi86], pp. 911-916, 1986.
[deG85] A.J. de Geus, W.Cohen, A rule-based system for optimizing combinational logic. *IEEE Design & Test*, 2, pp. 22-32, Aug. 1985.
[Deg89] M.G.R.Degrauwe *et al.*, Towards an analog system design environment. *IEEE Jour. Solid-State Circ.*, 24, pp. 659-671, June 1989.
[DeJ86] E.J.DeJesus, J.P.Callan, C.R.Whitehead, PEARL: An expert system for power supply layout. *23rd Des. Aut. Conf.*, Las Vegas, pp. 615-621, June/July 1986.
[DEJ86] G.DeJong, R.Mooney, Explanation-based learning: An alternative view. *Machine Learning*, 1, pp. 145-176, 1986.
[deK86] J. de Kleer, Problem solving with the ATMS. *Art. Intell.*, 28, pp. 197-224, 1986.
[deK80] J. de Kleer, G.J.Sussman, Propagation of constraints applied to circuit synthesis. *Circ. Th. Appl.*, 8, pp. 127-144, 1980.
[deK87] J. de Kleer, B.C.Williams, Diagnosing multiple faults. *Art. Intell.*, 32, pp. 97-130, 1987.
[deL91] A.A.J. de Lange, *Design and Implementation of Highly Parallel Pipelined VLSI Systems*. Doct. Diss., Delft Univ. Technology, 24 Jan. 1991.
[deL89] A.A.J. de Lange, A.J. van der Hoeven, E.F.Deprettere, P.M.Dewilde, HiFi: An object oriented system for the structural synthesis of signal processing algorithms and the VLSI compilation of signal flow graphs. In [Cla89], pp. 462-481, 1989.
[Del89] J.G.Delgado-Frias, W.R.Moore (eds), *VLSI for Artificial Intelligence*. Kluwer Acad., Boston, 1989.
[Del85] C.Delorme *et al.*, A functional partitioning expert system for test sequences generation. *22nd Des. Aut. Conf.*, Las Vegas, pp. 820-824, June 1985.
[DeM85] H.J. De Man, I.Bolsens, E.Vanden Meersch, J. Van Cleynenbreugel, DIALOG: An expert debugging system for MOS VLSI design. *IEEE Trans. CAD*, 4, pp. 303-311, July 1985.
[DeM86] H. De Man, J.Rabaey, P.Six, L.Claesen, Cathedral-II: A silicon compiler for digital signal processing. *IEEE Design & Test*, 3, pp. 13-25, Dec. 1986. H. De Man *et al.*, CATHEDRAL-II. A computer-aided synthesis system for digital signal processing VLSI systems. *Computer-Aided Eng. Jour.*, 5, pp. 55-66, April 1988.
[DeM87] G. De Micheli, A.Sangiovanni-Vincentelli, P.Antognetti (eds), *Design Systems for VLSI Circuits: Logic Circuits and Silicon Compilation*. Martinus Nijhoff, Dordrecht, 1987.
[Den85] P.B.Denyer, D.Renshaw, *VLSI Signal Processing. A Bit-Serial Approach*. Addison-Wesley, 1985.
[deS85] H. de Saram, *Programming in Micro-PROLOG*. Halsted Press, New York, 1985.
[Deu76] D.Deutsch, A "dogleg" channel router. *13th Des. Aut. Conf.*, San Francisco, pp. 425-433, June 1976.

REFERENCES

[Deu83] L.P.Deutsch, *The Dorado Smalltalk-80 Implementation: Hardware Architecture's Impact on Software Architecture*. Addison-Wesley, Reading, 1983.

[Deu90] O.Deux et al., The story of O_2. *IEEE Trans. Knowledge and Data Eng.*, 2, pp. 91-108, March 1990.

[Dev87] P.Deves, P.Dague, J.P.Marx, O.Raiman, DEDALE: An expert system for troubleshooting analogue circuits. *Int. Test Conf.*, Washington, pp. 586-594, Sept. 1987.

[Dew90] A.M.Dewey, S.W.Director, *Principles of VLSI System Planning: A Framework for Conceptual Design*. Kluwer Academic, Boston, 1990.

[Dew86] P.Dewilde (ed.), *The Integrated Circuit Design Book*. Delft Univ. Press, Delft, 1986.

[Dew88] P.Dewilde, Multiview and hierarchical VLSI design achievements of the ICD Project (ESPRIT 991), *Delft Progress Report*, 12, pp. 207-219, 1988. See also [ESP88], pp. 245-262, 1988.

[DEw90] P.Dewilde, Z.-Q.Ning, *Models for Large Integrated Circuits*. Kluwer Acad., Boston, 1990.

[Dic86] A.Dickinson, Floyd. A knowledge based floor plan designer. *Int. Conf. Comp. Des.*, Port Chester, pp. 176-179, Oct. 1986.

[Die90] J.Diederich (ed.), *Artificial Neural Networks. Concept Learning*. IEEE Comp. Soc. Press, Los Alamitos, 1990.

[Dil88] T.E.Dillinger, *VLSI Engineering*. Prentice-Hall, Englewood Cliffs, 1988.

[DiG89] J. Di Giacomo (ed.), *VLSI Handbook: Silicon, Gallium Arsenide, and Superconductor Circuits*. McGraw-Hill, New York, 1989.

[DiG90] J. Di Giacomo (ed.), *Digital Bus Handbook*. McGraw-Hill, New York, 1990.

[Dij68] E.W.Dijkstra, The structure of the T.H.E.-multiprogramming system. *Commun. ACM*, 11, pp. 341-346, My 1968.

[Din80] M.Dincbas, A knowledge-based expert system for automatic analysis and synthesis in CAD. *Information Proc. 80, IFIPS Proc.*, pp. 705-710, 1980.

[Dir81] S.W.Director, A.C.Parker, D.P.Siewiorek, D.E.Thomas, A design methodology and computer aids for digital VLSI systems. *IEEE Trans. Circ. Sys.*, 28, pp. 634-645, July 1981.

[Dis89] J.P.Dishaw, J.Y.C.Pan, AESOP: A simulation-based knowledge system for CMOS process diagnosis. *IEEE Trans. Semicond. Manuf.*, 2, pp. 94-103, Aug. 1989.

[DIs89] C.Dislis et al., Cost effective test strategy selection. In [Sau89], pp. 93-110, 1989.

[Dob87] T.Dobry, A coprocessor for AI; LISP, Prolog and data bases. *COMPCON Spring*, San Francisco, pp. 396-403, Feb. 1987.

[Dob90] T.P.Dobry, *A High Performance Architecture for Prolog*. Kluwer Acad., Boston, 1990.

[Dol88] S.B.Dolins, A.Srivastava, B.E.Flinchbaugh, Monitoring and diagnosis of plasma etch processes. *IEEE Trans. Semicond. Manuf.*, 1, pp. 23-27, Feb. 1988.

[Dou85] R.J.Douglass, A qualitative assessment of parallelism in expert systems. *IEEE Software*, 2, pp.70-81, May 1985.

[Dro85] P.J.Drongowski, A graphical, rule-based assistant for control graph - datapath dsign. *Int. Conf. Comp. Des.*, Port Chester, pp. 208-211, Oct. 1985.
[DSo89] D.F.D'Souza, KRAFT: A knowledge-base for CAD. In [Oda89], pp. 13-20, 1989.
[Dub90] P.F.Dubois, A.Puissochet, A.M.Tagant, A general and flexible switchbox router: CARIOCA. *IEEE Trans. CAD*, 9, pp. 1307-1317, Dec. 1990.
[Dud79] R.O.Duda, J.Gaschnig, P.E.Hart, Model design in the PROSPECTOR consultant system for mineral exploration. In [Mic79], pp. 153-167, 1979.
[Dun84] L.N.Dunn, IBM's Engineering Design System support for VLSI design and verification. *IEEE Design & Test*, 1, pp. 30-40, Feb. 1984.
[Dur89] M.D.Durand, Parallel simulated annealing. Accuracy vs. speed in placement. *IEEE Design & Test*, 6, pp. 8-34, June 1989.
[Edw83] M.D.Edwards, D.Aspinall, The synthesis of digital systems using ASM design techniques. In: T.Uehara, M.Barbacci (eds), *Computer Hardware Description Languages and Their Applications*, pp. 55-64, 1983.
[Eic65] E.B.Eichelberger, Hazard detection in combinational and sequential switching circuits. *BM Jour. Res. Develop.*, 9, pp. 90-99, March 1965.
[Eic91] E.Eichelberger, E.Lindbloom, J.Waicukauski, T.Williams, *Structured Logic Testing*. Prentice-Hall, Englewood Cliffs, 1991.
[Ein85] N.G.Einspruch (ed.), *VLSI Handbook*. Acad. Press, Orlando, 1985.
[Ein91] N.G.Einspruch, J.L.Hilbert (ed.), *Application Specific Integrated Circuit (ASIC) Technology*. Acad. Press, San Diego, 1991.
[Eli83] N.J.Elias, A.W.Wetzel, The IC module compiler, a VLSI system design aid. *20th Des. Aut. Conf.*, Miami Beach, pp. 46-49, June 1983.
[Elm89] R.Elmasri, S.B.Navathe, *Fundamentals of Database Systems*. Benjamin/Cummings, Redwood City, 1989.
[ElT89] F.El-Turky, E.E.Perry, BLADES: An Artificial Intelligence approach to analog circuit design. *IEEE Trans. CAD*, 8, pp. 680-692, June 1989.
[Enn84] J.R.Ennals, *Beginning Micro-Prolog*. Ellis Horwood, Chichester, (2nd ed.) 1984.
[Eno85] K.Enomoto et al., LORES-2: A logic reorganization system. *IEEE Design & Test*, 2, pp. 35-42, Oct. 1985.
[Erf91] S.Erfani, M.Malek, H.Sachar, An expert system-based approach to capacity allocation in a multiservice application environment. *IEEE Network Mag.*, 4, pp. 7-12, May 1991.
[Eri89] E.C.Ericson, L.Traeger Ericson, D.Minoli (eds), *Expert Systems Applications in Integrated Network Management*. Artech House, Norwood, 1989.
[Erm80] L.D.Erman et al., The Hearsay-II speech-understanding system: Integrating knowledge to resolve uncertainty. *Comp. Surv.*, 12, pp. 213-253, June 1980.
[ESP88] ESPRIT'88, *Putting the Technology to Use*. 5th Annual ESPRIT Conf., Brussels, Nov. 1988. North-Holland, Amsterdam, 1988.
[Est77] G.Estrin et al., Six papers on SARA. *Symp. Des. Aut. and Micropr.*, Palo Alto, pp. 54-94, Feb. 1977.
[Eur86] J.P.Eurich, A tutorial introduction to the Electronic Design Interchange Format. *23rd Des. Aut. Conf.*, Las Vegas, pp. 327-333, June/July 1986.

REFERENCES

[Eva83] D.J.Evans (ed.), *Parallel Processing Systems*. Cambridge Univ. Press, Cambridge, 1983.

[Eva84] W.H.Evans *et al.*, ADL: An Algorithmic Design Language for integrated circuit synthesis. *21st Des. Aut. Conf.*, Albuquerque, pp. 66-72, June 1984.

[Fah83] S.E.Fahlman, G.E.Hinton, Massively parallel architectures for AI: NETL, THISTLE and BOLTZMANN machines. *Nat. Conf. AI*, pp. 109-113, Aug. 1983.

[Fah87] S.E.Fahlman, G.E.Hinton, Connectionist architectures for artificial intelligence. *Computer*, 20, pp. 100-109, Jan. 1987.

[Fag82] R.Fagin, A.O.Mendelzon, J.D.Ullman, A simplified universal relation assumption and its properties. *ACM Trans. Database Sys.*, 7, pp. 343-360, Sept. 1982.

[Fal88] H.Falk, AI techniques enter the realm of conventional languages. *Comp. Design*, 27, pp. 45-49, 15 Oct. 1988.

[Fan88] A.Fanni, A.Giua, M.G.Manca, Automated diagnosis for digital circuits. In: T.O'Shea, V.Squrev (eds), *Artificial Intelligence III: Methodology, Systems, Applications*, North-Holland, Amsterdam, pp. 373-379, 1988.

[Far89] N.H.Farhat, Optoelectronic neural networks and learning machines. *IEEE Circ. Dev. Mag.*, 5, pp. 32-41, Sept. 1989.

[Fei63] E.A.Feigenbaum, J.Feldman (eds), *Computers and Thought*, McGraw-Hill, New York, pp. 109-133, 279-293, 1963.

[Fei83] E.A.Feigenbaum, P.McCorduck, *The Fifth Generation: Artificial Intelligence and Japan's Computer Challenge*. Addison-Wesley, Reading, 1983.

[Fen81] T.Y.Feng, A survey of interconnection networks. *Computer*, 14, pp. 12-27, Dec. 1981.

[Fic87] W.Fichtner, M.Morf (eds), *VLSI CAD Tools and Applications*. Kluwer Acad., Dordrecht, 1987.

[Fic83] S.Fickas, Automating software development: A small example. *Symp. Appl. Ass. Aut. Software Develop. Tools*, San Francisco, 1983.

[Fie86] R.D.Fiebrich, A supercomputer workstation for VLSI CAD. *IEEE Design & Test*, 3, pp. 31-37, June 1986.

[Fil88] B.Filipic, *Prolog User's Handbook: A Library of Utility Programs*. Wiley, New York, 1988.

[Fin84] T.Finin, J.McAdams, P.Kleinosky, FOREST: An expert system for automatic test equipment. *1st Conf. AI Appl.*, Denver, pp. 350-356, Dec.1984.

[Fin87] P.Fink, C.Lusth, Expert systems and diagnostic expertise in the mechanical and electrical domains. *IEEE Trans. Sys., Man, Cybernetics*, 17, pp. 340-349, May/June 1987.

[Fis84] A.L.Fisher, H.T.Kung, Special-purpose VLSI architectures. General discussions and a case study. In: P.R.Capello *et al.* (eds), *VLSI Signal Processing*, IEEE Press, New York, pp. 153-169, 1984.

[Fis88] A.S.Fisher, *CASE: Using Software Development Tools*. Wiley, New York, 1988.

[Fis89] D.Fishman *et al.*, Overview of the Iris DBMS. In [Kim89], pp. 219-150, 1989.

[Fla87] B.J.Fladung, *The XLISP Primer*. Prentice-Hall, Hemel Hempstead, 1987; D.Betz, XLISP: An experimental object-oriented language (version 1.4).

Manual, Manchester, 1985.

[Fla81] P.L.Flake, P.R.Moorby, G.Musgrave, HILO MARK 2 hardware desription language. In: M.A.Breuer, R.Hartenstein (eds), *Computer Hardware Description Languages and Their Applications*. North-Holland, Amsterdam, pp. 95-107, 1981.

[Fle85] H.Fleisher, Simulated annealing as a tool for logic optimization in a CAD environment. *Int. Conf. CAD*, Santa Clara, pp. 203-205, Nov. 1985.

[Flo82] R.W.Floyd, J.D.Ullman, The compilation of regular expressions into integrated circuits. *Jour. ACM*, 29, pp. 603-622, July 1982.

[Fly72] M.F.Flynn, Some computer organizations and their effectiveness. *IEEE Trans. Computers*, 21, pp. 948-960, Sept. 1972.

[Fod81] J.K.Foderaro, *The FRANZ LISP Manual*. Univ. of California, Berkeley, 1981.

[Fol82] J.D.Folay, A. van Dam, *Fundamentals of Interactive Computer Graphics*. Addison-Wesley, Reading, 1982.

[Foo90] S.Y.Foo, L.R.Anderson, Y.Takefuji, Analog components for the VLSI of neural networks. *IEEE Circ. Dev.*, 6, pp. 18-26, July 1990.

[Foo86] Y.P.S.Foo, H.Kobayashi, A knowledge-based system for VLSI module selection. *Int. Conf. Comp. Des.*, Port Chester, pp. 184-187, Oct. 1986.

[For82] C.L.Forgy, Rete: A fast algorithm for the many pattern / many object pattern match problem. *Art. Intell.*, 19, pp. 17-37, Sept. 1982.

[For84] C.L.Forgy, *The OPS83 Report*. Carnegie-Mellon Univ., Pittsburgh, May 1984; C.L.Forgy, *OPS83 User's Manual*. Production Systems Technologies, 1985.

[For77] C.Forgy, J.McDermott, OPS, a domain-independent production system language. *5th Joint Conf. AI*, Cambridge, pp. 933-939, Aug. 1977; C.L.Forgy, OPS5 User's Manual. Tech. Rep. CMU-CS-81-35, Carnegie-Mellon Univ., Pittsburgh, July 1981.

[For83] J.Forrest, M.D.Edwards, The automatic generation of programmable logic arrays from algorithmic state machine descriptions. In [Anc83], pp. 183-193, 1983.

[For89] R.Forsyth (ed.), *Machine Learning. Principles and Techniques*. Chapman and Hall, London, 1989.

[For86] R.Forsyth, R.Rada, *Machine Learning: Applications in Expert Systems and Information Retrieval*. Ellis Horwood, Chichester, 1986.

[Fos80] M.J.Foster, H.T.Kung, The design of special-purpose VLSI chips. *Computer*, 13, pp. 26-40, Jan. 1980.

[Foy84] G.Foyster, A knowledge-based approach to transistor sizing. Rep. HPP-84-3, Computer Sci. Dep., Stanford Univ., March 1984.

[Fre90] Y.Freundlich, Knowledge bases and databases: Converging technologies, diverging interests. *Computer*, 23, pp. 51-57, Nov. 1990.

[Fro86] R.A.Frost, *Introduction to Knowledge Base Systems*. Collins, London, 1986.

[Fro77] R.A.Frowerk *et al.*, Three articles on Signature Analysis. *Hewlett-Packard Jour.*, 28, pp. 2-21, May 1977.

[Fuj89] T.Fujii *et al.*, KAT: Knowledge acquisition tool for PWB layout expert system. In [Oda89], pp. 221-228, 1989.

[Fuj90] M.Fujita, Y.Matsunaga, T.Kakuda, Automatic and semi-automatic verification of switch-level circuits with temporal logic and binary decision diagrams. *Int.*

REFERENCES

[Fuj83] *Conf. CAD*, Santa Clara, pp. 38-41, Nov. 1990.
T.Fujita, S.Goto, A rule-based routing system. *Int. Conf. Comp. Des.*, Port Chester, pp. 451-454, Oct./Nov. 1983. See also: K.Mitsumoto *et al.*, *Int. Symp. Circ. Sys.*, Montreal, pp. 449-452, May 1984.

[Fuj87] T.Fujita, Y.Nakakuki, S.Goto, A new knowledge based approach to circuit design. *Int. Conf. CAD*, Santa Clara, pp. 456-459, Nov. 1987.

[Fuj86] T.Fujita, H.Mori, K.Mitsumoto, S.Goto, Artificial intelligence approach to VLSI design. In: S.Goto (ed.), *Design Methodologies*, pp. 441-464. North-Holland, Amsterdam, 1986.

[Fuj84] M.Fujita, H.Tanaka, T.Moto-oka, Specifying hardware in temporal logic and efficient synthesis of state-diagrams using Prolog. *Int. Conf. Fifth Gen. Comp. Sys.*, Tokyo, pp. 572-581, Nov. 1984.

[Fuj81] H.Fujiwara, K.Kinoshita, A design of programmable logic arrays with universal tests. *IEEE Trans. Comp.*, 30, pp. 823-828, Nov. 1981.

[FUj83] H.Fujiwara, T.Shimono, On the acceleration of test generation algorithms. *IEEE Trans. Comp.*, 32, pp. 1137-1144, Dec. 1983.

[Fuk82] K.Fukushima, S.Miyake, Neocognitron: A new algorithm for pattern recognition tolerant of deformations and shifts in position. *Pattern Recognition*, 15, pp. 455-469, 1982. See also: *Biol. Cybern.*, 36, pp. 193-202, 1980, and *Neural Networks*, 2, pp. 413-420, 1989.

[Fun85] S.Funatsu, M.Kawai, An automatic test generation system for large digital circuits. *IEEE Design & Test*, 2, pp. 54-60, 1985.

[Fun86] H.S.Fung, S.Hirschhorn, An automatic DFT system for the Silc silicon compiler. *IEEE Design & Test*, 3, pp. 45-57, Feb. 1986.

[Fun88] A.H.Fung *et al.*, Knowledge-based analog circuit synthesis with flexible architecture. *Int. Conf. Comp. Des.*, Rye Brook, pp. 48-51, Oct. 1988.

[Fur90] K.Furuya, S.Shinoda, K.Koguchi, A knowledge based system for embedding DFT-methodologies. *Electronics and Commun. in Japan*, Part 3, 73, pp. 22-30, 1990.

[Gaj84] D.D.Gajski, J.J.Bozek, ARSENIC: Methodology and implementation. *Int. Conf. CAD*, Santa Clara, pp. 116-118, Nov. 1984.

[Gaj86] D.Gajski, R.Kuhn, An expert silicon compiler. *Custom IC Conf.*, Rochester, pp. 116-119, May 1986.

[Gal88] S.I.Gallant, Connectionist expert systems. *Commun. ACM*, 31, pp.152-169, 1988.

[Gar84] K.Garrison, D.Gregory, W.Cohen, A. de Geus, Automatic area and performance optimization of combinatorial logic. *Int. Conf. CAD*, Santa Clara, pp. 212-214, Nov. 1984.

[Gas82] J.Gaschnig, Prospector: An expert system for mineral exploration. In [Mic82], pp. 47-64, 1982. See also: R.O.Duda, R.Reboh, AI and decision making: The PROSPECTOR experience. In [Rei84], pp. 111-147, 1984.

[Gas87] W.S.Gass *et al.*, Multiple digital signal processor environment for intelligent signal processing. *Proc. IEEE*, 75, pp. 1246-1259, Sept. 1987.

[Gen84] M.R.Genesereth, The use of design descriptions in automated diagnosis. *Art. Intell.*, 24, pp. 411-436, 1984.

REFERENCES

[Gen87] M.R.Genesereth, N.J.Nilsson, *Logical Foundations of Artificial Intelligence*. Morgan Kaufmann, San Mateo, 1987.

[Ger87] J.S.Gero (ed.), *Expert Systems in Computer-Aided Design*. IFIP WG5.2 Working Conf., Sydney, Feb. 1987. North-Holland, Amsterdam, 1987.

[Ger88] J.S.Gero (ed.), *Artificial Intelligence in Engineering: Design*. Elsevier, Amsterdam, 1988.

[GEr88] J.S.Gero (ed.), *Artificial Intelligence in Engineering: Diagnosis and Learning*. Elsevier, Amsterdam, 1988.

[Gev87] W.B.Gevarter, The nature and evaluation of commercial expert system building tools. *Computer*, 20, pp. 24-41, May 1987.

[Gho87] S.Ghosh, A distributed modeling approach for simulation and verification of digital designs. *IEEE Trans. Circ. Sys.*, 34, pp. 1171-1181, Oct. 1987.

[Gho88] S.Ghosh, Behavioral-level fault simulation. *IEEE Design & Test*, 5, pp. 31-42, June 1988.

[Gia86] F.Giannesini, H.Kanoui, R.Pasero, M. van Caneghem, *PROLOG*. Addison-Wesley, Wokingham, 1986.

[Gim87] C.E.Gimarc, V.M.Milutinović, A survey of RISC processors and computers of the mid-1980s. *Computer*, 20, pp. 59-69, Sept. 1987.

[Gla85] L.A.Glasser, D.W.Dobberpuhl, *The Design and Analysis of VLSI Circuits*. Addison-Wesley, Reading, 1985.

[Glo88] J.R.Glover et al., Knowledge-based signal understanding. *Int. Symp. Circ. Sys.*, Espoo, pp. 2367-2370, June 1988.

[Goe81] P.Goel, An implicit enumeration algorithm to generate tests for combinational logic circuits. *IEEE Trans. Comp.*, 30, pp. 215-222, March 1981.

[GOe81] P.Goel, B.C.Rosales, PODEM-X: An automatic test generation system for VLSI logic structures. *18th Des. Aut. Conf.*, Nashville, pp. 260-268, June 1981.

[Gol84] A.Goldberg, *Smalltalk-80. An Interactive Programming Environment*. Addison-Wesley, Reading, 1984.

[Gol89] A.Goldberg, D.Robson, *Smalltalk-80. The Language*. Addison-Wesley, Reading, 1989.

[GOl89] R.Goldman, R.P.Gabriel, QLisp: Parallel processing in Lisp. *IEEE Software*, 6, pp. 51-59, July 1989.

[Gol80] L.H.Goldstein, E.L.Thigpen, SCOAP: Sandia controllability/observability analysis program. *17th Des. Aut. Conf.*, Minneapolis, pp. 190-196, June 1980.

[Gon87] R.Gonzalez, Simplified in-circuit diagnosis of bus node failures using expert systems techniques. *Int. Test Conf.*, Washington, pp. 595-603, Sept. 1987.

[Got86] G.R.Gottschalk, R.M.Vandoorn, A rule-based system to diagnose malfunctioning computer peripherals. *Hewlett-Packard Jour.*, 37, pp. 48-53, Nov. 1986.

[Gra89] H.P.Graf, L.D.Jackel, Analog electronic neural network circuits. *IEEE Circ. Dev. Mag.*, 5, pp. 44-49, 55, July 1989; H.P.Graf, L.D.Jackel, W.E.Hubbard, VLSI implementation of a neural network model. *Computer*, 21, pp. 41-49, March 1988.

[Gra85] R.B.Grafton, T.Ichikawa et al., Visual programming. Special Issue of *Computer*, 18, pp. 6-94, Aug. 1985.

REFERENCES

[Gra91] A.J.Graham, The CAD Framework Initiative. *IEEE Design & Test*, 8, pp. 12-15, Sept. 1991.
[Gra90] I.Graham, T.King, *The Transputer Handbook*. Prentice-Hall, New York, 1990.
[GRa85] J.Granacki, D.Knapp, A.Parker, The ADAM Advanced Design Automation System: Overview, Planner and Natural Language Interface. *22nd Des. Aut. Conf.*, Las Vegas, June 1985.
[Gra82] J.P.Gray et al., Designing gate arrays using a silicon compiler. *19th Des. Aut. Conf.*, Las Vegas, pp. 377-383, June 1982.
[Gra83] J.P.Gray et al., Controlling VLSI complexity using high-level language for design description. *Int. Conf. Comp. Des.*, Port Chester, pp. 523-526, Oct./Nov. 1983.
[Gra91] J.Gray, D.P.Siewiorek, High-availability computer systems. *Computer*, 24, pp. 39-48, Sept. 1991.
[Gre89] A.D.P.Green, P.D.Noakes, A novel approach to VLSI routing using neural networks. *Eur. Conf. Circ. Th. Des.*, Brighton, pp. 68-72, Sept. 1989.
[GrE92] K.R.Green, J.G.Fossum, A pragmatic approach to integrated process/device/circuit simulation for IC technology development. *IEEE Trans. CAD*, 11, pp. 505-512, April 1992.
[Gre91] P.E.Green, The future of fiber-optic computer networks. *Computer*, 24, pp. 78-87, Sept. 1991.
[GRe92] P.E.Green, *Fiber Optic Communication Networks*. Prentice-Hall, Englewood Cliffs, 1992.
[Gre84] R.D.Greenblatt et al., The LISP machine. In: D.R.Barstow, H.E.Shrobe (eds), *Interactive Programming Environments*. McGraw-Hill, New York, pp. , 1984.
[Gre84] D.Gregory, K.Bartlett, A.J. de Geus, Automatic generation of combinatorial logic from a functional specification. *Int. Symp. Circ. Sys.*, Montreal, pp. 986-989, May 1984.
[Gre86] D.Gregory et al., SOCRATES: A system for automatically synthesizing and optimizing combinational logic. *23rd Des. Aut. Conf.*, Las Vegas, pp. 79-85, June/July 1986.
[Gre92] R.Grehan, S.Diehl, DOS databases at work. *Byte*, 17, pp. 226-248, Jan. 1992.
[Gri84] J.H.Griesmer et al., YES/MVS: A continuous real time expert system. *AAAI-84*, Austin, pp. 130-136, Aug. 1984.
[Gri80] M.R.Grinberg, A knowledge based design system for digital electronics. *AAAI-80*, Stanford, pp. 283-285, Aug. 1980.
[Gro91] P.Groeneveld, *Context-Driven Channel Routing*. Doct. Diss. Delft Univ. Tech., Delft, 1991.
[Gro82] S.Grossberg, *Studies of Mind and Brain*. Reidel, Dordrecht, 1982.
[Gro87] S.Grossberg, Competitive learning: From interactive activation to adaptive resonance. *Cogn. Science*, 11, pp. 23-63, 1987.
[Gro88] S.Grossberg (ed.), *Neural Networks and Natural Intelligence*. MIT Press, Cambridge, 1988. See also: G.A.Carpenter, S.Grossberg, The ART of adaptive pattern recognition by a self-organizing neural network. *Computer*, 21, pp. 77-88, March 1988.
[Gu86] J.Gu, K.F.Smith, KD2: An intelligent circuit module generator. *Int. Conf. Comp. Des.*, Port Chester, pp. 470-475, Oct. 1986.

[Gub88] P.Gubian, M.Zanella, AUSPICE. An expert system to assist simulations with SPICE. *Int. Symp. Circ. Sys.*, Espoo, pp. 889-892, June 1988.

[Gul85] E.Gullichsen, Heuristic circuit simulation using PROLOG. *Integration, VLSI Jour.*, 3, pp. 283-318, Dec. 1985.

[Gup88] A.Gupta, B.Welham, Functional test generation for digital circuits. In [GEr88], pp. 51-71, 1988.

[GUp88] N.K.Gupta, Digital system design partitioning aid. A knowledge-based approach. In [Ger88], pp. 57-71, 1988.

[Gup84] N.K.Gupta, R.E.Seviora, An expert system approach to real time system debugging. *1st Conf. AI Appl.*, Denver, Dec. 1984.

[Gup91] R.Gupta, E.Horowitz (eds), *Object-Oriented Databases with Applications to CASE, Networks, and VLSI CAD*. Prentice-Hall, Englewood Cliffs, 1991.

[GUp91] R.Gupta et al., The development of a framework for VLSI CAD. In: [Gup91], pp. 237-260, 1991. See also: *Computer*, 22, pp. 28-37, May 1989.

[GUP91] U.G.Gupta (ed.), *Validation and Verification of Expert Systems*. IEEE Comp. Soc. Press, Los Alamitos, 1991.

[Hac82] G.D.Hachtel, A.R.Newton, A.L.Sangiovanni-Vincentelli, Techniques for programmable array folding. *19th Des. Aut. Conf.*, Las Vegas, pp. 147-155, June 1982.

[Haf82] L.J.Hafer, A.C.Parker, Automated synthesis of digital hardware. *IEEE Trans. Comp.*, 31, pp. 93-109, Feb. 1982.

[Haf83] L.J.Hafer, A.C.Parker, A formal method for the specification, analysis, and design of register-transfer level digital logic. *IEEE Trans. CAD*, 2, pp. 4-18, Jan. 1983.

[Ham90] N.Hamada, K.Bekki, T.Yokota, VLSI logic design with logic programming and knowledge-base technology. *IEEE Trans. Industr. Electronics*, 37, pp. 1-5, Feb. 1990. Also in: *Int. Workshop AI Industr. Appl.*, Hitachi City, pp. 206-210, May 1988.

[Han77] W.Händler, The impact of classification schemes on computer architectures. *Int. Conf. Parallel Proc.*, New York, pp. 7-15, 1977. See also Chapter 1 of [Eva82].

[Har89] *Hardware Specification, Verification and Synthesis: Mathematical Aspects*. Springer, Berlin, 1989.

[HAr89] R.Harjani, R.A.Rutenbar, L.R.Carley, OASYS: A framework for analog circuit synthesis. *IEEE Trans. CAD*, 8, pp. 1247-1266, Dec. 1989.

[Har91] R.E.Harr,A.G.Stanculescu (eds), *Applications of VHDL to Circuit Design*, Kluwer Acad., Dordrecht, 1991.

[Har78] P.E.Hart, R.O.Duda, M.T.Einaudi, PROSPECTOR, a computer-based consultation program for mineral exploration. *Math. Geology*, 10, pp. 589-610, Oct. 1978.

[Har90] D.S.Harrison, A.R.Newton, R.L.Spickelmier, T.J.Barnes, Electronic CAD frameworks. *Proc. IEEE*, 78, pp. 393-417, Feb. 1990.

[Har84] R.T.Hartley, CRIB: Computer fault-finding through knowledge engineering. *Computer*, 17, pp. 76-83, March 1984.

[Has71] A.Hashimoto, J.Stevens, Wire routing by optimizing channel assignment within large apertures. *8th Des. Aut. Workshop*, Atlantic City, pp. 155-169, June 1971.

REFERENCES

[Hav86] B.L.Havlicsek, A knowledge based diagnostic system for automatic test equipment. *Int. Test Conf.*, Washington, pp. 930-938, Sept. 1986.

[Hay83] H.Hayashi, A.Hattori, H.Akimoto, ALPHA: A high-performance Lisp machine equipped with a new stack architecture and garbage collection system. *10th Int. Symp. Comp. Arch.*, Stockholm, pp. 342-348, June 1983. See also: *COMPCON85*, San Francisco, pp. 366-369, Feb. 1985.

[Hay91] F.Hayes-Roth, J.E.Davidson, L.D.Erman, J.S.Lark, Framework for developing intelligent systems. *IEEE Expert*, 6, pp. 30-40, June 1991.

[HAy83] F.Hayes-Roth, D.A.Waterman, D.B.Lenat (eds), *Building Expert Systems*. Addison-Wesley, Reading, 1983.

[Hay82] J.E.Hayes, D.Mitchie, Y.H.Pao (eds), *Machine Intelligence 10*. Horwood, Chichester, pp. 325-337, 1982.

[Hay87] J.P.Hayes, An introduction to switch-level modeling. *IEEE Design & Test*, 4, pp. 18-25, Aug. 1987.

[Heb49] D.O.Hebb, *The Organization of Behavior*. Wiley, New York, 1949.

[Hec90] R.Hecht-Nielsen, *Neurocomputing*. Addison-Wesley, Reading, 1990.

[Hek90] A.Hekmatpour, P.Chau, AI techniques and object-oriented technology for VLSI design-space representation, optimization and management. *SPIE, Vol. 1293, Appl. of AI VIII*, Orlando, pp. 85-94, April 1990.

[Hek89] A.Hekmatpour, A.Orailoglu, P.Chau, KINTESS: A knowledge-base expert system CAD for ASIC technology selection. *GOMAC Conf.*, Orlando, Nov. 1989.

[Hek91] A.Hekmatpour, A.Orailoglu, P.Chau, Hierarchical modeling of the VLSI design process. *IEEE Expert*, 6, pp. 56-70, April 1991.

[Hen82] J.Hennessy et al., The MIPS machine. *COMPCON Spring*, San Francisco, pp. 2-7, Feb. 1982.

[Hen86] M.Hennessy, Proving systolic algorithms correct. *ACM Trans. Progr. Lang. Sys.*, 8, pp. 344-387, 1986.

[Her92] J.Herrmann, M.Witthaut, LEDA. A learning apprentice system that acquires design plans for high-level synthesis of integrated circuits. *COMP EURO 92*, The Hague, pp. 430-435, May 1992.

[Hey80] W.Heyns, W.Sansen, H.Beke, A line-expansion algorithm for the general routing problem with a guaranteed solution. *17th Des. Aut. Conf.*, Minneapolis, pp. 243-249, June 1980.

[Hig69] D.W.Hightower, A solution to line-routing problems on the continuous plane. *6th Des. Aut. Workshop*, Miami Beach, pp. 1-24, June 1969.

[Hil91] D.D.Hill, A CAD system for the design of field programmable gate arrays. *28th Des. Aut. Conf.*, San Francisco, pp. 187-192, June 1991.

[Hil86] D.D.Hill, D.R.Coelho, *Multi-Level Simulation for VLSI Design*, Kluwer Acad., Dordrecht, 1986.

[Hil85] D.D.Hill, S.Roy, PROLOG in CMOS circuit design. *IEEE COMPCON Spring*, San Francisco, pp. 211-217, Feb. 1985.

[Hll86] M.Hill et al., Design decisions in SPUR. *Computer*, 19, pp. 8-22, Nov. 1986.

[Hil79] D.D.Hill, W. vanCleemput, SABLE: A tool for generating structured, multi-level simulations. *16th Des. Aut. Conf.*, San Diego, pp. 272-279, June 1979; SABLE: Multi-level simulation for hierarchical design. *Int. Symp. Circ. Sys.*,

Houston, pp. 431-434, April 1980.
[Hil85] W.D.Hillis, *The Connection Machine*, MIT Press, Cambridge, 1985.
[Hin86] G.E.Hinton, T.J.Sejnowski, Learning and relearning in Boltzmann machines. In [RUm86], Ch. 7, 1986.
[Hir86] S.Hirschhorn, R.Kulkarni, An automatic DFT system for the SILC silicon compiler. *IEEE Design & Test*, 3, pp. 45-57, 1986.
[Hit83] C.Y.Hitchcock, D.E.Thomas, A method of automatic data path synthesis. *20th Des. Aut. Conf.*, Miami Beach, pp. 484-489, June 1983.
[Ho86] W.P.C.Ho, A plan patching approach to switchbox routing. In [SRi86], pp. 947-958, 1986.
[Ho89] W.P.C.Ho, S.Cha, Performance analysis of an AI-based placement system. In [Oda89], pp. 197-204, 1989.
[Ho85] W.P.C.Ho, Y.H.Hu, D.Y.Y.Yun, An intelligent librarian for VLSI cell databases. *Int. Conf. Comp. Des.*, Port Chester, pp. 78-81, Oct. 1985.
[Hob84] L.C.Hobbs, Printing and storage peripherals: Past, present and future. *Computer*, 17, pp. 225-241, Oct. 1984.
[Hoe76] J.H.Hoel, Some variations of Lee's algorithm. *IEEE Trans. Comp.*, 25, pp. 19-24, Jan. 1986.
[Hol87] T.Holden, *Knowledge Based CAD and Microelectronics*, North-Holland, Amsterdam, 1987.
[Hom87] M.Homewood *et al.*, The IMS T800 transputer. *IEEE Micro*, 7, pp. 10-26, Oct. 1987.
[Hon74] S.J.Hong, R.G.Cain, D.L.Ostapko, MINI: A heuristic approach for logic minimization. *IBM Jour. Res. Develop.*, 18, pp. 443-458, Sept. 1974.
[Hon83] S.J.Hong, R.Nair, Wire-routing machines. New tools for VLSI physical design. *Proc. IEEE*, 71, pp. 57-65, Jan. 1983.
[Hon80] S.J.Hong, D.L.Ostapko, FITPLA: A programmable logic array for function independent testing. *Int. Symp. Fault-Tol. Comp.*, pp. 131-136, Oct. 1980.
[Hop82] J.J.Hopfield, Neural networks and physical systems with emergent collective computational abilities. *Proc. Nat. Acad. Sci.*, 79, pp. 2554-2558, Baltimore, April 1982.
[Hop84] J.J.Hopfield, Neurons with graded responses have collective computational properties like those of two-state neurons. *Proc. Nat. Acad. Sci.*, Baltimore, 81, pp. 3088-3092, 1984.
[Hop88] J.J.Hopfield, Artificial neural networks. *IEEE Circ. Dev. Mag.*, 4, pp. 3-10, Sept. 1988.
[Hop86] J.J.Hopfield, D.W.Tank, Computing with neural circuits: A model. *Science*, 233, pp. 625-633, 8 Aug. 1986.
[HOp86] F.R.A.Hopgood *et al.*, *Methodology of Window Management*. Springer, New York, 1986.
[Hor91] E.Horowitz, Q.Wan, An overview of existing object-oriented database systems. In [Gup91], pp. 101-116, 1991.
[Hor83] P.W.Horstmann, Design for testability using logic programming. *Int. Test Conf.*, Philadelphia, pp. 706-713, Oct. 1983.
[Hor84] P.W.Horstmann, A knowledge-based system using design for testability rules. *Int. Symp. Fault-Tol. Comp.*, Kissimmee, pp. 278-284, June 1984.

REFERENCES

[HOr84] P.W.Horstmann, E.P.Stabler, Computer Aided Design (CAD) using logic programming. *21st Des. Aut. Conf.*, Albuquerque, pp. 144-151, June 1984.
[Hos87] J.W.Hoskins, *IBM Personal System/2: A Business Perspective*. Wiley, New York, 1987.
[How86] M.M.How, B.Y.M.Pan, AMBER, a knowledge-based area estimation assistant. *Int. Conf. Computer Des.*, Port Chester, pp. 180-183, Oct. 1986.
[Hsi83] D.K.Hsiao (ed.), *Advanced Database Machine Architecture*. Prentice-Hall, Englewood Cliffs, 1983.
[Hsi91] P.Y.Hsiao, S.F.S.Chen, C.C.Tsai, W.S.Feng, A knowledge-based program for compacting mask layout of integrated circuits. *Computer-Aided Design*, 23, pp. 223-231, April 1991.
[HSi91] P.Y.Hsiao, C.C.Tsai, Expert compactor: A knowledge-based application in VLSI layout compaction. *IEE Proc.-E*, 138, pp. 13-20, Jan. 1991.
[HSI91] Y.C.Hsieh *et al.*, LiB: A CMOS cell compiler. *IEEE Trans. CAD*, 10, pp. 994-1005, Aug. 1991.
[Hsu87] A.Hsu, L.Hsu, HILDA: An integrated system design environment. *Int. Conf. Comp. Des.*, Port Chester, pp. 398-402, Oct. 1987.
[Hu90] Y.H.Hu, S.J.Chen, GM_Plan: A gate matrix layout algorithm based on Artificial Intelligence planning techniques. *IEEE Trans. CAD*, 9, pp. 836-845, Aug. 1990.
[Huf54] D.A.Huffman, The synthesis of sequential switching circuits. *Jour. Franklin Inst.*, 257, pp. 161-190, March 1954; pp. 257-303, April 1954.
[Hui88] L.M.Huisman, The reliability of approximate testability measures. *IEEE Design & Test*, 5, pp. 57-67, Dec. 1988.
[Hwa91] C.T.Hwang, J.H.Lee, Y.C.Hsu, A formal approach to the scheduling problem in high level synthesis. *IEEE Trans. CAD*, 10, pp. 464-475, April 1991.
[Hwa83] K.Hwang (ed.), Special Issue on Computer Architectures for Image Processing. *Computer*, 16, pp. 10-80, Jan. 1983.
[Hwa89] K.Hwang, D.DeGroot, *Parallel Processing for Supercomputers and Artificial Intelligence*. McGraw-Hill, New York, 1989.
[Hwa87] K.Hwang, J.Ghosh, Hypernet: A communication-efficient architecture for constructing massively parallel computers. *IEEE Trans. Comp.*, 36, pp. 1450-1466, Dec. 1987.
[HWa87] K.Hwang, J.Ghosh, R.Chowkwanyum, Computer architectures for Artificial Intelligence processing. *Computer*, 20, pp. 19-27, Jan. 1987.
[INM84] INMOS Ltd, *OCCAM Programming Manual*. Prentice-Hall, 1984.
[INM88] INMOS Ltd, *Transputer Reference Manual*. Prentice-Hall, New York, 1988.
[Int84] *The Knowledge Engineering Environment*. IntelliCorp, Menlo Park, 1984. See also: R.E.Filman, Reasoning with worlds and truth maintenance in a knowledge-based programming environment. *Commun. ACM*, 31, pp. 382-401, April 1988.
[Ish87] H.Ishikawa *et al.*, KID. Designing a knowledge-based natural language interface. *IEEE Expert*, 2, pp. 57-71, Summer 1987.
[Ish88] J.Ishikawa *et al.*, A rule based logic reorganization system LORES/EX. *Int. Conf. Comp. Des.*, Rye Brook, pp. 262-266, Oct. 1988.
[Ism90] M.Ismail, J.Franca (eds), *Introduction to Analog VLSI Design Automation*. Kluwer Acad., Boston, 1990.

[Iwa82] Y.Iwasaki, P.Friedland, SPEX: A second-generation experiment design system. *AAAI-82*, Pittsburgh, Aug. 1982.

[Jab90] M.A.Jabri, BREL. A Prolog knowledge-based system shell for VLSI CAD. *27th Des. Aut. Conf.*, Orlando, pp. 272-277, June 1990.

[Jab92] M.A.Jabri, X.Li, Predicting the number of contacts and dimensions of full-custom integrated circuit blocks using neural network. *IEEE Trans. Neural Netw.*, 3, pp. 146-153, Jan. 1992.

[Jab88] M.A.Jabri, D.J.Skellern, PIAF: A KBS/algorithmic IC floorplanner. In [Ger88], pp. 163-190, 1988.

[Jab89] M.A.Jabri, D.J.Skellern, PIAF: A knowledge-based/algorithmic top-down floorplanning system. *26th Des. Aut. Conf.*, Las Vegas, pp. 582-585, June 1989. See also: PIAF. Efficient IC floor planning. *IEEE Expert*, 4, pp. 33-45, Summer 1989.

[Jac90] R.J.K.Jacob, J.N.Froscher, A software engineering methodology for rule-based systems. *IEEE Trans. Knowl. Data Eng.*, 2, pp. 173-189, June 1990.

[Jac89] M.R.P.S.F.Jácome, M.J.A.Lança, An environment for the development of knowledge based expert systems for assistance to VLSI design. *MELECON89*, Lisbon, pp. 315-319, 1989.

[Jai83] S.K.Jain, V.D.Agrawal, Test generation for MOS circuits using D-algorithm. *20th Des. Aut. Conf.*, Miami Beach, pp. 64-70, June 1983.

[Jai85] S.K.Jain, V.D.Agrawal, Statistical Fault Analysis. *IEEE Design & Test*, 2, pp. 38-44, Feb. 1985.

[Jam85] R.Jamier, A.A.Jerraya, APOLLON, a data-path silicon compiler. *IEEE Circ. Dev. Mag.*, 1, pp. 6-14, May 1985. See also: *Int. Conf. Comp. Des.*, Port Chester, pp. 308-311, Oct. 1985.

[Jan87] G.Janac, C.Garcia, R.Davis, A knowledge-based GaAs design system. *VLSI Sys. Des.*, 8, pp. 68-75, April 1987.

[Jar84] M.Jarke, Y.Vassiliou, Coupling expert systems with database management systems. In [Rei84], pp. 65-85, 1984.

[Jer86] A.Jerraya et al., Principles of the SYCO compiler. *23rd Des. Aut. Conf.*, Las Vegas, pp. 715-721, June/July 1986.

[Jha90] N.K.Jha, S.Kundu, *Testing and Reliable Design of CMOS Circuits*. Kluwer Acad., Boston, 1990.

[Joh80] D.Johnson, Simulation, verification and test package for logic design. *Electronic Engng*, 52, pp. 81-89, Jan. 1980.

[Joh88] R.C.Johnson, C.Brown, *Cognizers. Neural Networks and Machines that Think*. Wiley, New York, 1988.

[Jon88] E.Jones, *dBase III Plus: Power User's Guide*. Osborne/McGraw-Hill, Berkeley, (2nd ed.) 1988.

[Jon87] N.A.Jones, K.Baker, Knowledge-based system tool for high-level BIST design. *Microproc. Microsys.*, 11, pp. 35-40, Jan./Feb. 1987.

[Jon89] O.Jones, *How to Use X Windows*. Prentice-Hall, Englewood Cliffs, 1989.

[Joo86] R.Joobbani, *An Artificial Intelligence Approach to VLSI Routing*. Kluwer Acad., Boston, 1986.

[Jos85] R.L.Joseph, An expert systems approach to completing partially routed printed circuit boards. *22nd Des. Aut. Conf.*, Las Vegas, pp. 523-528, June 1985.

REFERENCES

[Jud90] J.S.Judd, *Neural Network Design and the Complexity of Learning.* MIT Press, Cambridge, 1990.
[Kab88] A.M.Kabakcioglu, P.K.Varshney, C.R.P.Hartmann, An Artificial Intelligence Approach to PLA optimization. *Int. Symp. Circ. Sys.*, Espoo, pp. 1861-1864, June 1988.
[Kab85] W.C.Kabat, A.S.Wojcik, Automated synthesis of combinational logic using theorem-proving techniques. *IEEE Trans. Comp.*, 34, pp. 610-632, July 1985.
[Kae86] T.Kaehler, D.Patterson, *A Taste of Smalltalk.* W.W.Norton, New York, 1986.
[Kah85] H.J.Kahn, N.P.Filer, An application of knowledge based techniques to VLSI design. In [Mer85], pp. 307-321, 1985.
[Kah92] H.J.Kahn, R.F.Goldman, The Electronic Design Interchange Format EDIF: Present and future. *29th Des. Aut. Conf.*, Anaheim, pp. 666-671, June 1992.
[Kap91] G.Kaplan et al., Data Communications. Special Guide. *IEEE Spectrum*, 28, pp. 21-44, Aug. 1991.
[Kar88] A.Kara, R.Rastogi, K.Kawamura, An expert system to automate timing design. *IEEE Design & Test*, 5, pp. 28-40, Oct. 1988.
[Kat86] R.H.Katz, Design database. In: S.Goto (ed.), *Design Methodologies.* North-Holland, Amsterdam, 1986.
[Kaw85] N.Kawato, T.Saito, S.Hiroyuki, DDL/SX: a rule based expert system for logic circuit synthesis. *Int. Symp. Circ. and Sys.*, Kyoto, pp. 885-888, June 1985.
[Kaw82] N.Kawato, T.Uehara, S.Hirose, T.Saito, An interactive logic synthesis system based upon AI techniques. *19th Des. Aut. Conf.*, Las Vegas, pp. 858-864, June 1982.
[Kee86] M.Keefe, J.Kendall, An expert system for routing in VLSI. *Canadian Conf. VLSI*, Montreal, pp. 337-342, Oct. 1986.
[Kee89] S.Keene, *Object-Oriented Programming in Common LISP.* Addison-Wesley, Reading, 1989.
[Kee85] S.Keene, D.Moon, Flavors: Object-oriented programming on symbolic computers. *Common LISP Conf.*, Boston, 1985.
[Kel84] V.E.Kelly, The CRITTER system. Automated critiquing of digital circuit designs. *21st Des. Aut. Conf.*, Albuquerque, pp. 419-425, June 1984.
[Ken89] D.Kennett, K.A.E.Totton, Experience with an expert diagnostic system shell. In [Sau89], pp. 183-193, 1989.
[Ken91] K.B.Kenny, K.J.Lin, Building flexible real-time systems using the Flex language. *Computer*, 24, pp. 70-78, May 1991.
[Ker87] E.T.Keravnou, L.Johnson, NEOCRIB: An expert fault finding system that articulates the competence of field engineers. In [Bel87], pp. 107-118, Sept. 1987.
[Ker88] B.W.Kernighan, D.M.Ritchie, *The C Programming Language.* Prentice-Hall, Englewood Cliffs, (2nd ed.) 1988.
[Kim83] J.Kim, J. McDermott, TALIB: An IC layout design assistant. *AAAI-83*, Washington, pp. 197-201, Aug. 1983.
[Kim84] J.H.Kim, J.McDermott, D.P.Siewiorek, Exploiting domain knowledge in IC cell layout. *IEEE Design & Test*, 1, pp. 52-64, Aug. 1984.
[Kim86] J.Kim, J.McDermott, Computer aids for IC design. *IEEE Sofware*, 3, pp. 38-47, March 1986.

[Kim91] S.H.Kim, *Knowledge Systems Through Prolog*. Oxford Univ. Press, Oxford, 1991.

[Kim89] W.Kim, F.H.Lochovsky (eds), *Object-Oriented Concepts, Databases, and Applications*. Addison-Wesley, Reading, 1989.

[Kim90] W.Kim et al., Architecture of the ORION next-generation database system. *IEEE Trans. Knowl. Data Engng*, 2, pp. 109-124, March 1990.

[KIm89] Y.H.Kim, S.H.Hwang, A.R.Newton, Electrical-Logic simulation and its applications. *IEEE Trans. CAD*, 8, pp. 8-22, Jan. 1988.

[Kip91] J.R.Kipps, D.D.Gajski, Automating technology adaptation in design synthesis. In: N.G.Bourbakis (ed.), *Applications of Learning and Planning Methods*. World Scientific, Singapore, pp. 129-164, 1991.

[Kir88] T.Kirkland, M.R.Mercer, Algorithms for automatic test pattern generation. *IEEE Design & Test*, 5, pp. 43-55, June 1988.

[Kir83] S.Kirkpatrick, C.D.Gelatt, M.P.Vecchi, Optimization by simulated annealing. *Science*, 220, pp. 671-680, May 1983.

[Kit86] Y.Kitamura et al., Hardware engines for logic simulation. In: E.Hörbst (ed.), *Logic Design and Simulation*, North-Holland, Amsterdam, 1986.

[KIt86] C.T.Kitzmiller, J.Kowalik, Symbolic and numerical computing in knowledge-based systems. In [Kow86], pp. 3-17, 1986. See also: C.T.Kitzmiller, Simulation and AI: Coupling symbolic and numeric computing. In: T.Henson (ed.), *AI and Simulation*, Soc. Computer Simulation, pp. 3-7, 1988.

[Kle92] R.H.Klenke, R.D.Williams, J.H.Aylor, Parallel-processing techniques for automatic test pattern generation. *Computer*, 25, pp. 71-84, Jan. 1992.

[Kli92] R.Klinke et al., Rule-based analog circuit design. *Eur. Conf. Des. Aut.*, Brussels, pp. 480-484, March 1992.

[Kli88] G.L.Klir, T.A.Folger, *Fuzzy Sets, Uncertainty, and Information*. Prentice-Hall, Englewood Cliffs, 1988.

[Klj89] J.Kljaich, B.T.Smith, A.S.Wojcik, Formal verification of fault tolerance using theorem-proving tchniques. *IEEE Trans. Comp.*, 38, pp. 366-376, March 1989.

[Kna91] D.W.Knapp, A.C.Parker, The ADAM Design Planning Engine. *IEEE Trans. CAD*, 10, pp. 829-846, July 1991.

[Kno87] *Knowledge Craft CRL Technical Manual, Cersion 3.1*. Carnegie Group Inc., Pittsburgh, 1985.

[Koh84] T.Kohonen, *Self-Organization and Associative Memory*. Springer, Berlin, 1984.

[KOl85] P.W.Kollaritsch, N.H.E.Weste, TOPOLOGIZER: An expert system translator of transistor connectivity to symbolic cell layout. *IEEE Jour. Solid-State Circ.*, 20, pp. 799-804, June 1985.

[Kol86] P.W.Kollaritsch, B.Ackland, COORDINATOR: A complete design-rule enforced layout methodology. *Int. Conf. Comp. Des.*, Port Chester, pp. 302-307, Oct. 1986.

[Kol85] A.Kolodny, R.Friedman, T.Ben-Tzur, Rule-based static debugger and simulation compiler for VLSI schematics. *Int. Conf. CAD*, Santa Clara, pp. 150-152, Nov. 1985.

[Kon80] B.Könemann, J.Mucha, G.Zwiehoff, Built-in test for complex digital integrated circuits. *IEEE Jour. Solid-State Circ.*, 15, pp. 315-319, June 1980.

REFERENCES

[Kor84] J.Kornell, A VAX tuning expert built using automated knowledge acquisition. *1st Conf. AI Appl.*, Sheraton, pp. 38-41, Dec. 1984.
[Kor86] H.F.Korth, A Silberschatz, *Database System Concepts*. McGraw-Hill, New York, 1986.
[Kos88] T.Koschmann, M.W.Evens, Bridging the gap between object-oriented and logic programming. *IEEE Software*, 5, pp. 36-42, July 1988.
[Kos90] Y.Koseko et al., Rule-based testability rule check program. *Int. Conf. Comp. Des.*, Cambridge, pp. 95-98, Sept. 1990.
[Kos92] B.Kosko, *Neural Networks for Signal Processing*. Prentice-Hall, Englewood Cliffs, 1992.
[KOs92] B.Kosko, *Neural Networks and Fuzzy Systems: A Dynamical Systems Approach to Machine Intelligence*. Prentice-Hall, Englewood Cliffs, 1992.
[KOs88] A.P.Kostelijk, G.G.Schrooten, VERA, a rule based verification assistant for VLSI circuit design. In [ESP88], pp. 263-268, 1988.
[Kow86] J.S.Kowalik (ed.), *Coupling Symbolic and Numerical Computing in Expert Systems*. North-Holland, Amsterdam, 1986.
[Kow87] J.S.Kowalik (ed.), *Parallel Computation and Computers for Artificial Intelligence*, Kluwer Acad., Dordrecht, 1987.
[Kow79] R.Kowalski, *Logic for Problem Solving*. North-Holland, New York, 1979.
[Kow83] T.J.Kowalski, D.E.Thomas, The VLSI Design Automation Assistant: Learning to walk. *Int. Symp. Circ. Sys.*, Newport Beach, pp. 186-190, May 1983. T.J.Kowalski, D.E.Thomas, The VLSI Design Automation Assistant: Prototype system. *20th Des. Aut. Conf.*, Miami Beach, pp. 479-483, June 1983; T.J.Kowalski, D.E.Thomas, The VLSI Design Automation Assistant: An IBM System/370 design. *IEEE Design & Test*, 1, pp. 60-69, Feb. 1984.
[Kow85] T.J.Kowalski, *An Artificial Intelligence Approach to VLSI Design*. Kluwer Acad., Dordrecht, 1985; T.J.Kowalski, D.E.Thomas, The VLSI Design Automation Assistant: What's in a knowledge base. *22nd Des. Aut. Conf.*, Las Vegas, pp. 252-258, June 1985; T.J.Kowalski, D.J.Geiger, W.H.Wolf, W.Fichtner, The VLSI Design Automation Assistant: From algorithms to silicon. *IEEE Design & Test*, 2, pp. 33-43, Aug. 1985.
[Kra87] G.A.Kramer, Incorporating mathematical knowledge into design models. In [Ger87], pp. 229-265, 1987.
[Kri91] R.F.Krick, A.Dollas, The evolution of instruction sequencing. *Computer*, 24, pp. 5-15, April 1991.
[Kri89] E.V.Krishnamurthy, *Parallel Processing and Practice*. Addison-Wesley, Reading, 1989.
[Kro89] J.Krol, CIRCOR. An expert system for fault correction of digital NMOS circuits. *Eur. Conf. Circ. Th. Des.*, Brighton, pp. 674-676, Sept. 1989.
[Kun82] H.T.Kung, Why systolic architectures? *Computer*, 15, pp. 37-46, Jan. 1982.
[Kun80] H.T.Kung, P.L.Lehman, Systolic (VLSI) arrays for relational database operations. *ACM-SIGMOD Int. Conf. Management of Data*, pp. 105-116, May 1980.
[Kun88] S.Y.Kung, J.N.Hwang, Parallel architectures for artificial neural nets. *IEEE Int. Conf. Neural Netw.*, San Diego, pp. II 165-172, July 1988.

[Kuo89] Y.H.Kuo, L.Y.Kung, T.Chen, KMDS. An expert system for integrated hardware/software design of microprocessor-based digital systems. *IEEE TENCON*, Bombay, pp. 661-664, Nov. 1989.
[Kur90] Y.Kurosawa *et al.*, LSI logic synthesis expert system. *Electronics and Commun. in Japan*, Part 3, 73, pp. 10-21, 1990.
[Lac92] R.C.Lacher, S.I.Hruska, D.C.Kuncicky, Back-propagation learning in expert networks. *IEEE Trans. Neural Netw.*, 3, pp. 62-72, Jan. 1992.
[Laf84] T.J.Laffey, W.A.Perkins, O.Firschein, LES: A model-based expert system for electronic maintenance. *Joint Services Workshop on AI in Maintenance*, pp. 1-17, 1984.
[Lai87] F.Lai, S.M.Kang, T.N.Trick, V.B.Rao, iJADE: A rule-based hierarchical CMOS VLSI circuit optimizer. *Int. Conf. Comp. Des.*, Port Chester, pp. 38-41, Oct. 1987.
[LAi87] F.Lai, V.B.Rao, T.N.Trick, JADE: A hierarchical switch level timing simulator. *Int. Symp. Circ. Sys.*, Philadelphia, pp. 592-595, May 1987.
[Lai86] R.N.W.Laithwaite, An expert system to aid placement on gate arrays. *3rd Silicon Des. Conf.*, London, pp. 305-313, July 1986.
[LaL90] W.R.LaLonde, J.R.Pugh, *Inside Smalltalk*. Vol. 1. Prentice-Hall, Englewood Cliffs, 1990.
[Lam86] J.Lam, J.M.Delosme, Logic minimization using simulated annealing. *Int. Conf. CAD*, Santa Clara, pp. 348-351, Nov. 1986.
[Lan87] K.Lang, *The Writer's Guide to Desktop Publishing*. Acad. Press, London, 1987.
[Lar86] R.P.Larsen, Rules-based object clustering: A data structure for symbolic VLSI synthesis and analysis. *23rd Des. Aut. Conf.*, Las Vegas, pp. 768-777, June/July 1986.
[Lat86] D.O.Lathi, G.C.Chen-Ellis, PROSPECT: A Production System for Partitioning and Evaluating Chip Testability. *Int. Test Conf.*, Washington, pp. 360-367, Sept. 1986.
[Lat85] R.H.Lathrop, R.S.Kirk, An extensible object-oriented mixed-mode functional simulation system. *22nd Des. Aut. Conf.*, Las Vegas, pp. 630-636, June 1985.
[Lau80] U.Lauther, A min-cut placement algorithm for general cell assemblies based on a graph representation. *Jour. DA and FTC*, 4, pp. 21-34, 1980.
[Law76] E.L.Lawler, *Combinatorial Optimization*. Holt, Rinehart and Winston, New York, 1976.
[Lea88] S.M.Lea *et al.*, Expert system for the functional test program generation of digital electronic circuit boards. *Int. Test Conf.*, Washington, pp. 209-220, Sept. 1988.
[LeB89] G.T.LeBlond, W.B.LeBlond, B.Heslop, *dBase IV: The Complete Reference*. Osborne/McGraw-Hill, Berkeley, 1989.
[Lee61] C.Y.Lee, An algorithm for path connection and its application. *IRE Trans. El. Comp.*, 10, pp. 346-365, Sept. 1961.
[Lef91] A.Leff, C.Pu, A classification of transaction processing systems. *Computer*, 24, pp. 63-75, June 1991.
[Lei92] F.T.Leighton, *Introduction to Parallel Algorithms and Architetures: Arrays, Trees, Hypercubes*. Morgan Kaufmann, San Mateo, 1992.

REFERENCES

[Len82] D.B.Lenat, AM: Discovery in mathematics as heuristic search. In [Dav82], pp. 1-225, 1982.
[Len83] D.B.Lenat, EURISKO: A program that learns new heuristics and domain concepts. *Art. Intell.*, 21, pp. 61-98, March 1983.
[Let88] G.Letwin, *Inside OS/2*. Microsoft Press, Redmond, 1988.
[Lev82] Y.H.Levendel, P.R.Menon, Test generation algorithms for computer hardware description languages. *IEEE Trans. Comp.*, 31, pp. 577-588, July 1982.
[Lew77] D.Lewin, *Computer-Aided Design of Digital Systems*. Crane Russak, New York, 1977.
[Lew84] E.T.Lewis, Optimization of device area and overall delay for CMOS VLSI designs. *Proc. IEEE*, 72, pp. 670-689, June 1984.
[Lib89] R.Libeskind-Hadas, C.L.Liu, Solutions to the module orientation and rotation problems by neural computation networks. *26th Des. Aut. Conf.*, Las Vegas, pp. 400-405, June 1989.
[Lig86] M.M.Ligthart, E.H.L.Aarts, F.P.M.Beenker, Design-for-testability of PLA's using statistical cooling. *23rd Des. Aut. Conf.*, Las Vegas, pp. 339-345, June/July 1986.
[Lin92] R.B.Lin, E.Shragowitz, Fuzzy logic approach to placement problem. *29th Des. Aut. Conf.*, Anaheim, pp. 153-158, June 1992.
[Lin86] T.M.Lin, C.A.Mead, A hierarchical timing simulation model. *IEEE Trans. CAD*, 5, pp. 188-197, Jan. 1986.
[Lin87] Y.L.S.Lin, D.D.Gajski, H.Tago, A flexible-cell approach for module generation. *Custom Integr. Circ. Conf.*, Portland, pp. 9-12, May 1987.
[Lin89] Y.L.Lin, Y.C.Hsu, F.S.Tsai, SILK: A simulated evolution router. *IEEE Trans. CAD*, 8, pp. 1108-1114, Oct. 1989.
[Lin88] Y.L.Lin, D.D.Gajski, LES: A layout expert system. *IEEE Trans. CAD*, 7, pp. 868-876, Aug. 1988.
[Lip87] R.P.Lippmann, An introduction to computing with neural nets. *IEEE ASSP Mag.*, 4, pp, 4-22, April 1987.
[Llo90] P.Lloyd, H.K.Dirks, E.J.Prendergast, K.Singhal, Technology CAD for competitive products. *IEEE Trans. CAD*, 9, pp. 1209-1216, Nov. 1990.
[Lob85] C.Lob, A.R.Newton, RUBICC. A rule based expert system for VLSI integrated circuit critique. *CICC*, Portland, pp.379-383, May 1985.
[Loh90] F.Lohnert, F.Hoppe, An object-oriented multi-level VLSI simulator using POOL. In [Tre90], pp. 81-100, 1990.
[Lug81] G.F.Luger, Mathematical model building in the solution of mechanics problems: Human protocols and the MECHO trace. *Cogn. Science*, 5, pp. 55-77, 1981.
[Lug89] G.F.Luger, W.A.Stubblefield, *Artificial Intelligence and the Design of Expert Systems*. Benjamin/Cummings, Redwood City, 1989.
[Lug87] G.F.Luger, W.A.Stubblefield, Paradigm dependent human factors issues in expert system design. In [Roo87], pp. 1-11, 1987.
[Luk83] W.K.Luk, A greedy switch-box router. *Integration. VLSI Jour.*, 3, pp. 129-149, 1985.
[Lur84] C.Lursinsap, D.Gajski, Cell compilation with constraints. *21st Des. Aut. Conf.*, Albuquerque. pp. 103-108, June 1984.

[Lut87] D.Lutoff, C.Robach, Expert system for test strategy generation. *Int. Test Conf.*, Washington, pp. 576-585, Sept. 1987.

[Ma89] H.K.T.Ma, S.Devadas, R.S.Wei, A Sangiovanni-Vincentelli, Logic verification algorithms and their parallel implementation. *IEEE Trans. CAD*, 8, pp. 181-189, Feb. 1989.

[Mac91] R.MacGregor, M.H.Burstein, Using a description classifier to enhance knowledge representation. *IEEE Expert*, 6, pp.41-46, June 1991.

[Mac89] *Macintosh Programmer's Workshop Pascal 3.0 Reference*. Apple Computer, Cupertino, 1989.

[Mad89] J.C.Madre *et al.*, Formal verification and diagnosis of digital circuits using a propositional theorem prover. In [Oda89], pp. 101-108, 1989.

[Mag77] B.Magnhagen, *Probability Based Verification of Time Margins in Digital Designs*. Diss. Linköping Univ. (Sweden), Sept. 1977.

[Mag79] G.A.Mago, A network of microprocessors to execute reduction languages. *Int. Jour. Comp. Inf. Sci.*, 8, pp. 349-385, 435-471, 1979.

[Mal82] K.Maling, S.H.Mueller, W.R.Heller, On finding most optimal rectangular package plans. *19th Des. Aut. Conf.*, Las Vegas, pp. 663-670, June 1982.

[Man90] S.Manetti, M.C.Piccirilli, A.Liberatore, Automatic test point selection for linear analog network fault diagnosis. *Int. Symp. Circ. Sys.*, New Orleans, pp. 25-28, May 1990.

[Man83] T.E.Mangir, Design for testability. An integrated approach to VLSI testing. *Int. Conf. CAD*, Santa Clara, pp. 68-70, Sept. 1983.

[Man87] T.Mano *et al.*, A verifier tightly connected to synthesis expert system. *Int. Conf. CAD*, Santa Clara, pp. 414-417, Nov. 1987.

[MAn83] T.Manuel, Lisp and Prolog machines are proliferating. *Electronics*, 56, pp. 132-137, 3 Nov. 1983.

[Mar82] M.Marek-Sadowska, E.S.Kuh, A new approach to channel routing. *Int. Symp. Circ. Sys.*, Rome, pp. 764-767, May 1982.

[Mar90] A.J.Maren, C.T.Harston, R.M.Pap *et al.*, *Handbook of Neural Computing Applications*. Acad. Press, San Diego, 1990.

[MAr90] V.Margo, D.M.Etter, An expert system for digital filter design. *33rd Midwest Symp. Circ. Sys.*, Calgary, pp. 997-1000, Aug. 1990.

[Mar76] J.D.Markel, A.H.Gray, *Linear Prediction of Speech*. Springer, Berlin, 1976.

[Mar91] G.Martinelli, R.Perfetti, Synhesis of feedforward neural analogue-digital convertors. *IEE Proc.-G*, 138, pp. 567-574, Oct. 1991.

[Mar85] F.Maruyama, M.Fujita, Hardware verification. *Computer*, 18, pp. 22-32, Feb. 1985.

[Mar89] M.Marzouki, B.Courtois, Debugging integrated circuits: A.I. can help! *1st Eur. Test Conf.*, Paris, pp. 184-191, April 1989.

[Mar92] M.Marzouki, F.L.Vargas, Using a knowledge-based system for automatic debugging: Case study and performance analysis. *Microelectronic Engng*, 16, pp. 129-136, March 1992.

[Mas90] A.Masaki, Y.Hirai, M.Yamada, Neural networks in CMOS: A case study. *IEEE Circ. Dev.*, 6, pp. 12-17, July 1990.

[Mat83] T.G.Matheson *et al.*, Embedding electrical and geometric constraints in hierarchical circuit-layout generators. *Int. Conf. CAD*, Santa Clara, pp. 3-5,

Sept. 1983.
[Mat89] C.Matthäus, B.Krüger-Sprengel, H.T.Vierhaus, EXTEST, a knowledge based system for the design of testable logic circuits. *IEEE COMPEURO*, Hamburg, pp. 5.137-5.139, May 1989.
[Mau90] C.M.Maunder, R.E.Tulloss (eds), *The Test Access Port and Boundary-Scan Architecture*. IEEE Computer Soc. Press, Los Alamitos, Sept. 1990.
[May87] K.Mayaram, D.O.Pederson, CODECS: An object-oriented mixed-level circuit and device simulator. *Int. Symp. Circ. Sys.*, Philadelphia, pp. 604-607, May 1987.
[McA87] J.McAllister, *Artificial Intelligence and PROLOG on Microcomputers*. Arnold, London, 1987.
[McC87] J.V.McCanny, J.C.White (eds), *VLSI Technology and Design*. Acad. Press, London, 1987.
[McC65] J.McCarthy, P.W.Abrahams, D.J.Edwards, T.P.Hart, M.I.Levin, *LISP 1.5 Programmer's Manual*. MIT Press, Cambridge, (2nd ed.) 1965.
[McC88] J.L.McClelland, D.E.Rumelhart, *Explorations in Parallel Distributed Processing: A Handbook of Models, Programs, and Exercises*. MIT Press, Cambridge, 1988.
[McC89] C.McClure, *CASE is Software Automation*. Prentice-Hall, Englewood Cliffs, 1989.
[McC86] E.J.McCluskey, *Logic Design Principles with Emphasis on Testable Semicustom Circuits*. Prentice-Hall, Englewood Cliffs, 1986.
[McC43] W.McCulloch, W.Pitts, A logical calculus of the ideas imminent in nervous activity. *Bull. Math. Biophys.*, 7, pp. 113-146, 1943.
[McD83] D.McDermott, Duck: A Lisp-based deductive system. Yale Univ., Dept Computer Science, 1983.
[MCD82] D.McDermott, R.Brooks, ARBY: Diagnosis with shallow causal models. *AAAI-82*, Pittsburgh, pp. 370-372, Aug. 1982.
[McD82] J.McDermott, XSEL: A computer sales person's assistant. In [Ha82], pp. 325-337, 1982.
[McD84] J.McDermott, Building expert systems. In [Rei84], pp. 11-22, 1984.
[McF87] M.C.McFarland, T.J.Kowalski, Tools for VLSI synthesis. In [SAu87], pp. 97-108, 1987.
[McF90] M.C.McFarland, T.J.Kowalski, Incorporating bottom-up design into hardware synthesis. *IEEE Trans. CAD*, 9, pp. 938-950, Sept. 1990.
[McL83] D.McLeod, K.Narayanaswamy, K.V. Bapa Rao, An approach to information management for CAD/VLSI applications. *Conf. Database Week. Eng. Des. Appl.*, San Jose, pp. 39-50, May 1983.
[McW78] T.M.McWilliams, L.C.Widdoes, SCALD. *15th Des. Aut. Conf.*, Las Vegas, pp. 271-284, June 1978.
[Mea80] C.A.Mead, L.A.Conway, *Introduction to VLSI Systems*. Addison-Wesley, Reading, 1980.
[Mea88] C.A.Mead, *Analog VLSI and Neural Systems*. Addison-Wesley, Reading, 1988.
[Mea89] C.Mead, M.Ismail (eds), *Analog VLSI Implementation of Neural Systems*. Kluwer Acad., Boston, 1989.

[Mea55] G.H.Mealy, A method for synthesizing sequential circuits. *Bell Sys. Tech. Jour.*, 34, pp. 1045-1079, Sept. 1955.
[Meh87] Z.Mehmood et al., IDEAS: An integrated design automation system. *Int. Conf. Comp. Des.*, Rye Brook, pp. 407-412, Oct. 1987.
[Men90] *Explorer CheckMate*. Product description. Mentor Graphics, Beaverton, 1990.
[Mer89] D.Merritt, *Building Expert Systems in Prolog*. Springer, New York, 1989.
[Mer83] M.Merry, APEX3: An expert system shell for fault diagnosis. *GEC Jour. Research*, 1, pp. 39-47, 1983.
[Mer85] M.Merry (ed.), *Expert Systems 85*. Cambridge Univ. Press, Cambridge, 1985.
[Met91] W.Mettrey, A comparative evaluation of expert system tools. *Computer*, 24, pp. 19-31, Feb. 1991.
[Mic83] R.S.Michalski, J.G.Carbonell, T.M.Mitchell (eds), *Machine Learning. An Artificial Intelligence Approach*. Vol. I, Tioga Publ. Co., Palo Alto, 1983
[MIc83] R.S.Michalski, R.E.Stepp, Learning from observation: Conceptual clustering. In [Mic83], 1983.
[Mic86] R.S.Michalski, J.G.Carbonell, T.M.Mitchell (eds), *Machine Learning. An Artificial Intelligence Approach*. Vol. II, Morgan Kaufmann, Los Altos, 1986.
[MIc86] R.Michalski, I.Mozetic, J.Hong, N.Lavrac, The multi-purpose incremental learning system AQ15 and its testing application to three medical domains. *AAAI-86*, Philadelphia, pp. 1041-1045, Aug. 1986.
[Mic79] D.Michie (ed.), *Expert Systems in the Micro-electronic Age*. Edinburgh Univ. Press, Edinburgh, 1979.
[Mic82] D.Michie (ed.), *Introductory Readings in Expert Systems*. Gordon and Breach, New York, 1982.
[Mic88] D.Michie, Machine learning in the next five years. In: D.Sleeman (ed.), *3rd Eur. Working Session on Learning (EWSL-88)*, Pitman, London, pp. 107-122, 1988.
[Mil89] A.Miller, From expert assistant to design verification: Application of AI to VLSI design. *IEEE Southeastcon*, Columbia, pp. 406-408, April 1989.
[Mil92] D.D.Miller, *VAX/VMS. Operating System Concepts*. Digital Press, Bedford, 1992.
[Mil86] F.D.Miller, J.R.Rowland, E.M.Siegfried, ACE: An expert system for preventive maintenance operations. *AT&T Bell Lab. Rec.*, 64, pp. 20-25, Jan. 1986.
[Mil83] G.J.Milne, The correctness of a simple silicon compiler. In [Ueh83], pp. 1-12, 1983.
[Mil88] G.J.Milne (ed.), *The Fusion of Hardware Design and Verification*. North-Holland, Amsterdam, 1988.
[MIl86] G.J.Milne, P.A. Subrahmanyam (eds), *Formal Aspects of VLSI Design*. North-Holland, Amsterdam, 1986.
[MIL86] V.Milutinovic, GaAs microprocessor technology. *Computer*, 19, pp. 10-13, Oct. 1986.
[Mil91] K.Milzner, W.Brockherde, SILAS: A knowledge-based simulation assistant. *IEEE Jour. Solid-State Circ.*, 26, pp. 310-318, March 1991.
[Mil90] K.Milzner, F.Krohm, A knowledge based simulation environment. *IEEE Custom Integr. Circ. Conf.*, Boston, pp. 10.6.1-10.6.4, May 1990.
[Min75] M.Minsky, A framework for representing knowledge. In [Win75], pp. 211-277, 1975.

REFERENCES

[Min85] M.Minsky, *The Society of Mind*. Simon & Schuster, New York, 1985.
[Min69] M.L.Minsky, S.Papert, *Perceptrons*. MIT Press, Cambridge, 1969.
[Mir91] D.P.Miranker, B.J.Lofaso, The organization and performance of a TREAT-based production system compiler. *IEEE Trans. Knowl. and Data Engng*, 3, pp. 3-10, March 1991.
[Mir90] A.R.Mirzai (ed.), *Artificial Intelligence. Concepts and Applications in Engineering*. Chapman and Hall, London, 1990.
[Mit89] D.P.Mital, E.K.Teoh, A rule-based inspection system for printed circuit boards. *IEEE TENCON*, Bombay, pp. 665-668, Nov. 1989.
[Mit86] T.M.Mitchell, R.M.Keller, S.T.Kedar-Cabelli, Explanation-based generalization: A unifying view. *Machine Learning*, 1, pp. 47-80, 1986.
[Mit85] T.M.Mitchell, S.Mahadevan, L.I.Steinberg, LEAP: A learning apprentice for VLSI design. *9th Int. Joint Conf. AI*, Los Angeles, pp. 573-580, Aug. 1985.
[MIt85] T.M.Mitchell, L.I.Steinberg, J.S.Shulman, A knowledge-based approach to design. *IEEE Trans. Pattern Anal. Mach. Intell.*, 7, pp. 502-510, Sept. 1985.
[Mon82] M.Monachino, Design verification system for large-scale LSI design. *IBM Jour. Res. Dev.*, 26, pp. 89-99, Jan. 1982.
[Moo74] D.Moon, *Maclisp Reference Manual*. MIT Press, Cambridge, 1974.
[Moo86] D.Moon, Object-oriented programming with Flavors. *OOPSLA Proc.*, Portland, Sept./Oct. 1986.
[Moo87] D.A.Moon, Symbolics architecture. *Computer*, 18, pp. 43-52, Jan. 1987.
[Moo83] P.R.Moorby, Fault simulation using parallel value lists. *Int. Conf. CAD*, Santa Clara, pp. 101-102, Sept. 1983.
[Moo56] E.F.Moore, Gedanken-experiments on sequential machines. In: C.E.Shannon, J.McCarthy (eds), *Automata Studies*, Princeton Univ. Press, Princeton, pp. 129-153, 1956.
[Moo85] T.P.Moore, A.J. de Geus, Simulated annealing controlled by a rule-based expert system. *Int. Conf. CAD*, Santa Clara, pp. 200-202, Nov. 1985.
[Mor90] N.Morgan (ed.), *Artificial Neural Networks. Electronic Implementations*. IEEE Comp. Soc. Press, Los Alamitos, 1990.
[Mor88] T.Morgan, R.Engelmore (eds), *Blackboard Systems*. Addison-Wesley, London, 1988.
[MOr90] S.G.Morton, Electronic hardware implementations. In [Mar90], pp. 251-269, 1990.
[Mos85] B.Moszkowski, A temporal logic for multilevel reasoning about hardware. *Computer*, 18, pp. 10-19, Feb. 1985.
[Mos86] B.C.Moszkowski, *Executing Temporal Logic Programs*. Cambridge Univ. Press, New York, 1986.
[Mot82] T.Moto-oka (ed.), *Fifth Generation Computer Systems*. North-Holland, Amsterdam, 1982.
[Mou88] C.B.Mouleeswaran, A problem-solver for generating chip apportionments. In [Ger88], pp. 121-136, 1988.
[Mue89] K.D.Mueller-Glaser, J.Bortolazzi, An approach to intelligent assistance for the specification of ASIC design using objects and rules. *26th Des. Aut. Conf.*, Las Vegas, pp. 472-477, June 1989.

[Muk83] A.Mukhopadhyay, VLSI hardware algorithms. In: G.Rabbat (ed.), *Hardware and Software Concepts in VLSI.* Van Nostrand Reinhold, New York, pp. 72-94, 1983.

[Mul91] B.Müller, J.Reinhardt, *Neural Networks. An Introduction.* Springer, Berlin, 1991.

[Mur85] K.Murakami, K.Kakuta, R.Onai, N.Ito, Research on parallel machine architecture for fifth-generation computer systems. *Computer,* 18, pp. 76-92, June 1985.

[Mur96] M.C.Murphy Hoye, Artificial intelligence in semiconductor manufacturing for process development, functional diagnostics, and yield crash prevention. *Int. Test Conf.,* Washington, pp. 939-946, Sept. 1986.

[Mur89] A.Murray, A.Smith, L.Tarassenko, Fully programmable analogue VLSI devices for the implementation of neural networks. In [Del89], pp. 236-244, 1989.

[Mye86] G.J.Myers, A.Y.C.Yu, D.L.House, Microprocessor technology trends. *Proc. IEEE,* 74, pp. 1605-1622, 1986.

[Mye82] W.Myers, Lisp machines displayed at AI Conference. *Computer,* 15, pp. 79-82, Nov. 1982.

[Nag88] J.Naganuma *et al.,* High-speed CAM-based architecture for a Prolog machine (ASCA). *IEEE Trans. Comp.,* 37, pp. 1375-1383, Nov. 1988.

[Nag75] L.W.Nagel, *SPICE2: A computer program to simulate semiconductor circuits.* Memorandum UCB/ERL M75/520, Univ. of California, Berkeley, May 1975.

[Nag82] A.W.Nagle, R.Cloutier, A.C.Parker, Synthesis of hardware for the control of digital systems. *IEEE Trans. CAD,* 1, pp. 201-212, Oct. 1982.

[Nai88] P.Naish, P.Bishop, *Designing Asics.* Horwood, Chichester, 1988.

[Nak78] S.Nakamura, S.Murai, C.Tanaka, M.Terai, H.Fujiwara, K.Kinoshita, LORES. Logic Reorganization System. *15th Des. Aut. Conf.,* Las Vegas, pp. 250-260, June 1978.

[Nau83] D.S.Nau, Expert computer systems. *Computer,* 16, pp. 63-85, Feb. 1983.

[Nay87] C.Naylor, *Build Your Own Expert System.* Sigma, Wilmslow, (2nd ed.) 1987.

[Nes86] J.A.Nestor, D.E.Thomas, Behavioral synthesis with interfaces. *Int. Conf. CAD,* Santa Clara, pp. 112-115, Nov. 1986.

[New72] A.Newell, H.A.Simon, *Human Problem Solving.* Prentice-Hall, Englewood Cliffs, 1972.

[New79] W.M.Newman, R.F.Sproull, *Principles of Interactive Computer Graphics.* McGraw-Hill, New York, 1985.

[New87] A.R.Newton, A.L.Sangiovanni-Vincentelli, CAD tools for ASIC design. *Proc. IEEE,* 75, pp. 765-776, June 1987.

[Ng89] C.K.Ng, K.P.Chow, An expert system for diagnosis of electronic equipment using structural model and trouble-shooting heuristics. *IEEE TENCON,* Bombay, pp. 669-673, Nov. 1989.

[Nii86] H.P.Nii, Blackboard systems: The blackboard model of problem solving and the evolution of blackboard architectures. *AI Mag.,* 7, pp. 38-53, Summer 1986. H.P.Nii, Blackboard systems: Blackboard application systems, blackboard systems from a knowledge engineering perspective. *AI Mag.,* 7, pp. 82-106, Aug. 1986.

REFERENCES

[Nil82] N.J.Nilsson, *Principles of Artificial Intelligence*. Springer, Berlin, 1982.
[Nin86] T.H.Ning, D.D.Tang, Bipolar trends. *Proc. IEEE*, 74, pp. 1669-1677, 1986.
[Nin88] Z.Q.Ning, P.M.Dewilde, SPIDER: Capacitance modelling for VLSI interconnections. *IEEE Trans. CAD*, 7, pp. 1221-1228, Dec. 1988.
[Noo87] A.Noore, R.S.Nutter, R.E.Swartwout, Creating a test knowledge base for VLSI testing. *Int. Conf. Comp. Des.*, Port Chester, pp. 288-291, Oct. 1987.
[Nor86] D.A.Norman, S.W.Draper, *User Centered System Design: New Perspectives on Human-Computer Interaction*. Lawrence Erlbaum, Hillsdale, 1986.
[Nye83] W.Nye et al., DELIGHT.SPICE: An optimization-based system for the design of integrated circuits. *IEEE Trans. CAD*, 7, pp. 501-519, April 1988.
[Oda89] G.Odawara (ed.), *CAD Systems Using AI Techniques*. North-Holland, Amsterdam, 1989.
[Oda84] G.Odawara, J.Sato, M.Tomita, A symbolic functional description language. *21st Design Aut. Conf.*, Albuquerque, pp. 73-80, June 1984.
[Oda86] G.Odawara, M.Tomita, O.Okuzawa, T.Ohta, Z.Q.Zhuang, A logic verifier based on Boolean comparison. *23rd Des. Aut. Conf.*, Las Vegas, pp. 208-214, Junei/July 1986.
[Odr85] P.Odryna, A.J.Strojwas, PROD: A VLSI fault diagnosis system. *IEEE Design and Test*, 2, pp. 27-35, Dec. 1985.
[Oht86] T.Ohtsuki (ed.), *Layout Design and Verification*. North-Holland, Amsterdam, 1986.
[Ora86] A.Orailoglu, D.D.Gajski, Flow graph representation. *23rd Des. Aut. Conf.*, Las Vegas, pp. 503-509, June 1986.
[Orc84] E.Orciuch, J.Frost, ISA: Intelligent scheduling assistant. *1st Conf. AI Appl.*, Denver, pp. 314-320, Dec. 1984.
[Org84] E.I.Organick et al., Transforming an Ada program unit to silicon and verifying its behavior in an Ada environment: A first experiment. *IEEE Software*, 1, pp. 31-49, Jan. 1984.
[Ost79] D.L.Ostapko, S.J.Hong, Fault analysis and test generation for programmable logic arrays (PLA's). *IEEE Trans. Comp.*, 28, pp. 617-626, Sept. 1979.
[Pal88] J.C.Palais, *Fiber Optic Communications*. Prentice-Hall, Englewood Cliffs, (2nd ed.) 1988.
[Pan89] J.Y.C.Pan, J.M.Tenenbaum, J.Glicksman, A framework for knowledge-based computer-integrated manufacturing. *IEEE Trans. Semicond. Manuf.*, 2, pp. 33-46, May 1989.
[Pan86] B.M.Pangrle, D.D.Gajski, State synthesis and connectivity binding for microarchitecture compilation. *Int. Conf. CAD*, Santa Clara, pp. 210-213, Nov. 1986.
[Pan87] B.M.Pangrle, D.D.Gajski, Slicer: A state synthesizer for intelligent silicon compilation. *Int. Conf. Comp. Des.*, Rye Brook (NY), pp. 42-45, Oct. 1987.
[PAn87] B.M.Pangrle, D.D.Gajski, Design tools for intelligent compilation. *IEEE Trans. CAD*, 6, pp. 1098-1112, Nov. 1987.
[Pan88] B.M.Pangrle, Splicer: A heuristic approach to connectivity binding. *25th Des. Aut. Conf.*, Anaheim, pp. 536-540, June 1988.
[Pap88] A.C.Papaspyridis, A Prolog-based connectivity verification tool. *25th Des. Aut. Conf.*, Anaheim, pp. 523-527, June 1988; The use of Prolog for connectivity

verification. *Int. Symp. Circ. Sys.*, Espoo, pp. 1433-1436, June 1988.
[Pap80] S.Papert, *Mindstorms, Children, Computers and Powerful Ideas*. Basic Books, New York, 1980.
[Par76] G.J.Parasch, R.L.Price, Development and application of a designer oriented cyclic simulator. *13th Des. Aut. Conf.*, San Francisco, pp. 48-53, June 1976.
[Par87] A.C.Parker, S.Hayati, Automating the VLSI design process using expert systems and silicon compilation. *Proc. IEEE*, 75, pp. 777-785, June 1987.
[Par86] P.S.Parry, I.A.Guyler, J.S.Bayliss, An application of artificial intelligence techniques to VLSI test. *3rd Silicon Des. Conf.*, London, pp. 325-329, June 1986.
[Par89] K.Parsaye, M.Chignell, S.Khoshafian, H.Wong, *Intelligent Databases. Object-Oriented, Deductive Hypermedia Technologies*. Wiley, New York, 1989.
[Pat84] A.Patel, C.Yeh, L.C.Cote, Theoretical and experimental wirability analysis system (TEWAS). *Int. Conf. CAD*, Santa Clara, pp. 69-71, Nov. 1984.
[Pat85] D.A.Patterson, Reduced Instruction Set Computers. *Commun. ACM*, 28, pp. 8-21, Jan 1985.
[Pau86] P.G.Paulin, J.P.Knight, E.F.Girczyc, HAL: A multi-paradigm approach to automatic data path synthesis. *23rd Des. Aut. Conf.*, Las Vegas, pp. 263-270, June 1986. See also *24th Des. Aut. Conf.*, Miami Beach, pp. 195-202, June/July 1987.
[Pea84] J.Pearl, *Heuristics. Intelligent Search Strategies for Computer Problem Solving*. Addison-Wesley, Reading, 1984.
[Pea88] J.Pearl, *Probabilistic Reasoning in Intelligent Systems: Networks of Plausible Inference*. Morgan Kaufmann, San Mateo, 1988.
[Pel91] D.Pellerin, M.Holley, *Practical Design Using Programmable Logic*. Prentice-Hall, New York, 1991.
[Pen86] J.M.Pendleton et al., A 32-bit microprocessor for Smalltalk. *IEEE Jour. Solid-State Circ.*, 21, pp. 741-749, Oct. 1986.
[Per78] L.M.Pereira, F.Pereira, D.H.D.Warren, *User's Guide to DEC System-10 Prolog*. Artificial Intelligence Unit, Univ. of Edinburgh, Edinburgh, 1978.
[Per77] G.Persky, D.N.Deutsch, D.G.Schweikert, LTX. A minicomputer-based system for automated LSI layout. *Jour. DA and FTC*, 1, pp. 217-255, May 1977.
[Pet88] G.Peterson, *Object-Oriented Computing. Vol.1: Concepts. Vol.2: Implementations*. IEEE Comp. Soc. Press, Los Alamitos, 1988.
[PEt88] C.Petzold, *Programming Windows*. Microsoft Press, Redmond, 1988.
[Pin88] L.J.Pinson, R.S.Wiener, *An Introduction to Object-Oriented Programming and Smalltalk*. Addison-Wesley, Reading, 1988.
[Pin83] R.Y.Pinter, River routing: Methodology and analysis. In [Bry83], pp. 141-163, 1983.
[Pip86] F.Pipitone, The FIS electronic troubleshooting system. *Computer*, 19, pp. 68-76, July 1986.
[Pit83] K.Pitman, *The Revised Maclisp Manual*. MIT/LCS/TR-295, Massachusetts Inst. Tech., Cambridge, May 1983.
[Ple87] A.R.Pleszkun, M.J.Thazhuthaveetil, The architecture of Lisp machines. *Computer*, 20, pp. 35-44, March 1987.

REFERENCES

[Poh90] H.Pohjonen, Expert system for parameterized analog cells. In [Ism90], pp. 29-43, 1990.
[Pou85] D.Pountain, Parallel processing: A look at the ALICE hardware and HOPE language. *Byte*, 10, pp. 385-395, May 1985.
[Pow85] P.A.D.Powell, M.I.Esmary, ICEWATER: A procedural design language for VLSI. Univ. of Waterloo, Waterloo, 1985.
[Pre86] B.T.Preas, P.G.Karger, Automatic placement. A review of current techniques. *23rd Des. Aut. Conf.*, Las Vegas, pp. 622-629, June/July 1986.
[Puc88] D.A.Pucknell, K.Eshraghian, *Basic VLSI Design. Systems and Circuits.* Prentice-Hall, New York, 1988.
[Pur88] E.T.Purcell, Fault diagnosis assistant. *IEEE Circ. Sys. Mag.*, 4, pp. 47-59, Jan. 1988.
[Qua86] T.Quarles et al., SPICE3A5 User's Guide. Dept EECS, Univ. of California, Berkeley, April 1986; T.L.Quarles, *The SPICE3 Implementation Guide*, Univ. of California, Memorandum UCB/ERL M89/44, 24 April 1989.
[Qui68] M.R.Quillian, Semantic memory. In: M.Minsky, *Semantic Information Processing.* MIT Press, Cambridge, pp. 227-270, 1968.
[Qui83] J.R.Quinlan, Learning efficient classification procedures and their application to chess end games. In [Mic83], pp. 463-480, 1983.
[Qui87] J.R.Quinlan, P.J.Compton, K.A.Horn, L.Lazarus, Inductive knowledge acquisition: A case study. In: J.R.Quinlan (ed.), *Applications of Expert Systems.* Addison-Wesley, Sydney, pp. 157-173, 1987.
[Qui91] P.Quinton, Y.Robert, *Systolic Algorithms and Architectures.* Prentice-Hall, Hemel Hempstead, 1991.
[Rad92] Z.Radenković et al., A knowledge-based layout compaction procedure. *Microelectronics Jour.*, 23, pp. 121-132, April 1992.
[Rad82] G.Radin, The 801 minicomputer. *Symp. Arch. Supp. Progr. Lang. Op. Sys.*, Palo Alto, pp. 39-47, March 1982.
[Rae90] P.G.Raeth (ed.), *Expert Systems: A Software Methodology for Modern Applications.* IEEE Comp. Soc. Press, Los Alamitos, 1990.
[Ram91] U.Ramacher, U.Rückert (eds), *VLSI Design of Neural Networks.* Kluwer Acad., Boston, 1991.
[Rat82] I.M.Ratiu, A.Sangiovanni-Vincentelli, D.O.Pederson, VICTOR: A fast VLSI testability analysis program. *Int. Test Conf.*, Philadelphia, pp. 397-401, Nov. 1982.
[Ree85] J.Reed, A.Sangiovanni-Vincentelli, M.Santomauro, A new symbolic router: YACR2. *IEEE Trans. CAD*, 4, pp. 208-219, July 1985.
[Rei82] S.P.Reiss, J.E.Savage, SLAP. A methodology for silicon layout. *Int. Conf. Circ. Comp.*, New York, pp. 281-285, Sept./Oct. 1982.
[Rei84] W.Reitman (ed.), *Artificial Intelligence Applications for Business.* Ablex, Norwood, 1984.
[Ric85] E.Rich, *Artificial Intelligence.* McGraw-Hill, New York, 1985.
[Riv83] R.L.Rivest, C.M.Fiduccia, A 'greedy' channel router. *Computer-Aided Design*, 15, pp. 135-140, May 1983.
[Rob89] C.Robach, D.Lutoff, N.Garcia, Knowledge-based functional specification of test and maintenance programs. *IEEE Trans. CAD*, 8, pp. 1145-1156, Nov. 1989.

[Rob84] C.Robach, P.Malecha, G.Michel, CATA: A computer-aided test analysis system. *IEEE Design & Test*, 1, pp. 68-79, May 1984.
[Rob83] G.D.Robinson, HITEST. Intelligent test generation. *Int. Test Conf.*, Philadelphia, pp. 311-323, Oct. 1983.
[Rog89] B.Rogel-Favila, P.Y.K.Cheung, Deep reasoning approach to sequential circuit fault diagnosis. *Eur. Conf. Circ. Th. Des.*, Brighton, pp. 665-669, Sept. 1989.
[Rog85] C.D.Rogers et al., VIVID. Four papers in *22nd Des. Aut. Conf.*, Las Vegas, pp. 62-87, June 1985; C.D.Rogers, The VIVID symbolic design system: Current overview and future directions. *IEEE Design & Test*, 3, pp. 75-81, Feb. 1986.
[ROg85] D.F.Rogers, *Procedural Elements for Computer Graphics*. McGraw-Hill, New York, 1985.
[Roo87] R.W.Root (ed.), *Expert Systems in Telecommunications*. Symp. Human Factors Soc., New York, March 1987.
[Ros83] S.Rosenberg, HPRL: A language for building expert systems. *8th Int. Joint Conf. AI*, Karlsruhe, pp. 215-217, Aug. 1983.
[Ros62] F.Rosenblatt, *Principles of Neurodynamics. Perceptrons and the Theory of Brain Mechanisms*. Spartan, New York, 1962.
[Rot66] J.P.Roth, Diagnosis of automata failures: a calculus and a method. *IBM Jour. Res. Dev.*, 10, pp. 278-291, July 1966.
[Rot67] J.P.Roth, W.G.Bouricius, P.R.Schneider, Programmed algorithms to compute tests to detect and distinguish between failures in logic circuits. *IEEE Trans. El. Comp.*, 16, pp. 567-580, Oct. 1967.
[Roy85] C.Roy, L.P.Demers, E.Cerny, J.Gecsei, An object-oriented switch-level simulator. *22nd Des. Aut. Conf.*, Las Vegas, pp. 623-629, June 1985.
[Rub74] F.Rubin, The Lee path connection algorithm. *IEEE Trans. Comp.*, 23, pp. 907-914, Sept. 1974; 25, p. 208, Feb. 1976.
[Rub87] S.M.Rubin, *Computer Aids for VLSI Design*. Addison-Wesley, Reading, 1987.
[Rub83] J.Rubinstein, P.Penfield, M.A.Horowitz, Signal delay in RC tree networks. *IEEE Trans. CAD*, 2, pp. 202-211, July 1983.
[Rum86] D.E.Rumelhart, G.E.Hinton, R.J.Williams, Learning internal representations by error propagation. In [RUm86], pp. 318-362, 1986. See also: *Nature*, 323, pp. 533-536, 1986.
[RUm86] D.E.Rumelhart, J.L.McClelland (eds), *Parallel Distributed Processing: Explorations in the Microstructure of Cognition. Vol.1: Foundations*. MIT Press, Cambridge, 1986.
[Rus89] L.Ruston, P.Sen, Rule-based network design: Application to packet radio networks. *IEEE Netw. Mag.*, 3, pp. 31-39, July 1989.
[Sab89] S.K.Sabat, S.Thanawastien, An AI (Artificial Intelligence) tool for reliable VLSI synthesis. In [Oda89], pp. 161-169, 1989.
[SAC88] Knowledge-Based Systems for Communications. Special Issue of *IEEE Jour. Select. Areas Commun.*, 6, pp. 781-898, June 1988.
[Sai81] T.Saito et al., A CAD system for logic design based on frames and demons. *18th Des. Aut. Conf.*, Nashville, pp. 451-456, June/July 1981.
[Sai86] T.Saito, H.Sugimoto, M.Yamazaki, N.Kawato, A rule-based logic circuit synthesis system for CMOS gate arrays. *23rd Des. Aut. Conf.*, Las Vegas, pp. 594-600, June/July 1986.

REFERENCES

[Sal83] K.K.Saluja, K.Kinoshita, H.Fujiwara, An easily testable design of programmable logic arrays for multiple faults. *IEEE Trans. Comp.*, 32, pp. 1038-1046, Nov. 1983.

[Sam86] M.A.Samad, J.A.B.Fortes, DEFT. A design for testability expert system. *Fall Joint Comp. Conf.*, Dallas, pp. 899-908, Nov. 1986.

[SAm86] M.A.Samad, J.A.B.Fortes, Explanation capabilities in DEFT. A design-for-testability expert system. *Int. Test Conf.*, Washington, pp. 954-963, Sept. 1986.

[Sam91] M.Sami, J.Calzadilla-Daguerre (eds), *Silicon Architectures for Neural Nets*. North-Holland, Amsterdam, 1991.

[San92] E.Sánchez-Sinencio, R.W.Newcomb (eds), Neural Network Hardware. Special Issue of *IEEE Trans. Neur. Netw.*, 3, pp. 345-518, May 1992.

[Sar92] T.M.Sarfert et al., A hierarchical tets pattern generation system based on high-level primitives. *IEEE Trans. CAD*, 11, pp. 34-44, Jan. 1992.

[Sar90] A.Sarkar, S.Bandyopadhyay, G.A.Jullien, Bit-level designer's assistant. A knowledge based approach to systolic processor design. *33rd Midwest Symp. Circ. Sys.*, Calgary, pp. 1001-1004, Aug. 1990.

[SAr92] M.A.Sartori, K.M.Passino, P.J.Antsaklis, A multilayer perceptron solution to the match phase problem in rule based Artificial Intelligence systems. *IEEE Trans. Knowl. Data Engng*, 4, pp. 290-297, June 1992.

[Sau89] G.Saucier, A.Ambler, M.A.Breuer (eds), *Knowledge Based Systems for Test and Diagnosis*. North-Hlland, Amsterdam, 1989.

[Sau87] G.Saucier, M.Crastes de Paulet, P.Sicard, ASYL: A rule-based system for controller synthesis. *IEEE Trans. CAD*, 6, pp. 1088-1097, Nov. 1987.

[Sau86] G.Saucier, S.Hanriat, Rule based logical synthesis for silicon compilers. *Int. Conf. CAD*, Santa Clara, pp. 166-168, Nov. 1986.

[SAu87] G.Saucier, E.Read, J.Trilhe (eds), *Fast-Prototyping of VLSI*. North-Holland, Amsterdam, 1987.

[SAU87] G.Saucier, P.Genestier, C.Jay, Rule based synthesis environment application to a dedicated microprocessor design. In [SAu87], pp. 69-89, 1987.

[Sav83] J.Savir, Good controllability and observability do not guarantee good testability. *IEEE Trans. Comp.*, 32, pp. 1198-1200, Dec. 1983.

[Sch86] R.W.Scheifler, J.Gettys, The X Window System. *ACM Trans. Graphics*, 5, pp. 79-109, April 1986.

[Sch83] J.P.Schoellkopf, LUBRICK: A silicon assembler and its application to data-path design for FISC. In [Anc83], pp. 435-445, 1983.

[SCh86] M.I.Schor, Declarative knowledge programming: Better than procedural? *IEEE Expert*, 1, pp. 36-43, Spring 1986.

[Sch88] M.H.Schulz, E.Trischler, T.M.Sarfert, SOCRATES: A highly efficient automatic test pattern generation system. *IEEE Trans. CAD*, 7, pp. 126-137, Jan. 1988.

[SCh87] D.Schutzer, *Artificial Intelligence. An Applications-Oriented Approach*. Van Nostrand Reinhold, New York, 1987.

[Sch92] D.G.Schwartz, G.J.KLir, Fuzzy logic flowers in Japan. *IEEE Spectrum*, 29, pp. 32-35, July 1992.

[Sch87] A.F.Schwarz, *Computer-Aided Design of Microelectronic Circuits and Systems. Fundamentals, Methods and Tools*. Two volumes. Acad. Press, London, 1987.

[Sea91] M.S.Seadle, *Automated Mainframe Management: Using Expert Systems with Examples from VM and MVS*. McGraw-Hill, New York, 1991.

[Sec88] C.Sechen, *VLSI Placement and Global Routing Using Simulated Annealing*. Kluwer Acad., Dordrecht, 1988.

[Sei91] C.Seitz, *Developments in Concurrency and Communication*. Addison-Wesley, Reading, 1991.

[Sej87] T.J.Sejnowski, C.R.Rosenberg, Parallel networks that learn to pronounce English text. *Complex Systems*, 1, pp. 145-168, 1987.

[Set89] R.Sethi, *Programming Languages, Concepts and Constructs*. Addison-Wesley, Reading, 1989.

[Set90] D.E.Setliff, R.A.Rutenbar, *Automatic Programming Applied to VLSI CAD Software: A Case Study*. Kluwer Acad. Publ., Boston, 1990. See also *IEEE Trans. CAD*, 10, pp. 783-801, June 1991.

[Sha89] R.Shankar, E.B.Fernandez, *VLSI and Computer Architecture*. Acad. Press, San Diego, 1989.

[Sha90] S.C.Shapiro (ed.), *Encyclopedia of Artificial Intelligence*. Two volumes. Wiley, New York, (2nd ed.) 1990.

[Sha86] S.C.Shapiro, S.N.Srihari, M.R.Taie, J.Geller, VMES: A network-based versatile maintenance expert system. In [SRi86], pp. 925-936, 1986.

[SHa89] R.Sharma, VLSI electronic neural networks. In [Wat89], pp. 413-433, 1989.

[Sha85] J.A.Sharp, *Data Flow Computing*. Ellis Horwood, Chichester, 1985.

[She88] B.J.Sheu, A.H.Fung, Y.N.Lai, A knowledge-based approach to analog IC design. *IEEE Trans. Circ. Sys.*, 35, pp. 256-258, Feb. 1988.

[She90] B.J.Sheu, J.C.Lee, A.H.Tung, Flexible architecture approach to knowledge-based analogue IC design. *IEE Proc. G*, 137, pp. 266-274, Aug. 1990.

[SHe88] P.C.Y.Sheu, VLSI design with object-oriented knowledge bases. *Computer-Aided Design*, 20, pp. 272-280, June 1988.

[She85] J.P.Shen, W.Maly, F.J.Ferguson, Inductive fault analysis of MOS integrated circuits. *IEEE Design & Test*, 2, pp. 13-26, Dec. 1985.

[Shi91] P.H.Shih, K.E.Chang, W.S.Feng, Neural computation network for global routing. *Computer-Aided Design*, 23, pp. 539-547, Oct. 1991.

[Shi83] T.Shimizu, K.Sakamura, MIXER: An expert system for microprogramming. *16th ACM Micropr. Workshop*, pp. 168-175, Oct. 1983.

[Shi87] H.Shin, A.Sangiovanni-Vincentelli, A detailed router based on incremental routing modifications: Mighty. *IEEE Trans. CAD*, 6, pp. 942-955, Nov. 1987.

[Shi89] K.Shirai, T.Takezawa, Interactive refinement of architecture using knowledge in special-purpose circuit design system. In [Oda89], pp. 59-66, 1989. See also: K.Shirai, T.Takezawa, Expert system for designing digital signal processor architectures. *Microproc. Microsys.*, 12, pp. 83-91, March 1988; K.Shirai, T.Takezawa, S.Ueno, Knowledge based techniques on a development system of application specific integrated circuits. *Int. Workshop AI Industr. Appl.*, Hitachi City, pp. 200-205, May 1988.

[SHi89] I.Shirakawa *et al.*, An expert system for mask pattern generator of CMOS logic cells. *Eur. Conf. Circ. Th. Des.*, pp. 324-328, Sept. 1989.

[Shi92] N.Shiratori *et al.*, Using Artificial Intelligence in communication system design. *IEEE Software*, 9, pp. 38-46, Jan. 1992.

REFERENCES

[SHi87] M.Shirley, P.Wu, R.Davis, G.Robinson, A synergistic combination of test generation and design for testability. *Int. Test Conf.*, Washington, pp. 701-711, Sept. 1987.

[Shn87] B.Shneiderman, *Designing the User Interface: Strategies for Effective Human-Computer Interaction*. Addison-Wesley, Reading, 1987.

[Sho75] H.A.Sholl, S.C.Yang, Design of asynchronous sequential networks using READ-ONLY memories. *IEEE Trans. Comp.*, 24, pp. 195-206, Feb. 1975.

[Sho76] E.H.Shortliffe, *Computer-Based Medical Consultations: MYCIN*. Amer. Elsevier, New York, 1976.

[Sho84] E.H.Shortliffe et al., An expert system for oncology protocol management. In [Buc84], 1984.

[Shu82] H.Shubin, J.W.Ulrich, IDT: An intelligent diagnostic tool. *AAAI-82*, Pittsburgh, pp. 290-295, Aug. 1982.

[Shu91] C.B.Shung et al., An integrated CAD system for algorithm-specific IC design. *IEEE Trans. CAD*, 10, pp. 447-463, April 1991.

[Sia87] P.Siarry, L.Bergonzi, G.Dreyfus, Thermodynamic optimization of block placement. *IEEE Trans. CAD*, 6, pp. 211-221, March 1987.

[Sil89] Silicon Neural Networks. Special Issue of *IEEE Micro*, 9, pp. 5-76, Dec. 1989.

[Sim89] E.Simoudis, A knowledge-based system for the evaluation and redesign of digital circuit networks. *IEEE Trans. CAD*, 8, pp. 302-315, March 1989.

[Sim90] E.Simoudis, Learning redesign knowledge. *IEEE Trans. CAD*, 9, pp. 1047-1062, Oct. 1990.

[Sim85] E.Simoudis, S.Fickas, The application of knowledge-based design techniques to circuit design. *Int. Conf. CAD*, Santa Clara, pp. 213-215, Nov. 1985.

[Sim87] A.Simpson, *Advanced Techniques in dBase III Plus*. Sybex, San Francisco, 1987.

[Sin87] N.Singh, *An Artificial Intelligence Approach to Test Generation*. Kluwer Academic, Boston, 1987.

[Sle82] D.Sleeman, J.S.Brown (eds), *Intelligent Tutoring Systems*. Acad. Press, London, 1982.

[Sma81] Smalltalk-80. Special Issue of *Byte Magazine*, 6, pp. 6-387, Aug. 1981.

[Sma86] *Smalltalk/V. Tutorial and Programming Handbook*. Digitalk Inc., Los Angeles, 1986.

[Smi84] D.E.Smith, J.E.Clayton, Another look at frames. In [Buc84], pp. 441-452, 1984.

[Smi82] G.L.Smith, R.J.Bahnsen, H.Halliwell, Boolean comparison of hardware and flowcharts. *IBM Jour. Res. Dev.*, 26, pp. 106-116, Jan. 1982.

[Sno78] E.A.Snow, D.P.Siewiorek, D.E.Thomas, A technology-relative computer-aided design system: Abstract representations, transformations, and design tradeoffs. *15th Des. Aut. Conf.*, Las Vegas, pp. 220-226, June 1978.

[Son85] K.Son, Rule based testability checker and test generator. *Int. Test Conf.*, Philadelphia, pp. 884-889, 1985.

[Son84] K.Son, J.Y.O.Fong, Table driven behavioral test generation. *Int. Symp. Circ. Sys.*, Montreal, pp. 718-722, May 1984.

[Sou89] B.Soucek, *Neural and Concurrent Real-Time Systems: The Sixth generation*. Wiley, New York, 1989.

[Sou91] B.Soucek et al., *Neural and Intelligent Systems Integration. Fifth and Sixth Generation Integrated Reasoning Information Systems.* Wiley, New York, 1991.

[Sou83] J.R.Southard, MacPitts: An approach to silicon compilation. *Computer*, 16, pp.74-82, Dec. 1983.

[Spi85] R.L.Spickelmier, A.R.Newton, Connectivity verification using a rule-based approach. *Int. Conf. CAD*, Santa Clara, pp. 190-192, Nov. 1985.

[Spi88] R.L.Spickelmier, A.R.Newton, Critic: A knowledge-based program for critiquing circuit designs. *Int. Conf. Comp. Des.*, Rye Brook, pp. 324-327, Oct. 1988.

[Spr91] J.D.Spragins et al., *Telecommunications: Protocols and Design.* Addison-Wesley, Reading, 1991.

[Sri88] N.C.E.Srinivas, V.D.Agrawal, Formal verification of digital circuits using hybrid simulation. *IEEE Circ. Dev. Mag.*, 4, pp. 19-27, Jan. 1988.

[Sri86] N.C.E.Srinivas, A.S.Wojcik, Y.H.Levendel, An Artificial Intelligence based implementation with the P-algorithm for test generation. *Int. Test Conf.*, Washington, pp. 732-739, Sept. 1986.

[SRi86] D.Sriram, R.Adey (eds), *Applications of Artificial Intelligence in Engineering Problems.* Springer, Berlin, 1986.

[Sta88] W.Stallings, *Computer Organization and Architecture: Principles of Structure and Fundamentals.* MacMillan, New York, 1987.

[STa88] W.Stallings (ed.), *Integrated Services Digital Networks (ISDN).* IEEE Comp. Soc. Press, Los Alamitos, (2nd ed.) 1988.

[Sta90] W.Stallings (ed.), *Reduced Instruction Set Computers (RISC).* IEEE Comp. Soc. Press, Los Alamitos, (2nd ed.) 1990.

[Sta91] W.Stallings, *Local Networks.* Macmillan, New York, (3rd ed.) 1991.

[Sta92] W.Stallings, *ISDN and Broadband ISDN.* Macmillan, New York, (2nd ed.) 1992.

[Sta77] R.M.Stallman, G.J.Sussman, Forward reasoning and dependency-directed backtracking in a system for computer-aided circuit analysis. *Art. Intell.*, 9, pp. 135-196, Aug. 1977. See also [Win79], Vol. 1, pp. 33-91, 1979.

[STa90] J.Staunstrup (ed.), *Formal Methods for VLSI Design.* North-Holland, Amsterdam, 1990.

[Ste84] G.L.Steele et al., *Common LISP: The Language*, Digital Press, Burlington, 1984.

[Ste79] M.J.Stefik, An examination of a frame-structured representation system. *6th Int. Joint Conf. AI*, Tokyo, pp. 845-852, Aug. 1979.

[Ste81] M.J.Stefik, Planning with constraints, MOLGEN: Part 1; Planning and meta-planning, MOLGEN: Part2. *Art. Intell.*, 16, pp. 111-139, pp. 141-169, 1981.

[Ste82] M.Stefik, J.Aikins, R.Balzer, J.Benoit, L.Birnbaum, F.Hayes-Roth, E.Sacerdoti, The organization of expert systems. A tutorial. *Art. Intell.*, 18, pp. 135-173, 1982.

[Ste83] M.Stefik, D.G.Bobrow, S.Mittal, L.Conway, Knowledge programming in LOOPS: Report on an experimental course. *Art. Intell.*, 4, pp. 3-14, Fall 1983. See also: Object-oriented programming: Themes and variations. *AI Mag.*, 6, pp. 40-62, Winter 1986.

REFERENCES

[Ste85] L.I.Steinberg, T.M.Mitchell, The Redesign system: A knowledge-based approach to VLSI CAD. *IEEE Design & Test*, 2, pp. 45-54, Feb. 1985. See also *21st Des. Aut. Conf.*, Albuquerque, pp. 412-418, June 1984.

[Ste86] L.Sterling, E.Shapiro, *The Art of Prolog: Advanced Programming Techniques*. MIT Press, Cambridge, 1986.

[Sto87] S.J.Stolfo, Initial performance of the DADO2 prototype. *Computer*, 20, pp. 75-83, Jan. 1987.

[Sto90] H.S.Stone, *High-Performance Computer Architecture*. Addison-Wesley, Reading, (2nd ed.) 1990.

[Sto91] H.S.Stone, J.Cocke, Computer architecture in the 1990s. *Computer*, 24, pp. 30-38, Sept. 1991.

[Sto76] M.Stonebraker, The design and implementation of INGRES. *ACM Trans. Database Sys.*, 1, pp. 189-222, Sept. 1976.

[Sto88] M.Stonebraker *et al.*, The POSTGRESS rule manager. *IEEE Trans. Software Engng*, 14, pp. 897-907, July 1988; M.Stonebraker *et al.*, The implementation of PROGRESS. *IEEE Trans. Knowl. Data Engng*, 2, pp. 125-142, March 1990.

[Str88] A.J.Strojwas, The Process Engineer's Workbench. *IEEE Jour. Solid-State Circ.*, 23, pp. 377-386, April 1988.

[Str91] B.Stroustrup, *The C++ Programming Language*. Addison-Wesley, Reading, (2nd ed.) 1986.

[Su82] S.Y.H.Su, Y.Hsieh, Testing functional faults in digital systems described by register transfer language. *Jour. Dig. Sys.*, 6, pp. 161-183, Summer/Fall 1982. Also in: *Int. Test Conf.*, Philadelphia, pp. 447-457, Oct. 1981.

[Su88] S.Y.W.Su, *Database Computers: Principles, Architectures, and Techniques*. McGraw-Hill, New York, 1988.

[Sub83] P.A.Subrahmanyam, Synthesizing VLSI circuits from behavioral specifications: A very high level silicon compiler and its theoretical basis. In [Anc83], pp. 195-210, 1983.

[Sub86] P.A.Subrahmanyam, The algebraic basis of an expert system for VLSI design. In [Mil86], pp. 59-81, 1986.

[SUb86] P.A.Subrahmanyam, Synapse: An expert system for VLSI design. *Computer*, 19, pp. 78-89, July 1986.

[Suc87] L.A.Suchman, *Plans and Situated Actions. The Problem of Human-Machine Communication*. Cambridge Univ. Press, Cambridge, 1987.

[Sul90] T.E.Sullivan, B.P.Butz, R.J.Schuhl, Expert system diagnosis of CMOS VLSI fabrication problems. *SPIE, Vol.1293, Applications of AI (Part 1)*, Orlando, pp. 62-71, April 1990.

[Sus81] G.J.Sussman, J.Holloway, G.L.Steele, A.Bell, Scheme-79. Lisp on a chip. *Computer*, 14, pp. 10-21, July 1981.

[Sus75] G.J.Sussman, R.M.Stallman, Heuristic techniques in computer-aided circuit analysis. *IEEE Trans. Circ. Sys.*, 22, pp. 857-865, Nov. 1975.

[Sus82] G.J.Sussman *et al.*, Tools for the design of large VLSI circuits. *Int. Conf. Circ. Sys.*, Rome, pp. 256-259, May 1982.

[Sut81] R.S.Sutton, A.G.Barto, Towards a modern theory of adaptive networks: Expectation and prediction. *Psychol. Rev.*, 88, pp. 135-170, 1981.

[Suz85] N.Suzuki, Concurrent Prolog as an efficient VLSI design language. *Computer*, 18, pp. 33-40, Feb. 1985.
[Sva84] D.Svanaes, E.J.Aas, Test generation through logic programming. *Integration. VLSI Jour.*, 2, pp. 49-67, 1984.
[Swa89] E.E.Swartzlander (ed.), *Wafer Scale Integration*. Kluwer Acad., Dordrecht, 1989.
[Szu90] H.H.Szu, Optical neuro-computing. In [Mar90], pp. 271-293, 1990.
[Tab90] D.Tabak, *RISC Systems*. Wiley, New York, 1990.
[Tak88] M.Takashima et al., A circuit comparison system with rule-based functional isomorphism checking. *25th Des. Aut. Conf.*, Anaheim, pp. 512-516, June 1988.
[Tan86] H.Tanaka, A parallel inference machine. *Computer*, 19, pp. 48-54, May 1986.
[Tan88] T.Tanaka, Structural analysis of electronic circuits in a deductive system. In: L.Bolc, M.J.Coombs (eds), *Expert System Applications*. Springer, Berlin, pp. 257-308, 1988.
[Tan89] A.S.Tanenbaum, *Computer Networks*. Prentice-Hall, Englewood Cliffs, (2nd ed.) 1989.
[Tan87] S.L.Tanimoto, *The Elements of Artificial Intelligence*. Computer Sci. Press, Rockville, 1987.
[TAn86] D.W.Tank, J.J.Hopfield, Simple "neural" optimization networks: An A/D coverter, signal decision circuit and a linear programming circuit. *IEEE Trans. Circ. Sys.*, 36, pp. 533-541, May 1986.
[Tat87] D.Tatar, *Programmer's Guide to Common Lisp*. Digital Press, Bedford, 1987.
[Tef89] L.Teft, *Programming in Turbo Prolog with an Introduction to Knowledge-Based Systems*. Prentice-Hall, Englewood Cliffs, 1989.
[Tei81] W.Teitelman, L.Masinter, The Interlisp programming environment. *Computer*, 14, pp. 25-33, April 1981.
[Tel89] E.R.Tello, *Object-Oriented Programming for Artificial Intelligence*. Addison-Wesley, Reading, 1989.
[Tem88] K.H.Temme, A.Nitsche, Chip-architecture planning: An expert system approach. In [Ger88], pp. 137-161, 1988.
[TEm88] K.H.Temme, R.Brück, Chip-architecture planning based on expert knowledge. *Int. Workshop AI Industr. Appl.*, Hitachi City, pp. 188-193, May 1988.
[ten90] K.O. ten Bosch, P.Bingley, P. van der Wolf, Design flow management in the NELSIS CAD framework. *28th Des. Aut. Conf.*, San Francisco, pp. 711-716, June 1991.
[ter86] J.H. ter Bekke, OTO-D: Object type oriented data modeling. Report 86-02, Fac. Tech. Math. and Informatics, Delft Univ. of Technology, 1986.
[ter91] J.H. ter Bekke, *Semantic Data Modeling*. Prentice-Hall, Hemel Hempstead, 1991. See also: *Semantic Data Modeling in Relational Environments*, Doct. Diss., Delft Univ. of Technology, Delft, June 1991.
[Ter85] C.J.Terman, Timing simulation for large digital MOS circuits. In A.Sangiovanni-Vincentelli (ed.), *Computer-Aided Design of VLSI Circuits and Systems*. JAI Press, London, 1985.
[Tew89] S.K.Tewksbury, *Wafer-Level Integrated Systems: Implementation Issues*. Kluwer Acad., Hingham, 1989.

REFERENCES

[Tho83] D.E.Thomas, G.W.Leive, Automating technology relative logic synthesis and module selection. *IEEE Trans. CAD*, 2, pp. 94-105, April 1983.

[Tho91] D.E.Thomas, P.R.Moorby, *The Verilog Hardware Description Language*. Kluwer Acad., Dordrecht, 1991.

[THo83] D.E.Thomas, J.A.Nestor, Defining and implementing a multilevel design representation with simulation applications. *IEEE Trans. CAD*, 2, pp. 135-145, July 1983.

[Tho81] D.E.Thomas, D.P.Siewiorek, Measuring designer performance to verify design automation systems. *IEEE Trans. Comp.*, 30, pp. 48-61, Jan. 1981.

[Tho89] D.E.Thomas et al., *Algorithmic and Register-Transfer Level Synthesis: The System Architect's Workbench*. Kluwer Acad., Dordrecht, 1989. Anaheim, pp. 337-343, June 1988.

[Tic89] E.Tick, Comparing two parallel logic-programming architectures. *IEEE Software*, 6, pp. 71-80, July 1989.

[Tim83] C.Timoc et al., Logical models of physical failures. *Int. Test Conf.*, Philadelphia, pp. 546-553, Oct. 1983.

[To85] E.A.Torrero, *Next-Generation Computers*. IEEE Press, New York, 1985.

[Tou82] F.N.Tou et al., RABBIT: An intelligent database assistant. *AAAI-82*, pp. 314-318, 1982.

[Tow86] C.Townsend, D.Feucht, *Designing and Programming Personal Expert Systems*. TAB Books, Blue Ridge Summit, 1986.

[Tra80] Translator Writing Systems. Special Issue of *Computer*, 13, pp. 9-49, Aug. 1980.

[Tre81] N.Tredenick, How to flowchart for hardware. *Computer*, 14, pp. 87-102, Dec. 1981.

[Tre82] P.C.Treleaven, Parallel models of computation. In [Eva82], pp. 348-380, 1983.

[Tre83] P.C.Treleaven, Decentralised computer architectures for VLSI. In: B.Randell, P.C.Treleaven (eds), *VLSI Architecture*. Prentice-Hall, Englewood Cliffs, pp. 348-380, 1983.

[Tre90] P.C.Treleaven, *Parallel Computers. Object-Oriented, Functional, Logic*. Wiley, Chichester, 1990.

[Tre85] R.Treuer, H.Fujiwara, V.K.Agarwal, Implementing a built-in self-test PLA design. *IEEE Design & Test*, 2, pp. 37-48, April 1985.

[Tre91] A.Trew, G.Wilson (eds), *Past, Present, Parallel: A Survey of Available Parallel Computing Systems*. Springer, New York, 1991.

[Tri83] H.W.Trickey, Good layouts for pattern recognizers. *IEEE Trans. Comp.*, 31, pp. 514-520, June 1983.

[Tri84] E.Trischler, An integrated design for testability and automatic test pattern generation system: An overview. *21st Des. Aut. Conf.*, Albuquerque, pp. 209-215, June 1984.

[Tsa88] C.Tsareff, T.M.Cesear, E.Iodice, An expert system approach to parameterized module synthesis. *IEEE Circ. Dev. Mag.*, 4, pp. 28-36, Jan. 1988.

[Tse81] C.J.Tseng, D.P.Siewiorek, The modeling and synthesis of bus systems. *18th Des. Aut. Conf.*, Nashville, pp. 471-478, June/July 1981.

[Tse86] C.J.Tseng, D.P.Siewiorek, Automated synthesis of data paths in digital systems. *IEEE Trans. CAD*, 5, pp. 379-395, July 1986.

[Tuc88] L.W.Tucker, G.G.Robertson, Architecture and applications of the connection machine. *Computer*, 21, pp. 26-38, Aug. 1988.

[Tui88] P.W.Tuinenga, *A Guide to Circuit Simulation and Analysis Using PSpice*. Prentice-Hall, Englewood Cliffs, 1988.

[Tur88] *Turbo Prolog 2.0 User's Guide* and *Turbo Prolog 2.0 Reference Guide*. Borland International, Scotts Valley, 1988.

[Ueh85] T.Uehara, A knowledge-based logic design system. *IEEE Design & Test*, 2, pp. 27-34, Oct. 1985.

[Ueh83] T.Uehara, M.Barbacci (eds), *Computer Hardware Description Languages and Their Applications*. North-Holland, Amsterdam, 1983.

[UEh83] T.Uehara, T.Saito, F.Maruyama, N.Kawato, DDL verifier and temporal logic. In [Ueh83], pp. 91-102, 1983.

[Ueh81] T.Uehara, W.M. van Cleemput, Optimal layout of CMOS functional arrays. *IEEE Trans. Comp.*, 30, pp. 305-312, May 1981.

[Uhr87] L.Uhr, *Multi-Computer Architectures for Artificial Intelligence. Toward Fast, Robust, Parallel Systems*. Wiley, New York, 1987.

[Ull84] J.D.Ullman, *Computational Aspects of VLSI*. Computer Science Press, Rockville, 1984.

[Ull88] J.D.Ullman, *Principles of Database and Knowledge-Base Systems. Vols 1 & 2*. Computer Sci. Press, Rockville, 1988, 1989.

[Ung87] D.Ungar, D.Patterson, What price Smalltalk? *Computer*, 20, pp. 67-74, Jan. 1987. See also: D.Ungar et al., Architecture of SOAR: Smalltalk on a RISC. *11th Int. Symp. Comp. Arch.*, Ann Arbor, pp. 188-197, 1984.

[Vai87] M.K.Vai, M.A.Shanblatt, An improved building block model for placement using simulated annealing. *Int. Symp. Circ. Sys.*, Philadelphia, pp. 572-575, May 1987.

[Vak88] D.Vakil, M.R.Zargham, K.J.Danhof, A knowledge-based system for multi-layer channel routing. *IEEE COMPSAC*, Chicago, pp. 427-424, Oct. 1988.

[Van88] E. Vanden Meersch, L.Claesen, H. De Man, Accurate timing verification algorithms for synchronous MOS circuits. In [ESP88], pp. 283-302, 1988.

[van92] A.J. van der Hoeven, *Concepts and Implementation of a Design System for Regular Processor Arrays*. Doct. Diss., Delft Univ. Technology, 6 Oct. 1992.

[van88] N. van der Meijs, A.J. van Genderen, Space: an accurate and efficient extractor for submicron integrated circuits. *Delft Progress Report*, 12, pp. 260-279, 1988.

[van87] N. van der Meijs, T.G.R. van Leuken, P. van der Wolf, I.Widya, P.Dewilde, A Data Management Interface to facilitate CAD/IC software exchanges. *Int. Conf. Comp. Des.*, Port Chester, pp. 403-406, Oct. 1987.

[vAn87] P.S. van der Meulen, INSIST: Interactive simulation in Smalltalk. *ACM SIGPLAN Not.*, 22, pp. 366-376, 1987.

[van93] P. van der Wolf, *Architecture for an Open and Efficient CAD Framework*. Doct. Diss., Delft Univ. Technology, 1993.

[van90] P. van der Wolf, P.Bingley, P.Dewilde, On the architecture of a CAD framework: The NELSIS approach. *Eur. Des. Aut. Conf.*, Glasgow, pp. 29-33, March 1990.

[vAn90] P. van der Wolf, G.W.Sloof, P.Bingley, P.Dewilde, Meta data management in the NELSIS CAD framework. *27th Des. Aut. Conf.*, Orlando, pp. 142-145, June

1990.

[vAn88] P. van der Wolf, T.G.R. van Leuken, Object type oriented data modeling for VLSI data management. *25th Des. Aut. Conf.*, Anaheim, pp. 351-356, June 1988. See also: T.G.R.M. van Leuken, *Data Management for VLSI Design in an Open and Distributed Environment*. Doct. Diss., Delft Univ. Technology, 24 March 1988.

[van91] A.J. van Genderen, *Reduced Models for the Behavior of VLSI Circuits*. Doct. Diss., Delft Univ. Technology, 3 Oct. 1991.

[van86] A.J. van Genderen, A.C. de Graaf, SLS: A switch-level timing simulator. In [Dew86], pp. 2.93-2.145, 1986; A.J. van Genderen, SLS: A switch-level timing simulator. *Delft Progress Report*, 12, pp. 280-290, 1988.

[vAN88] A.J. van Genderen, N.P. van der Meijs, Extracting simple but accurate RC models for VLSI interconnect. *Int. Symp. Circ. Sys.*, Espoo, pp. 2351-2354, June 1988.

[VAn88] J. Vanhoof, J.Rabaey, H. De Man, A knowledge-based CAD system for synthesis of multi-processor digital signal processing chips. In: C.H.Séquin, *VLSI Design of Digital Systems*, North-Holland, Amsterdam, pp. 73-88, 1988.

[vAn91] T.G.R. van Leuken, P. van der Wolf, P.Bingley, Standardization concepts in the NELSIS CAD framework. In: F.J.Rammig, R.Waxman (eds), *Electronic Design Automation Frameworks*. North-Holland, Amsterdam, pp. 57-70, 1991.

[Van91] P.Vanoostende, P.Six, H.J. De Man, DARSI: RC data reduction. *IEEE Trans. CAD*, 10, pp. 493-500, April 1991.

[Van92] P. Van Roy, A.M.Despain, High-performance logic programming with the Aquarius Prolog compiler. *Computer*, 25, pp. 54-68, Jan. 1992.

[Var88] P.Varma, Y.Tohma, A knowledge-based test generator for standard cell and iterative array logic circuits. *IEEE Jour. Solid-State Circ.*, 23, pp. 428-436, April 1988.

[Vem88] V.Vemuri (ed.), *Artificial Neural Networks. Theoretical Concepts*. IEEE Comp. Soc. Press, Los Alamitos, 1988.

[Ver89] M.Verleysen *et al.*, Neural networks for high-storage content-addressable memory: VLSI circuit and learning algorithm. *IEEE Jour. Solid-State Circ.*, 24, pp. 52-568, June 1989.

[Ver87] G.L.Vernazza, S.B.Serpico, S.G.Dellepiana, A knowledge-based system for biomedical image processing and recognition. *IEEE Trans. Circ. Sys.*, 34, pp. 1399-1416, Nov. 1987.

[Ves83] G.T.Vesonder, S.J.Stolfo, J.E.Zielinski, F.D.Miller, D.H.Copp, ACE: An expert system for telephone cable maintenance. *IJCAI-83*, Karlsruhe, pp. 116-121, Aug. 1983.

[Via89] T.Viacroze, M.Lequeux, An expert system to assist in diagnosis of failures on VLSI memories. *IEEE Custom Integr. Circ. Conf.*, San Diego, pp. 26.4.1-26.4.4, May 1989.

[Vie87] H.T.Vierhaus, Rule-based design for testability. The EXTEST approach. *COMPEURO87*, Hamburg, pp. 949-952, May 1987.

[Voi90] Voice in Computing. Special Issue of *Computer*, 23, pp. 8-80, Aug. 1990.

[Vor89] R.S.Voros, D.J.Hillman, D.R.Decker, G.D.Blank, Frame-based approach to database management. *SPIE, vol. 1095, Appl. AI, VII*, Orlando, pp. 86-91,

March 1989.
- [Wad88] S.Wada, Y.Koseki, T.Nishida, An expert system architecture for switching system diagnosis. *Int. Workshop AI Industr. Appl.*, Hitachi City, pp. 123-128, May 1988.
- [Wad85] W.W.Wadge, E.A.Ashcroft, LUCID. *The Dataflow Programming Language.* Acad. Press, Orlando, 1985.
- [Wah86] B.Wah, G.J.Li (eds), *Computers for Artificial Intelligence Applications.* IEEE Comp. Soc. Press, Washington, 1986.
- [Wal91] I.Walker, A Smalltalk/V VLSI CAD application. *Computer-Aided Eng. Jour.*, 8, pp. 47-53, April 1991.
- [Wal83] B.Walter, Timed Petri-nets for modelling and analyzing protocols with real-time characteristics. In: H.Rudin, C.H.West (eds), *Protocol Specification, Testing and Verification III.* North-Holland, Amsterdam, 1983.
- [Wan91] F.C.Wang, *Digital Circuit Testing. A Guide to DFT and Other Techniques.* Acad. Press, San Diego, 1991.
- [Wan88] J.S.Wang, R.C.T.Lee, An A. algorithm to yield an optimal solution for the channel routing problem in VLSI. *Int. Workshop AI Industr. Appl.*, Hitachi City, pp. 181-187, May 1988.
- [Wan84] L.T.Wang, E.Law, DTA: Daisy Testability Analyzer. *Int. Conf. CAD*, Santa Clara, pp. 143-145, Nov. 1984.
- [War89] F.Warkowski, J.Leenstra, J.Nijhuis, L.Spaanenburg, Issues in the test of artificial neural networks. *Int. Conf. Comp. Des.*, Cambridge, pp. 487-490, Oct. 1989.
- [War83] D.H.D.Warren, An abstract Prolog instruction set. Tech. note 309. AI Center, SRI International, Menlo Park, Oct. 1983.
- [War88] D.H.D.Warren, The Aurora OR-parallel Prolog system. *Int. Conf. Fifth Gen. Comp.*, Tokyo, 1988.
- [Was89] P.D.Wasserman, *Neural Computing. Theory and Practice.* Van Nostrand Reinhold, New York, 1989.
- [Wat87] H.Watanabe, B.Ackland, FLUTE. An expert floorplanner for full-custom VLSI design. *IEEE Design & Test*, 4, pp. 32-41, Feb. 1987.
- [Wat86] T.Watanabe, T.Masuishi, T.Nishiyama, N.Horie, Knowledge-based optimal IIL circuit generator from conventional logic circuit descriptions. *23rd Des. Aut. Conf.*, Las Vegas, pp. 608-614, June/July 1986.
- [WAt86] D.A.Waterman, *A Guide to Expert Systems.* Addison-Wesley, Reading, 1986.
- [Wat89] R.K.Watts (ed.), *Submicron Integrated Circuits.* Wiley, New York, 1989.
- [Waw89] K.Wawryn, Prolog-based active filter synthesis. *Eur. Conf. Circ. Th. Des.*, Brighton, pp. 670-673, Sept. 1989.
- [Waw91] K.Wawryn, A.Guzinski, Circuit verification based on diagnostic expert system methodology. *Electronics Lett.*, 27, pp. 958-960, 23 May 1991.
- [WAw89] K.Wawryn, W.Zinka, A prototype expert system for fault diagnosis in electronic device. *Eur. Conf. Circ. Th. Des.*, Brighton, pp. 677-680, Sept. 1989.
- [Wax86] R.Waxman *et al.*, The VHSIC Hardware Description Language. *IEEE Design & Test*, 3, pp. 10-65, April 1986.
- [Wei84] S.M.Weiss, C.A.Kulikowski, *A Practical Guide to Designing Expert Systems.* Rowman and Allanhold, Totowa, 1984.

REFERENCES

[Wei91] S.M.Weiss, C.A.Kulikowski, *Computer Systems that Learn. Classification and Prediction Methods from Statistics, Neural Nets, Machine Learning, and Expert Systems*. Kaufmann, San Mateo, 1991.

[Wen89] J.Wenin *et al.*, Rule-based VLSI verification system constrained by layout parasitics. *26th Des. Aut. Conf.*, Las Vegas, pp. 662-667, June 1989.

[Wes81] N.H.E.Weste, MULGA, an interactive symbolic layout system for the design of integrated circuits. *Bell Sys. Tech. Jour.*, 60, pp. 823-857, July/Aug. 1981.

[Wes85] N.H.E.Weste, K.Eshraghian, *Principles of CMOS VLSI Design*. Addison-Wesley, Reading, 1985.

[Wha83] D.J.Wharton, The HITEST test generation system. Overview. *Int. Test Conf.*, Philadelphia, pp. 302-310, Oct. 1983.

[Whi92] B.A.White, M.I.Elmasry, The digi-neocognitron: A digital neocognitron neural network model. *IEEE Trans. Neural Netw.*, 3, pp. 73-85, Jan. 1992.

[Wid60] B.Widrow, M.E.Hoff, Adaptive switching circuits. *IRE WESCON Conv. Rec.*, pp. 96-104, 1960.

[Wie83] G.Wiederhold, *Database Design*. McGraw-Hill, New York, (2nd ed.) 1983.

[Wil79] P.Wilcox, Digital logic simulation at the gate and functional level. *16th Des. Aut. Conf.*, San Diego, pp. 242-248, June 1979.

[Wil87] D.C.Wilkins, Knowledge base refinement using apprenticeship learning techniques. In: K.Morik (ed.), *Knowledge Representation and Organization in Machine Learning*. Springer, Berlin, pp. 247-257, 1987.

[Wil85] A.J.Wilkinson, MIND: An inside look at an expert system for electronic diagnosis. *IEEE Design & Test*, 3, pp. 69-77, Aug. 1985.

[Wil90] K.Wilkinson, P.Lyngbaek, W.Hasan, The Iris architecture and implementation. *IEEE Trans. Knowl. and Data Engng*, 2, pp. 63-75, March 1990.

[Wil84] C.Williams, ART: The advanced reasoning tool. Report Inference Corp., Los Angeles, 1984.

[Wil73] M.J.Y.Williams, J.B.Angell, Enhancing testability of large-scale integrated circuits via test point and additional logic. *IEEE Trans. Comp.*, 22, pp. 46-60, Jan. 1973.

[Wil83] T.L.Williams, P.J.Orgren, C.L.Smith, Diagnosis of multiple faults in a nationwide communications network. *IJCAI-83*, pp. 179-181, 1983.

[Win75] P.H.Winston (ed.), *The Psychology for Computer Vision*. McGraw-Hill, New York, 1975.

[Win84] P.H.Winston, *Artificial Intelligence*. Addison-Wesley, (2nd ed.) 1984.

[Win79] P.H.Winston, R.H.Brown (eds), *Artificial Intelligence: An MIT Perspective*. 2 volumes, MIT Press, Cambridge, 1979.

[Win88] P.H.Winston, B.K.P.Horn, *LISP*. Addison-Wesley, Reading, (3rd ed.) 1988.

[Wir77] N.Wirth, Modula: A language for modular multiprogramming. *Software Pract. Exp.*, 7, pp. 3-35, Jan./Feb. 1977.

[Wir88] N.Wirth, *Programming in Modula-2*, Springer, Berlin, (4th ed.) 1988.

[Woj84] A.S.Wojcik, J.Kljaich, N.Srinivas, A formal design verification system based on an automated reasoning system. *21st Des. Aut. Conf.*, Albuquerque, pp. 641-647, June 1984.

[Wol83] W.Wolf *et al.*, Dumbo, a schematic-to-layout compiler. In [Bry83], pp. 379-393, 1983.

[Wol89] W.H.Wolf, How to build a hardware description and measurement system on an object-oriented programming language. *IEEE Trans. CAD*, 8, pp. 288-301, March 1989.
[Won87] D.F.Wong, H.W.Leong, C.L.Liu, PLA folding by simulated annealing. *IEEE Jour. Solid-State Circ.*, 22, pp. 208-215, April 1987.
[Won88] D.F.Wong, H.W.Leong, C.L.Liu, *Simulated Annealing for VLSI Design*. Kluwer Academic, Boston, 1988.
[Woo85] N.S.Woo, A Prolog based verifier for the functional correctness of logic circuits. *Int. Conf. Comp. Des.*, Port Chester, pp. 203-207, Oct. 1985.
[Wos84] L.Wos, R.Overbeek, E.Lusk, J.Boyle, *Automated Reasoning. Introduction and Applications*. Prentice-Hall, Englewood Cliffs, 1984.
[Wu87] C.F.E.Wu, A.S.Wijcik, L.M.Ni, A rule-based circuit representation for automated CMOS design and verification. *24th Des. Aut. Conf.*, Miami Beach, pp. 786-792, June/July 1987.
[Wu90] J.G.Wu, Automatic knowledge acquisition in a digital circuit design system. *Eur. Des. Aut. Conf.*, Glasgow, pp. 180-184, March 1990.
[WU90] J.G.Wu, Y.H.Hu, W.P.C.Ho, D.Y.Y.Yun, A model-based expert system for digital system design. *IEEE Design & Test*, 7, pp. 24-40, Dec. 1990.
[Wya84] J.L.Wyatt, Q.J.Yu, Signal delay in RC meshes, trees and lines. *Int. Conf. CAD*, Santa Clara, pp. 15-17, Nov. 1984.
[Yau86] C.W.Yau, Concurrent test generation using AI techniques. *Int. Test Conf.*, Washington, pp. 722-731, Sept. 1986.
[Yih90] J.S.Yih, P.Mazumder, A neural network design for circuit partitioning. *IEEE Trans. CAD*, 9, pp. 1265-1271, Dec. 1990, Also in: *26th Des. Aut. Conf.*, Las Vegas, pp. 406-411, June 1989.
[Yin87] K.M.Yin, D.Solomon, *Using Turbo Prolog*. Que Corp., Indianapolis, 1987.
[Yoo90] M.S.Yoo, A.Hsu, DEBBIE: A configurable user interface for CAD frameworks. *Int. Conf. Comp. Des.*, Cambridge, pp. 135-140, Sept. 1990.
[Yor86] B.W.York, KBTA: An expert aid for chip test. In [SRi86], pp. 959-969, 1986.
[Yos86] T.Yoshimura, S.Goto, A rule-base and algorithmic approach for logic synthesis. *Int. Conf. CAD*, Santa Clara, pp. 162-165, Nov. 1986.
[Yos82] T.Yoshimura, E.S.Kuh, Efficient algorithms for channel routing. *IEEE Trans. CAD*, 1, pp. 25-35, Jan. 1982.
[Yu87] M.L.Yu, Fork. A floorplanning expert for custom VLSI design. *Int. Conf. Comp. Des.*, Rye Brook, pp. 34-37, Oct. 1987.
[Yu89] M.L.Yu, A study of the applicability of Hopfield decision neural nets to VLSI CAD. *26th Des. Aut. Conf.*, Las Vegas, pp. 412-417, June 1989.
[Zad83] L.A.Zadeh, Commonsense knowledge representation based on fuzzy logic. *Computer*, 16, pp. 61-65, Oct. 1983.
[Zad88] L.A.Zadeh, Fuzzy logic. *Computer*, 21, pp. 83-93, April 1988.
[Zad89] L.A.Zadeh, Knowledge representation in fuzzy logic. *IEEE Trans. Knowl. Data Engng*, 1, pp. 89-100, March 1989.
[Zak91] M.Zak, An unpredictable-dynamics approach to neural intelligence. *IEEE Expert*, 6, pp. 4-10, Aug. 1991.
[Zan90] M.Zanella, P.Gubian, A learning scheme for SPICE simulations diagnostics. *Int. Symp. Circ. Sys.*, New Orleans, pp. 510-513, May 1990.

REFERENCES

[Zan86] C.Zaniolo et al., Object oriented database systems and knowledge systems. In: L.Kerschberg (ed.), *Expert Database Systems*. Benjamin/Cummings, Menlo Park, 1986.

[Zdo90] S.Zdonik, Object-oriented type evolution. In [Ban90], pp. 277-288, 1990.

[Zhu88] X.A.Zhu, M.A.Breuer, A knowledge-based system for selecting test methodologies. *IEEE Design & Test*, 5, pp. 41-59, Oct. 1988.

[Zim79] G.Zimmermann, The MIMOLA design system: A computer-aided digital processor design method. *16th Des. Aut. Conf.*, San Diego, pp. 53-58, June 1979; P.Marwedel, The MIMOLA design system: Detailed description of the software system. *16th Des. Aut. Conf.*, pp. 59-63, June 1979.

[Zim87] H.J.Zimmermann, *Fuzzy Sets, Decision Making, and Expert Systems*. Kluwer Acad., Boston, 1987.

[Zip83] R.Zippel, An expert system for VLSI design. *Int. Symp. Circ. Sys.*, Newport Beach, pp. 191-193, May 1983.

[Zub89] M.Zubair, B.B.Madan, Systolic implementation of neural networks. *Int. Conf. Comp. Des.*, Cambridge, pp. 479-482, Oct. 1989.

INDEX

ABE, 508
Abu-Hanna/Gold system, 361-362
Access time, 94
ACE, 213
ACT2, 97
Actor computer, 96
Ada, 101, 201, 244, 310
Adaline, 500
ADAM, 281, 397
ADFT, 402
Adjacency graph, 420
ADL, 310
ADLIB, 305
ADVICE, 347
AESOP, 214
AI (Artificial Intelligence), 32
AI machines, 198
AI-based placement, 428
AIX, 54, 93
ALAP, 251
ALEA, 342
ALGOL, 83, 97
Algorithm, 84, 209
Algorithmic State Machine, 233, 234, 258
ALPHA, 123, 199
ALICE, 199
ALU, 27
Alvey VLSI-BIST system, 402
AM, 216, 510
AMBER, 428
Ambiguity region, 312
Analog circuit, 18
Analog-circuit diagnosis, 343
Analog-circuit experts, 340-348
Analog-digital converter, 501
Analogical reasoning, 282
AND/OR graph, 167-168, 384
ANSI, 290
ANSYS, 347
APEX3, 356
APOLLON, 293
Application generator, 91
Application language, 91
Application program, 26
Applicative language, 98
Apprenticeship learning, 485
AQ Algorithm, 482
Aquarius Prolog, 146
ARBY, 191, 210
Architecture, 29
Area splitter, 430
Arity Prolog, 187, 202
ARPANET, 78
Array, 152
ARS, 215
ARSENIC, 293
ART, 191, 210, 214
ART-FUL, 214
ART-IM, 192
Artificial Intelligence, 32, 108-117
ASAP, 251
ASCA, 199
ASCAP, 306
ASIC, 4, 13
ASIC architectures, 299
ASIC design, 285-302
ASIC design experts, 297-302
ASIC design specification, 297-298
ASP, 302
ASPOL, 304
Assembly language, 82
Assignable-delay model, 311-312
Assignment statement, 97
Associative architecture, 489
Associative memory, 200
ASTAP, 19
ASYL, 300
ATEX, 357
Atom, 118
ATMS, 360
Attribute, 155
Attribute-value pairs, 278

INDEX

Auto-associative architecture, 489
Automated logic synthesis, 258-268
AV pair, 161
Avance, 157, 158
Axon, 485

B & D router, 458
Backtrace, 365, 372
Backtrack(ing), 133, 137, 143, 145, 373
Backtrack search, 137
Backward chaining, 165, 166, 187, 191, 192, 206, 207, 214
Backward error propagation, 491
Backward implication, 369
BASIC, 86, 98
Bayes's Theorem, 170, 207
BDL/CS, 259
BDS, 210, 213
BEAVER, 454
Behavior, 242, 243
Behavioral level, 303
Behavioral-level fault simulation, 382
B-ISDN, 473
BILBO, 394, 398, 400, 404, 406
BIMOS, 319
BIM Prolog, 146
Binary-decision diagram, 275
BIOS, 54, 144
Bipolar technology, 16, 457
BIST, 402
Bit-serial architecture, 294
Blackboard architecture, 191, 192-193, 264, 281, 433, 435, 439, 441, 455, 458, 459, 461
BLADES, 347
BLINK, 262
BLISS, 272, 442
Block-channel graph, 417
Block placement, 429
Block structures, 416
Blockage recovery, 461
BLOREC, 461
BNF, 86
Boltzmann machine, 200, 492
Boolean comparison, 302
Borland C++ 3.0, 149

Boundary scan, 392
Brahme-Abraham test generator, 384
Breadth-first search, 167
Bridging fault, 351
Browser, 266, 267
BUD, 274
Bus, 25
Bus failures, 362
Bus standards, 290
Butterfly, 200

C, 52, 83, 98, 109, 116, 128, 149, 157, 191, 201, 217, 244, 357, 405, 434, 461
C++, 149
C4, 483
C algorithm, 384
CAD Framework Initiative, 264
CAD Frameworks, 263-267
CADCAS, 2
CADHELP, 218
CADR, 122
CADRE, 431-433
Cadweld, 264
CAM, 200
CAMELOT, 355
CAP/DSDL, 404
CAPRI, 293
CARIOCA, 461
CASE, 31
CASNET, 207-208
CASS, 479
CATA, 385, 396
CATHEDRAL, 295
Cbase, 157, 158
CC-TEGAS, 378
CCL, 386
CCOMP/EX, 464
CDC 6600, 95
Cellular reduction machine, 103
Certainty factor, 170, 206, 207, 209, 210
CGEN, 275
Chaining, 251
Chan's channel router, 448
Channel router, 442, 444, 446-452
CHARM, 281
Checking of DFT rules, 403-404

INDEX

Chip-area estimation, 428, 429
Chip Layout Assistant, 429
Chip partitioning, 281
Chip Planner, 431
Chippe, 273
CIF, 24, 279, 294, 406, 429, 441
CIP, 259
CIRCOR, 360
Circuit abstractors, 464
Circuit extraction, 299, 463
Circuit simulation, 334-337
CISC, 28, 93, 146
CGA, 41
Class, 147, 155
CLASS, 429
Class abstraction, 147, 150
Class hierarchy, 155
Class variables, 148, 152
Classification of architectures, 95-96
Classification of expert systems, 202-203
Clausal form, 127
Clause, 127, 128
Client/server architecture, 71
CLIPS, 192
Clique partitioning, 262
Clock buffering, 289
Clock period, 94
Clock skew, 288, 323
Clocking schemes, 288
CLOS, 149
Clustering, 279, 429
CMOS mask pattern generator, 462
CMOS technology, 16-17, 214, 240, 292, 299, 435, 438, 442, 443, 459, 462, 503
CMU-DA System, 260-263
CNAPS, 498
CODASYL, 64
Combinatorial explosion, 163, 167
Command interpreter, 186
Command language, 26
Common Framework, 265
Common LISP, 123, 149, 157, 199, 282, 346, 359, 360
Compactor, 430
Compiler, 28, 83, 85, 86
Complex objects, 155, 156

Compound object, 140
Compression ridge, 419, 441
Computation-intensive algorithm, 105
Computer architectures, 91-100, 505
Computer communication network, 75-80, 473
Computer performance, 94
Concept learning, 481
Concurrent fault simulation, 380-381
Concurrent languages, 98-99
Concurrent Pascal, 99, 101
Conflict resolution, 183
Conjunction, 132, 177
Connection machine, 199
Connectionist system, 200, 487
Connectivity graph, 420, 433
Constraint propagation, 208, 215, 340, 341
Content-addressable memory, 199, 200
Control allocator, 262
Control-flow computer, 101
Control flow-graph, 245, 246, 249, 279
Control path, 233
Control section, 232
Control synthesis, 245, 261
Controllability, 353
Controlled switch, 328
Coordinating expert, 193
COORDINATOR, 433, 436
CORAL, 66, 210
CORDIC processor, 269
Cost function, 433, 440
Coupling architectures, 202
CP/M, 54
C-Prolog, 279, 361
CPU, 27
CRAS, 213
Cray, 17, 94, 95, 304
CRIB, 211
Critic, 346
Critical input, 375
Critical path, 375
Critical-path analysis, 431
Critical-Path Test Generation, 375-376
Critical path tracing, 381
Critical race, 315

CRITTER, 330, 313
CSG, 276
Cube, 366
CV (Connectivity Verificaton), 465
CWL, 386
CYBER 205, 95
Cyclic constraint, 447
Cyclic simulator, 306
Cytocomputer, 200

D algorithm, 366-371, 387
D cube, 366
D drive, 369
D intersection, 369
DA, 2
DAA (Design Automation Asssistant), 271-274, 431
DADO2, 199
DAP, 95
DARPA, 508
DARSI, 465
DART, 359
Data abstraction, 147-148
Data clause, 128
Data flow-graph, 278
Data/control flow-graph, 250
Data-dependence graph, 102, 268
Data dependency, 250
Data dictionary, 88
Data driven, 96, 102
Data-driven reasoning, 165
Data-flow computers, 96, 101-102
Data-flow graph, 102, 245, 246, 249
Data handling, 96
Data independence, 63, 71
Data/memory allocator, 261, 262
Data models, 63-68
Data path, 233
Data-path graph, 261
Data-path synthesis, 245, 261
Data section, 232
Data processing, 104
Data token, 101
Database, 62, 160
Database computer, 74
Database language, 136

Database-management system, 68
Database machine, 74
Database schema, 157
Database system, 62-74
Data-flow computer
DAV, 307
dBase, 73
DBMS, 62, 68
DDL, 70, 157
DDL/SX, 270
DDL Verifier, 307, 330
DEBBIE, 264
Decision diagram, 371
Decision-support language, 90
Declarative language, 83, 136, 201
Decoder, 221-222, 270
DECtalk, 491
DEDALE, 343
Dedicated hardware for design, 291
Deductive fault simulation, 378-379
Deductive principle, 163
Deep knowledge, 204, 299, 356
Default option, 89
DEFT, 405
DEGAS, 331
Delay analysis, 319
Delay calculation, 322
Delay fault, 351
Delay models, 310-313
Delft Placement and Routing System, 267
DELIGHT.SPICE, 347
Delta rule, 490
Density column, 448
DFT Expert, 405
Demand driven, 96
DEMO, 403
Demon, 169, 180, 181
Demultiplexer, 224
DENDRAL, 110, 123, 205
Dendrite, 485
Depth-first search, 145, 167, 187
DESCART, 299
Descriptive language, 136
Design automation, 2
Design Automation Conference, 2
Design flow, 284

INDEX

Design fault, 21
Design for testability, 22, 389-406
Design frameworks, 263-267, 284
Design methodology, 11
Design partitioning tool, 428
Design quality, 247
Design-rule checking, 463
Design rules, 462
Design specification, 10, 303
Design strategy, 248
Design style, 14, 247, 261
Design systems, 215-217
DESIGNER, 438
DFT (Design for Testability), 389-395
DFT system, 396
Diagnostic signal, 366
Diagnostic systems, 210-214
DIALOG, 332, 465
Dialog interaction, 89
Dialog system 1800, 200
DIANA, 328
Digi-neocognitron, 501
Digital, 153
Digital-circuit verification experts, 329-333
Digital filters, 281, 283
Digital logic circuit, 19
Digital neurochips, 498
Digital signal processing, 104, 107-108
Digitalk, 153
Digitizer, 468
DIGSIM, 313
Disjunction, 177
Distributed computer network, 198
Distributed data management, 75
Distributed computing, 75
Distributed-memory system, 99
DMI, 266
DML, 70, 157
DNNA, 498
Dogleg router, 448
Domain, 32
Domain expert, 159, 172-173
DOOM, 150, 338
Dorado, 199
DOS, 55, 153

DOS 5, 60
DOS Shell, 154
DQDB, 79
DSUI, 267
DTA, 355
DTAS, 271
DTP (Desktop publishing), 44
Duck, 342
DUMBO, 294
Dynamic binding, 148
Dynamic hazard, 314, 316

E-Crystal, 329
Edge triggering, 319
EDIF, 290
Edinburgh Prolog, 129, 138
EGA, 42
EL, 215, 340-341
ELF, 443
ELogic, 329
ELOSIM, 329
ELSYN, 341
EMERALD, 263
EMUCS, 263
EMYCIN, 166, 186, 190
Encapsulation, 148, 156
Encore, 158
Engineering workstation, 475-477
Entity-relationship model, 66
Error model, 351
ESCHER, 267
ESCON, 29
ESPRIT, 6, 265
ESTA, 404
ESTC, 363
ESTEPS, 348
ET, 212
ETA, 95
ETHERNET, 78
EURISKO, 216
EURONET, 78
Event, 310
Event-driven simulation, 277, 316
EXCIRSIZE, 217
Excitation table, 230
Exhaustive testing, 364

Expert Essence, 345
Expert networks, 502
Expert systems, 2, 15, 32, 110-114, 159-218, 502
Expert-system tools, 191, 192
Explanation-based learning, 481, 484-485
Explorer, 123, 199, 284
Explorer CheckMate, 123, 338
EXTEST, 405

FACET, 262
F-LOGIC, 378
Facts, 109, 112, 129, 513
FAIM-1, 199
FAN, 373-374
Fault analysis, 351-352
Fault coverage, 350, 364
Fault detection, 349
Fault diagnosis, 349
Fault insertion, 378
Fault list, 378
Fault location, 349
Fault model, 349-351
Fault simulation, 377-386
FDA, 360
FDDI, 79
Feedback architecture, 489, 493
Feedforward architecture, 489, 493
Feedback shift register, 393
FFSIM, 381
FG502-TASP, 212
FGCS, 114, 507
Fiber-optic network, 473
Field-Programmable Logic Array, 241
Field-Programmable Logic Sequencer, 241
Fifth-generation computer system, 114-115, 507-508
Fifth-generation language, 83, 114-154
File, 149
File handling, 139
File system, 62
Filter synthesis, 342
Finite-element method, 267
Finite-state machine, 225-231
FIRST, 294

FIS, 212
FISC, 293
FLAIR, 435
Flavors, 150, 189, 199, 295, 298
Flex, 149, 440
F/LOGIC, 313, 315
Floorplan evaluator, 433
Floorplan experts, 431, 432-435
Floorplanning, 243, 287, 420, 429, 431, 432
Flowchart method, 258
Flow-graph, 249
FLOYD, 433, 434
FLUTE, 433
FOREST, 210
FORK, 433-434
FORTRAN, 83, 117, 199, 215, 217, 304, 460
Forward chaining, 165-166, 214, 215
Forward tracing, 365
Fourth generation, 81, 83
Fourth-generation language, 81-91
Frame-based representation, 209, 210, 220
Frame-based system, 189, 191, 299, 300, 362
Frames, 180-181, 270, 278, 281, 398, 429, 437, 464
Franz LISP, 123, 210, 211, 215, 217, 406, 461, 331
Fred, 278
Function, 126
Functor, 129
Functional language, 97-98, 119
Functional test generation, 389
FUTURE, 374
Fuzzy logic, 171, 510
Fuzzy logic application, 437

GaAS overlay shell, 442
GaAS technology, 17, 442
Garbage collection, 117-122, 137, 146
GASP, 304
Gate array, 270, 286, 287, 430
Gate assignment, 188
Gate-level simulation, 310-319

INDEX

Gate matrix, 442, 443
Gateway, 80
Gaussian machine, 493
GDE, 359
GemStone, 157
General routers, 411
Generate-and-test, 168
Geometry Engine, 94
Geometric layout, 433, 436, 440
GKS, 290, 471
Glitter, 466
Global clause, 128
Global domains, 141
Global predicates, 141
Global routing, 410-414, 429, 432, 440, 458
Global routing optimizer, 429
Global transformation, 259
Global variable, 152
GM_Plan, 442-443
Goal clause, 128
Goal-driven reasoning, 165, 206
Golden Lisp, 123
GPSS, 304
Graph reduction, 96, 103, 180
Graphics displays, 469-470
Graphics software, 470
Graphics supercomputer, 477
Graphics tablet, 468
Graphics tools, 467-471
Greedy algorithm, 384
Greedy channel router, 449-451
Greedy switchbox router, 452-454
Gridless router, 414

HAL, 278
Hardware-verification experts, 306-310
Harpy, 200
HAT, 395
Hazard, 313-314
Heap, 145
HEARSAY, 194, 200, 283
HEICLAS, 431
Helix, 278
Hephaestas, 465
Heuristic search, 168

Heuristics, 109, 215
Hierarchical block design, 287
Hierarchical data model, 64
Hierarchical levels, 7-9, 242
HIFI, 268
High-level system design, 241-268
Hightower's routing algorithm, 413
HILDA, 264, 284
HILO-2, 386
HILO simulator, 318
HITEST, 386, 404
HITSIM, 339
Hold time, 319
HOPE, 430
Hopfield network, 491-492
Hopfield-Tank model, 496-497
Horizontal block graph, 416
Horizontal constraints, 447
Horn clause, 127, 128
HPRL, 182, 210, 346
Human-computer interaction, 467-474
Human-computer interface, 86
Human factors, 474
Hyperclass, 210
Hypermedia, 195, 472
Hypernet, 200
HyperPIES, 214
Hypertext, 195, 472

IBIS, 219
IBM 801, 93
IBM Nat. Task, 200
IC, 1, 22
I-CAT, 212
ICD, 265
ICDL, 435
ICEWATER, 292
ICOMP, 438, 441
Icon, 50, 89
ID3 Algorithm, 482
IDAC3, 19
IDAS, 355
IDEAS, 284
Identifier, 97
IDT, 211
IEC, 290

IFS, 307
IGES, 290, 471
IIL-circuit generation, 345
iJADE, 345
ILLIAC, 95
Imperative language, 84, 97
Implementation technologies, 16
Implication, 365, 372
IMS T800 Transputer, 101
IN-ATE, 212
Inductive learning, 282, 481-484
Inductive principle, 163
Inertial delay, 313
Inference, 109
Inference engine, 112, 137, 165, 183
Infix notation, 142
Info Center/1, 90
INGRES, 72, 157
INGRES/MENU, 73
Inheritance, 147, 148, 150, 264
INMOS transputer, 101
Inner-product-step processor, 106
Instance, 154
Instantiation, 131, 154, 155
Instruction set, 28, 37, 145
Instruction Set Architecture, 146
Integer-linear programming, 258
Integrated layout verification, 465
Integration density, 6
Intelligence, 31, 108
Intelligent databases, 196
Intelligent interface, 111, 200
Intelligent Librarian, 280
Intelligent machines, 508-509
Intelligent Network, 474
Intelligent signal processing, 283
Intelligent silicon compilers, 296-297
Interactive operation, 86
Interactive Theorem Prover, 384
Interlisp, 123, 181, 199, 205, 207, 215, 217, 276, 331
Intermittent faults, 214, 351
Interpreter, 85, 86
IOS, 290
IPC, 290
IPDA, 461

IPT, 214
IQCLISP, 123
IQLISP, 123, 192
Iris, 157
ISA, 219
ISDN, 77
ISPS, 260, 261, 262, 273
IS-SPICE, 337
Iterative-improvement method, 409

JADE, 345
JESSI, 265

KAT, 429
Karnaugh map, 231, 240
KBTA, 388
KD2, 299
KEE, 191, 210
KES, 190, 192
Keyword message, 151
KINDEN, 300
KINFIDA, 301
KINTESS, 301
KL-ONE, 209
KL-TWO, 192
KMDS, 301
Knowledge, 109, 513
Knowledge acquisition, 172, 174, 429
Knowledge base, 109, 112
Knowledge-based management, 194, 197-198
Knowledge-based system, 109, 199
Knowledge Craft, 192
Knowledge engineer(ing), 159, 172
Knowledge processing, 116, 159, 183-194
Knowledge representation, 159, 172-182
Knowledge source, 173, 193
KOCOS, 441
KRAFT, 401
KRL, 182

LAGER, 296
LAMBDA, 123, 199
Lambda calculus, 119
LAN (Local-Area Network), 78
Language-based AI machines, 198-199

INDEX

Laser printer, 42
Layout, 243
Layout analysis, 463
Layout compaction, 299, 418-419
Layout-compaction expert, 441
Layout design, 22-25
Layout experts, 428-444
Layout generator, 430
Layout structures, 286-288
Layout-to-circuit extractor, 267
Layout verification, 462-466
Layout-verification experts, 464-466
LDSS, 259
Leaf-cell layout tools, 435-437
LEAP, 282, 331, 484
Learning, 493-496
Learning assistant, 271
Learning by examples, 494-495
Learning machines, 510-511
Least-commitment approach, 430, 435
LEDA, 282
Lee's routing algorithmn 412
Left-edge channel router, 469
LES (Layout Expert System), 157, 213, 434, 461, 465
LES/PL1, 210
Level5, 192
Level graph, 294
LEX, 183
LEXTOC, 332
LiB, 294
LIFO (Last-In First-Out) stack, 373
Light pen, 468
LISP, 83, 116, 117-122, 157, 178, 180, 190, 191, 198, 201, 206, 215, 217, 293, 308, 362, 397, 434, 461, 465
LISP machine, 122-123, 278, 284
List, 135, 140
LMLisp, 199
Lists, 118
Local routing, 411, 432, 444-462
Local tratnsformations, 259, 269, 270
LOGAN, 319
Logic connectives, 124
Logic-design experts, 269-284
Logic fault, 21

Logic-level synthesis, 246
Logic minimization, 270
Logic modules, 221
Logic programming, 124-146
Logic programming language, 116
Logic programming system, 187, 207
Logic simulation, 310-329
Logic synthesis, 270
Logic verification, 306
Logo, 122
LOLA, 271
LOOPS, 182, 189, 191, 215, 276, 277, 278, 281 402
LORES-2, 269
Lotus 1-2-3, 90
LOVE, 259
LSS, 259, 269
LSSD, 381, 391, 392, 398, 404
LTS, 260
LUBRICK, 293
Lucid Common Lisp, 202, 264

M1, 275
Machine language, 82
Machine learning, 480-485
Macintosh, 40, 46-47, 153, 212
MacLisp, 123, 199, 215, 388
MACLOG, 264
MacPitts, 292
MAGIC, 306
Manhattan structure, 408, 411
MARS, 278, 291
Mask artwork, 23
Matching, 127, 131, 183
Maximum cardinality matching, 448, 449
Maze-running router, 412, 443
MBESDSD, 279
McCulloch-Pitts neuron model, 489-490
MEA, 183
Mealy machine, 227
MECHO, 207
Memory bandwidth, 95
Memory failures, 352-353, 361
Menu, 40, 89
Message, 97, 148, 278
Messaging, 147-148, 150

Metaknowledge, 161
Metarule, 161, 175
Method, 147, 155
Mflops, 94
MICON, 216, 275
Microarchitecture, 252, 257, 278
Micro IN-ATE, 212
Micro-Prolog, 138
Microprocessor, 37
Microprogramming, 28
Mighty, 452
MIMD, 95, 200
MIMOLA, 305
Min-cut placement, 409
MIND, 357
MINI, 225
Minimal Rectangular Steiner Tree, 457
Minimal spanning tree, 411
Min-max model, 312
MISD, 95
MIPS, 93
Mi/s, 94
MIT system, 357
MIXER, 219
Mixed-integer linear programming, 263
Mixed-mode simulation, 337
MNA, 334
MODEL, 293
Modem, 75
MODULA, 99
Modular program development, 140
Modularity, 141
Module binder, 261
Module compilers, 293
Module database, 261
Modus ponens, 125
Modus tollens, 125
MOLGEN, 208
Moore machine, 227
MOSSIM, 323
MOS technology, 221, 224
Mouse, 39, 86, 89, 283, 468
MRS, 187, 276, 277, 359, 388
MS-DOS, 39, 54-60
MS-Windows, 47, 60
MTA, 218

MULGA, 419, 431, 435
MuLisp, 123
Multicycling, 251
Multilevel simulation, 304
Multiplan, 90
Multiple faults, 210, 214, 350, 359
Multiple-valued logic, 315
Multiplexer, 222, 240, 273, 275
Multiprogramming, 27, 49
Multiprocessor ssytem, 93
Multitasking, 476
Multiwindowing, 476-477
MVISA-DA system, 405
MYCIN, 123, 166, 171, 190, 206, 471

NAP2, 19
NAPLPS, 471
NDS, 213
NEC DP-100, 200
NELSIS, 265-267, 284
Neocognitron, 501
NEOCRIB, 211
NEPTUNE, 300
Net, 445
Net list, 270
NETL, 199
NETSIM, 498
NETtalk, 491
Network data model, 64
Network topology, 76
Neural computers, 501, 509-510
Neural network architecture, 488-489
Neural network applications, 500-502
Neural networks, 100, 200, 284, 485-502, 505
Neuro-ASIC, 498
Neurocomputing, 100, 506
Neuron, 485
Next-event simulation, 316-317
Ng-Chow troubleshooter, 363
NMD-CAD, 265
NMOS technology, 16, 223, 276, 430, 465
Nonprocedural language, 84, 87
Non Von Neumann architecture, 91-108, 504

INDEX

NOP, 458-459
NP complete, 19

O_2, 158
OASE, 344
OASYS, 347
OAV triplet, 161
Object, 147, 150, 178, 278
Object identity, 155, 156
Object Manager, 157
Object-oriented approach, 278
Object-oriented representation, 283, 298
Object-oriented simulation, 339
Object-oriented systems, 202, 210, 264, 284
Object Pascal, 150
Object Server, 157
Objective C, 150
Object-oriented data model, 155
Object-oriented database, 154-158
Object-oriented language, 116, 147, 149
Object-Oriented LISP, 150
Object-oriented programming, 189, 191
Object-oriented Prolog, 150
Object-oriented (OO) systems, 188, 197, 202, 210, 264
Objectworks/Smalltalk, 153
Observability, 353
OCCAM, 101
OCR, 460
Odyssey, 284
ONCOCIN, 471
OO database management system (OO-DBMS), 156-158
OP-1, 344
OPAL, 147, 157
Operating system, 27, 49-61
OPS5, 116, 176-177, 185-186, 190, 210, 215, 217, 272, 301, 401, 425, 429, 441, 442, 444, 455, 462, 465
OPS83, 178, 187, 264, 284, 360, 404, 405, 438
Optical computer, 506
ORBS, 332, 466
ORION, 157
Orthogonal layout, 408, 429

OSI (Open Systems Interconnection), 78
OTO-D, 266

P algorithm, 384
Packet switching, 80
PAL, 234, 241
Palladio, 182, 217, 276-278
PAMS, 299
PARAID, 428
Parallel computer, 90, 100
Parallel fault simulation, 378
Parallel processing, 100
Parallel test generation, 377
Parallel Value List method, 386
Parallelism, 169
Parameterized module, 288, 294, 295
Parameterized version, 155
Parametric fault, 21, 214, 350
ParcPlane, 153
PA-RISC, 478
Partitioning, 407-408, 428-431
Partitioning expert system, 383
Pascal, 83, 97, 109, 128, 179, 201, 244, 461
Pass transistor, 277, 294
Path sensitizing, 365
Pattern, 177
Pattern-invoked program, 175
Pattern matching, 127, 131, 164
Pattern recognizer, 430
PC (Personal Computer), 37
PC-DOS, 39, 54
PC Scheme, 123
PC table, 366
PDES, 290
PEACE, 210, 213, 216
PEARL, 442
PEDX, 220
Penalty-credit function, 401
Perceptron, 490
Persistent object, 157
Personal computer, 37-49
PESTICIDE, 361
Petri net, 284, 308
PEX, 219
PHIGS, 471

Physical defect, 21, 349
PIAF, 437
PIE, 199
PIM, 199
Pipelining, 91
PISCES-II, 339
PL/1, 83
PLA (Programmable Logic Array), 224-225, 234, 237, 240, 258, 261, 294
PLA folding, 225
PLA testing, 395
Placement, 407-410
Placement experts, 428-431
Planar Original Graph, 422
PLA-TSS, 400-401
PLEX, 292
Plotters, 468-469
PMLS, 338
PODEM, 372-373, 387
PODEM-X, 385
Polynomial-time complexity, 19, 35
POOL, 150, 338
POP-2, 356
Portable, 83
POSTGRESS, 157
PPL, 300
PPRG, 319
Precision Architecture, 93
Predicate, 125, 129
Predicate calculus, 125-126
Predicate logic, 124, 187
Prefix notation, 142
PRIAM, 309
Primary inputs, 310
Primary outputs, 310
Prime implicant, 371
Primitive cube, 366
Printer, 42
Probabilities, 212
Probability theory, 169
Problem reduction, 164, 168-169
Problem solving strategies, 163-171
Procedural language, 83, 93, 201
Processing element, 105
Processor array architecture, 104
PROD, 360

Production rule, 175-177, 183
Production system, 175, 183-185, 205, 272
Programmable Logic Devices, 286
Programmable technologies, 286-287
Programmer productivity, 86
Programming environment, 116
Programming language, 81
Prolog 83, 86, 116, 127-146, 144, 171, 187-188, 190, 192, 199, 201, 210, 215, 217, 224, 298, 300, 309, 356, 357, 362, 404, 430, 436, 442, 456, 457, 458, 464
Prolog1, 144
Prolog2, 138, 202
Prolog V, 144
Prolog-based channel router, 457
Prolog-based CMOS design, 443
Prolog-based expert systems, 188
Prolog-based silicon compiler, 302
Prolog-based testing, 394, 395, 403, 405, 406
Prolog-based test generation, 383
Prolog compiler, 144
Prolog interpreter, 144
Prolog machine, 146
Prolog rule, 132
ProLogic, 279
Propagation D cubes, 368
Proposition, 124
Propositional calculus, 124-125
PROSAIC, 347
PROSPECT, 403-404
PROSPECTOR, 206-207
Protean, 383
ProTest, 404
Prototyping, 162, 175, 190
PROVE, 309
PS/2, 153
PSPICE, 337
PTRANS, 219
Pull-down menu, 40, 89, 281
Pull-up menu, 40
Pumps, 200
PUSH and POP, 156

QBE, 69, 90

INDEX

QCritic, 346
Qlisp, 123
Quantifier, 126
Query clause, 129
Query language, 69, 90, 136
Quintus Prolog, 144, 146, 190, 202, 361

R 3000, 93
RABBIT, 209
Race, 314
RAM (Random-Access Memory), 26, 37, 38, 352
Random Access Scan, 392
Random pattern, 377
RAPS, 381
RDV, 330
Reasoning, 125
Recognize-act cycle, 183
Reconvergent fanout, , 313-314, 370
Rectangular graph, 434
Rectangular Dual Graph, 423
Rectangular layout block, 420
Recursive computer, 97
REDESIGN, 282, 331
Reductio ad absurdum, 125
Reduction computer, 102-103
Redundant gate, 270
Referential transparency, 98
Register-transfer synthesis, 245
Regular-expression compiler, 296
Relational data model, 65
Relationship, 130
Rete algorithm, 186
RIP, 42, 46
Rip-up and reroute, 294, 452
RISC, 28, 93
River routing, 454
RL, 296
RLL-1, 216
ROM (Read Only Memory), 28, 224, 234, 241, 258, 261, 393
Routability, 408, 414, 430
Routing channel, 444
Routing experts, 455-461
Routing problem, 411
RSIM, 328

RUBICC, 346
Rule, 109
Rule base, 112, 160, 183
Rule-based language, 116
Rule-based system, 175, 207
Rule clause, 160
Rule interpreter, 184

S-expression, 118
SABLE, 305
SADD, 216
SALOG-IV, 315
SAPTES, 348
SARA, 304
Sass, 296
SATURN, 387
SCAI, 474
SCALD, 304, 395
SCAMPS, 306
Scan path, 390
Scan-path system, 392
Scan-set logic, 392
SCHEMA, 298
Schematic entry, 267, 478
SCHEME 79, 93, 123
Scheme-81, 293
SCOAP, 354, 355
SDL, 305
Sea-of-gates routing, 267, 287-288, 438
Search strategies, 167
Selective trace, 316
Self-testing, 392
Self-timed system, 107
Semantic network, 178-180, 207, 208, 220, 281, 363
Semiconductor technologies, 16, 503
Semicustom design, 13, 14
Sensitized path, 365
Sequent Balance 8000, 457
Sequential circuit synthesis, 231-232
Sequential function, 226
Serial adder, 230
Setup time, 319
SFDL, 259
Shallow knowledge model, 204, 356
Shared-memory system, 99, 199

Shearing method, 419
SHEDIO, 438
Shell, 161, 190-191
SHIELD, 306
Shift-register mode, 391
SHOOTX, 213
Shove aside, 452
Side effect, 97
Sidetracking, 146
SIGGRAPH, 470
SIGMA, 265
Signal flow, 188
Signal processing, 104
Signature analysis, 393
Silage, 296
SILAS, 344
Silc silicon compiler, 295, 402
Silc TESTPERT, 402
Silicon compiler, 209, 309-310
Silicon foundry, 12, 285
Silicon assembler, 291, 293
Silicon compiler, 291-297
SILK, 452
SIMD, 95
SIMSCRIPT, 304
SIMULA, 149, 304
Simulated annealing, 406, 414, 493
Simulation, 19, 21, 310
Simulation compiler, 332
Simulation language, 303
Singular code, 221
SISD, 95
Sixth generation, 503-513
Sixth-generation machine, 511-513
SL1, 304
SLA, 300
SLAP, 294
Slicer, 251
SLS, 267, 324-327
Smalltalk, 83, 116, 149, 150-154, 157, 189, 199, 202, 207, 210
Smalltalk V, 150, 153
SMART, 214
SMPL, 304
SNA, 80
SNEL, 327

SNePS, 362
SOAR, 199
SOCRATES, 275, 333, 374, 383
Software engineering, 30
Solution path, 165
SOPHIE, 217
SPACE, 267, 464
SPARC, 93, 288
Specification language, 91
Speech analysis, 108
Speech synthesis, 108
SPEX, 208
SPICE, 19, 84, 332, 335-336
SPIDER, 464
Spike, 313
Splicer, 253
Spreadsheet, 48, 461
SPUR, 199
SQL, 69, 157, 158, 198
SRTG, 381
Stack, 145, 156, 316
STAFAN, 355, 382
Standard-cell approach, 286
Standardization, 290
STARAN, 95
Start-Small approach, 396
State assignment, 231
State (transition) diagram, 229
State (transition) table, 227
State-graph, 251, 278
State-space search, 164-165
Static debugger, 332
Steiner tree, 411, 457, 458
STEP, 290
Stick diagram, 294
STONN, 498
Stored-program computer, 232
String, 152
String handling, 139
String reduction, 96, 102
Structural level, 303
Structure, 243
Stuck-at-0 fault, 350
Stuck-at-1 fault, 350
Stuck-open fault, 351
Stuck-type fault, 350

INDEX

Subclass, 155
SUD2, 305
SUPERCAT, 384-386
Superclass, 155
Supercomputer, 94-95
Supervised learning, 495
SUPREM-3, 339
SWIFT, 406
Switchbox, 446
Switchbox router, 446, 458, 461
Switch-level simulation, 323-324
SYCO, 295
Symbolic design, 279
Symbolic language, 117
Symbolic layout, 417-418, 431, 435, 436
Symbolic simulation, 397, 403
Symbolics 3600, 123, 157, 199, 295, 388
Symbolics 3670, 214
Symbolics LM2, 122
SYN, 341
Synapse, 486
SYNAPSE, 271, 301
Synchronous design, 288
Synthesis, 244
System8000, 79
System Architect's Workbench, 263
System architecture, 14
System planning, 280
Systems programming, 50
Systems analyst, 84
Systolic array, 94, 105-107, 499

TACOS, 284
Tagged architecture, 145
TALIB, 429-430
Target system, 20
TDES system, 282, 333, 397-400
Technology CAD, 339
TEGAS, 312, 315
TEIRESIAS, 192
TEMPIC, 213
Temporal logic, 307
TEST, 400
Test, test pattern, 350
Test set, 350
Testability measures, 353-355

Testability programs, 355
Testable Design Methodology, 398
Testing, 21
Test generation, 364-381
Test-set verification, 377
Test-strategy planner, 402
TEXAS, 396
Text processing, 43
Third-generation language, 82, 85-86
Three-valued logic, 313
Throughput, 49
TIGER, 401
Timing analysis, 438
Timing simulation, 327-329
Timing verification, 319-323
TIMM-Tuner, 218
TInMANN, 498
TIP, 355
TLTS, 332
Top-down design, 9-11
Topological evaluator, 431
Topological placer, 430
TOPOLOGIZER, 433, 435, 437
Tospics, 200
Trace, 140, 142
Track, 445
Traffic-light controller, 236-240
Trail, 145
Training, 488
Transaction, 71
Transistor sizing, 217, 479
Transputer, 101
Trigger, 155
TSS, 217
Turbo C++, 149
Turbo Pascal, 150
Turbo Prolog, 139-144, 344, 348
Turnaround time, 2, 20, 50, 285, 297
Turtle graphics, 122
Two-phase clock, 245, 288
Type hierarchy, 155

ULTRIX, 54
ULYSSES, 274
Unate function, 390
Uncertainty, 169-171

Unification, 128, 137, 164
Unit-delay model, 311
UNITS, 191
UNIX, 48, 52-54, 208
Unsupervised learning, 495
User interfaces, 89, 471-473

Value, 176
Value Trace, 260, 273
Value of an attribute, 155
Variable, 125, 131
VAX OPS5, 185, 186, 192
Vbase, 158
VDI, 471
VDM, 471
VERA, 346
VERCON, 465
VERIFY, 308
Verilog, 290
Vertical block graph, 416
Vertical constraints, 447
Vertical-constraint graph, 447
Version, 154, 155
VEXED, 271, 484
VHDL, 282, 290
Via, 445
VICTOR, 355
VIEWSIM/AD, 337
Virtual grid, 419, 452
VISTA, 306
VIVID, 419-420
VLSI, 1, 6
VLSI testing, 349-406
VMES, 362
VM/PROLOG, 343
Vocoder, 108
Von Neumann bottleneck, 100
Von Neumann computer, 25, 97
VT bodies, 260, 273

Wafer-scale integration, 30, 503
WAM (Warren Abstract Machine), 145
Wavefront processor array, 107
Wavefront router, 412
WEAVER, 429, 455-457
WEDS, 363

Westinghouse diagnostic system, 358
WHEEZE, 209
Widrow-Hoff rule, 490
WIMP, 40, 474
Windows, 40, 89, 139, 281
WIREX PROLOG, 459-460
Wiring congestion, 430
Word length, 25, 37
Word processing, 43
Wordperfect, 43
Wordstar, 43
Working memory, 183
Working-memory element, 176, 272, 274
Workstations, 475-479
WORM, 479
WYSIWYG, 44, 474

XCON, 215, 219, 471
XENIX, 54
Xenologix X-1, 199
Xerox 1100 series, 123, 199
XLISP, 429
X-path check, 372
XSEL, 219
X-window system, 472-473

YACR2, 452
YES-MVS, 218-219
Yield, 428

Zero-assignment language, 98
Zetalisp, 123, 199
Zone, 447